Diffusion of Chloride in Concrete

Modern concrete technology series

A series of books presenting the state-of-the-art in concrete technology

Series Editors

Arnon Bentur
National Building Research Institute
Faculty of Civil and Environmental Engineering
Technion-Israel Institute of Technology
Technion City
Haifa 32 000
Israel

Sidney Mindess
Department of Civil Engineering
University of British Columbia
6250 Applied Science Lane
Vancouver, B.C. V6T 1Z4
Canada

1. **Fibre Reinforced Cementitious Composites**
   A. Bentur and S. Mindess
2. **Concrete in the Marine Environment**
   P.K. Mehta
3. **Concrete in Hot Environments**
   I. Soroka
4. **Durability of Concrete in Cold Climates**
   M. Pigeon and R. Pleau
5. **High Performance Concrete**
   P.C. Aïtcin
6. **Steel Corrosion in Concrete**
   A. Bentur, S. Diamond and N. Berke
7. **Optimization Methods for Material Design of Cement-based Composites**
   Edited by A. Brandt
8. **Special Inorganic Cements**
   I. Odler
9. **Concrete Mixture Proportioning**
   F. de Larrard
10. **Sulfate Attack on Concrete**
    J. Skalny, J. Marchand and I. Odler
11. **Fundamentals of Durable Reinforced Concrete**
    M.G. Richardson
12. **Pore Structure of Cement-Based Materials: Testing, Interpretation and Requirements**
    K.K. Aligizaki
13. **Aggregates in Concrete**
    M.G. Alexander and S. Mindess
14. **Diffusion of Chloride in Concrete**
    L. Mejlbro and E. Poulsen

Diffusion of Chloride in Concrete

Theory and Application

Ervin Poulsen
Professor Emeritus of Civil Engineering
Technical University of Denmark

Leif Mejlbro
Professor Emeritus of Mathematics
Technical University of Denmark

CRC Press
Taylor & Francis Group
Boca Raton London New York

CRC Press is an imprint of the
Taylor & Francis Group, an **informa** business
A TAYLOR & FRANCIS BOOK

CRC Press
Taylor & Francis Group
6000 Broken Sound Parkway NW, Suite 300
Boca Raton, FL 33487-2742

First issued in paperback 2019

ISBN-13: 978-0-419-25300-6 (hbk)
ISBN-13: 978-0-367-39152-2 (pbk)

This book has been produced from camera-ready copy supplied by the authors.

British Library Cataloguing in Publication Data
A catalogue record for this book is available from the British Library

Library of Congress Cataloging in Publication Data
Mejlbro, Leif.
Diffusion of chloride in concrete : theory and application / Leif Mejlbro, Ervin Poulsen.
p. cm. — (Modern concrete technology ; 14)
Includes bibliographical references and index.
ISBN 0-419-25300-9
1. Concrete—Deterioration. 2. Reinforced concrete—Deterioration. 3. Concrete—Effect of salt on.
I. Poulsen, Ervin. II. Title. III. Series.
TA440.M394 2005
620.1′36′2—dc22 2005050889

Visit the Taylor & Francis Web site at
http://www.taylorandfrancis.com

and the CRC Press Web site at
http://www.crcpress.com

Contents

List of Figures

List of Tables

List of Examples

Preface

This book provides information on chloride ingress into concrete in marine (seawater) and road environments (de-icing salt). It reviews the present state of knowledge of chloride ingress into concrete and presents the information ready for analysis and design of *reinforced concrete structures* (RC *structures* for short in the following) exposed to chloride laden environment.

The aim of the book is to bridge the gap between concrete technologists and mathematicians. Therefore, the book could be of interest to

- Students and researchers of concrete technology with an interest in chloride ingress into concrete by diffusion and mathematical modelling of the diffusion process and the effect on concrete.

- Students and researchers of applied mathematics with an interest in the application of diffusion processes.

- Structural engineers who want to apply methods of chloride ingress into concrete based upon diffusion in the analysis and design of RC structures with respect to durability.

The book deals with testing, analysis of chloride ingress into existing RC structures, estimation of chloride ingress into concrete and prediction of initiation period and service lifetime of RC structures under design. Special problems are also considered like effects of e.g. prediction of reinforcement corrosion and coating.

In the literature of applied mathematics Fick's second law is mostly known as the *heat equation*, because it was first applied when describing heat transfer. A better name would be the *diffusion equation*, because it describes the first linear approximation of diffusion of something, e.g. heat, or chloride ingress into concrete. Now, one cannot expect any physical process to be truly linear. In the case of the actual fire the nonlinearity of the problem was very striking. In case of chloride ingress, however, the linear model will in quite a few cases give us a good approximation of the real world. For that reason we have focussed so much on the Fick's second law in this book.

We believe that there exist other applications, because the diffusion equation contains some not well defined random element. Furthermore, unlike the elliptic differential equation and the wave equation, one can never reconstruct the past for the diffusion equation. In some sense the diffusion equation obeys the law of entropy, which the former two mentioned types of equations do not. Hence, the diffusion equation is closer to the truth when describing the real world.

Unfortunately, from a mathematical point of view the diffusion equation is much harder to solve than the elliptic equation or the wave equation, so even the rich classical literature on the heat equations may seem poor compared with the literature on the elliptic equations or the wave equations.

The solution of Fick's second law is not an easy matter. We have tried to extend the solution formulæ to more realistic cases than already known in the classical literature. Occasionally we have been forced to start from scratch. We do not know whether these results exist elsewhere.

Although the nature in principle behaves in a nonlinear way, the book only treats linear models. There are many reasons for this limitation. One is that chloride ingress into concrete is such a slow process that linearization usually gives fair approximate results, because higher order terms are small compared with the dominating linear term.

Another reason is that even the linear model contains many unsolved problems, or useful results which are not well-known outside a small group of experts. By 'useful' we here mean 'applicable in practical engineering problems'. Many of the formulæ in this book require much less computer time than the traditional finite element method which usually is applied.

A third reason is that the linear theory is much easier to handle than a full nonlinear model (i.e. non-Fickean diffusion). It would be outside the scope of this book to present extremely complicated solution formulæ for more generalized cases, as long as even the linear model can give so many new results, which have not been tested yet. The philosophy is that one should only consider the non-linear theory, when a linear model has been tested by experiments, and the experiments deviate too much from the prediction given by the linear model.

Finally, the lazy aspect, the linear theory has been known for centuries by mathematicians, so there exists a rich literature which we could use.

However, this last non-serious reason did not help us much. The mathematical solution formulæ of Fick's second law are extremely useful from a theoretical point of view, because one can derive general properties of the solution (e.g. the maximal principle, the solution is of class C^∞, except possibly on the boundary, etc.). However, except for only very special cases, which cannot be expected in general, the usual mathematical solution formula cannot successfully be applied in practical calculations. Convolution integrals involving $\operatorname{erfc}(x)$ are not particular nice, to put it mildly.

We have tried to develop alternative methods which can be applied on a computer, or even on an advanced pocket calculator. The main guideline in writing the mathematical part of this book has been the following: Whenever a model and a solution formula have been developed, they have been tested on a pocket calculator (TI–92). Only in the case of coating in Chapter 7 the calculations became too big for a pocket calculator, so we have used MAPLE instead. Here, the calculations concerned with coating lie on the edge of what can be performed on a pocket calculator (it can be done, but it will take a fairly long time), while the problem of cracks in concrete is far more complicated. This is the only example where a pocket calculator could not be applied. Therefore we have not included these calculations, though we could do something using MAPLE instead.

As mentioned above, the linear theory has been known for centuries. It was first introduced by *Fourier* in his treatment of heat flow in 1822. For that reason it is usually called the *heat equation.* It is only in civil engineering that it is called *Fick's second law*, so readers interested in more theoretical works should search for the heat equation as well.

It should be mentioned that Fourier solved the heat equation by means of what is known as the theory of *Fourier series.* It should therefore be no surprise that much of the book contains formulæ, which has the structure of a generalized Fourier series.

Traditionally, the mathematicians mainly study three types of classical partial differential equation, or their generalizations, namely elliptic, hyperbolic and parabolic equations. (There are other types, but they are of less importance.) Elliptic equations are potential (static) equations. Hyperbolic equations are wave equations, while parabolic equations, or heat equations, make up a limiting case lying in some sense between the two other cases, so they take some properties from the elliptic case and some from the hyperbolic case, and then they have their own important properties, which neither the elliptic not the hyperbolic equations share with them. We shall in the following briefly discuss a couple of these differences, because they will describe

- why parabolic equations are better than elliptic and hyperbolic ones in modelling, and
- why the practical implementation often is so difficult.

Since the solution formulæ of the elliptic and hyperbolic equations are very different in their structure, and the parabolic equations can be obtained as a limiting case of both elliptic and hyperbolic equations, we may expect a rather messy solution. This is indeed true. Since elliptic and hyperbolic equations are nicer, most mathematical papers are dealing with these two cases.

On the other hand, elliptic equations are static and they cannot directly describe an impact in time. They are therefore useless in describing e.g. chloride profiles in time.

Hyperbolic equations describe typically the propagation of a wave, so here we have a candidate. Unfortunately, it can be proved that one can always reconstruct the past and predict the future for the wave equation, and this violates the law of entropy.

For parabolic equations one can predict the future, but never reconstruct the past, so the law of entropy is not violated (the time variable can only increase, so there is only one time direction, which cannot be reversed). At the same time the solution of the heat equation inherits the same nice property as the solutions of the elliptic equations, namely that the solution is always smooth in the interior of the domain. Solutions of the hyperbolic equations do not have this property. A discontinuity on the boundary will always propagate into the domain and it will never disappear.

From this point of view, the parabolic equations can be considered as the most important modelling equations in any engineering science (and not just Fick's second law for chloride ingress), and therefore one has to live with the fact that the solution may cause some numerical difficulties.

This can be put in the following way: Whenever a stochastic process is at hand, one should look for a suitable parabolic equation. This has recently been done by the introduction of the so-called *stochastic differential equations*, which e.g. are used to predict the behaviour of the stock market. Roughly speaking, one considers the heat equation and then add a term of 'white noise', calling the result a stochastic differential equation. The solution of the heat equation (here called the diffusion equation) gives the trend of the market, while the additional white noise takes care of unpredictable fluctuations of the market.

For the time being we do not know how far one has come in predicting the stock market by using stochastic differential equations, except for the fact that it still cannot handle the effect of sudden disasters like wars, terror actions, earthquakes, floods, etc. on the world economy. The point here is that the same mathematical equation is used under different names to describe such diverse subjects as heat transfer, chloride ingress into concrete and the trend of the stock market. There may be many other applications not known to us.

The work has been divided between us, such that Ervin Poulsen has written Chapter 1, Chapter 2 and Chapter 5, while Leif Mejlbro has written the remaining chapters. In the draft, Chapter 6 consisted of 70 pages, but we judged that it was far too difficult in that version, compared with the rest of the book, so it was cut down to its present size.

Ervin Poulsen

Leif Mejlbro

January 2005

Acknowledgement

The authors have freely drawn on the data, information and conclusions from the Danish HETEK project and other projects carried by the AEClaboratory. Especially thanks to *Jens Mejer Frederiksen*, B.Sc.Eng., for fruitful discussion on chloride ingress into concrete.

The authors also wish to acknowledge the valuable advice and guidance provided by Prof. Emer. *Per W. Karlsson*, and to Ass. Prof. *Gunnar Mohr*, Department of Mathematics, Technical University of Denmark, for constructive and helpful comments in the preparation of the present book.

Chapter 1

Introduction and Reader's Guide

Chloride is present everywhere. However, the knowledge that chloride (e.g. in marine environment) is very aggressive against reinforcement steel bars has not been a hindrance for the development of marine *reinforced concrete structures* (RC *structures* in the following). Early in the history of RC structures, concrete has been applied for marine RC structures and other chloride exposed RC structures like e.g. jetties, piers, breakwaters, bridges, pavements, ships, floating platforms and other off-shore structures.

According to previous technical papers about marine RC structures it is seen that even if the transport mechanism of chloride in concrete was not known, 'seawater attack' became an institution. Several field studies and examinations of reinforced concrete structures exposed to chloride were carried out and reported in the literature, e.g. by Atwood *et al.* (1924) and Feld (1968). The knowledge at the time about the consequences of seawater attack upon RC structures was very practical orientated and based on the engineers' practical experiences – something which we need today.

Many RC structures are exposed to chloride. It is not only marine RC structures which are exposed to chloride (from seawater). Also road RC structures suffer from chloride attack due to de-icing with chloride-laden agents. In some geographical areas the soil may contain chloride (raised marine soil). From a chloride-laden soil the chloride may penetrate concrete in contact with the soil.

Furthermore, several of our industries apply chloride in their production (e.g. abattoirs). As a special chloride exposure for an RC structure is the chloride attack on a structure with building materials of PVC, which during a fire will develop chloride as hydrochloric acid (HCl).

The chloride from the various sources are divided into two main groups

- *Chloride incorporated* in the (fresh) concrete when it was mixed, e.g.

from salty aggregates, salty mixing water and admixtures containing chloride.

- *Chlorides penetrating* into the (hardened) concrete from the environment, e.g. from seawater, salty pool-water, salty groundwater, sea spray, de-icing salts, from processing or storage of halides, and from other industries like the food-industries (slaughterhouses, osterias etc.).

Only two decades ago it was believed that the chlorides incorporated in the concrete when it was mixed were bound (i.e. insoluble) chlorides and should therefore not be risky as long as the content of chloride was kept at a maximum of two per cent by mass binder. Now, several cases have shown (laboratory tests and field inspections) that the bound chlorides will be released by the carbonation and leaching, leading to corrosion of reinforcement. Thus, it is now specified by codes, e.g. the Eurocode ENV 1992-1-1 and EN 206-1 (1999), that there must be an upper limit of chloride in reinforced and prestressed concrete. This limit is a certain fraction of the threshold value of chloride in concrete.

1.1 The process of chloride ingress

Concrete is a kind of material to which liquids may penetrate. When concrete is exposed to chloride, the penetration of chloride occurs via the system of capillary pores, cracks and defects of the concrete.

1.1.1 Types of chloride transport in concrete

The transport mechanisms are usually divided into the following groups

- *Diffusion*, where the transport of chloride is driven by the difference of the concentration of chloride in various zones. The chloride always diffuses into zones with smaller chloride concentration.

- *Permeation*, where the chloride transport is driven by a difference of the hydraulic pressure in various zones. Chloride always moves into zones with smaller hydraulic pressure. The transport of chloride in concrete by means of permeation may occur in concrete with a high intensity of cracks and defects.

- *Migration*, where the chloride transport is driven by a difference in electrical potential. Chloride always moves into zones with less electrical potential, cf. Mays *et al.* (1992). Migration of chloride in concrete may occur when exposed to stray current e.g. on Metro tunnels, cf. Kirsh (1981). The negative chloride ions (e.g. from soil in connection with the concrete) will migrate towards the (positive) anode zones of the reinforcement and there increase the corrosion rate and cause pitting corrosion. Migration is applied by electro-chemical chloride extraction, cf. NCT (1991), and by various test methods for the determination of the chloride diffusion coefficient, cf. the test method ASTM C 1202-94 (1994).

- *Convection*, where the chloride transport is driven by a difference in moisture content (pressure). Water (with chloride) always moves towards zones with less moisture content, all other parameters being equal. A special case is drying of concrete from a surface. Then the pore liquid of the concrete moves towards a surface exposed to evaporation. In case the concrete contains chloride, the evaporation involves concentration of chloride at the surface where the evaporation takes place, the so-called wick action. Another special case is the repeated wetting and drying which may involve very high concentration of chloride at the surface.

1.1.2 Equations of diffusion

The theory of diffusion is mainly based upon the mathematical models by *Adolph Eugen Fick*. Diffusion is the predominant transport mechanism of chloride in concrete. Thus, it is important for the understanding of the mathematical models to know the conditions and background for diffusion.

Adolph Eugen Fick, 1829–1901, had a special interest in mathematics and wanted to become a mathematician. However, his family persuaded him to study medicine at the University of Marburg. Here he became convinced that medicine fundamentally must have its basis in mathematics, physics and

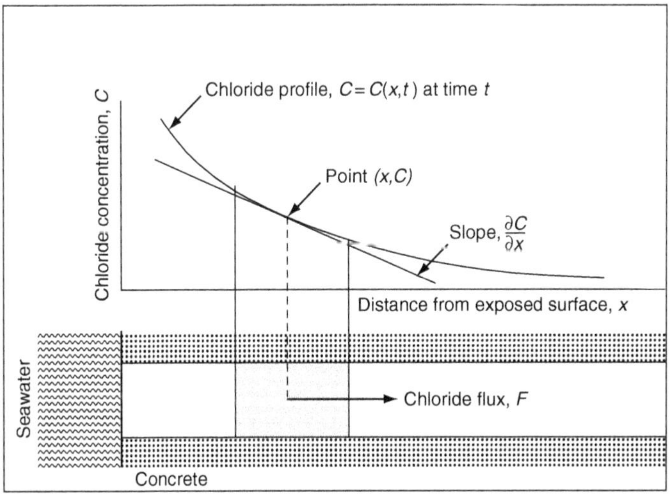

Figure 1.1: Fick's first general law of diffusion expresses that the flux of chloride ions is proportional to the chloride concentration gradient normal to the section.

chemistry. *Fick* had many interests, but they were all based upon his great interest for mathematics.

In his first paper on diffusion, Fick (1855) brought system and theory to many observations. In fact, he knew that the same mathematical theory was valid for diffusion as well as for heat movement (as stated by Fourier in 1822).

The application of Fick's laws on chloride diffusion in concrete appears many years later in Collepardi *et al.* (1970, 1972). Thus, the study of chloride diffusion processes in concrete technology is still fairly young.

Fick's second general law of diffusion

When chloride diffuses into concrete, a change of the chloride concentration C occurs at any time t in every point x of the concrete, i.e. it is a non-steady state of diffusion.

In order to simplify the problems we first deal with one-dimensional diffusion, i.e. there is a gradient of chloride concentration C only along the x-axis, cf. Figure 1.1 and Figure 1.2. Many problems of chloride ingress can be solved by the applications of this simplification in practice.

Consider an element parallel to the chloride diffusion into a semi-infinite volume of concrete. For convenience the cross-section of the element is taken as $dA = 1$. Thus, between the two consecutive sections at a distance of dx the volume is $dV = dx$. Per unit time, $dt = 1$, the amount of chloride which diffuses into this volume dV is by definition the flux F, cf. Figure 1.2. In the same way, the amount of chloride which diffuses out from the volume dV is the flux at the abscissa $x + dx$. Therefore, the flux changes along the x-axis

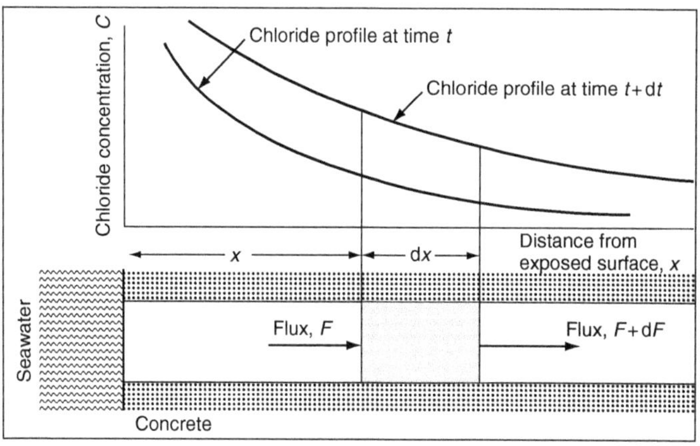

Figure 1.2: Fick's second general law of diffusion expresses that change in chloride ion content per unit time (on the t-axis) is equal to the change of flux per unit length (on the x-axis).

into the concrete with $\partial F/\partial x$ per unit of the x-axis. Thus, during the time $dt = 1$ the increase of chloride in the volume dV becomes:

$$\frac{\partial C}{\partial t}\,dx = F - \left\{F + \frac{\partial F}{\partial x}\,dx\right\} = -\frac{\partial F}{\partial x}\,dx, \tag{1.1}$$

or

$$\frac{\partial C}{\partial t} = -\frac{\partial F}{\partial x}. \tag{1.2}$$

This law, Fick's second law of diffusion, simply says that the change in chloride content per unit time is equal to the change of flux per unit length.

Equation (1.2) is also called the mass balance equation, since it describes that the change in chloride content per unit time of an infinitesimal volume is equal to the difference in chloride flux to and from the infinitesimal volume per unit length.

Fick's first law of diffusion

The assumption of Fick's diffusion theory is that the transport in concrete of chloride ions through an unit area of a section of the concrete per unit of time (the flux F) is proportional to the concentration gradient of the chloride ions measured normally to the section, i.e. $\partial C/\partial x$, i.e.

$$F = -D \cdot \frac{\partial C}{\partial x}. \tag{1.3}$$

The negative sign in Fick's first law of diffusion, Equation (1.3), arises because the diffusion of chloride ions occurs in the direction opposite to that of increasing concentration of chloride ions, cf. Figure 1.1. In Equation (1.3) the constant of proportionality D is called the chloride diffusion coefficient. In general D is not a constant, but depends on many parameters like the time for which diffusion has taken place, location in the concrete, composition of the concrete, etc.

If the chloride diffusion coefficient is a constant, $D = D_0$, Equation (1.3) is usually referred to as Fick's first law of diffusion. If this is not the case, the relation is usually referred to as Fick's first general law of diffusion. There are cases where such a simple relation as Equation (1.3) should not be applied.

The diffusion process may be irreversible or have a history-dependence. In such cases Fick's law of diffusion is not valid, and the diffusion process is referred to as a 'non-Fickean' or 'anomalous' diffusion process. However, no observation so far indicates that the chloride diffusion in concrete should be characterized as an 'anomalous' diffusion.

The differential equation of diffusion

One-dimensional problems. By applying Fick's first general law of diffusion, i.e. Equation (1.3), the mass balance equation Equation (1.2) becomes

$$\frac{\partial C}{\partial t} = \frac{\partial}{\partial x}\left\{D\,\frac{\partial C}{\partial x}\right\}. \tag{1.4}$$

In order to apply Fick's second law of diffusion in this form for concrete exposed to chloride during a longer period of time, one ought to know the development of the chloride diffusion coefficient in time, i.e. the function $D = D(t)$. If only few observations exist in a specific case it is possible to estimate an upper and lower bound for the variation of $D = D(t)$.

To solve this partial differential equation the initial condition and the boundary condition have to be known, cf. Section 1.1.3.

In general, the chloride diffusion coefficient depends on the variables of the differential equation x, t and C, i.e. $D = D(x, t, C)$. However, in many cases it is possible to ignore such dependency, except the time dependency.

The special case where the chloride diffusion coefficient is independent of location x, time t and the chloride concentration C, i.e. $D = D_0$, is in particular interesting. In this case *Fick's second law* can be written in the simpler form

$$\frac{\partial C}{\partial t} = D_0 \cdot \frac{\partial^2 C}{\partial x^2}. \tag{1.5}$$

Examples of solving Equation (1.5) are e.g. shown in Sections 4.3.1 and 4.3.2.

Three-dimensional problems. Certain problems of chloride ingress have to be solved in two or three dimensions. The diffusion of chloride in concrete, when assuming the concrete to be an isotropic medium, obeys the following equation, corresponding to Equation (1.2),

$$\frac{\partial C}{\partial t} + \frac{\partial F_x}{\partial x} + \frac{\partial F_y}{\partial y} + \frac{\partial F_z}{\partial z} = 0. \tag{1.6}$$

Here, F_x, F_y and F_z are the flows of chloride ions in the x, y and z-axis respectively. By inserting Fick's first law of diffusion, Equation (1.6) becomes

$$\frac{\partial C}{\partial t} = \frac{\partial}{\partial x}\left\{D\,\frac{\partial C}{\partial x}\right\} + \frac{\partial}{\partial y}\left\{D\,\frac{\partial C}{\partial y}\right\} + \frac{\partial}{\partial z}\left\{D\,\frac{\partial C}{\partial z}\right\}, \tag{1.7}$$

which is the corresponding equation to Equation (1.4).

The special case where the chloride diffusion coefficient is independent of location x, time t and the chloride concentration C, i.e. $D = D_0$ is of interest. In this case *Fick's second law* is written in the simpler form

$$\frac{\partial C}{\partial t} = D\left\{\frac{\partial^2 C}{\partial x^2} + \frac{\partial^2 C}{\partial y^2} + \frac{\partial^2 C}{\partial z^2}\right\}. \tag{1.8}$$

Application and solution of Equation (1.8) are shown in Sections 4.3.3 and 4.3.4.

Three-dimensional problems in semi-polar (or *cylindrical*) *coordinates.* The equation for the diffusion of chloride in a concrete cylinder results from a transformation of the coordinates in Equation (1.7) by

$$\begin{cases} x = r \cos\varphi, \\ y = r \sin\varphi, \\ z = z. \end{cases} \tag{1.9}$$

This gives the equation

$$\frac{\partial C}{\partial t} = \frac{1}{r}\left\{\frac{\partial}{\partial r}\left(r\,D\,\frac{\partial C}{\partial r}\right) + \frac{\partial}{\partial \varphi}\left(\frac{D}{r}\,\frac{\partial C}{\partial \varphi}\right) + \frac{\partial}{\partial z}\left(r\,D\,\frac{\partial C}{\partial z}\right)\right\}. \tag{1.10}$$

In case the chloride diffusion coefficient remains constant, $D = D_0$, i.e. is independent of the parameters r, φ and z, Equation (1.10) may be written as

$$\frac{\partial C}{\partial t} = D_0\left\{\frac{\partial^2 C}{\partial r^2} + \frac{1}{r}\,\frac{\partial C}{\partial r} + \frac{1}{r^2}\,\frac{\partial^2 C}{\partial \varphi^2} + \frac{\partial^2 C}{\partial z^2}\right\}. \tag{1.11}$$

The solution of this model is e.g. shown in Section 4.3.9 and Section 4.3.11.

Two-dimensional problems in polar coordiates are described by the model above by deleting every term, which contains the third variable z.

Effect of cracks on chloride ingress

The assumption of the diffusion equations given above is that the concrete can be considered as being a quasi-homogeneous medium. Therefore, chloride ingress into concrete through cracks cannot be covered directly by the diffusion equations. However, by assuming that the sides of a crack represent concrete surfaces through which the chloride can penetrate, it becomes possible, even though it is difficult, to determine the effect of cracks on the chloride ingress.

1.1.3 Initial and boundary conditions

In order to solve Fick's second law of diffusion the initial conditions and the boundary conditions must be known. Depending on the level of knowledge about these conditions, it is convenient to define some special cases.

The initial condition is the statement of the chloride content C_i of the concrete when exposure starts. Here it is assumed that C_i is a constant.

The Collepardi model

Old RC structures. For a concrete which is fairly old (e.g. more that approximately 20 years) the chloride diffusion coefficient may be considered to remain constant, $D = D_0$. Furthermore, the chloride content of the chloride exposed concrete surface C_{sa} may have stabilized to a constant value $C_{sa} = C_0$.

This leads to a simple diffusion equation, cf. Equation (1.5) with a simple initial condition C_i = constant (i.e. equally distributed) and a simple boundary condition $C_{sa} = C_0$. Section 2.4 deals with this problem.

Determination of chloride parameters. When the chloride diffusivity is described one usually determines the chloride ingress and one determines the chloride parameters as if they were constants. The chloride diffusivity of concrete is determined under two circumstances

1. Determination of the chloride parameters of laboratory specimens, kept under specified exposure, e.g. NT Build 443, cf. Nordtest (1996a).

2. Determination of the chloride parameters of RC structures (e.g. marine RC structures) or concrete specimens (e.g. at a marine exposure station) exposed to a chloride-laden environment.

When obtaining a chloride profile there is no information about the initial, boundary and physical conditions. Thus, in order to make it as simple as possible it is assumed that the chloride parameters remain constant during the entire period of chloride exposure. However, this is not true, but since only one chloride profile is determined versus time, no information is available on these conditions.

Section 2.1 deals with the determination of the chloride parameters from a chloride profile.

The LIGHTCON model

Gradually as the numbers of chloride profiles from the same concrete and same environment, but at various times (ages) increased, it was recognized that the achieved chloride diffusion coefficient of concrete, D_a, was time-dependent, i.e. $D_a = D_a(t)$, as shown by Takewaka *et al.* (1988). The achieved chloride diffusion coefficient of concrete is defined as the constant value of the diffusion coefficient which leads to the measured chloride ingress (at a constant surface chloride content D_a).

Takewaka *et al.* (1988) proposed to use a power function to model the achieved diffusion coefficient. This dependency is later proved by e.g. Mangat *et al.* (1995), and Maage *et al.* (1993). Section 5.1 deals with the LIGHTCON model applied to RC structures in chloride-laden environment.

The HETEK model

The increasing number of observations from inspection of marine RC structures, especially from Japan, demonstrated that the achieved surface chloride content must be time-dependent. Uji *et al.* (1990) proposed on basis of their observations to model C_{sa} as being proportional to the square root of the time. This was convenient since Fick's second law of diffusion in this case has a simple solution, cf. Crank (1956). This solution was applied for the first time by Purvis *et al.* (1994) and Gautefall *et al.* (1994).

Later Swamy *et al.* (1995) showed by a comprehensive investigation that C_{sa} is not always increasing with the square root of time. However, Swamy *et al.* (1995) showed that it was possible to assume that C_{sa} is increasing with time as a power function.

It was not possible at that time to determine the influence of such parameters as the w/c-ratio, types of cement and other binders like fly ash, silica fume and slag, but laboratory studies showed that there was an effect, cf. Byfors (1987). Many inspections and field investigations at that time in Scandinavia have also shown a great influence.

At that time there was no general solution to Fick's second law of diffusion, when the surface chloride content versus time was a power function. However, Mejlbro (1996) modelled the solution and tabulated the necessary functions. There is a certain scepticism of the increase of the surface chloride content versus time. Not too many observations have been made on old marine RC structures. Section 5.2 deals with the HETEK model, and it covers also other relations for surface chloride than the power function.

1.1.4 Chloride binding

Chloride which diffuses through the pores of the cement paste will be exposed to various chemical compounds. Thus, there is a possibility of the chloride to be bound to the cement paste, either physically to the cement gel or chemically. Only the free chloride will diffuse and therefore the binding of chloride is important to the chloride profile. The relationship between the free and the bound chlorides will vary with the type of binder and has to be determined experimentally.

Some of the chloride bound to the cement gel may be released when such parameters of the pore liquid change as e.g. pH-value, temperature and pressure. Therefore, it is most convenient to determine the chloride of the concrete as acid soluble chloride.

In this book it is assumed that chloride in concrete is determined or referred to as the acid soluble chloride content of the concrete or its binder.

1.2 Chloride-laden environments

A very important parameter to the chloride ingress into concrete is the environment. When two identical concretes are exposed to e.g. a marine environment at various positions of a marine RC structure, the chloride ingress into the concrete varies with its position. Therefore, it is not possible to describe the concrete just by its composition. The dissimilarities of the chloride-laden environment in various positions of the structure play an important role.

The chloride ingress into concrete depends on an interaction of mainly the local environment (i.e. the intensity of chloride), the composition of the concrete (i.e. the binder and the w/c-ratio), defects of the concrete and time.

1.2.1 Sources of chloride

Chloride may be added to the concrete as impurities of the constituent materials or added to the concrete by means of admixtures. However, the chloride content of any concrete admixture is limited today. In former times calcium chloride used as an accelerator to concrete has caused so many defects that calcium chloride is no longer accepted in concrete. Thus, left as predominant chloride sources are seawater, de-icing salts and industrial processes.

Seawater

Seawater and brackish water contain substances, which are aggressive against concrete and its steel reinforcement (i.e. mainly chloride). Chloride involves heavy pitting corrosion of reinforcement in concrete, if the conditions are present, i.e. a suitable concentration of chloride in concrete in touch with the reinforcement, suitable humidity of the concrete and a possibility of an oxygen supply.

The presence of magnesium sulphate in the seawater may influence the diffusibility of the concrete by forming a coating of brucite, $Mg(OH)_2$. Other reactions may take place in forming aragonite, $CaCO_3$. In such cases the diffusibility of concrete exposed to seawater and in the laboratory to water containing chloride may differ, even when all other parameters are equal. Also, marine growth and formation of shells on the concrete surface of marine RC structures and attack of marine organisms may be of some importance, cf. Hoff (1986).

Chloride containing de-icing salts

Not all de-icing salts contain chloride, but the de-icing salt containing chloride is the cheapest and the most efficient. Especially sodium chloride, NaCl, is cheap and therefore commonly used. Not only RC structures suffer from de-icing salts; so do plantations along roads.

When removing ice from RC structures by means of chloride containing de-icing salts, the structure will be exposed to chloride through the melt water as well as from traffic splash. The melt water often has a very high concentration of chloride. The chloride is transported from the road and to e.g. car parks. In a multi-storey car park where the remaining chloride is not removed, the chloride ingress into the concrete may be very serious.

PVC fire

In general, concrete has good properties to resist fire. However, when an RC concrete component is covered by panels made of PVC and exposed to fire, then there is a risk of corrosion. When PVC burns it releases gaseous hydrochloric acid, HCl. In reaction with concrete it forms calcium chloride, $CaCl_2$, which is water soluble. Thus the chloride will penetrate the concrete.

Industrial processes

A part of industrial processes involves chloride in high concentration, e.g. abattoirs and osterias. Thus, there is a risk for reinforcement corrosion, if the RC structures are not sufficiently protected against chloride ingress into the concrete.

1.2.2 Marine environments

It is convenient to define the following three environmental zones of marine exposure, cf. Frederiksen *et al.* (1997),

- *Marine atmosphere, ATM.* Concrete placed 3 m or more above the highest maximum water level incl. waves, cf. Figure 1.3. Concrete exposed to marine atmosphere can, if relevant, be subdivided into leeward and windward marine atmospheres.

- *Marine splash, SPL.* Concrete placed between 3 m above the highest maximum water level incl. waves and 3 m below the lowest minimum water level inclusive of waves, cf. Figure 1.3.

- *Submerged in seawater, SUB.* Concrete placed 3 m or more below the lowest minimum water level inclusive of waves, cf. Figure 1.3.

Governing parameters of marine structures

The governing parameters depend on the composition of the concrete as well as the environmental zones. Their values have to be estimated from in situ observations achieved by the marine structure in the relevant environment. A set of at least two observations per parameter is necessary for this estimation, and the observations ought to cover a time interval as wide as possible.

As the first set of values, the HETEK model uses the achieved values after exactly one year of exposure. These values are obtained from the analysis of a large number of concrete specimens exposed to various environments established in Träslövsläge Marine Exposure Station (on the west coast of Sweden), cf. Sandberg (1996) on the Swedish research project BMB and cf. Frederiksen *et al.* (1997) on the research project HETEK. At this exposure station several concrete specimens representing fifteen different types of concrete were exposed. The following parameters were studied

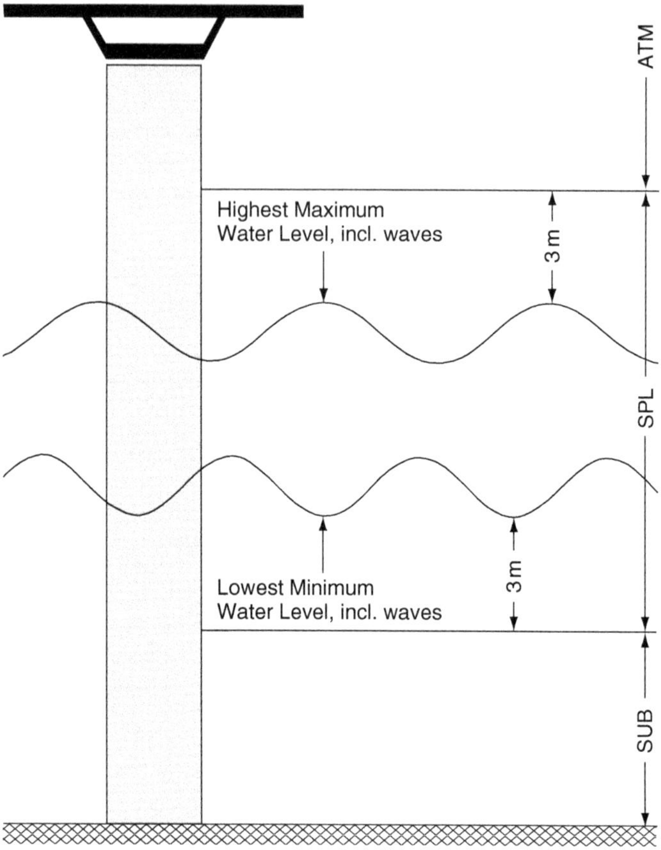

Figure 1.3: A schematic illustration of a bridge pier to visualize the definition of the various local marine environments according to the HETEK model, cf. Frederiksen *et al.* (1997) and (2000):

• *Concrete exposed to marine atmosphere*, ATM, is concrete placed 3 m or more above the highest maximum water level, incl. waves.

• *Concrete submerged in seawater*, SUB, is concrete placed at most 3 m or more below the lowest minimum water level, incl. waves.

• *Concrete exposed to marine splash*, SPL, is concrete placed in between ATM and SUB.

- 4 types of Portland cement,
- 2 types of silica fume,
- 2 types of fly ash,
- w/c-rations from 0.25 to 0.75 by mass.

Figure 1.4: Seawater increases the frost action on concrete without adequate entrained air. Especially, in the constructions joints.

Concrete submerged in seawater versus marine atmosphere

Calculation of the governing parameters for concrete exposed to marine splash, submerged in seawater and to marine atmosphere principally takes place in the same way as described above. Since the temporary conclusion of the HETEK project and the BMB project, some observations of chloride ingress into concrete exposed to marine splash or submerged in seawater have indicated that the chloride content of the exposed concrete surface seems to stabilize after approximately two years, cf. Maage *et al.* (1999) and Mangat *et al.* (1995). However, the two to five years observations for concrete specimens at Träslövsläge Marine Exposure Station do not show this phenomenon.

In the case of an existing marine structure, it is recommended to choose the 1-year values of the parameters, cf. Equation (5.2) and Equation (5.78). As a proper date for the later observation one can choose to carry out an inspection in such a way that it is possible to obtain several chloride profiles of the concrete in question. The achieved chloride diffusion coefficient and the achieved chloride content of the concrete surface can be determined from the observed chloride profiles by nonlinear regression analysis.

However, in the case of a marine structure under design one is forced to apply the estimation of the 100-years value found by the HETEK project (based on the observations from Träslövsläge Marine Exposure Station) or a similar kind of information.

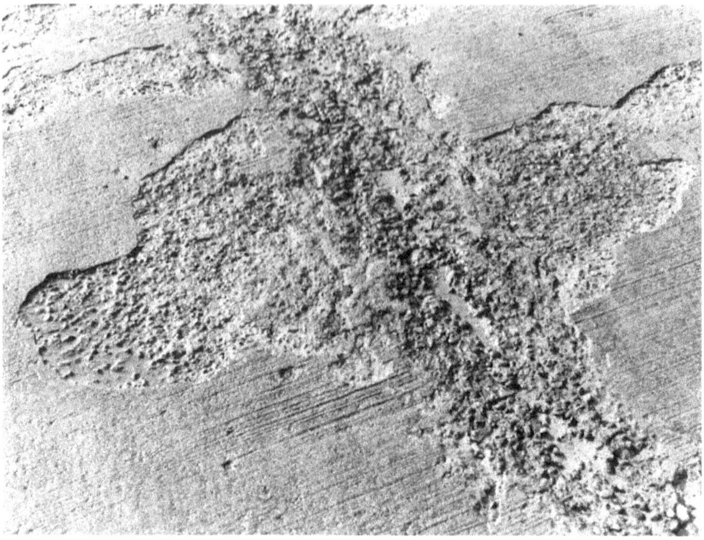

Figure 1.5: A near-by-surface layer of concrete exposed to seawater (breakwater) is spalled and eroded, especially at locations of construction joints.

Figure 1.6: De-icing salt creates a very hostile local environment. When a concrete staircase is exposed to de-icing salts and the concrete is not designed to resist such an exposure (by air entrainment) the concrete will spall.

Figure 1.7: Even in cases where the concrete has a higher strength, but does not contain a sufficiently amount of entrained air, the concrete will deteriorate due to frost action.

1.2.3 Road environments

Special environments are found for road bridges or equivalent RC structures, where the chloride exposure originates from the use of de-icing salt containing chloride. Not all de-icing salts contain chloride, but the de-icing salts containing chloride are the cheapest and the most efficient ones. Therefore, they are the most common ones in use. A detailed description of the road environments is given by Frederiksen *et al.* (1997).

Parameters of the road environment

- *Chloride containing de-icing salt.* The most important parameter for the road environment is the de-icing salt, NaCl, which may be spread as a NaCl solution or as dry salt. In Scandinavia de-icing salts are spread 30 to 60 times each winter season. Each time 10 to 25 g NaCl is spread per m^2 of the road surface, depending on the conditions.
- *Traffic splash.* When the de-icing salts are spread they melt ice and snow on the road. Thus, when the road surface gets wet a driving car will generate a splash of chloride-laden melt water, depending on the speed of the car. When the traffic splash hits the concrete of an RC structure, a chloride ingress will take place.
- *Airborne chloride.* Some of the traffic splash consists of salted water drops so small that drag from the car and wind may deposit the chloride in the neighbourhood of the road.

Figure 1.8: The marine environment is among the most corrosive natural environment for concrete structures. However, the road environment is the most hostile man-made and accepted environment for RC structures. The degree of which reinforced concrete will be able to resist the environmental exposure is dependent on the concrete (i.e. constituent materials, composition, casting, compaction and curing), the overall and detailing structural design (e.g. drainage), the reinforcement detailing (e.g. spacer and cover).
The photo shows an edge beam of an RC bridge. The surfaces of the edge beam are affected by 'off-scaling' where water soaked. It should be noted that damage does not occur just below the pilasters, acting as shelters for the water passing.
The bridge was constructed before the invention of air-entrainment of concrete to resist frost action. Therefore, the concrete detailing is of great importance, e.g. the horizontal soffit of the edge beam will absorb water from puddles and become water saturated. Concrete, which is not entrained with sufficient amount of air, will not be able to resist frost actions without 'off-scaling' especial when de-iced.
In this case the composition of the concrete is unknown.

- *Rain.* In Scandinavia it also rains during the winter season. This means that if a part of the concrete of the road RC structure is exposed to the rain, the chloride on the concrete surface is washed out to a certain stage. Therefore, the road environments are divided into two local environments, namely the wet and the 'dry' road environment.

Figure 1.9: Concrete blocks of a breakwater showing surface spalled and erosion (by salt crystallization) from evaporation of seawater puddles.

Figure 1.10: Close-up of the photo shown on Figure 1.9. Notice the corroded lifting hooks across the slot.

- *Drainage.* The drainage system of the road structure is of importance to the chloride ingress into the concrete. When the drainage is not working satisfactorily the surrounding concrete may suffer from frost damage as well as the reinforcement may show corrosion (sooner or later). Therefore, maintenance is of great importance to road structures.

Figure 1.11: Deterioration of a marine RC structure (jetty), where the tidal zone covers a great deal of the structure. Notice that the deterioration takes place in the splash zone or the atmospheric zone, while the concrete exposed to the tidal zone shows almost no deterioration. It has been saturated with water during high water level so that the reinforcement has been protected from corrosion by lack of oxygen.

- *Height and distance from the road.* The distance from the road is the most important parameter of the road environment. When the distance decreases from the road where the chloride exposure takes place, the response on the concrete with respect to chloride ingress reduces drastically.

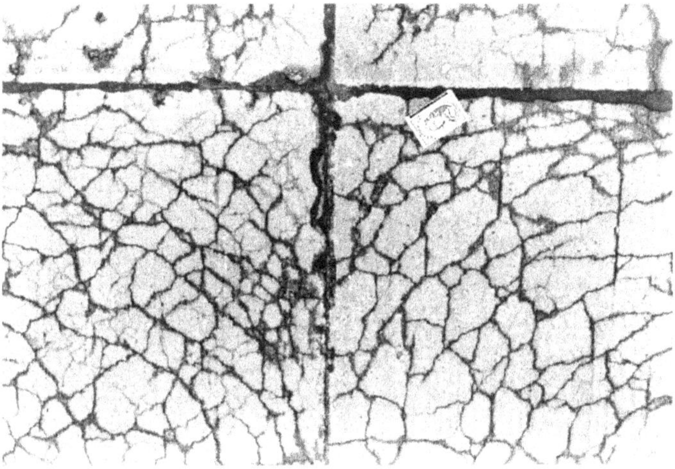

Figure 1.12: An example of a severely cracked concrete pavement due to shrinkage. In such cases the chloride from de-icing salt has an easy access through the concrete to the reinforcement.

Figure 1.13: When de-iced a spacer of plastic material forms an easy access for the chloride to migrate along the interface between the plastic spacer and the concrete and right to the reinforcement.

Road traffic zones

On the basis of the parameters mentioned above, it is convenient to define the following three environmental zones of road exposure of a road bridge or an equivalent RC structure, cf. Frederiksen *et al.* (1997),

Figure 1.14: A badly delaminated concrete soffit of a walkway of RC concrete. The concrete has contained chloride from the aggregates. Therefore, the concrete cover of the reinforcement has not been able to protect the reinforcement. Besides chloride attack the concrete suffers from carbonation.

- *Open traffic splash*, OTS, or wet road environment covers RC components which are exposed to direct driven rain and direct traffic splash from a wet road surface. The maximum chloride exposure takes place at ground level and is decreasingly linearly to the exposure of the distant traffic atmosphere, DTA, defined below. An example may be an edge beam of a bridge deck.

- *Sheltered traffic splash*, STS, or 'dry' road environment covers RC components which are exposed to airborne chloride, but are sheltered from rain and situated close to relatively dry road surfaces (i.e. sheltered from rain, e.g. by the bridge deck). The maximum chloride exposure takes place at ground level and is decreasing linearly to the exposure of DTA. Examples may be columns and the surroundings of a bridge.

- *Distant traffic atmosphere*, DTA. This covers RC components which are exposed to airborne chloride, but placed more than 3 m above or more than 5 m from the road surface. Examples may be the soffit of a bridge deck and the wing walls of a bridge.

Figure 1.15: Failure of a pilaster of the walkway shown on Figure 1.14. When the reinforcing bar is exposed to dry air and sunshine, the corrosion stops, while the embedded reinforcement corrodes since the concrete contains chloride and holds the humidity.

1.3 Corrosion

Steel (mild steel and high strength steel) forms a 'passive' layer and will not corrode when embedded in wet concrete, which is alkaline and free from chloride. Chloride is a catalyst to the corrosion, if the content of chloride in the concrete up to a reinforcing bar is higher than the threshold value of chloride in the concrete. Corrosion of steel is a process, where the steel oxidizes and changes to rust when reacting with the oxygen and water, cf. Broomfield (1997), Bentur *et al.* (1997) and Fontana (1967).

1.3.1 Corrosion and its consequences

Corrosion of the reinforcement will cause defects. The concrete cover of the reinforcement will spall off and the cross section of the reinforcement will be reduced, i.e. the bearing capacity may be reduced.

Anodes, cathodes and incipient anodes

When the chloride content of the concrete up to the reinforcement exceeds the threshold value of chloride in concrete, the passive layer of the reinforcement breaks down. Thus, the reinforcement is no longer protected. This means that defects and other favourable conditions for corrosion will be able to start

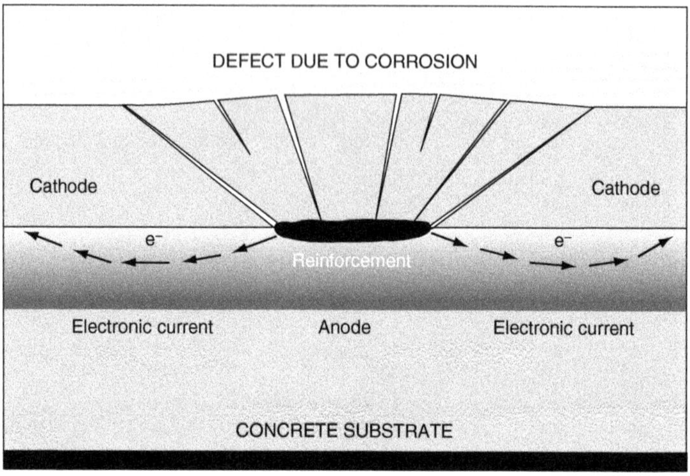

Figure 1.16: When the chloride reaches the reinforcement and breaks down the passive layer, the reinforcement starts to corrode where the conditions are favourable (e.g. optimum of defects and humidity). Corrosion is an anodic dissolution in the concrete pore liquid and a liberation of electrons (which forms an electric current).

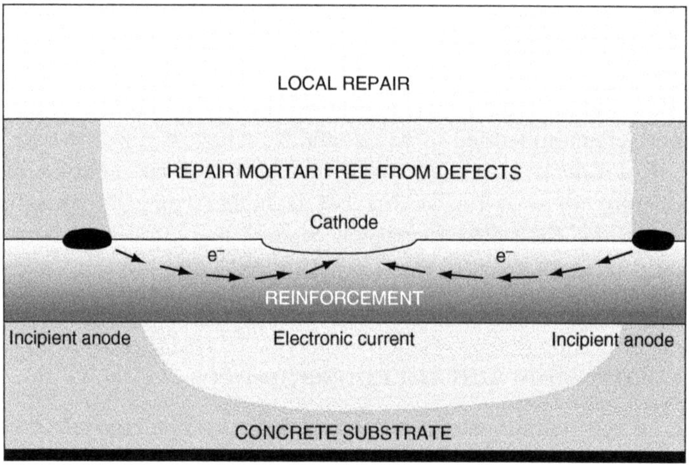

Figure 1.17: By a traditional repair a difference between the 'perfect' repair mortar and the concrete substrate (full of defects) may occur, especially in pH-value, micro cracks and chloride content. The electronic current changes direction and incipient anodes are developed, especially if there is a shrinkage crack at the interface between the substrate and the repair mortar (repair mortar has to be shrinkage compensated).

the corrosion. When the corrosion takes place the steel rusts, i.e. the steel dissolves in the pore liquid under liberation of electron. Therefore, the location of the corrosion is called an anode. In order to stay neutral the reinforcement also creates cathodes. One could say that the difference between the anodic zone and the cathodic zone (chloride content, defects, pH-value etc.) is responsible for corrosion of the reinforcement.

When repairing the corrosion zone the former anodic zone may be in a better condition than the former cathodic zone. This may change the situation so that the former cathodic zone becomes an anode and the former anodic zone turns into a cathodic zone. In other words, the steel reinforcement around the repair starts to corrode. This phenomenon is called incipient anodes after repair.

Corrosion current and corrosion rate

When the couple of an anode and a cathode is formed, an electronic current takes place in the steel reinforcement. The unit of the electronic current I_{corr} is usually μA. It is convenient to define the corrosion rate of a corroding reinforcing bar where the corrosion zone has a length ℓ_s and a diameter d_s as

$$i_{\text{corr}} = \frac{I_{\text{corr}}}{\pi \ell_s d_s}. \tag{1.12}$$

Thus, the unit of the corrosion rate i_{corr} is $\mu\text{A/cm}^2$. When a rebar corrodes there is a rust growth (rate of corrosion) of $r = 0.116$ mm/yr when the corrosion rate is $10\,\mu\text{A/cm}^2$ or a rust growth of $r = 11.6\,\mu\text{m/yr}$ for a corrosion rate of $1\ \mu\text{A/cm}^2$, cf. Broomfield (1997).

In order to define the corrosion activity of reinforcement in concrete Broomfield *et al.* (1993), (1994), defines the following under the assumption that the corrosion rate is kept constant

- Passive condition: $r < 1\,\mu\text{m/yr}$
- Low to moderate corrosion rate: $1\,\mu\text{m/yr} \le r < 5\,\mu\text{m/yr}$
- Moderate to hich corrosion rato: $5\,\mu\text{m/yr} < r < 10\,\mu\text{m/yr}$
- High corrosion rate: $r \ge 10\,\mu\text{m/yr}$.

According to Clear (1989), the consequences are

- No corrosion expected: $r < 6\,\mu\text{m/yr}$
- Corrosion possible in 10–15 yr: $6\,\mu\text{m/yr} \le r < 30\,\mu\text{m/yr}$
- Corrosion expected in 2–10 yr: $30\,\mu\text{m/yr} \le r < 300\,\mu\text{m/yr}$
- Corrosion expected in 2 yr or less: $r \ge 300\,\mu\text{m/yr}$.

Since these tables are only valid for a constant corrosion rate they are mainly useful for a rough estimate because the corrosion rate in practice is time-dependent.

Figure 1.18: An example of typical cracks caused by chloride ingress into concrete of a marine RC structure (quay). When the concrete, besides chloride attack, is exposed to carbonation the critical value of chloride in concrete yields zero. This will accelerate the corrosion of the reinforcement. The reinforcement bars placed at the corners of a column will receive chloride from two sides while reinforcing bars placed at the centre of the sides of the column only receive chloride from one direction.

1.3.2 Threshold value of chloride in concrete

The threshold value of chloride in concrete is here defined as the total acid soluble chloride content of the concrete which just starts corrosion of steel reinforcing bars in the concrete exposed to the chloride-laden environments in question. By this model, the threshold value depends on the concrete composition in the following way

Table 1.1: Efficiency factors and environmental factors.

Binder	Efficiency factors, f	Environmental factors, k_{cr}
Portland cement, PC	+ 1.0	1.25
Silica fume, SF	– 4.7	1.25
Fly ash, FA	– 1.4	3.35

$$C_{cr} = k_{cr} \times \exp\left(-1.5 \times \text{eqv}\left\{w/c\right\}\right) \qquad \text{unit: \% mass binder,} \tag{1.13}$$

where the equivalent water/cement ratio (by mass) is

$$\text{eqv}\left\{w/c_{cr}\right\} = \frac{W}{PC + f_{fa} \times FA + f_{sf} \times SF}. \tag{1.14}$$

Here W, PC, FA and SF denote the mass of water, Portland cement, fly ash and silica fume per m^3 of the concrete. The environmental factors k_{cr} and the efficiency factors f_{fa} f_{sf} are shown in Table 1.1.

The threshold value of chloride in concrete estimated in this way is only valid for concrete free from macro cracks with a crack width greater than 0.1 mm and a rebar cover greater than 25 mm. Therefore, the formulæ given above are not valid for the determination of the initiation time in concrete with crack widths greater than 0.1 mm. The influences of cracked concrete on the threshold value of chloride in concrete are dealt with by Frederiksen *et al.* (1997) and especially by Sandberg (1998).

The threshold value of chloride in concrete is applied in several numerical examples in this book, cf. Sections 2.4.1 and 5.1 and 5.2.

Notice that when corrosion inhibitors are added to the concrete the threshold value of chloride in concrete increases, cf. Section 1.3.3.

1.3.3 Corrosion inhibitors

The definition of a corrosion inhibitor is: 'a small amount of chemical substances which reduces or stops the rate of corrosion without reducing the concentration of the corrosive agents'. cf. EN ISO 8044 (1989). For a corrosion inhibitor for reinforced concrete the definition especially reads: 'a small amount of an admixture added to concrete which increases the onset of corrosion and reduces the rate of corrosion, once the corrosion starts'.

Effects of inhibitors

Corrosion inhibitors have been applied since the 1970s in e.g. metal manufacturing treatments and packaging, in the process industries and in high performance coatings. Now, corrosion inhibitors are used for concrete mainly in connection with concrete repair.

Corrosion inhibitors can either influence the anodic or the cathodic reactions, or both. One must distinguish between corrosion inhibitors and other admixtures or additives to concrete that improve the corrosion stability by reducing the chloride ingress, e.g. silica fume.

Thus, corrosion inhibitors are normally divided into five groups, cf. Fontana (1967), namely:

1. *Anodic corrosion inhibitors* which inhibit corrosion associated with the stabilization of the passivity film which tends to break down when chloride ions are present at the steel surface. Calcium nitrite is e.g. an inhibitor, and it acts to raise the threshold value of chloride in concrete to initiate corrosion. Calcium nitrite is being offered to improve e.g. the efficiency of concrete repairs (patch repairs). However, calcium nitrite is a set accelerator, and measures ought to be taken to prevent 'flash setting', cf. Broomfield (1997).

2. *Cathodic corrosion inhibitors* which adsorb on the steel surface and thus act as a barrier to the reduction of oxygen which is the primary cathodic reaction for steel in concrete. However, most cathodic inhibitors, such as amines, phosphates, zincates, and phosphonates severely increase set retardation of concrete, cf. Bentur *et al.* (1997).

3. *Organic corrosion inhibitors* or adsorption-type corrosion inhibitors, which appear to affect as well the anodic as the cathodic processes and thus being double-active. Typical of these inhibitors are organic amines, e.g. the amino alcohol type.

4. *Vapour phase corrosion inhibitors* are similar to the adsobtion-type corrosion inhibitor, but they possess a very high vapour pressure. Therefore, vapour phase inhibitors are suitable to inhibit atmosphere corrosion without being in direct contact with the metal surface to be protected.

5. *Oxidizing inhibitors.* Such substances as chromates are acting as corrosion inhibitors in many systems. They are primarily used to inhibit the corrosion of metals and alloys that demonstrate active-passive transitions, such as iron and its alloys and stainless steel.

The organic corrosion inhibitors are of great interest for the repair of reinforced concrete since they can migrate through hardened concrete via diffusion. Upon the contact with reinforcing steel the organic corrosion inhibitor forms a monomolecular protective layer which reduces corrosion dramatically by a protection of as well the anodic areas as the cathodic areas of the reinforcing rebars in concrete, i.e. double-acting.

It is important to add enough inhibitor to the concrete since some inhibitors accelerate corrosion when present in small concentrations (may cause

Figure 1.19: A sixty years old reinforced concrete column of a swimming pool. The column has been hidden by a wall structure and not maintained since constructed. The reinforcing bars have been severely corroded. At several locations the reinforcing bars have been totally corroded. Figure 1.21 shows such a corrosion, where all the metal has been dissolved.

pitting corrosion), cf. Fontana (1967). Therefore, too little inhibitor added to concrete is less desirable than none inhibitors added. On the other hand it must be ensured that excess of corrosion inhibitors added will not damage the concrete itself, cf. Hope *et al.* (1995). To avoid these phenomena it should be documented that the inhibitors added do not cause corrosion of the rebars in any possible concentration in the concrete and do not effect the durability of the hardened concrete as well as the workability of fresh concrete, i.e. they act as set accelerators.

Figure 1.20: Reinforcing bars collected from a walkway of RC concrete. The bars show various degree of pitting corrosion.

When two or more inhibitors are added to a corrosive system there is sometimes a synergistic effect, i.e. the inhibiting effect is greater that would be achieved by either of the inhibitors, cf. Fontana (1967). At present, the mechanism of this synergistic effect is not fully understood.

There are limitations in the use of some of the corrosion inhibitors, since they are toxic or they may contaminate the environment. Furthermore, some inhibitors may have a serious effect on the properties of fresh and hardened, and a few corrosion inhibitors lose their efficiency as the temperature of the environment increases, cf. Fontana (1967).

Corrosion tests with migrating inhibitors

Application of organic corrosion inhibitors in concrete began approximately 20 years ago. At the same time laboratory tests had been carried out. Obviously, only short term tests had been carried out. These tests demonstrate that organic corrosion inhibitors may migrate through concrete, cf. Phanasgaonkar *et al.* (1997), and Ukrainczyk *et al.* (1993). They cause a significant delay of the onset of corrosion of reinforcing bars in concrete exposed to chlorides, e.g. by the test method ASTM G 109 (1992). Furthermore, once the corrosion starts the rate is significantly reduced, cf. Nagayama *et al.* (1998).

Figure 1.21: The rust produced by corrosion caused by chlorides is typically black and liquid so that it penetrates the concrete and will be absorbed rather than spalling the concrete cover.

Field tests

Most tests with corrosion inhibitors are carried out as laboratory tests, but a few field tests exist on the efficiency of corrosion inhibitors. Here, the most comprehensive ones are the field tests carried out as part of the SHRP program, cf. SHRP–S–658 (1993).

1.3.4 Initiation period of time and service lifetime

The initiation period of time is defined as the time between the complete mixing of concrete until the time when the reinforcing bar starts to corrode. Since corrosion of the reinforcement in a structural member does not start at the same time, the initiation period of time is a stochastic variable.

The service lifetime is defined as the initiation time plus the time, when the reinforcement left from corrosion reaches the requirements in strength (i.e. the bearing capacity) and deformation required by the Code of Practice.

Examples of calculation of the initiation period of time are illustrated by Examples in Sections 2.4, 5.1 and 5.2. The service lifetime is treated in Section 4.2.4.

1.3.5 Corrosion multi-probes

Chloride-induced corrosion of reinforced steel is one of the major causes of deterioration of marine RC structures and de - iced infrastructures. When

Figure 1.22: A severe pitting corrosion of reinforcement of a bridge deck. Notice, that the second layer of reinforcement shows no corrosion, indicating that the aggressive substances (chlorides) have penetrated from the surface.

corrosion of reinforcing steel bars due to chloride attack occurs, the concrete will spall. However, this 'warning' is often too late for an optimum repair and maintenance of the structure. The rehabilitation of RC structures due to corroding steel reinforcing bars is quite expensive compared with preventive maintenance. Therefore, a warning system is needed so that preventive maintenance can be chosen at the right time and repair expenses minimized.

There are several such multi-probes on the market. The end of the initiation period of time may be predicted from a regression analysis of observations by the multi-probes. The mathematical models of chloride ingress, such as the LIGHTCON model and the HETEK model are applied for the interpretation of observations from installed multi-probes in marine RC structures, cf. Sections 2.4, 5.1, 5.2 and 7.1.

1.3.6 Design against corrosion

Design of RC structures against corrosion from exposure of marine or road environments is not a specific task of the Code of Practice. There are some requirements, but many questions are left to the designer's decision.

The resistance of an RC structure to a chloride-laden environment is mainly determined by the following parameters

- The thickness of the concrete cover of the reinforcing bars.

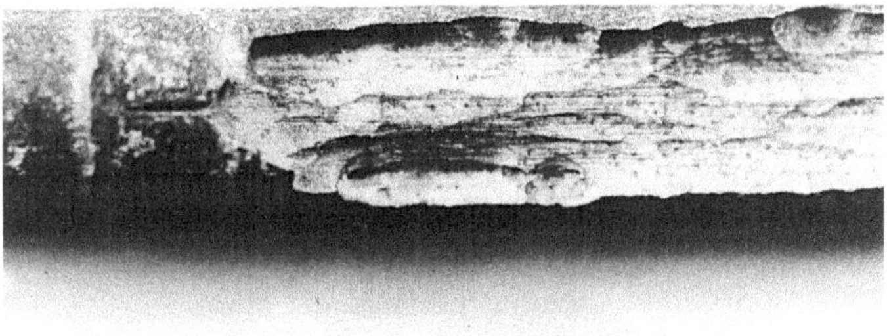

Figure 1.23: The pitting of reinforcing bars can be very severe. Here more than 50 % of the cross section of the reinforcing bar has disappeared by corrosion.

- The structural detailing of the structure, and appropriate practical design to facilitate inspection, maintenance and repair.
- The diffusivity of the concrete cover (i.e. mainly determined by the water/binder-ratio, compaction and cracks (i.e. intensity and widths).
- The types and the composition of the binder of the concrete.
- The types of reinforcement.

Class of environment and method of safety

As stated in Section 1.2 there are several classes and subclasses of a chloride-laden environment, while the Code of Practice often deals with one or two classes. The classes of environment must be carefully chosen, cf. Sections 1.2.2 and 1.2.3.

It must be realized that the chloride-laden environmental exposure is a stochastic variable. Therefore, proper measures must be taken in order to ensure the required initiation period of time. This book describes the probabilistic analysis, the method of Cornell's reliability index and the determination of the characteristic initiation period of time, cf. Sections 2.4.6 and 5.2.6.

Composition of the concrete

A necessary but not sufficient condition is that the concrete has a sufficient resistance against chloride ingress. At present this can be achieved by

- The binder of the concrete should be composed of Portland cement, silica fume and fly ash.

- The equivalent water/cement-ratio must be sufficiently low.
- The compaction of the concrete shall be appropriate.
- The curing condition must ensure a high degree of hydration.

Structural design and design of rebar cover

The geometry of the structure has a significant influence on the rate of chloride ingress into concrete. Horizontal concrete surfaces ought to be avoided. Just a small inclination of a concrete surface may dry out the surface. Also details of a joint are important. As a rule of thumb the details ought to be designed so that they can dry out as soon as possible.

It should also be considered that the structure is open for inspection, maintenance and repair.

The thickness and quality of the concrete cover of the reinforcement is very important for the length of the initiation period of time. This book deals with rebar cover design in Section 5.2.

1.4 Test methods

There are several laboratory test methods and in-situ test methods for chloride ingress into concrete, cf. Tang (1993), and Andrade (1993). It is characteristic for a majority of the test methods on diffusion in concrete that the observations obtained have to be evaluated before applied in practice. Therefore, these test methods are mainly for making comparison between various types of concrete.

1.4.1 Analysis of chloride content of concrete

The various types of chlorides in concrete (free and bound) are determined by measurements in different ways. The total amount of chloride is found as acid-soluble chloride while the free chloride is found as water-soluble chloride. Thus the bound chloride is the difference between the total and the free chloride.

It is the free chlorides that are able to cause corrosion of embedded steel reinforcement and inserts of corrosive metals. Therefore, it is of interest to measure especially the free chlorides, since the ratio bound to free chloride varies with the types of binders. E.g. concrete made with fly ash, silica fume and slag as part of the binder has a high binding capacity. Even concretes made with the various types of cement have different binding capacities. Concrete made with sulphate resisting cement, for example, has a lower binding capacity that concrete made with another type of Portland cement, all other parameters made equal.

When in the USA the chloride content of concrete became an important parameter to control, ACI required in 1977 that the free (i.e. water-soluble)

Figure 1.24: Drilling concrete cores from the bottom of a swimming pool. In this case the drilling machine is operating on compressed air. The water of the swimming pool contains chloride and the chloride profile of the core has to be determined.

chloride of the concrete was kept at a low level. Later this was changed to a requirement to the total (acid-soluble) chloride content of concrete, determined by the test method of ASTM C 114, due to the lack of reproductiveness of the content of the free chlorides. This book only deals with the total chloride.

Sources of uncertainties

There are several sources of uncertainties when determining the chloride content of concrete. Even mistakes can take place as shown by several robin tests. There are four main sources of uncertainties

1. *Exposure conditions*, in field as well as in laboratory.
2. *Preparing samples for analysis*, in-situ as well as in laboratory.
3. *Analysis of chloride content*, the test methods chosen in-situ as well as in the laboratory.
4. *Interpretation of observations*, when rejecting or accepting observations for curve-fitting.

A few comments on these main sources are given as follows.

Figure 1.25: The diver has collected the concrete core. The chloride profile is determined by the profile grinding technique, cf. Figure 1.26.

Exposure conditions in field

The local environment plays an important role for the chloride ingress, cf. Sand (1993). A predominant direction for strong winds leads to a difference in chloride ingress at windward and leeward parts of a structure exposed to airborne chloride, e.g. in coastal regions. Generally speaking the concrete at the leeward part of the structure will contain more chloride than concrete at the windward side, other parameters being equal. The concrete at the leeward side is often sheltered from driven rain which at the windward side tend to wash down most of the chlorides deposited at the concrete surface by sea spray, salty fogs etc.

Besides this shelter effect the concentration of airborne chloride is not constant but varies with the level above sea and distance from the beach.

When sampling concrete in order to determine its chloride content one has to take the above mentioned influences into account. However, no official guidance, recommendation or standard exists so far for the sampling of concrete for this purpose.

Laboratory exposure conditions

In order to characterize the intrinsic chloride diffusivity of concrete, samples of concrete are exposed to a standard solution of chloride as for the test

Figure 1.26: Equipment for the grinding and collecting of concrete powder from thin layers, each from 0.7 mm to 2 mm when applying a turning lathe. The concrete powder is collected and analyzed by a chemical test method, e.g. Nordtest (1984), NT Build 208, in order to determine the chloride content of the powder. The diameter of the core and the thickness of the layers milled away should be chosen in such a way that the cement content of a powder sample for an analysis fulfils the requirement given by NT Build 208 or similar test methods used for the analysis of the chloride content.

Figure 1.27: The concrete core is fixed into the lathe and rotates while very thin layers of concrete powder are milled away by a rotation diamond tool which moves horizontally.

Figure 1.28: In-situ equipment for the grinding and collecting of concrete powder from thin layers, each by a depths increment of 0.5 mm and upwards. The Profile Grinding Kit is fixed to a concrete surface and the principle of milling corresponds to that of Figure 1.26. On horizontal surfaces the powder is collected with a dust buster with a re-usable filter. On a vertical surface the collection of powder is made automatically in a plastic bag attached to the grinding plate. The grinding area is circular with a diameter of 73 mm. The maximum depth of grinding is 50 mm. One powder sample from a 2 mm thick layer of concrete may be collected in approximately 5 minutes.

method NT 443, cf. Nordtest (1996a), or a similar test method. By keeping important parameters like temperature, chloride concentration and exposure period constant and by cutting off a 10 mm thick slice of the concrete from the exposed surface, the deviation of the observation is kept at a minimum.

However, the deviations found by robin tests are not negligible and the sources of the uncertainties may be found in minor differences in test conditions, the microstructure of the concrete (degree of compaction and intensity of defects), and the processing of the powder samples for analysis more than the application of titration procedures.

Preparation of powder samples for analysis

As explained later several techniques are used in order to prepare samples of the exposed concrete. The various sampling methods involve different types of deviation and it is important for the application of the obtained chloride profiles that this is taken into account. Furthermore, the choice of the numbers of observations to determine one chloride profile is also important, cf. Pedersen *et al.* (1993).

There is a need for several techniques for the preparation of the powder samples for analysis. However, one should realize the difference in the accuracy when applying a chloride profile determined by analysis of three powder samples prepared by hammer drilling compared with a chloride profile determined by a dozen observations found by powder grinding. Both test methods are needed but the accuracy should be taken into account when applying the profiles.

Analysis of chloride content

Several test methods exist suitable for measurements in-situ as well as for laboratory purposes. Applied test methods in the Scandinavia are mainly

- *RCT-method*, the Rapid Chloride Test method.
- *Quantab-method.*
- *Volhard titration*, e.g. NT Build 208, cf. Nordtest (1984).
- *Potentiometric titration.*

The conclusion from a round robin test reported by Reknes (1994), is that sufficient accuracy is obtained if

- Skilled personal should have fixed routines for the test method applied.
- The test equipment is calibrated and well-kept.
- In-situ methods are not applied in laboratory testing.
- Blind test samples are incorporated to reveal any mistakes.

Other test methods than the above mentioned exist, like the colourimetric spray indicators, cf. Collepardi (1995), Engell (1960), and Schöppel (1988). These test methods are based upon a spray of silver nitrate, $AgNO_3$. When free chloride is present, silver chloride, AgCl, is formed. When exposed to UV light (sun light or artificial UV) $AgNO_3$ changes its colour into black or dark brown. There are several additives which improve the determination of the chloride ingress by making the border line between the chloride-laden area and the area free from chloride as sharp as possible.

Figure 1.29: Chamber of the salt spray test method, cf. Vennesland (1993). Concrete specimens are exposed to a cycling fog of a solution of NaCl. The chloride ingress into the concrete is determined using the profile grinding technique, cf. Figure 1.26.

1.4.2 Chloride profiles

The basis of how to characterize the chloride diffusivity of concrete is normally the chloride profile, i.e. the graph of the chloride content of the concrete versus the distance from the chloride exposed concrete surface. A theoretical chloride profile is found by solving the diffusion equation, Equation (1.5) for the relevant initial and boundary conditions.

In order to be able to handle the information represented by the chloride profile in a simple way, a chloride profile is described by the following four parameters assuming the chloride ingress into the concrete to obey Fick's second law of diffusion with constant chloride parameters.

1. The ordinate C_{sa} of the chloride profile at the exposed surface $x = 0$, i.e. the achieved surface chloride concentration.

2. The asymptote C_i of the chloride for $x \to \infty$, i.e. the chloride concentration of the undisturbed (non-exposed concrete).

3. The achieved chloride diffusion coefficient D_a.

4. The period of time t where the concrete has been exposed.

Figure 1.30: The Träslövsläge Marine Exposure Station at the west coast of Sweden (near Varberg). Concrete specimens of various concrete with w/c-ratio from 0.35 to 0.70 were exposed to seawater. The chloride profiles were determined several times and the time-dependency of the chloride parameters was observed.

Chloride profile of concrete exposed to seawater

Figure 1.31 shows some typical chloride profiles from a specimen corresponding to various periods of seawater exposure (No. 1–50 at the Träslövsläge Marine Station). The specimen was exposed to seawater when it was 14 days old (i.e. $t_{\text{ex}} = 0.038$ yr) at the Träslövsläge Marine Exposure Station. The concrete contains Swedish Anläggs Cement of 370 kg/m^3 concrete with $w/c = 0.50$ by mass and aggregates having a maximum size of $d_{\text{max}} = 20$ mm. The exposed surface of the concrete was cast against formwork. The following is learned from this example:

- C_{sa}, which is the ordinate of the chloride profile at the abscissa $x = 0$, is time-dependent.
- The chloride profiles show a hump at a distance of approximately $x = 0.5\,d_{\text{max}}$.

This can be explained in the following way: Concrete is a heterogeneous medium, and the chloride from the seawater penetrates concrete through the cement matrix assuming that the aggregates do not allow the chloride to pass. This is the common case of marine concrete. When concrete is cast against formwork, the void fraction of the coarse aggregates (filled with cement matrix) ranges from 100 % at the surface to a minimum at $x = 0.5\,d_{\text{max}}$, cf. Figure 1.32.

Figure 1.33 shows a chloride profile from a marine RC structure which has been submerged in seawater for 20 years. The unit of the ordinate of

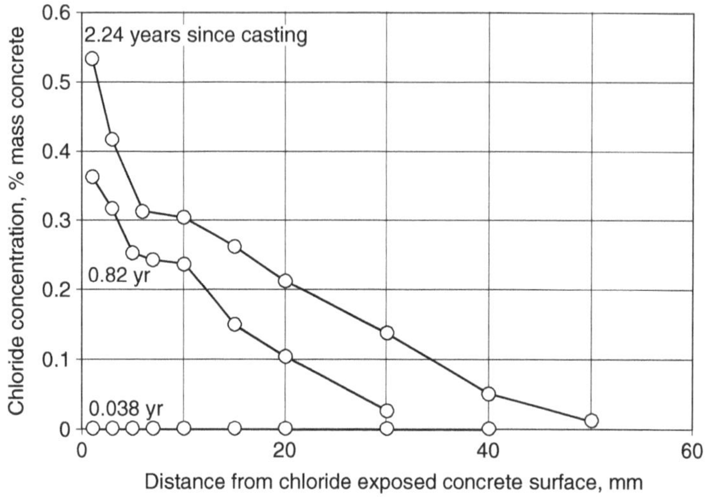

Figure 1.31: Typical chloride profiles of concrete submerged in seawater at various periods of exposure. It is seen that the chloride content of the concrete surface and the diffusion coefficient varies with the exposure time. The chloride profiles show a hump at a distance from the exposed surface of approximately 50 % of the maximum aggregate size.

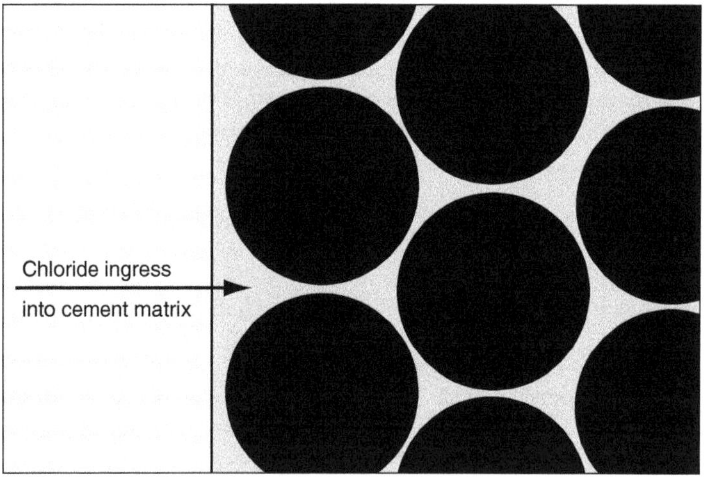

Figure 1.32: When concrete is cast against form work the void fraction of the coarse aggregates (filled with binding matrix) ranges from 100 % at the surface to a minimum at 50 % of the maximum aggregate size.

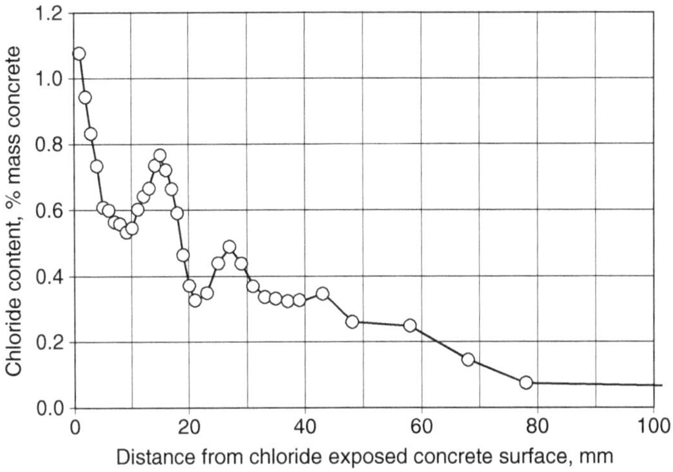

Figure 1.33: A chloride profile from a marine RC structure which has been submerged in seawater (at Nuuk, Greenland) for 20 years. The unit of the ordinate is % by mass concrete. It is seen that the profile reflects the presence of the coarse aggregate in the concrete. The detailed shape was made possible by the analysis of 1 mm thick layers.

the chloride profile is % by mass concrete. The chloride has penetrated the concrete to about 80 mm from the exposed surface. This made it possible to obtain the detailed shape of the profile by grinding and analyzing layers having a thickness of 1 mm. It is seen that the profile reflects the presence of the aggregates in the concrete.

The aggregates used did not contain calcite (checked by thin section analysis) and the Portland cement used at the time of casting (Danish Rapid Hardening Portland cement) contained 65 % by mass of CaO. Thus, by the calcium-profile, cf. Figure 1.34, the chloride profile showed in Figure 1.35 was obtained. Here the unit of the ordinate of the chloride profile is % by mass binder (Portland cement).

1.4.3 Determination of chloride parameters

Chloride ingress into the surface of a concrete structure can be determined and described by the achieved chloride profile. The total information about the environment, the exposure time and the achieved chloride profile will give a clear but also an inconvenient picture of the chloride exposure and the response from the concrete.

The information given by the exposure time and the achieved chloride profile of the concrete may be simplified by a few parameters, sufficient to

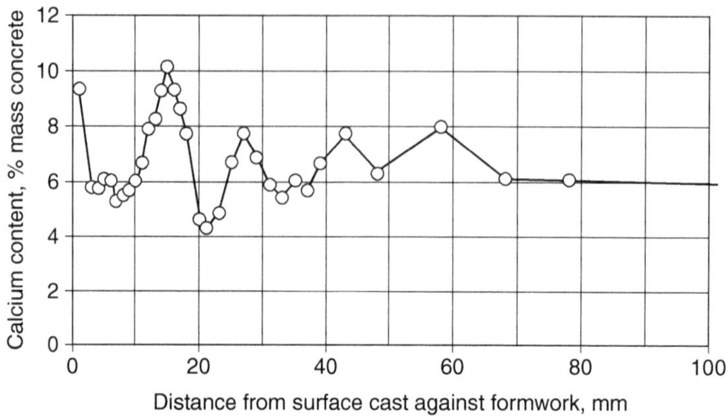

Figure 1.34: The calcium profile of the concrete determined from the same powder samples as the chloride profile shown in Figure 1.26. Assuming that the cement contains 65 % by mass of CaO it is possible to determine the cement profile. Thus the calcium profile is proportional to the distribution of the cement matrix since the aggregates contain no calcite.

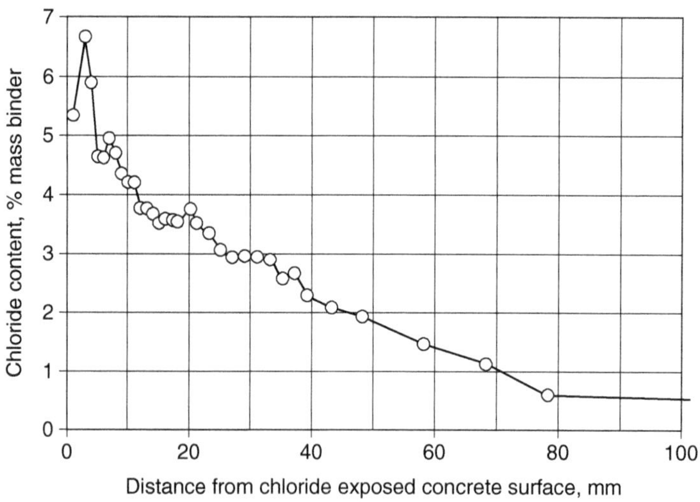

Figure 1.35: The chloride profile shown in Figure 1.33 transformed by means of the calcium profile shown in Figure 1.34 in such a way that the unit of the ordinate of the chloride profile yields % by mass binder (cement). An effect of leaching is visible at the near-to-surface layer of the concrete.

determine the shape of the chloride profile from a mathematical point of view. The values of these parameters, their stochastic distributions and their development in time will give a convenient way of handling the chloride ingress into concrete.

As mentioned previously, the transport mechanism may vary with the composition and the compaction of the concrete and the intensity and widths of its cracks. Here, it is assumed that diffusion is the predominant mechanism of the chloride ingress and that the concrete is a quasi homogeneous material. These assumptions have proved to be valid for test specimens of uncracked high performance concrete having water/binder-ratios from 0.35 to 0.70 by mass and tested by field exposure at the Träslövsläge Marine Exposure Station in Sweden.

Chloride profiles for diffusion

It is convenient to speak about two different chloride profiles, the *achieved chloride profile* and the *potential chloride profile.* These chloride profiles may determine bound chlorides as well as free chlorides according to the purpose of testing. The achieved chloride profile is found in a concrete structure or a concrete specimen when field exposed to chloride. It could be said that the achieved chloride profile is the response of the concrete to the chloride exposure from a given environment.

The potential chloride profile is found in a concrete specimen when it is exposed to a standard solution of chloride at a standard temperature and during a standard length of time. It could be said that the potential chloride profile is the intrinsic parameter of the concrete, i.e. a materials constant.

Section 2.1.1 deals with chloride profiles in details and Section 2.1.4 deals with the determination of chloride parameters by regression analysis.

Achieved chloride profile. According to the assumptions given above the achieved chloride profile for chloride diffusion will be unambiguously determined by the following three parameters

1. The initial chloride content C_i of the concrete.

2. The ordinate C_{sa} of the chloride profile at the surface of the concrete.

3. The chloride diffusion coefficient D_a of the concrete.

As mentioned earlier these three parameters are not materials constants but they refer to the exposure load, including the exposure time and the age of the concrete. The achieved chloride profiles are determined by the test method NT Build 443, cf. Nordtest (1996a), or a similar test method.

Potential chloride profile. According to the assumptions given above the potential chloride profile for chloride diffusion will be unambiguously determined by the following three parameters

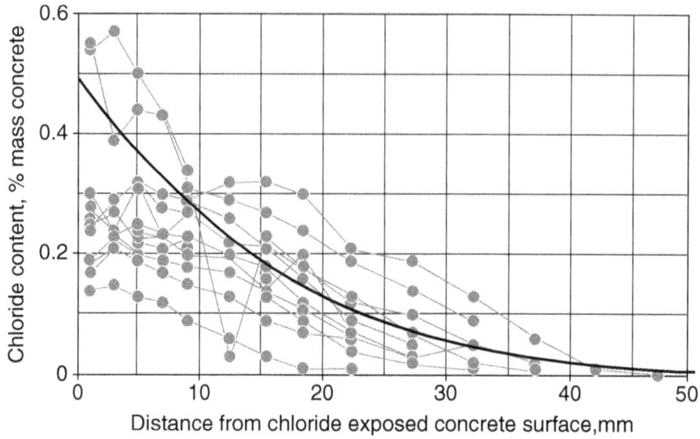

Figure 1.36: Measured chloride profiles of concrete after 10 yr of exposure by marine atmosphere. The curve shown is an estimate according to the HETEK model of chloride ingress, cf. Section 5.2.

1. The initial chloride content C_i of the concrete.
2. The ordinate C_{sp} of the chloride profile at the surface of the concrete.
3. The chloride diffusion coefficient D_p of the concrete.

These three parameters are 'materials constants', but they refer to the test method. Potential chloride profiles are determined by the test method NT Build 208, cf. Nordtest (1984), or any similar test method.

Deviations from the ideal shape

It is common knowledge that the theoretical shape found by solving the mathematical equation for chloride ingress into a quasi homogeneous material sometimes deviates from the shape of a determined chloride profile. Several possible reasons for such deviations could be mentioned. However, in each case the concrete should be examined (micro-structural examination) before jumping to conclusions.

Two main deviations are observed, namely at the exposed concrete surface and at a small distance from the surface. The observed types of deviations are not the same for achieved and for potential chloride profiles. Some reasons are listed below.

- A carbonated surface layer of chloride exposed concrete will have a much smaller binding than the non-carbonated concrete. This means that the concrete cannot be considered as a homogeneous material but rather as two layers of different materials. This will not be relevant to the potential chloride profile since the concrete is not exposed to carbonation.

- A leached surface layer of chloride exposed concrete will have a much smaller binding than a non-leached concrete. This means that the concrete cannot be considered as a quasi homogeneous material but rather as two layers of different materials. This will not be relevant to the potential chloride profile, since the concrete normally is only exposed for a short period (35 days) to a chloride solution and care is taken to minimize any leaching effect.

- A significant part of the penetrating chloride is bound to the C–S–H particles in the cement matrix of the concrete. Since the cement matrix is not uniformly distributed through the concrete (highest content at the surface) the concrete is not a quasi homogeneous material. This phenomenon could be taken into account by another mathematical than that given above. It is, however, also possible to neglect the observations of chloride content in the outermost layer corresponding to approximately half the size of the coarse aggregate, i.e. approximately 10 mm. When determining the potential chloride profile of a concrete this is done physically by cutting away the outermost layer of a thickness of 10 mm.

- The exposure of concrete specimens to a solution of chlorides, as applied by the test method NT Build 443 or similar methods, often results in a higher chloride concentration at the surface of the concrete ($x = 0$) than it appears from the mathematical solution. It is generally believed that the chloride concentrates in cavities and voids of the exposed surface, and this phenomenon has nothing to do with the diffusion mechanism (it is rather a wick action).

Other methods for the determination of the chloride parameters

There are several other methods by which chloride parameters (mainly the chloride diffusion coefficient) may be determined, e.g.

- *Diffusion cell test method*, where a sample (a slice) of the concrete is placed in between two cells. One of the cells contains chloride. Since the total amount of chloride is constant, the diffusion coefficient can be calculated from the flow of the concrete. Cf. Bigas (1994) and Nilsson *et al.* (1994).

- *Electrical field migration test method* is in principle an accelerated diffusion cell test method. Non-steady state migrating test methods like AASTO T277, cf. ASTM standards (1994), and the CTH test method, cf. Tang *et al.* (1992), are quick in use and widely used. However, steady state migrating test methods are time consuming and not so commonly used.

- *Salt spray test method* is a method where the concrete specimens are exposed to a cycling fog of a solution of sodium chloride, NaCl. Such a method is the NTH spray method, cf. Vennesland (1993). In between fog periods of four hours the concrete specimens are allowed to dry for four hours. The chloride ingress into the specimens are determined by chloride profiles after six weeks of exposure.

- *Resistivity test method* is a method where the resistivity of the concrete is measured, cf. Klinghoffer *et al.* (1995). The resistivity is dependent on the temperature, the resistivity of the pore liquid and the degree of saturation. The resistivity test method is mainly used for comparisons between concretes.

1.4.4 Characteristic value of observations

According to the EN 1504 standards (1997), and the Eurocode EN 206-1 the lower characteristic value shall be the 5 % fractile and the upper characteristic value shall be the 95 % fractile.

According to prEN 14358, (2002), concerning the characteristic values of observations obtained from testing concrete or similar building materials, the determination may be carried out applying the following assumptions

- The lower characteristic value is defined as the 5 % fractile.
- The upper characteristic value is defined as the 95 % fractile.
- The characteristic value shall be determined from observations at a level of confidence of $\alpha = 84.1$ %.
- The observations from the testing are assumed statistically to be logarithmic normally distributed.

The coefficient of variation is unknown. In the case of $n \geq 3$ observations (in practice $n \geq 5$ observations) from one single section of inspection, calculation of the characteristic value of the following observations

$$D_1, D_2, \ldots, D_n$$

are carried out as follows: First calculate the mean value $M_{\ln D}$ and the standard deviation $S_{\ln D}$ of the Napierian (or natural) logarithm of the observations, i.e. the values

$$\ln D_1, \ln D_2, \ln D_3, \ldots, \ln D_n.$$

The easiest way is to apply a spreadsheet, e.g. Excel, cf. Example 1.4.1. Then the lower characteristic value (5 % fractile) is

$$D_{k\ell} = \exp\left(M_{\ln D} - k_n \times S_{\ln D}\right), \tag{1.15}$$

Table 1.2: Values of the factor k_n in Equations (1.15) and (1.16).

n	3	4	5	6	7	8	9	10
k_n	4.11	3.28	2.91	2.70	2.57	2.47	2.40	2.34
n	11	12	15	20	30	50	100	
k_n	2.29	2.25	2.16	2.07	1.98	1.89	1.81	

Table 1.3: Calculation of the upper characteristic value of observed chloride contents in Example 1.4.1.

		Chloride, mass concrete	$\ln C$
C_1	%	0.160	–1.833
C_2	%	0.154	–1.871
C_3	%	0.185	–1.687
C_4	%	0.176	–1.737
C_5	%	0.192	–1.650
C_6	%	0.174	–1.749
Mean value	%	0.1735	–1.755
Standard deviation	%	0.0144	+0.084
Coefficient of variation	%	8.30	—
Upper characteristic value	%	0.2171	—

and the upper characteristic value (95 % fractile) is

$$D_{ku} = \exp\left(M_{\ln D} + k_n \times S_{\ln D}\right). \tag{1.16}$$

The factor k_n is based upon the non-central t-distribution and it takes on the values shown in Table 1.2. It is seen that the characteristic value of n observations stabilizes from about $n = 20$ Nos.

Example 1.4.1 The chloride measurements by RCT (Rapid Chloride Test) in a 10×25 m bridge slab at a depth of 20–25 mm below the exposed concrete surface gave the following data

0.160, 0.154, 0.185, 0.176, 0.192, 0.174 % chloride by mass concrete.

Calculation of the upper characteristic value (95 % fractile) is carried out in the following way, applying a spreadsheet, cf. Table 1.3.

In Table 1.3 the mean value and the standard deviation of the logarithms of the chloride contents of the concrete are determined as $M_{\ln C} = -1.755$ and $S_{\ln C} = 0.084$, respectively. Thus, the upper characteristic value of the

Table 1.4: Calculation of the upper characteristic value of observed diffusion coefficients in Example 1.4.2.

		Diffusion coefficient, D	$\ln D$
D_1	mm^2/yr	56	4.025
D_2	mm^2/yr	48	3.871
D_3	mm^2/yr	44	3.784
D_4	mm^2/yr	47	3.850
D_5	mm^2/yr	56	4.025
Mean value	mm^2/yr	50.2	3.911
Standard deviation	mm^2/yr	5.5	0.109
Coefficient of variation	%	11.0	—
Upper characteristic value	mm^2/yr	68.6	—

chloride contents of the concrete (95 % fractile) is given by

$$
\begin{aligned}
C_{ku} &= \exp\left(M_{\ln C} + k_n \times S_{\ln C}\right) \\
&= \exp\left(-1.755 + 2.70 \times 0.0840\right) \\
&= 0.217 \text{ \% chloride by mass concrete.} \quad \Diamond
\end{aligned}
$$

Example 1.4.2 A concrete (typeno. L32F19M5) has been tested by means of NT Build 403, cf. Sørensen (1996). The upper characteristic value (the 95 % fractile) of the diffusion coefficient is determined. The diffusion coefficients are given by the following data

$$56 \quad 48 \quad 44 \quad 47 \quad 56 \quad mm^2/yr.$$

Calculation of the upper characteristic value (95 % fractile) is carried out in the following way, applying a spreadsheet, cf. Table 1.4.

In Table 1.4 the mean value and the standard deviation of the logarithms of the diffusion coefficients of the concrete are determined as $M_{\ln D} = 3.911$ and $S_{\ln D} = 0.109$, respectively. Thus, the upper characteristic value of the chloride contents of the concrete (95 % fractile) is

$$
\begin{aligned}
D_{k\ell} &= \exp\left(M_{\ln D} + k_n \times S_{\ln D}\right) \\
&= \exp\left(3.911 + 2.91 \times 0.109\right) = 68.6 \ mm^2/yr. \quad \Diamond
\end{aligned}
$$

1.4.5 Examination of concrete

An examination of chloride ingress into concrete covers a determination of chloride profiles as well as an analysis of the microstructure, especially the types and intensity of internal defects (e.g. cracks, voids, honeycombing, set of aggregates, carbonation, lack of hydration, leaching).

Guidance for the examination of concrete by means of the petrographic method is found in ASTM (1990), and by Concrete Society (1989).

This book does not cover the examination of concrete besides the background of the chloride profiles and their evaluation.

Sampling

The preferred sampling technique today is core drilling. If a core diameter of 100 mm to 75 mm is chosen the examination of the core will provide a chloride profile as well as several thin sections. If a thin section analysis is not necessary an in-situ chloride profile grinding is also available, cf. Germann Petersen (2003).

1.5 Maintenance and renovation of RC structures

It is a necessary condition that the cause of reinforcement corrosion is found and removed before or by the concrete repair and that the condition is retained. The Standard EN 1504 part 9 (1997) describes five principles of restoring reinforcement corrosion. Here, the following four methods of maintenance and repair are mentioned.

1.5.1 Repair of corrosion with corrosion inhibitors

In order to avoid incipient anodes after repairing reinforcement steel it is recommended to apply corrosion inhibitor to the concrete and repair, cf. EN 1504, part 9 (1997).

Choice of corrosion inhibitor for repair of RC structures

The properties of a chosen corrosion inhibitor for repair of reinforced concrete structures shall at least fulfil the following requirements:

- Be non-toxic and will not contaminate the environment when applied.
- Be able to migrate through hardened concrete.
- Be long lasting and efficient at temperatures relevant for the RC structures exposed to the environment concerned.
- Shall inhibit corrosion independent of changes in concentrations in the concrete.
- Must have a minor effect on properties of fresh and hardened concrete.

Application of corrosion inhibitors in concrete repairs

A traditional repair of a corroding reinforcing bar in concrete due to chloride attack (and even carbonation) of the concrete often shows new corrosion of the reinforcing bar at the edges of the repair, i.e. just outside the reinforcement protection (incipient anodes), cf. Figure 1.17. Therefore, the repair technique applying migrating corrosion inhibitors is performed in the following way:

- Break out the concrete.
- Clean the reinforcement (rust) and the prepared surface of the concrete (defects) by grit-blasting or water jetting.
- Apply a cementitious reinforcement protection containing inhibitor.
- Spray or brush the migrating corrosion inhibitors onto the prepared concrete surface and its surroundings areas until it soaks into the concrete in order to reduce the incipient anode effect.
- Apply a cementitious bonding aid (containing corrosion inhibitor) to the prepared and cleaned concrete surface.
- Reinstate a repair mortar, which contains migrating corrosion inhibitor.
- Give the structural component and its repair a surface protection.

The reinforcement protection, the bonding aid and the repair mortar all should contain migrating corrosion inhibitors. In this way the corrosion inhibitor migrates through the concrete around the repair and its near surroundings to provide protection to the neighbour reinforcing steel not treated and will extend the service life of the structure.

1.5.2 Electro-chemical chloride removal

Electro-chemical removal of chlorides from contaminated concrete is theoretically simple. Chlorides are negatively charged ions. The concrete is porous and the pores are more or less filled with pore liquid in which the chloride ions are dissolved. Thus, the free chloride ions may migrate when an electrical field is supplied. The idea of electro-chemical removal is that the free chloride ions is repelled by the reinforcement when negatively charged and attached to a surface electrode when positively charged.

The surface electrode may be a mesh of titanium (will not corrode) or a mesh of steel (which will corrode and therefore is not recommended since the concrete surface may be discoloured). The surface electrode is placed in an electrolyte which may consist of sprayed cellulose fibre or felt cloth saturated with an electrolyte (usually water). Also coffer tanks containing the electrolyte may be applied.

Figure 1.37: Corrosion of reinforcing bars due to chloride attack will cause cracks in the concrete. It is not possible to repair such cracks unless the cause of the corrosion is removed. Therefore, the chloride has to be removed, e.g. by replacing contaminated concrete before repair, and by injection of cracks.

When the electrical continuity in the reinforcement is documented, the electrical field is applied by approximately 1 A/m^2 concrete surface. Depending on the circumstances the chloride removal will last up to eight weeks.

Physically and chemically bound chlorides are not easy to remove. However, by making a suitable break in the process some of the bound chlorides are released to free chlorides.

By taken concrete samples and determining the chloride content it may be decided when to stop the electro-chemical chloride removal.

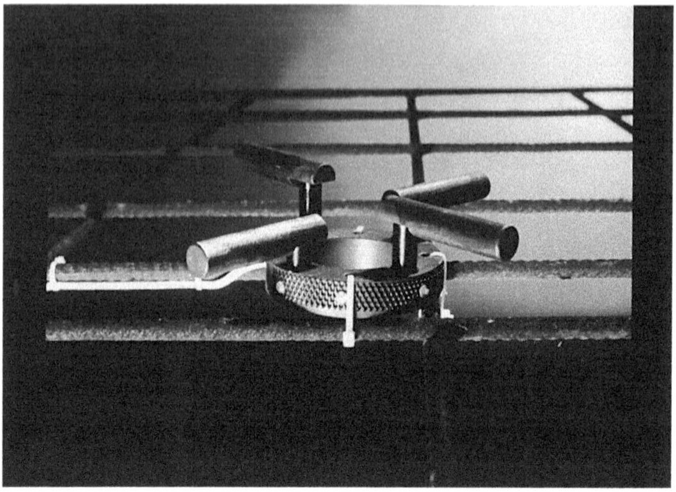

Figure 1.38: The multi-probe *CorroWatch* from FORCE Technology. This multi-probe has four anodes. The initiation period of time can be estimated by the LIGHTCON model, cf. Section 5.1.5, as well as by the HETEK model, cf. Section 5.2.

Detailed information on electro-chemical removal of chlorides from reinforced concrete is reported in e.g. Broomfield (1997), Chess *et al.* (1998), and NCT (1991).

1.5.3 Cathodic protection

Corrosion of steel reinforcement at the anode involves a liberation of electrons. There are two main ways to go when applying cathodic protection, namely

1. *Sacrificial anode cathodic protection.* The reinforcing bars of the concrete component which has to be protected are electrically connected to an external anode of zinc, aluminium, magnesium or their alloys. On condition that the whole system is exposed to a wet environment, e.g. a marine environment, the corrosion takes place on the external anodes. Thus, the external anodes have to be replaced from time to time. The advantage of this method is that no electrical power is needed.

2. *Impressed current cathodic protection.* The reinforcing bars of the concrete component which has to be protected is made by impressing an electric current on the cathode. The external anode is placed on the concrete surface and various types are used, like conductive coating systems (e.g. paint) or nets of noble metal, placed in saw cut slots with an overlay of good quality mortar or concrete.

Detailed information on cathodic protection of reinforced concrete is reported in e.g. Broomfield (1997), Chess *et al.* (1998) and Mailvaganam (1992).

1.5.4 Surface protection

When the resistance to chloride ingress is inadequate a surface treatment may retard the chloride ingress into the concrete. Protection of concrete against chloride ingress may be carried out either as part of the maintenance or after the repair or an electro-chemical treatment. The surface treatment may be hydrophobic impregnation, concrete sealers, paints and membranes. It is of great importance that the surface treatment has a suitable low water vapour diffusivity, especially in a cold climate, in order to avoid moisture and frost damages to the concrete substrate.

To calculate the chloride ingress into a coated concrete substrate one has to deal with chloride ingress through a two-layer material. This is not an easy task. In Section 7.2.1 the method of Crank (1956) is given for constant diffusivity. In Section 7.2.2 a more realistic model is presented. This method involves more calculation than that of Crank. A numerical example is presented.

1.6 Design of chloride exposed RC structures

The object of design of chloride exposed reinforced concrete structures, e.g. marine or road structures, is to achieve a prescribed service lifetime.

1.6.1 Service life

British Standard (1993), defines the service life as the period of time during which no excessive expenditure is required on operation, maintenance or repair of a component or construction.

A chloride exposed RC component may have various local types of environments, the concrete may have various resistance to chloride ingress (chloride diffusivity) and the reinforcing bars may have various thickness of concrete cover etc. Thus, corrosion of the steel reinforcing bars does not start at the same time. It may be in the spirit of the Code of Practice of today to define the service life as the period of time which has elapsed since the concrete was cast until 5 % of the surface of the reinforcing bars have corroded, i.e. the characteristic value.

1.6.2 Methods of design

In order to obtain a required service lifetime, several methods of design are available. These methods all deal with the determination of the thickness of the concrete cover of reinforcing bars.

Deterministic design

The calculation of the service lifetime is shown in Section 2.4.3. By means of this method there are two ways to obtain a suitable safety. The RC structure is estimated, i.e. the dimensions and the concrete.

1. The service lifetime is calculated on the basis of the mean values of the governing parameters of the environment and the concrete. The structure is accepted in case the determined service lifetime is 25–50 years greater than the required lifetime (approximately 100 years). This method of design is simple when only few observations are available on the environment and the concrete, but the method is not unambiguous.

2. The service lifetime is calculated on the basis of the characteristic values of the governing parameters of the environment and the concrete (95 % and 5 % fractiles; notice that the diffusion coefficient has to be the 95 % fractile). The structure is accepted in case the determined service lifetime is equal to or greater than the required lifetime (approximately 100 years). The method requires enough observations or data so that the characteristic values can be determined or estimated.

Stochastic design

In case a sufficient amount of observation and data exists it becomes possible to use stochastic design. At present two methods are available:

1. *Cornell's method of reliability index.* A method of design may be adapted from the structural design. It is the Cornell's method of reliability index, which makes it possible to take some knowledge about the stochastic distribution into account. Only the expectation values and the standard deviation of the governing parameters ought to be known. The method is presented in Section 2.4.7 and an numerical example is given.

2. *Probabilistic method of design.* In case it is possible to estimate the stochastic distribution of the variables involved, the stochastic distribution of the service time can be calculated. This knowledge may be based upon observations, by engineering judgements or both. This method requires a fairly large computer and will only be relevant for large structures. The method is not illustrated in this book.

Chapter 2

Constant Chloride Diffusivity

When a concrete component is exposed to chloride it has become normal practice to describe the response of the concrete to the chloride exposure by its chloride profile, i.e. the distribution of the chloride content of the concrete (or its binder) in its near-to-surface layer versus the distance from the chloride exposed concrete surface and at the time since the start of chloride exposure of the concrete. A chloride profile is also named the chloride concentration-distance graph, cf. Section 1.4.2.

This chapter discusses the basic evaluation of a chloride profile obtained either from laboratory testing, e.g. by the test method NT Build 443, cf. Nordtest (1996), or from inspection and testing of a chloride exposed concrete structure, e.g. a marine RC structure.

The aim of this chapter is to inform the reader about the basic mathematical treatment of the chloride profile and its application when having as simple assumptions as possible, i.e. constant chloride diffusivity.

2.1 Parameters of the chloride profile

Chloride ingress into the near-to-surface layer of a chloride exposed concrete structure or a concrete specimen can, at any time, be determined and described by its chloride profile. The total information about the environment, the length of the exposure period, and the chloride profile will give a clear but also an inconvenient picture of the chloride exposure and the response of the concrete.

The information given by the exposure time and the chloride profile of the concrete may be simplified by a few parameters, sufficient to determine the shape of the chloride profile from a mathematical point of view, while the number of observations of a (good) chloride profile often counts 7–10

Figure 2.1: A specimen made of concrete with the same mixture proportions as concrete for the Great Belt Link. The specimen is exposed to water of the Great Belt (at Nyborg) and the chloride profiles were determined versus time, cf. Vincentsen *et al.* (1999).

observations. The values of the chloride parameters of the concrete, their stochastic distribution and their development in time will give a convenient way of handling the chloride ingress into a chloride exposed concrete component or specimen.

The transport mechanism may vary with the composition, the compaction and the intensity and the widths of the cracks of the concrete. Here, it is assumed that chloride diffusion is the predominant mechanism of the chloride ingress, and that the concrete is a quasi homogeneous material.

These assumption have proved to be valid for test specimens of uncracked high performance concrete having water/binder-rations from 0.35 to 0.70 by mass and tested by field exposure, e.g. at Träslövsläge Marine Exposure Stationnear Varberg on the Swedish west coast, cf. Frederiksen *et al.* (1997). A similar agreement is found by the Great Belt Link for a test specimen placed on the beach of Nyborg in Denmark, cf. Vincentsen *et al.* (1999).

2.1.1 Chloride profile

The equation of a chloride profile created by chloride diffusion into the concrete is found by solving Fick's second law of diffusion for a semi-infinite medium, i.e. the dimension of the concrete member or specimen must not be smaller than the maximum chloride ingress. When obtaining a chloride profile there is no information about the initial, boundary and physical conditions.

Thus, in order to make it as simple as possible it is assumed that during the entire period of chloride exposure

- the surface chloride content has remained constant
- the chloride diffusion coefficient has remained constant
- the initial chloride content of the concrete was equally distributed.

This is *not* true, but since only one chloride profile is determined versus time, no information is available about these conditions. When two or more chloride profiles are determined at different periods of time it is possible to determine the variation of the chloride parameters in time. We shall, however, not deal with this here, cf. Chapter 5.

Equation of a chloride profile

The equation of a chloride profile is here obtained under the following assumptions and conditions

- The concrete is cast at time $t = -t_{\text{ex}}$.
- The concrete is exposed to chloride, e.g. seawater, for the first time at time $t = 0$.
- The chloride-laden environment of the concrete is assumed to remain constant.
- The chloride profile is determined (inspected) at time $t = t_{\text{in}}$.
- The concrete is ($x \geq 0$) assumed to be a quasi-homogeneous, semi-infinite medium.
- The transport of chloride in the concrete is assumed to obey Fick's second law of diffusion, cf. Section 1.1.2 and Equation (2.2).
- The chloride diffusion coefficient ($x \geq 0$) is a constant $D = D_0$ (i.e. the physical condition), cf. Equation (2.1).
- The chloride content of the concrete surface ($x = 0$) is constant $C_s = C_0$ (i.e. the boundary condition), cf. Equation (2.4).
- The initial chloride content of the concrete (or binder) C_i is assumed to be equally distributed throughout the concrete (i.e. the initial condition), cf. Equation (2.3).

Under the given conditions above the necessary and sufficient equations for determining the chloride profile $C = C(x)$, $x \geq 0$ and at time $t = t_{\text{in}}$ are given by

$$D = D_0 \qquad \text{for } x \geq 0 \text{ and } t \geq 0, \tag{2.1}$$

$$\frac{\partial C}{\partial t} = D_0 \frac{\partial^2 C}{\partial x^2} \qquad \text{for } x > 0 \text{ and } t > 0, \tag{2.2}$$

$$C(x,t) = C_i \qquad \text{for } x \geq 0 \text{ and } t = 0, \tag{2.3}$$

$$C(x,t) = C_0 \qquad \text{for } x = 0 \text{ and } t \geq 0. \tag{2.4}$$

The solution of Fick's second law of diffusion under these conditions, cf. Equation (3.13), is

$$C\left(x, t_{\text{in}}\right) = C_i + \left(C_0 - C_i\right) \text{erfc}\left(\frac{0.5\,x}{\sqrt{t_{\text{in}}\,D_0}}\right). \tag{2.5}$$

The complementary error function $y = \text{erfc}(u)$ is tabulated in Section 8.1.

The slope of the chloride profile at the point (x, C) at the time of inspection t_{in} is the partial derivative of the function Equation (2.5), which is modelling the chloride profile. It therefore follows that the slope of the tangent at the point of the chloride profile $(x, C) = (0, C_0)$ is given by

$$\left[\frac{\partial C}{\partial x}\right]_{x=0} = -\left[\frac{C_0 - C_i}{\sqrt{\pi\,t_{\text{in}}\,C_0}} \exp\left(-\frac{x^2}{4\,t_{\text{in}}\,D_0}\right)\right]_{x=0}, \tag{2.6}$$

which is reduced to

$$\left[\frac{\partial C}{\partial x}\right]_{x=0} = -\frac{C_0 - C_i}{\sqrt{\pi\,t_{\text{in}}\,D_0}}. \tag{2.7}$$

Thus, the tangent of the chloride profile at the point $(0, C_s)$ determines a line segment a of the asymptote from the ordinate axis to the intersection between the tangent and the asymptote. The length of a is

$$a = \sqrt{\pi\,t_{\text{in}}\,D_0}. \tag{2.8}$$

It follows that the chloride profile may be expressed in the following way

$$\frac{C - C_i}{C_0 - C_i} = \text{erfc}\left(\frac{x\sqrt{\pi}}{2a}\right). \tag{2.9}$$

The 'first year chloride ingress'

The location of a given reference chloride content C_{re} is of interest for the evaluation of the diffusivity of the concrete according to the test method NT Build 443, cf. Nordtest (1996a). Often the reference chloride content is chosen as the threshold value of chloride in concrete, cf. Section 1.3.2, though arbitrary values as $C_{\text{re}} = 0.1$ % mass concrete may occur.

The location $x_{\rm re}$ of a reference chloride content can be determined by solving the following equation

$$C_{\rm re} = C_i + (C_0 - C_i)\ \mathrm{erfc}\left(\frac{0.5\, x_{\rm re}}{\sqrt{t_{\rm in}\, D_0}}\right). \tag{2.10}$$

The inverse complementary error function $u = \mathrm{inv\ erfc}(y)$ is tabulated in Section 8.2. Thus, by introducing the parameter

$$\xi_{\rm re} = 2\,\mathrm{inv\,erfc}\left(\frac{C_{\rm re} - C_i}{C_0 - C_i}\right), \tag{2.11}$$

the location $x_{\rm re}$ of the reference chloride content $C_{\rm re}$ is

$$x_{\rm re} = \xi_{\rm re}\sqrt{t_{\rm in}\, D_0} = \frac{\xi_{\rm re} a}{\pi} = k_1\sqrt{t_{\rm in}}, \tag{2.12}$$

where a is the length of the segment of the asymptote from the ordinate axis to the intersection between the asymptote and the tangent of the chloride profile at point $(0, C_s)$, cf. Equation (2.8). It is seen that k_1 could be explained as the depth of the penetration of $C_{\rm re}$, when the chloride exposure times is $t_{\rm in} = 1$ year. Therefore,

$$k_1 = \xi\sqrt{D_0} = 2\sqrt{D_0} \times \mathrm{inv\,erfc}\left(\frac{C_{\rm re} - C_i}{C_0 - C_i}\right) \tag{2.13}$$

is called the 'first year chloride ingress'. However, the chloride parameters do not remain constant during the exposure of one year. Therefore, Equation (2.13) has only theoretical interest. However, Equation (2.13) is useful when the chloride diffusivity of different types of concrete are to be compared, since k_1 is a function of all the chloride parameters.

Flux of chlorides

According to Fick's first law of diffusion the flux of chlorides (i.e. the rate of transfer of chlorides through a unit area of the concrete perpendicular to the velocity vector of the chloride) F at location x at time t is

$$F = -D_0\frac{\partial C}{\partial x} = -D_0\frac{\partial}{\partial x}\left\{C_i + (C_0 - C_i)\,\mathrm{erfc}\left(\frac{0.5\, x}{\sqrt{t\, D_0}}\right)\right\}. \tag{2.14}$$

By introducing the derivative of the complementary error function the flux becomes

$$F = -\frac{D_0\,(C_0 - C_i)}{\sqrt{\pi\, t\, D_0}}\ \exp\left(-\frac{x^2}{4tD_0}\right). \tag{2.15}$$

Thus, through the concrete surface ($x = 0$) the flux is

$$F_s = -\frac{D_0\,(C_0 - C_i)}{\sqrt{\pi\, t\, D_0}} = -\frac{C_0 - C_i}{a}\, D_0, \tag{2.16}$$

where a is the length of the segment of the asymptote from the ordinate axis to the intersection between the asymptote and the tangent of the chloride profile at point $(x, C) = (0, C_s)$, cf. Equation (2.8).

Intensity of penetrating chloride

During the exposure period from 0 to t_{in} the total amount M_{in} of chloride which during the period of time from $t = 0$ to $t = t_{\text{in}}$ penetrates from the environment through the concrete surface per unit area may be determined from the chloride flux Equation (2.16) in the following way,

$$M_{\text{in}} = -\int_{t=\text{tex}}^{t_{\text{in}}} F_s \, dt = \frac{D_0 \, (C_0 - C_i)}{\sqrt{\pi \, D_0}} \int_{t=0}^{t_{\text{in}}} \frac{dt}{\sqrt{t}}. \tag{2.17}$$

For $t > 0$ Equation (2.17) is reduced to

$$M_{\text{in}} = (C_0 - C_i) \sqrt{\frac{4 t_{\text{in}} D_0}{\pi}} = \frac{2a}{\pi} \, (C_0 - C_i) \,, \tag{2.18}$$

where

$$a = \sqrt{\pi \, t_{\text{in}} \, D_0} \tag{2.19}$$

is the length of the segment of the asymptote from the ordinate axis to the intersection between the asymptote and the tangent of the chloride profile at point $(0, C_s)$.

Diffusion rate of chloride

Through the time the chloride content of the concrete is increasing. Its rate of chloride diffusion depends on the location x and the time t_{in}. At position x for $t > 0$ the rate of chloride diffusion is

$$\frac{\partial C}{\partial t} = \frac{\partial}{\partial t} \left\{ C_i + (C_0 - C_i) \text{ erfc} \left(\frac{0.5 \, x}{\sqrt{t \, D_0}} \right) \right\}. \tag{2.20}$$

By introducing the derivative of the complementary error function the rate of chloride diffusion can be written

$$\frac{\partial C}{\partial t} = \frac{(C_0 - C_i) \, x}{\sqrt{16 \, \pi \, t_{\text{in}}^3 \, D_0}} \exp \left(- \frac{x^2}{4 t_{\text{in}} \, D_0} \right). \tag{2.21}$$

Equation (2.21) may also be written as

$$\frac{\partial C}{\partial t} = \frac{x}{4a} \frac{C_0 - C_i}{t_{\text{in}}} \exp \left(- \frac{\pi}{4} \left\{ \frac{x}{a} \right\}^2 \right), \tag{2.22}$$

where a is the length of the segment of the asymptote from the ordinate axis to the intersection between the asymptote and the tangent of the chloride profile at point $(0, C_s)$. At the concrete surface $x = 0$ the rate of chloride ingress is zero at any time $t > 0$ in agreement with the assumption that the surface chloride content $C_s = C_0$ remains constant. The rate of chloride diffusion increases with the distance x from the exposed concrete surface to a maximum.

Summary of 2.1.1

Chloride ingress into concrete by diffusion is described at time $t = t_{\text{in}}$ by a chloride distribution $C = C(x,t)$, which obeys

$$C = C_i + (C_0 - C_i)\ \text{erfc}\left(\frac{0.5\,x}{\sqrt{t_{\text{in}}\,D_0}}\right)$$

under the following assumptions

- The chloride parameters: diffusion coefficient $D = D_0$ and the surface chloride content $C_s = C_0$, remain constant, i.e. independent of location x, time t, and chloride concentration C.
- The concrete is exposed to chloride for the first time at time $t = 0$.
- The initial chloride content of the concrete C_i is constant, i.e. independent of location x, of time t and of chloride concentration C.

A simple estimate of the complementary error function is, cf. Equation (3.73) for $p = 0$,

$$\text{erfc}(u) = \left(1 - \frac{u}{1.66}\right)^2,$$

which is valid for $u \leq 1.1$ with a maximum deviation of less that ± 10 %. Other estimations and approximations are available, like e.g. Equation (3.18), Equation (3.21), Equation (3.22), Equation (3.23) and Equation (3.24). The position x_{re} of a reference chloride content C_{re} is given by

$$x_{\text{re}} = k_1\,\sqrt{t_{\text{in}}},$$

where the 'first year chloride ingress' is

$$k_1 = 2\,\sqrt{D_0} \times \text{inv erfc}\left(\frac{C_{\text{re}} - C_i}{C_0 - C_i}\right) \approx 3.32 \times \sqrt{D_0} \times \left(1 - \sqrt{\frac{C_{\text{re}} - C_i}{C_0 - C_i}}\right).$$

The length of the segment a of the asymptote $C = C_i$ from the ordinate axis to the intersection between the asymptote and the tangent of the chloride profile at point $(0, C_s)$, is given by

$$a = \sqrt{\pi\,t_{\text{in}}\,D_0}.$$

Table 2.1: Determination of the chloride profiles specified in Example 2.1.1.

u	$\mathrm{erfc}(u)$	$C(x,t)$	x_{35}
0.00	1.0000	6.500	0.0
0.10	0.8875	5.774	1.0
0.20	0.7773	5.064	2.0
0.30	0.6714	4.381	3.0
0.40	0.5716	3.737	4.0
0.50	0.4795	3.143	5.0
0.60	0.3961	2.605	6.0
0.70	0.3222	2.128	7.1
0.80	0.2579	1.714	8.1
0.90	0.2031	1.360	9.1
1.00	0.1573	1.065	10.1
1.25	0.0771	0.547	12.6
1.50	0.0339	0.269	15.1
2.00	0.0047	0.080	20.2
2.50	0.0004	0.053	25.2

During the exposure time t_{in} the total amount M_{in} of chloride which penetrates from the environment into the concrete is

$$M_{\mathrm{in}} = \frac{2}{\pi}\,(C_0 - C_i)\,a.$$

The rate of chloride diffusion in (x,t) is

$$\frac{\partial C}{\partial t} = \frac{x}{4a}\,\frac{C_0 - C_i}{t}\,\exp\left(-\frac{\pi}{4}\left\{\frac{x}{a}\right\}^2\right).$$

Example 2.1.1 A concrete specimen having a binder content of $B = 370$ kg/m^3 concrete is exposed to a 10 % NaCl solution according to NT Build 443, cf. Nordtest (1996), i.e. an exposure of 35 days at temperature 23°C. The following conditions are assumed

- $C_i = 0.05$ % mass binder
- $C_0 = 6,50$ % mass binder
- $t_{\mathrm{in}} = 35$ days $= 0.09582$ yr
- $D_0 = 265$ mm^2/yr
- $B = 370$ kg/m^3 concrete.

The 'first year chloride ingress' k_1 of $C_{\mathrm{re}} = 0.40$ % mass binder, the location x_{re} of C_{re} at 35 days of exposure, and the intensity M_{35} of chloride ingress into the concrete for this period should be calculated.

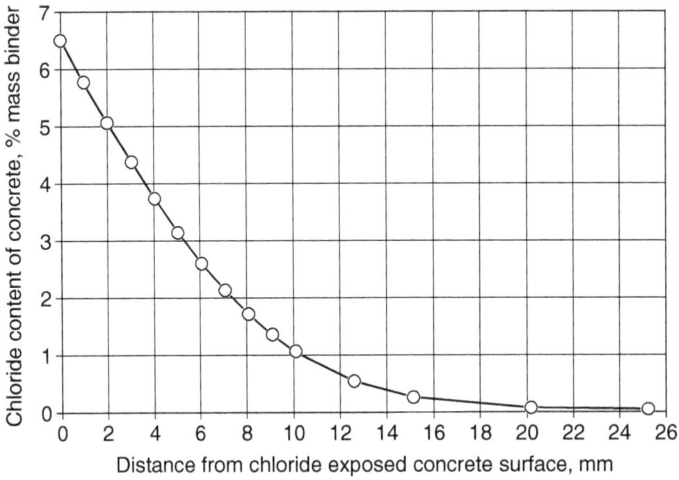

Figure 2.2: Calculated chloride profile on the conditions given in Example 2.1.1.

The estimated chloride profile is determined by the following spread sheet, cf. Table 2.1, where u denotes

$$u = \frac{0.5\,x}{\sqrt{t_{\text{in}}\,D_0}}.$$

For a 35 days of exposure, cf. Figure 2.2, we get

$$x_{35} = u\,\sqrt{4\,t_{\text{in}}\,D_0} = u\,\sqrt{4 \times 0.09582 \times 265} = 10.08\,u \text{ mm}.$$

Calculation of the ordinates of the chloride profile is carried out by means of the spread sheet shown in Table 2.1. Its graph is shown in Figure 2.2. The parameter ξ_{re} is given by, cf. Summary 2.1.1,

$$\xi_{\text{re}} \approx 3.32 \times \left(1 - \sqrt{\frac{0.40 - 0.05}{6.50 - 0.05}}\right) = 2.55.$$

Thus, the 'first year chloride ingress' k_1 into the concrete of $C_{\text{re}} = 0.40$ % mass binder, cf. Equation (2.13) is

$$k_1 = \xi_{\text{re}}\,\sqrt{D_0} = 2.55 \times \sqrt{265} = 41.5 \text{ mm}/\sqrt{\text{yr}}.$$

The increase of the chloride content of the surface chloride per m^2 of the surface yields

$$C_0 - C_i = \frac{6.50 - 0.05}{100} \times 370 = 23.9 \text{ kg/m}^3 \text{ concrete}.$$

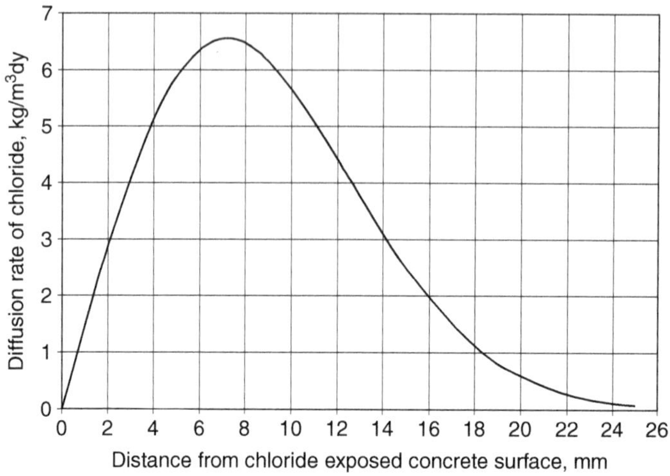

Figure 2.3: Calculated diffusion rate of chloride versus the distance from the exposed surface on the conditions given in Example 2.1.1. The diffusion rate of chloride has a maximum at the distance 7 mm from the chloride exposed surface.

The length of the segment of the asymptote from the ordinate axis to the intersection between the asymptote and the tangent of the chloride profile at point $(0, C_s)$, cf. Equation (2.19), is

$$a = \sqrt{\pi\, t_{\text{in}}\, D_0} = \sqrt{\pi \times 0.09582 \times 265} = 8.93 \text{ mm}.$$

Thus, the chloride ingress M_{35} of chloride into the concrete specimen per m^2 of the surface, cf. Equation (2.18), becomes

$$M_{35} = \frac{2}{\pi}\,(C_0 - C_i)\, a = \frac{2}{\pi} \times 6.45 \times 8.93 \times 10^{-3} = 0.0372 \text{ kg/m}^2 \text{ surface.}$$

The rate of chloride diffusion, cf. Equation (2.22) and Figure 2.3, is

$$\begin{aligned}\frac{\partial C}{\partial t} &= \frac{x}{4a}\,\frac{C_0 - C_i}{t_{\text{in}}}\,\exp\left(-\frac{\pi}{4}\left\{\frac{x}{a}\right\}^2\right)\\ &= \frac{x \times 6.46}{4 \times 8.93 \times 35} \times \exp\left(-\frac{\pi}{4}\left\{\frac{x}{8.93}\right\}^2\right)\\ &= \frac{x}{52.3} \times \exp\exp\left(-\frac{x^2}{101.5}\right) \text{ kg/m}^3\text{dy}.\end{aligned}$$

The rate of chloride diffusion has a maximum at the position

$$x = 8.93\sqrt{\frac{2}{\pi}} = 7.125 \text{ mm.} \qquad \Diamond$$

2.1.2 Types of chloride profiles

It is convenient to speak about two different chloride profiles, the achieved chloride profile and the potential chloride profile. These chloride profiles may determine bound chlorides as well as free chlorides according to the purpose of testing, cf. Section 1.4.

One basic difference between a potential and an achieved chloride profile is that a potential chloride profile is the response of a short-term chloride exposure (accelerated by increased temperature and chloride concentration) while the achieved chloride profile is the response of a long-term chloride (natural) exposure, cf. Section 1.4.

Achieved chloride profile

The achieved chloride profile is found in a concrete structure or a concrete component when field exposed to chloride from a chloride-laden environment. It could be said that the achieved chloride profile is the response of the concrete to the chloride exposure from a given environment.

According to the assumptions given above the achieved chloride profile for chloride diffusion will be unambiguously determined by the following three parameters

1. the initial chloride content C_i of the concrete,
2. the achieved ordinate C_{sa} of the chloride profile at the surface of the concrete,
3. the achieved chloride diffusion coefficient D_a of the concrete,

when the exposure time t_{in} is known. As mentioned earlier these three parameters are not materials constants. Besides being dependent on the composition of the concrete they refer to the exposure load, including the exposure time and the age of the concrete.

Potential chloride profile

The potential chloride profile is found in a concrete specimen when exposed to a standard solution of chloride in water at a standard temperature and during a standard length of time. It could be said that the chloride parameters obtained from a potential chloride profile are the intrinsic parameters, i.e. the materials constants, of the concrete.

According to the assumptions given above, the potential chloride profile for chloride diffusion will be unambiguously determined by the following three parameters

1. The initial chloride content C_i of the concrete.
2. The potential ordinate C_{sp} of the chloride profile at the surface of the concrete.

3. The potential chloride diffusion coefficient D_p of the concrete.

These three parameters are 'materials constants', but they refer to the test method. Potential chloride profiles are determined by the test method NT Build 443, cf. Nordtest (1996a), or any similar test method.

Deviations from the ideal shape

Sometimes it is observed that the ideal shape of the chloride profile given by Equation (2.5) deviates from the shape of a measured chloride profile. Several possible reasons for such deviations could be mentioned. However, in each case the concrete should be investigated, e.g. by microstructural examination (i.e. thin section analysis) before jumping to conclusions.

- A carbonated surface of chloride exposed concrete will have a much smaller binding effect than the non-carbonated concrete, other conditions being equal. This means that the concrete cannot be considered as a quasi-homogeneous material but rather as consisting of two layers of 'different' materials. This will not be relevant to the potential chloride profile, since the concrete is not exposed to carbonation by the test method NT Build 443, cf. Nordtest (1996a), and similar potential test methods.

- A leached surface layer of chloride exposed concrete will have a much smaller binding effect than the non-leached concrete. This means that the concrete cannot be considered as a quasi-homogeneous material but rather as two layers of 'different' materials. This will not be relevant to the potential chloride profile since the concrete is only exposed for a short period of time (NT Build 443: 35 days) to a chloride solution and care is taken to minimize any leaching effect.

- A significant part of the penetration chloride is bound to the C–S–H particles in the cementing matrix (binder) of the concrete. Since the cementing matrix of the concrete is not uniformly distributed through the concrete (i.e. highest content of matrix at the surface) the concrete is not a quasi-homogeneous material. This phenomenon could be taken into account by another mathematical solution than Equation (2.5). However, it is also a possible way to neglect the observations of chloride content in the outermost layer corresponding to approximately half the size of the coarse aggregate, i.e. approximately 10 mm. When determining the potential chloride profile of a concrete according to NT Build 443 this is done physically by cutting away the outermost layer of a thickness of 10 mm. The same procedure is not possible for an achieved chloride profile.

- The exposure of concrete specimens to a (rather) concentrated solution of chlorides, as applied by the test method NT Build 443 or similar test methods, often results in a higher chloride concentration at the exposed concrete surface (i.e. $x = 0$) than appears from the mathematical solution Equation (2.5). It is generally believed that chloride concentrates in cavities and voids of the exposed surface, especially caused by the wick action. This phenomenon has nothing to do with the diffusion mechanism. Thus, the observation is ignored when the ideal shape Equation (2.5) is fitted to the obtained observations.

- The potential chloride profiles are determined in the laboratory at the temperature specified by the test method. This temperature is often the standard temperature of the laboratory (23 °C ± 2 °C) or higher in order to accelerate the diffusion of the chloride in the concrete specimen. It has not been shown that increase of temperature has any influence on the shape of the chloride profile.

2.1.3 Chloride parameters determined by approximation

It is a typical engineering method to plot the observed chloride profile and to estimate the chloride parameters from the shape of the graph. There are several such engineering methods the use of which is for a first estimate of the parameters of the chloride profile, e.g. for a more accurate estimation by one of the iterative methods described in Section 2.1.4.

Test of heterogeneity

In order to estimate the value of the chloride parameters as defined by the error function solution, cf. Equation (2.5), an obtained (i.e. measured) chloride profile must follow the ideal shape within its tolerances. A systematic deviation from the ideal shape is an indication of that the assumptions made are not valid. Therefore, detailed examination of the concrete core is needed, e.g. by plane (polished) section analysis and thin section analysis, cf. Hansen (1995), in order to find the reasons for the observed deviations. If possible, the assumptions should be changed to fit the concrete in question. Typical reasons for changes of the assumptions may be separation of aggregates, cracks and delamination etc.

A first assumption is normally that the chloride diffusion coefficient and the surface chloride content are constant, i.e. independent of time t, location x and chloride content C. Let

$$\mathbf{x} = \{x_1, x_2, x_3, \ldots, x_n\} \tag{2.23}$$

be the locations at which the observed chloride contents (measurements)

$$\text{obs}\{\mathbf{C}\} = \{C_1, C_2, C_3, \ldots, C_n\} \tag{2.24}$$

determine a chloride profile, and let

$$\text{est}\{\mathbf{C}\} = \text{est}\,\{C_1, C_2, C_3, \ldots, C_n\} \tag{2.25}$$

be the ideal shape of the chloride profile found by means of the chloride parameters as described in the following methods of approximation, e.g. the non-linear regression analysis. Thus, the residuals, res{**C**}, are defined as the values which have to be added to the measurements, obs{**C**}, in order to obtain the ideal chloride contents given by the ideal shape of the chloride profile, est{**C**}. Hence,

$$\mathrm{res}\{\mathbf{C}\} = \mathrm{est}\{\mathbf{C}\} - \mathrm{obs}\{\mathbf{C}\}. \tag{2.26}$$

If the residual Equation (2.26) versus the locations Equation (2.23) show a significant and systematic variation, the assumptions made for the calculation of the ideal shape of the chloride profile are not valid (it is assumed that the observations Equation (2.24) do not show systematic deviations). If there is a systematic deviation further examination of the concrete ought to be carried out. Causes of systematic deviation could be heterogeneity of the concrete, e.g. by segregation of the aggregates, carbonation of the near-to-surface layer of the concrete and delamination of the concrete. Such heterogeneity of the concrete will cause the diffusion coefficient D of the concrete to depend on the location x at variance with the assumption that $D = D_0$ remains constant.

Microstructural examination of the concrete (e.g. by thin section analysis) may disclose heterogeneity of the concrete, but it will not be able to determine how the heterogeneity of the concrete influences the distribution of the chloride diffusion coefficient.

By the use of chloride indications, cf. Section 1.4.1, it is possible to let the influence of location on the chloride diffusion coefficient become visible. By cutting a specimen from concrete perpendicular to the concrete surface and expose the cutting surface to chloride (like test method NT Build 443) the penetration of the chloride will depend on the value of the chloride diffusion coefficient. After an exposure of approximately 35 days or more the chloride penetration can be made visible by a chloride indicator or by a special grinding technique.

The influence of the distribution of the chloride content (concentration) of the concrete on the chloride diffusion coefficient cannot be determined in a similar simple way.

Example 2.1.2 A 37 years old concrete pile has been exposed to chloride during its entire life by seawater. The chloride profile shows systematic deviation from the ideal shape of the error-function solution, cf. Figure 2.4 and Figure 2.5. An examination of a plane section of a core of the concrete disclosed that the aggregates were segregated in such a way that the concrete actually appears like a two-layer medium, i.e. one layer of mortar and another layer of concrete. ◊

Method of surface tangent

From the graph of a chloride profile it is possible to give a fairly accurate estimate of the initial chloride content C_i and the surface chloride content

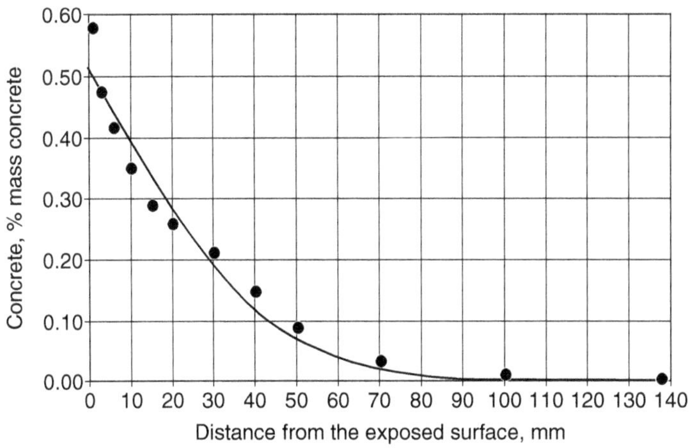

Figure 2.4: An observed chloride profile of concrete from a pile, submerged 37 yr in seawater in Esbjerg harbour. The ideal chloride profile is drawn on the basis of the chloride parameters found by a non-linear regression analysis: $D_0 = 14.7$ mm^2/yr, $C_0 = 0.512$ % mass of concrete, and $C_i = 0.002$ % mass of concrete. Notice that the ideal chloride profile seems to deviate systematically from observations with a hump 40 mm from the surface.

C_s while the estimate of the chloride diffusion coefficient D_0 is not possible without a geometrical estimate of the position of the tangent of the chloride profile at the exposed concrete surface.

Some chloride profiles deviate too much from the ideal shape near the concrete surface, and in such cases the following method is not valid. Especially achieved chloride profiles of concrete which have been exposed to carbonation or leaching besides chloride exposure deviate rather much from the ideal shape at the near-to-surface layer of the concrete.

The slope of the chloride profile at the point (x, C) at time of inspection t_{in} is the partial derivative of the function Equation (2.5) modelling the chloride profile. The tangent of the chloride profile at the point $(0, C_s)$ determines a line segment a of the asymptote from the ordinate axis to the intersection between the tangent and the asymptote, cf. Equation (2.19). From this equation it is seen that the chloride diffusion coefficient D_0 could be expressed as

$$D_0 = \frac{a^2}{\pi \, t_{\text{in}}}. \tag{2.27}$$

This equation leads to a simple rule: From the graph of the chloride profile the asymptote C_i is estimated. Then the tangent to the chloride profile at the point $(0, C_s)$ is drawn and the distance a of the asymptote from the ordinate axis to the intersection between the tangent and the asymptote is measured. Thus, the chloride diffusion coefficient is calculated from Equation (2.27) where t_{in} is the time of inspection.

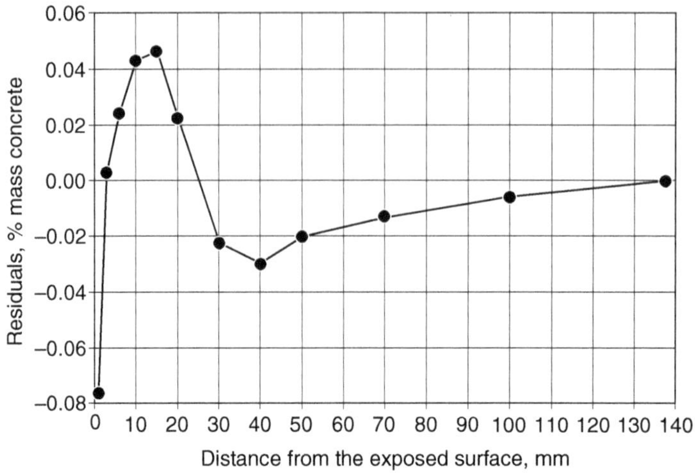

Figure 2.5: The residuals, i.e. the ideal chloride profile minus the observed, versus the distance from the exposed surface.

Three sets of observation

Theoretically, a chloride profile is determined by three points. In practice three points are never sufficient to determine the chloride parameters due to the uncertainties. Uncertain measurements (i.e. the measurement of the chloride content) will cause gross uncertainties in the calculation of the chloride parameters. However, from at theoretical point of view, formulæ for the chloride parameters are formulated in the following when we know just three points of the chloride profile.

From a practical point of view 'the method of three points' has some interest. It is a quick way to determine the chloride parameters from at given chloride profile: Choose three representative points of the chloride profile and determine D_0, C_0 and C_i from 'the method of three points' as described in the following.

Let (x_1, C_1), (x_2, C_2) and (x_3, C_3) be three points of a chloride profile where $0 < x_1 < x_2 < x_3$ and $C_1 > C_2 > C_3$. It is assumed that the location x_3 has a depth beneath the concrete surface so that C_3 is measured in the virgin concrete, i.e. no chloride ingress has reached position x_3. From this follows that $C_i = C_3$. Thus, the remaining chloride parameters $D = D_0$ and $C_s = C_0$ are determined from the two simultaneous equations

$$C_1 = C_3 + (C_0 - C_3)\ \text{erfc}\left(\frac{0.5\, x_1}{\sqrt{t_{\text{in}}\, D_0}}\right), \tag{2.28}$$

$$C_2 = C_3 + (C_0 - C_3)\ \text{erfc}\left(\frac{0.5\, x_2}{\sqrt{t_{\text{in}}\, D_0}}\right). \tag{2.29}$$

In order to solve these two equations we introduce a new set of variables

$$u = \frac{0.5\,x_1}{\sqrt{t_{\text{in}}\,D_0}}, \tag{2.30}$$

$$v = \frac{C_1 - C_3}{C_0 - C_3}. \tag{2.31}$$

Hence, the Equations (2.28) and (2.29) are transformed into the equivalent problem

$$v = \text{erfc}(u), \tag{2.32}$$

$$v = B\,\text{erfc}(A\,u), \tag{2.33}$$

where the constants A and B are given by

$$A = \frac{x_2}{x_1} > 1, \tag{2.34}$$

$$B = \frac{C_1 - C_3}{C_2 - C_3} > 1. \tag{2.35}$$

Thus, the solution u of the Equations (2.34) and (2.35) is determined by the following equation

$$f(x) = B\,\text{erfc}(A\,u) - \text{erfc}(u) = 0. \tag{2.36}$$

The solution of Equation (2.36), u and $v = \text{erfc}(u)$, may be found graphically from Figure 2.6 as shown in Example 2.1.3. From the solution (u, v) the chloride parameters are calculated to be

$$t_{\text{in}}\,D_0 = \frac{1}{4}\left\{\frac{x_1}{u}\right\}^2 \qquad \text{or} \qquad D_0 = \frac{(x_1/u)^2}{4t_{\text{in}}}, \tag{2.37}$$

$$C_0 = \frac{C_1 - (1-v)C_3}{v} \qquad \text{and} \qquad C_i = x_3. \tag{2.38}$$

The accuracy of solving Equation (2.36) is limited. However, the accuracy of the solution u to Equation (2.36) might be improved from a theoretical point of view by a complementary iteration method.

Let u_0 and u_1 be the results of the graphical solution in such a way that $f(u_0) > 0$ and $f(u_1) < 0$, cf. Equation (2.36). Then a more accurate solution by the method of *regula falsi* is given by

$$u_2 = \frac{f(u_1)u_0 - f(u_0)u_1}{f(u_1) - f(u_0)}. \tag{2.39}$$

The iteration process may be extended. A result obtained by Equation (2.36) will, however, in most cases be satisfactory. A case of iteration applying Equation (2.39) is shown in Example 2.1.3.

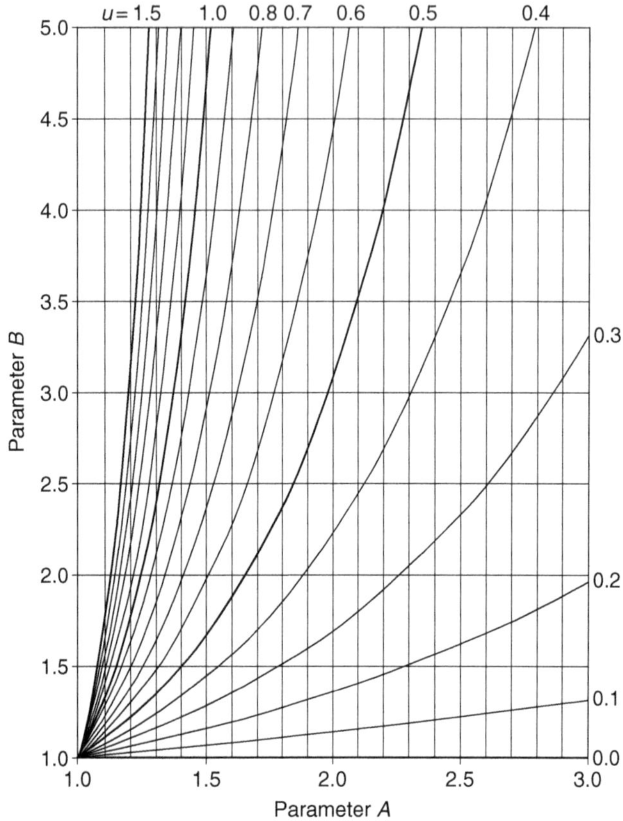

Figure 2.6: Diagram for solving the equation: $B\,\mathrm{erfc}(A\,u) - \mathrm{erfc}(u) = 0$, cf. Equation (2.36), when the parameters A and B are known. The solution is $u = 0.4$ when $A = 2.50$ and $B = 3.64$, cf. Example 2.1.3. The solution may be improved by iteration, cf. Equation (2.39).

Example 2.1.3 A chloride profile is determined by three points Nos. 1, 2 and 3, cf. Figure 2.7 and Table 2.2. The concrete has been exposed to chloride in $t_{\mathrm{in}} = 35$ days.
From Equation (2.34) and Equation (2.35) respectively, A and B are given by

$$A = \frac{x_2}{x_1} = \frac{10.0}{4.0} = 2.5,$$

$$B = \frac{C_1 - C_3}{C_2 - C_3} = \frac{0.576 - 0.008}{0.164 - 0.008} = 3.641.$$

Figure 2.8 shows the graphs of $v = \mathrm{erfc}(u)$ and $v = 3.641 \times \mathrm{erfc}(2.5u)$ based upon the principle of calculation given in Table 2.2. It is seen that the

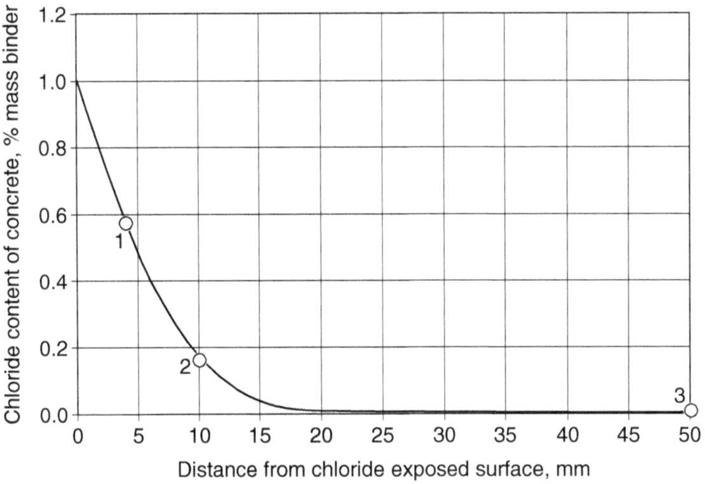

Figure 2.7: A chloride profile determined from three sets of observations, given in Example 2.1.3.

intersection of the two graphs is approximately $u \approx 0.4$. From the diagram in Figure 2.8 it also appears that the solution is $u \approx 0.4$. Hence, the chloride parameters imply that

$$v = \mathrm{erfc}(u) = \mathrm{erfc}(0.4) = 0.572.$$

Let $u_0 = 0.40$ and $u_1 = 0.42$. Then follows, cf. Equation (2.36), that

$$\begin{aligned} f(u_0) &= B\,\mathrm{erfc}(A\,u_0) - \mathrm{erfc}(u_0) \\ &= 3.641 \times \mathrm{erfc}(2.5 \times 0.40) - \mathrm{erfc}(0.40) \\ &= 3.641 \times 0.1573 - 0.5716 = 0.001129 > 0, \\ f(u_1) &= B\,\mathrm{erfc}(A\,u_1) - \mathrm{erfc}(u_1) \\ &= 3.641 \times \mathrm{erfc}(2.5 \times 0.42) - \mathrm{erfc}(0.42) \\ &= 3.641 \times 0.1376 - 0.5525 = -0.0515 < 0. \end{aligned}$$

Thus, an improved solution, cf. Equation (2.39) and Equation (2.32), is

$$\begin{aligned} u_2 &= \frac{f(u_1)u_0 - f(u_0)u_1}{f(u_1) - f(u_0)} = \frac{-0.0515 \times 0.40 - 0.001129 \times 0.42}{-0.0515 - 0.001129} \\ &= 0.40043, \\ v_2 &= \mathrm{erfc}(u_2) = \mathrm{erfc}(0.40043) = 0.57121. \end{aligned}$$

The calculation of erfc(0.40043) is carried out by the approximation formula Equation (3.73) in order to obtain a more accurate value than given by tables.

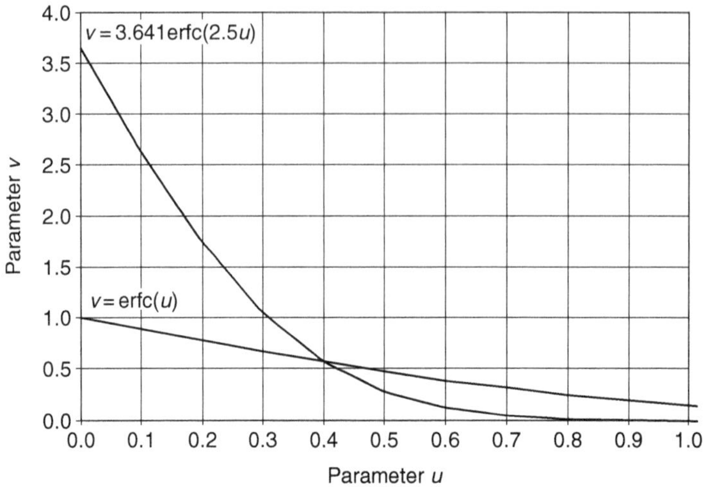

Figure 2.8: The solution of the equations $v = \text{erfc}(u)$ and $v = 3.641 \times \text{erfc}(2.5u)$ is determined graphically in this diagram. The solution is made more accurate by the method of regula falsi, cf. Example 2.1.3.

Hence, the chloride parameters are

$$t_{\text{in}}\, D_0 = \frac{1}{4}\left\{\frac{x_1}{u_2}\right\}^2 = \frac{1}{4}\left\{\frac{4.0}{0.40043}\right\}^2 = 24.946 \text{ mm}^2,$$

$$\begin{aligned} C_0 &= \frac{C_1 - (1 - v_2)C_3}{v_2} = \frac{0.576 - (1 - 0.57121) \times 0.008}{0.57121} \\ &= 1.0024 \text{ \% mass concrete.} \quad \Diamond \end{aligned}$$

2.1.4 Chloride parameters by regression analysis

The determination of the chloride parameters by the methods described above is just for rough (and quick) estimates. There is, however, also a certain need

Table 2.2: Chloride analysis of concrete in Example 2.1.3.

Point No.	Depth x mm	Chloride C % mass concrete
1	$x_1 = 4.0$	$C_1 = 0.576$
2	$x_2 = 10.0$	$C_2 = 0.164$
3	$x_3 = 50.0$	$C_3 = 0.008$

for approximative methods of estimation which apply to all the observations. Let the set of observations

$$(x_1, C_1), (x_2, C_2), (x_3, C_3), \ldots, (x_n, C_n) \tag{2.40}$$

determine an achieved chloride profile for which the chloride parameters (D_0, C_0, C_i) must be determined in such a way that the ideal chloride profile of the concrete

$$C = C_i + (C_0 - C_i)\ \text{erfc}\left(\frac{0.5\,x}{\sqrt{t_{\text{in}}\,D_0}}\right) \tag{2.41}$$

fits the observations as correctly as possible.

In Equation (2.41) the depth below the exposed concrete surface, x, is the independent variable and the chloride content C at location x is the dependent variable. The accuracy by which x is determined when using a grinding technique, cf. Section 1.4.1, is far better than the accuracy of C. This is, however, not the case when a hammer-drill technique is applied, though documentation has not been reported in detail. Here, the use of the grinding technique is assumed. In such cases the chloride parameters can be determined by a (non-linear) regression analysis, i.e. by minimizing the sum of the squares of the residuals (i.e. the method of least squares).

It is possible to apply the weighed residuals, but in connection with chloride profiles it has been common practice to apply either the weight 1 or 0. When we apply the weight 0 we ignore the observation in question. This is done when it is documented that a near-to-surface layer of the concrete is carbonated. The observations of a chloride profile within the carbonated zone are ignored when carrying out a regression analysis.

Non-linear curve-fitting

For any final estimation of the parameters of a chloride profile a non-linear curve-fitting should be applied. Here, the choice is whether the parameter C_i should be estimated by the curve-fitting (i.e. three parameters to estimate) or the observations should be planned in such a way that the observations are divided into two groups — i.e. one for the estimation of D_0 and C_0 and another one for the estimation of C_i. It has become general practice to use the last choice, so only this is described in detail in the following.

It is supposed that a (potential or achieved) chloride profile has been determined by the following observations

$$(x_1, C_1), (x_2, C_2), (x_3, C_3), \ldots, (x_n, C_n). \tag{2.42}$$

From these observations C_0 and D_0 should be determined in the following equation

$$C = C_i + (C_0 - C_i)\ \text{erfc}\left(\frac{0.5\,x}{\sqrt{t_{\text{in}}\,D_0}}\right), \tag{2.43}$$

assuming that the values of t_{in} and C_i are known. This will give n equations and two unknowns C_0 and D_0. Thus, the determination of C_0 and D_0 is not unambiguous unless further assumptions are made. If the observations are plotted into a Cartesian coordinate system it is unlikely that a chloride profile described by Equation (2.43) for any chosen value of C_0 and D_0 will pass through all the observations.

However, if it is required that a measure of the deviations between the (ideal) chloride profile and the observations is at a minimum, the determination of C_0 and D_0 will become unambiguous.

Assuming that the inaccuracy of C play the predominant role and x should be considered deterministic, the method of least squares will lead to an unambiguous determination of C_0 and D_0, i.e. it is required that

$$(\Delta C_1)^2 + (\Delta C_2)^2 + (\Delta C_3)^2 + \cdots + (\Delta C_n)^2 = \text{minimum}, \tag{2.44}$$

cf. Figure 2.5. Here, ΔC_1 is the residual at x_1, i.e. the deviation between the ordinate est$\{C_1\}$ of Equation (2.43) at x_1 and the observation obs$\{C_1\}$. The residuals ΔC_2 to ΔC_n are defined in a similar way.

2.2 Chloride ingress into prismatic specimens

The systematic collection of data from chloride ingress into concrete specimens exposed to various marine environment (marine atmosphere, marine splash and submerged in seawater) has mainly taken place at Marine Exposure Stations. When the water/cement-ratio of the concrete is relatively high (e.g. $0.5 < w/c < 0.7$) and the specimens allow the chloride to penetrate from more that one side of the specimens the chloride profiles cannot be determined in the same way as described in Section 2.1.

In case the specimens have a rectangular or especially a quadratic cross-section it is also possible to calculate the chloride profiles as shown in the following.

Structural components like concrete columns with square or circular cross-sections are rather common among concrete structures. The chloride ingress into the concrete of such a structural component cannot be determined in the same way as described in Section 2.1. It is, however, possible to calculate the chloride profiles as shown in the following.

2.2.1 Chloride ingress into walls from opposite sides

Often specimens tested at exposure stations have shapes like a panel. The chloride ingress into the panel from the opposite sides predominates. Therefore, it is possible to calculate the chloride profiles as chloride ingress into a concrete wall.

Structural components like thin concrete walls have often just one layer of reinforcement (at the centre plane). When such walls are exposed to chloride from both sides the chloride content of the concrete at the layer of the reinforcement will receive chloride from both sides of the wall. Therefore, the chloride profiles of thin concrete walls cannot be determined as described in Section 2.1. Instead, it is recommended to calculate the chloride ingress (chloride profiles) as chloride ingress into a concrete wall as shown in the following, cf. also Sections 5.1 and 5.2.

Fick's second law of diffusion

The chloride profile of a wall is here determined by means of Fick's second law of diffusion assuming the chloride diffusion coefficient to remain constant. The equation of the chloride profile is here obtained for the following assumptions and conditions

- The concrete is exposed to chloride, e.g. seawater, for the first time at time $t = 0$.
- The chloride-laden environment of the concrete is assumed to remain constant.
- The chloride profile is determined (inspected) at time $t = t_{\text{in}}$.
- The concrete (for $x \geq 0$) is assumed to be a quasi-homogeneous, semi-infinite medium.
- The transport of chloride in the concrete is assumed to obey Fick's second law of diffusion, cf. Section 1.1 and Equation (2.2).
- The chloride diffusion coefficient (for ≥ 0) is a constant $D = D_0$. This is the physical condition, cf. Equation (2.1).
- The chloride content of the concrete surface at one of the sides of the wall ($x = 0$) is constant $C_s = C_{00}$. The chloride content of the concrete surface at the opposite side of the wall ($x = d$) is also constant, $C_s = C_{0d}$, where the thickness (or depth) of the wall is d. This is the boundary condition, cf. Equation (2.4).
- The initial chloride content of the concrete (or binder) C_i is assumed to be equally distributed throughout the concrete. This is the initial condition, cf. Equation (2.3).

Under the given conditions above the necessary and sufficient equations for determining the chloride profile $C = C(x)$, $x \geq 0$ and at time $t = t_{\text{in}}$ become:

$$D = D_0, \qquad \text{for } x \geq 0 \text{ and } t > 0, \tag{2.45}$$

$$\frac{\partial C}{\partial t} = D_0 \frac{\partial^2 C}{\partial x^2} \qquad \text{for } x > 0 \text{ and } t > 0 \tag{2.46}$$

$$C(x,t) = C_i \qquad \text{for } x \geq 0 \text{ and } t = 0 \tag{2.47}$$

$$C(x,t) = C_{00} \qquad \text{for } x = 0 \text{ and } t > 0 \tag{2.48}$$

$$C(x,t) = C_{0d} \qquad \text{for } x = d \text{ and } t > 0 \tag{2.49}$$

Chloride profiles

The solution of Fick's second law of diffusion under these conditions, cf. Example 4.3.1 and Section 3.5, is

$$C(x, t_{\text{in}}) = C_i + (C_{00} - C_i) \times H_0(\xi, \tau) + (C_{0d} - C_i) \times H_0(1 - \xi, \tau), \quad (2.50)$$

where the function $H_0(\xi, \tau)$ is defined by

$$H_0(\xi, \tau) = 1 - \xi - \frac{2}{\pi} \sum_{n=1}^{\infty} \frac{1}{n} \exp\left(-n^2 \pi^2 \tau\right) \sin(n\pi\xi), \quad (2.51)$$

and is tabulated in Section 8.8. The parameters ξ and τ are respectively

$$\xi = \frac{x}{d} \quad \text{and} \quad \tau = \frac{tD}{d^2}, \quad (2.52)$$

where t is the time since the exposure started. In general the chloride content of the surfaces, C_{00} and C_{0d}, are different. However, in the special case where $C_{00} = C_{0d} = C_0$, the chloride profile is reduced to

$$\begin{aligned} C(x, t) &= C_i + (C_0 - C_i) \times \quad (2.53) \\ &\times \left\{ 1 - \frac{4}{\pi} \sum_{n=0}^{\infty} \frac{\exp\left(-(2n+1)^2 \pi^2 \tau\right) \sin((2n+1)\pi\xi)}{2n+1} \right\}. \end{aligned}$$

Example 2.2.1 A test specimen of concrete has the shape of a wall and has the thickness $d = 100$ mm. The following conditions are assumed

- C_i = 0.05 % mass binder
- C_{00} = 6.50 % mass binder
- C_{0d} = 5.50 % mass binder
- t_{in} = 1.00 yr
- D_0 = 400 mm^2/yr.

It is assumed that the concrete has a constant chloride diffusivity. The chloride profile after 1 year of exposure is predicted according to the following formula, cf. Equation (2.50),

$$\begin{aligned} C(x, t_{\text{in}}) &= C_i + (C_{00} - C_i) \times H_0(\xi, \tau) + (C_{0d} - C_i) \times H_0(1 - \xi, \tau) \\ &= 0.05 + 6.45 \times H_0(\xi, \tau) + 5.45 \times H_0(1 - \xi, \tau). \end{aligned}$$

Calculation of the ordinates of the chloride profile is carried out by means of the spread sheet shown in Table 2.3. The value of the parameter τ is

$$\tau = \frac{t_{\text{in}} D_0}{d^2} = \frac{1 \times 400}{100^2} = 0.04.$$

The chloride profile is shown in Figure 2.9. ◊

Table 2.3: Determination of the chloride profile specified in Example 2.2.1.

x mm	ξ non-dim	$1-\xi$ non-dim	$H_0(\xi,\tau)$ non-dim	$H_0(1-\xi,\tau)$ non-dim	$C(x,t)$ % mass binder
0	0.00	1.00	1.00000	0.00000	6.50
10	0.10	0.90	0.72367	0.00136	4.73
20	0.20	0.80	0.47950	0.00466	3.17
30	0.30	0.70	0.28884	0.01332	1.99
40	0.40	0.60	0.15730	0.03387	1.25
50	0.50	0.50	0.07710	0.07710	0.97
60	0.60	0.40	0.03387	0.15730	1.13
70	0.70	0.30	0.01332	0.28884	1.71
80	0.80	0.20	0.00466	0.47950	2.69
90	0.90	0.10	0.00136	0.72367	4.00
100	1.00	0.00	0.00000	1.00000	5.50

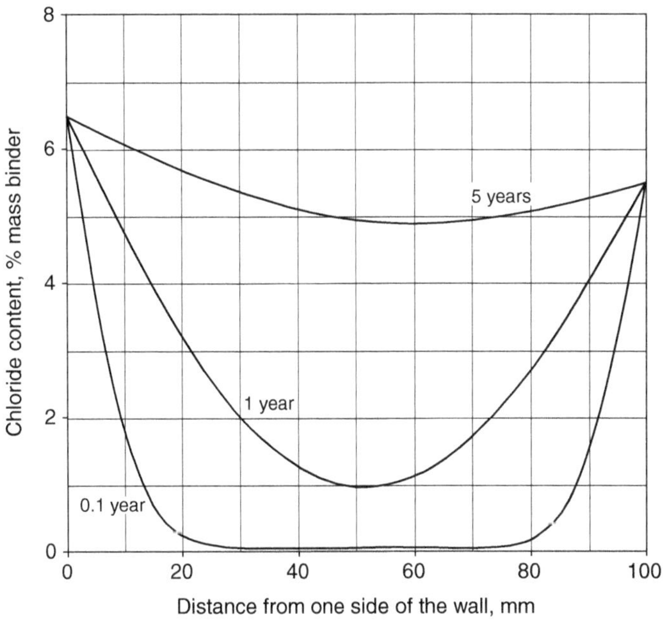

Figure 2.9: Chloride profiles in a 100 mm thick wall of concrete at a chloride exposure of 0.1 year, 1 year and 5 years under the assumptions that the concrete obeys a constant chloride diffusivity. The chloride diffusion coefficient is $D_0 = 400\ mm^2/year$, cf. Example 2.2.1 for further details. Notice: The data given in this example are all chosen for the purpose of illustration; none of the data is found by laboratory tests or field inspection.

2.2.2 Chloride ingress into specimens having square cross-sections

In case the test specimens have a rectangular or especially a quadratic cross-section it is also possible to calculate the chloride profiles as shown in the following.

Structural components like concrete columns with square (or circular) cross-sections are rather common among concrete structures. The chloride ingress into the concrete of such a structural component cannot be determined as described in Section 2.1. It is, however, nevertheless possible to calculate the chloride profiles as shown in the following.

Fick's second law of diffusion

The chloride profile of a prismatic specimen having a quadratic cross-section, $a \times a$, is here determined by means of Fick's second law of diffusion assuming that the chloride diffusion coefficient remains constant, i.e. $D = D_0$. Thus, the equation of the chloride profile is here obtained under the following assumptions and conditions

- The concrete is exposed to chloride, e.g. seawater, for the first time at time $t = 0$.
- The chloride-laden environment is assumed to remain constant.
- The chloride profile is determined (inspected) at time $t = t_{\text{in}}$.
- The concrete (for $x \geq 0$) is assumed to be a quasi-homogeneous, semi-infinite medium.
- The transport of chloride in the concrete is assumed to obey Fick's second law of diffusion, cf. Section 1.1 and Equation (2.55).
- The chloride diffusion coefficient (for $x \geq 0$) is a constant $D = D_0$. This is the physical condition, cf. Equation (2.54).
- The chloride content of the concrete surfaces at the sides (boundary) of the specimen is constant in time. In general, the surface chloride contents vary along the boundary for such specimens. However, in order to make this introduction simple it is here assumed that the surface chloride contents are constant along the boundary, $C_s = C_0$. This is the boundary condition, cf. Equation (2.57).
- The initial chloride content of the concrete (or binder) C_i is assumed to be equally distributed throughout the concrete. This is the initial condition, cf. Equation (2.56).

Under the given conditions above the necessary and sufficient equations for determining the chloride profile $C = C(x, y)$, x and $y \geq 0$ and at time $t = t_{\text{in}}$ are

$$D = D_0 \quad \text{for } x \geq 0,\ y \geq 0 \text{ and } t > 0, \tag{2.54}$$

$$\frac{\partial C}{\partial t} = D_0 \left\{ \frac{\partial^2 C}{\partial x^2} + \frac{\partial^2 C}{\partial y^2} \right\} \quad \text{for } x > 0,\ y > 0 \text{ and } t > 0, \tag{2.55}$$

$$C(x, y, t) = C_i \quad \text{for } 0 < x < a,\ 0 < y < a \text{ and } t = 0 \tag{2.56}$$

$$C(x, y, t) = C_0 \quad \text{for all } (x, y) \text{ at the boundary and } t > 0. \tag{2.57}$$

Chloride profiles

The solution of Fick's second law of diffusion for a quadratic cross-section under these conditions is

$$C\left(x, y, t_{\text{in}}\right) = C_i + (C_0 - C_i) \times \left\{1 - H\left(\xi_a, \tau_a\right) \times H\left(\eta_a, \tau_a\right)\right\}, \tag{2.58}$$

where the functions $H(\xi, \tau)$ and $H(\eta, \tau)$ respectively are defined by

$$\begin{aligned} H(\xi, \tau) &= \frac{4}{\pi} \sum_{n=0}^{\infty} \frac{\exp\left(-(2n+1)^2 \pi^2 \tau\right) \sin((2n+1)\pi\xi)}{2n+1}, \\ H(\eta, \tau) &= \frac{4}{\pi} \sum_{n=0}^{\infty} \frac{\exp\left(-(2n+1)^2 \pi^2 \tau\right) \sin((2n+1)\pi\eta)}{2n+1}, \end{aligned} \tag{2.59}$$

for $0 < \xi < 1$ and $0 < \eta < 1$ and $\tau \geq 0$. The function $H(\xi, \tau)$ is tabulated in Section 8.7. The notations ξ_a, η_a and τ_a in Equation (2.58) stand for

$$\xi_a = \frac{x}{a} \quad \text{and} \quad \eta_a = \frac{y}{a} \quad \text{and} \quad \tau_a = \frac{t_{\text{in}} D_0}{a^2}, \tag{2.60}$$

respectively. The chloride profile in the centre of one of the sides of the quadratic cross-section, e.g. $\eta_a = 0.5$, is described by

$$C\left(x, t_{\text{in}}\right) = C_i + (C_0 - C_i) \times \{1 - H(\xi, \tau) \times H(0.5, \tau)\}, \tag{2.61}$$

where

$$\xi = \frac{x}{a} \quad \text{and} \quad \tau = \frac{t_{\text{in}} D_0}{a^2}. \tag{2.62}$$

Example 2.2.2 A prismatic specimen of concrete is exposed to a marine environment. The specimen has a quadratic cross-section 100×100 mm^2. The predicted chloride profile in the centre of the side of the specimen shall be estimated under the following conditions

Table 2.4: Determination of the chloride profiles specified in Example 2.2.2.

x mm	ξ non-dim	$H(\xi,\tau)$ non-dim	$C(x,t)$ % mass binder
0	0.00	0.00000	4.50
10	0.10	0.27496	3.47
20	0.20	0.51584	2.56
30	0.30	0.69783	1.87
40	0.40	0.80881	1.46
50	0.50	0.84580	1.32
60	0.60	0.80881	1.46
70	0.70	0.69783	1.87
80	0.80	0.51584	2.56
90	0.90	0.27496	3.47
100	1.00	0.00000	4.50

- $C_i = 0.05$ % mass binder.
- $C_0 = 4.50$ % mass binder.
- The concrete is 28 days old when exposed to chloride.
- $t_{\text{in}} = 1.00$ yr.
- $D_0 = 400$ mm^2/yr.

It is assumed that the concrete has a constant chloride diffusivity. The chloride profile after 1 year of exposure is predicted according to the following formula, cf. Equation (2.61),

$$C(x, t_{\text{in}}) = C_i + (C_0 - C_i) \times \{1 - H(\xi, \tau) \times H(0.5, \tau)\},$$

where the value of the parameter τ is

$$\tau = \frac{t_{\text{in}} D_0}{d^2} = \frac{1 \times 400}{100^2} = 0.04.$$

Since $H(0.5, \tau) = 0.8458$, cf. Table 8.7, the chloride profile yields

$$C(x, t) = 0.05 + 4.45 \times \{1 - 0.8458 \times H(\xi, \tau)\}.$$

Calculation of the ordinates of the chloride profile is carried out by means of the spread sheet shown in Table 2.4. ◊

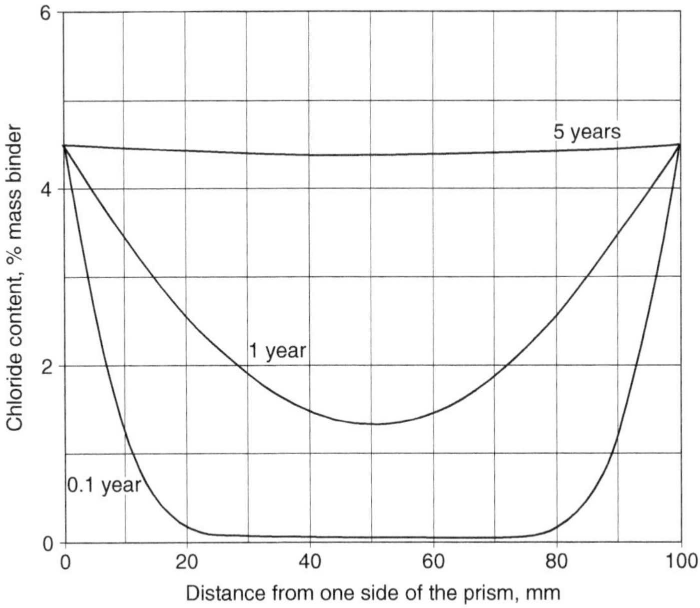

Figure 2.10: Chloride profiles in a prismatic concrete specimen with quadratic cross-section 100×100 mm, corresponding to a chloride exposure of 0.1 year, 1 year and 5 years. The assumptions are that the concrete obeys a constant chloride diffusivity. The chloride diffusion coefficient is $D_0 = 400$ mm^2/year, cf. Example 2.2.2 for further details. Notice: The data given in this example are all chosen for the purpose of illustration; none of the date is found by laboratory tests or field inspection.

2.3 Chloride ingress from de-icing salt

Concrete exposed to de-icing salt containing chloride during the winter and exposed to leaching during the summer from traffic splash and driving rain during the summer season cannot be modelled by the simple error function solution, previously shown. Very few observations of the chloride content of concrete surfaces exposed to de-icing salt (particularly during the summer season) versus time exist. From these very few observations, the chloride content of the exposed concrete surface versus time cannot be estimated with great accuracy. Thus, in order to make it simple, the chloride content of the exposed concrete surface is approximated by a step function versus time. Using such a function as the boundary condition, Fick's second law of diffusion can be solved by means of the principle of superposition, cf. Section 4.1.

When exposed to de-icing salt during the wintertime the chloride content of the concrete surface is here assumed to be a constant, $C_s = C_0$. When the de-icing stops and the chloride content is leached from the concrete surface it is here assumed that the chloride content is decreasing momentarily to

$C_s = 0$. Therefore, the chloride content of the near-to-surface layer of concrete exposed to de-icing salt shows a cyclic variation with time. Furthermore, it is here assumed that the chloride diffusion coefficient of the concrete remains constant $D = D_0$. Thus, it is possible to set up the solution of the problem by superposition of a series of error function solutions.

The conditions given above are valid when determining the chloride parameters from a chloride profile of a de-iced concrete structure. For more advanced conditions, cf. Section 5.1.3 assuming the LIGHTCON model and Section 5.2.3 assuming the HETEK model.

2.3.1 Chloride content of concrete surface

The boundary conditions are here given in the following way

- The first de-icing period runs from $t = t_0$ to $t = t_1$. During this period the chloride content of the surface is assumed to be constant, i.e. $C_s = C_0$.
- The first summer period runs from $t = t_1$ to $t = t_2$. During this period the chloride content of the surface is assumed to be zero, i.e. $C_s = 0$.
- The following de-icing periods and summer periods are repeated so that

  $$t_2 = t_0 + 12 \text{ months},\ t_3 = t_1 + 12 \text{ months},\ \ldots,\ t_{n+1} = t_n + 12 \text{ months},$$

  etc.

The chloride contents of the concrete surface given above may be written by means of the Heaviside's function

$$H(t - t_j) = \begin{cases} 0 & \text{for } t < t_j, \\ 1 & \text{for } t \geq t_j. \end{cases} \tag{2.63}$$

Thus, by means of Heaviside's function the chloride content of the concrete surface is determined by

$$C(0,t) = 0 \quad \text{for } 0 \leq t < t_0, \tag{2.64}$$

$$C(0,t) = C_0 \times H(t - t_0) \quad \text{for } t_0 \leq t < t_1, \tag{2.65}$$

$$C(0,t) = C_0 \times \{H(t - t_0) - H(t - t_t)\} \quad \text{for } t_1 \leq t < t_2, \tag{2.66}$$

$$C(0,t) = C_0 \times \{H(t - t_0) - H(t - t_1) + H(t - t_2)\},\ t_2 \leq t < t_3, \tag{2.67}$$

etc. It is seen that the chloride content of the concrete surface found above may be written as

$$C(0,t) = C_0 \times \sum_{j=0}^{n} (-1)^j \times H(t - t_j) \quad \text{for } t_n \leq t < t_{n+1}. \tag{2.68}$$

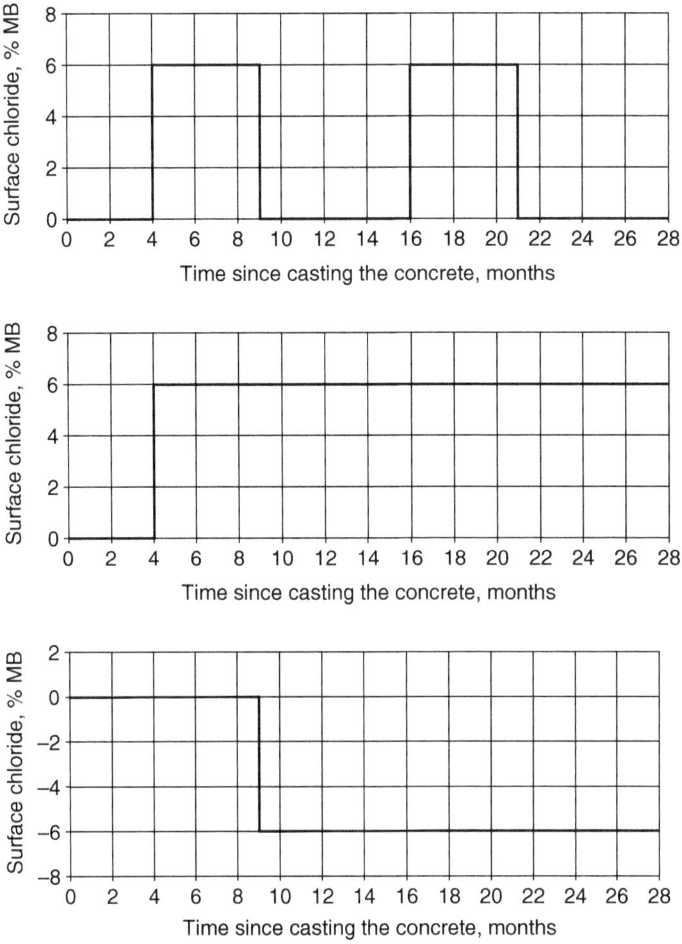

Figure 2.11: On top: Diagram of the chloride content of the concrete surface versus time in case where the concrete surface is exposed to de-icing salt according to Example 2.3.1. This diagram may be composed of the sum of the two step-functions shown below. Fick's second law of diffusion is simple when the boundary condition is a step-function and by superposition the chloride ingress into the concrete versus the simple is simple for the first nine months. Then by repeating this procedure the solution may be written as long as wanted, cf. Example 2.3.1.

Fick's second law of diffusion

Under the above given conditions the necessary and sufficient equations for determining the chloride profile $C = C(x)$, $x \geq 0$ and at time t_{in} are

$$D = D_0 \quad \text{for } x \geq 0 \text{ and } t \geq 0, \tag{2.69}$$

$$\frac{\partial C}{\partial t} = D_0 \frac{\partial^2 C}{\partial x^2} \quad \text{for } x > 0 \text{ and } t > 0, \tag{2.70}$$

$$C(x,t) = C_i \quad \text{for } x \geq 0 \text{ and } t = 0, \tag{2.71}$$

$$C(0,t) = C_0 \times \sum_{j=0}^{n} \left\{ (-1)^j \times H(t - t_j) \right\} \quad \text{for } 0 \leq t_n \leq t < t_{n+1}. \tag{2.72}$$

Chloride profiles of de-iced concrete

The chloride profile is found by solving Equation (2.64) to Equation (2.68),

$$C(x,t) = 0 \quad \text{for } t < t_0, \tag{2.73}$$

$$C(x,t) = C_0 \operatorname{erfc}(u_{t,t_0}) \quad \text{for } t_0 \leq t < t_1, \tag{2.74}$$

$$C(x,t) = C_0 \left\{ \operatorname{erfc}(u_{t,t_0}) - \operatorname{erfc}(u_{t,t_1}) \right\} \quad \text{for } t_1 \leq t < t_2, \tag{2.75}$$

$$C(x,t) = C_0 \left\{ \operatorname{erfc}(u_{t,t_0}) - \operatorname{erfc}(u_{t,t_1}) + \operatorname{erfc}(u_{t,t_2}) \right\} \quad \text{for } t_2 \leq t < t_3, \tag{2.76}$$

etc., where

$$u_{t,t_j} = \frac{0.5\,x}{\sqrt{(t - t_j)\,D_0}} \quad \text{for } t_j \leq t < t_{j+1}. \tag{2.77}$$

Example 2.3.1 A motorway bridge was cast and ready in service August 1st. The initial chloride content of the concrete is $C_i = 0$. On December 1st the de-icing of the bridge starts and continues until May 1st. Until December 1st there is no supply to the concrete of chloride. The exposure continues and the chloride content of the concrete surface is assumed to vary with time in the following way the first 16 months, cf. Figure 2.11,

- $0 \leq t < 4$ months: duration 4 months with $C_s = 0$ % mass binder,
- $4 \leq t < 9$ months: duration 4 months with $C_s = 6$ % mass binder,
- $9 \leq t < 16$ months: duration 4 months with $C_s = 0$ % mass binder.

An area of the motorway bridge has to be inspected for the chloride ingress. A rough estimate of the chloride ingress has to be established during the first 16 months. It is assumed that the chloride diffusion coefficient remains constant at a value of $D_0 = 85$ mm^2/mo.

Note. *The data given in this example are all chosen for the purpose of illustration. None of the data is found by laboratory tests nor field inspection and testing.*

Period 0: First summer season, i.e. $0 \leq t < 4$ months
The bridge is finished at time $t = 0$. During the period $0 \leq t < 4$ months the concrete is not exposed to chloride. Thus, the chloride content is zero, i.e.

$$C(x,t) = 0 \quad \text{for } 0 \leq t < 4 \text{ months.}$$

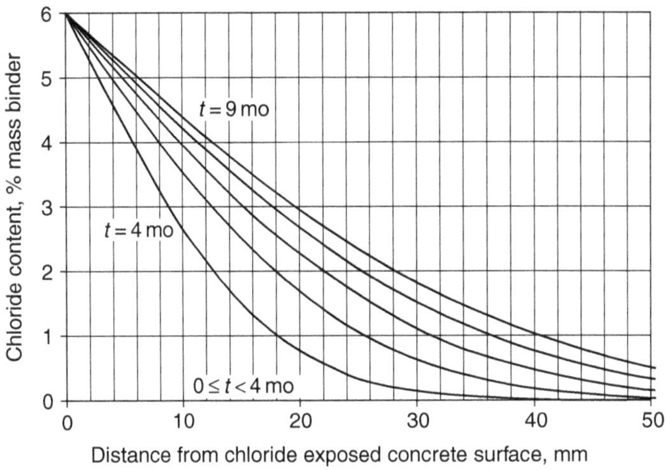

Figure 2.12: The development of the chloride profiles from month to month until the de-icing season stops, cf. Example 2.3.1.

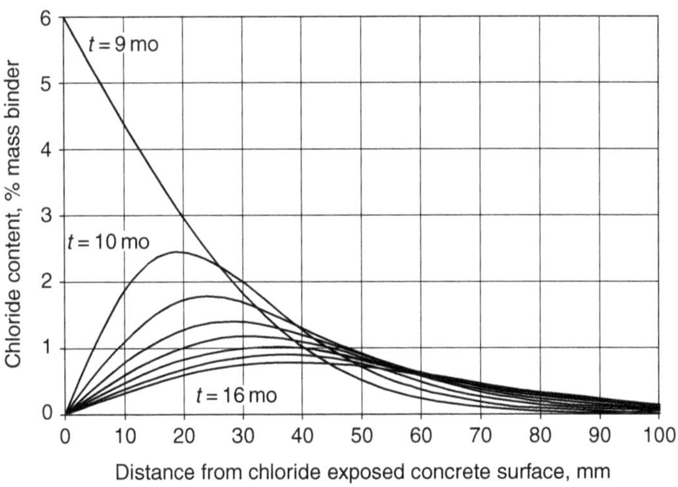

Figure 2.13: The development of the chloride profiles from month to month from the end of the de-icing season until the de-icing season starts, cf. Example 2.3.1.

Period 1: First winter season, i.e. $4 \leq t < 9$ months

After the first summer period the de-icing of the bridge starts on December 1st, i.e. at time $t = 4$ months. During this period the chloride of the concrete surface is increased (momentarily) so that the chloride content of the surface is $C_s = 6.0$ % mass binder during $4 \le t < 9$ months. During the first period the chloride profiles are month by month

$$t = 5 \text{ mo}: \quad C(x,t) = 6.0 \times \text{erfc}\,(u_{5,4}) \quad \text{where } u_{5,4} = \frac{0.5\,x}{\sqrt{(5-4)\times 85}} = 0.05423\,x,$$

$$t = 6 \text{ mo}: \quad C(x,t) = 6.0 \times \text{erfc}\,(u_{6,4}) \quad \text{where } u_{6,4} = \frac{0.5\,x}{\sqrt{(6-4)\times 85}} = 0.03835\,x,$$

$$t = 7 \text{ mo}: \quad C(x,t) = 6.0 \times \text{erfc}\,(u_{7,4}) \quad \text{where } u_{7,4} = \frac{0.5\,x}{\sqrt{(7-4)\times 85}} = 0.03131\,x,$$

$$t = 8 \text{ mo}: \quad C(x,t) = 6.0 \times \text{erfc}\,(u_{8,4}) \quad \text{where } u_{8,4} = \frac{0.5\,x}{\sqrt{(8-4)\times 85}} = 0.02712\,x,$$

$$t = 9 \text{ mo}: \quad C(x,t) = 6.0 \times \text{erfc}\,(u_{9,4}) \quad \text{where } u_{9,4} = \frac{0.5\,x}{\sqrt{(9-4)\times 85}} = 0.02425\,x.$$

Period 2: Second summer season, i.e. $9 \le t < 16$ months
The de-icing stops when the summer starts, i.e. at time $t = 9$ months. During this period the chloride of the concrete surface is washed out (momentarily) so that the chloride content of the surface is $C_s = 0$ during the time interval $9 \le t < 16$ months. During the second period the chloride profiles are month by month

$$t = 10 \text{ mo}: \quad C(x,t) = 6.0 \times \text{erfc}\,(u_{10,4}) - 6.0 \times \text{erfc}\,(u_{10,9}),$$

$$t = 11 \text{ mo}: \quad C(x,t) = 6.0 \times \text{erfc}\,(u_{11,4}) - 6.0 \times \text{erfc}\,(u_{11,9}),$$

$$t = 12 \text{ mo}: \quad C(x,t) = 6.0 \times \text{erfc}\,(u_{12,4}) - 6.0 \times \text{erfc}\,(u_{12,9}),$$

$$t = 13 \text{ mo}: \quad C(x,t) = 6.0 \times \text{erfc}\,(u_{13,4}) - 6.0 \times \text{erfc}\,(u_{13,9}),$$

$$t = 14 \text{ mo}: \quad C(x,t) = 6.0 \times \text{erfc}\,(u_{14,4}) - 6.0 \times \text{erfc}\,(u_{14,9}),$$

$$t = 15 \text{ mo}: \quad C(x,t) = 6.0 \times \text{erfc}\,(u_{15,4}) - 6.0 \times \text{erfc}\,(u_{15,9}),$$

$$t = 16 \text{ mo}: \quad C(x,t) = 6.0 \times \text{erfc}\,(u_{16,4}) - 6.0 \times \text{erfc}\,(u_{16,9}),$$

where

$$u_{10,4} = \frac{0.5\,x}{\sqrt{(10-4)\cdot 85}} = 0.02214x \text{ and } u_{10,9} = \frac{0.5\,x}{\sqrt{(10-9)\cdot 85}} = 0.05423\,x,$$

$$u_{11,4} = \frac{0.5\,x}{\sqrt{(11-4)\cdot 85}} = 0.02050x \text{ and } u_{11,9} = \frac{0.5\,x}{\sqrt{(11-9)\cdot 85}} = 0.03835\,x,$$

$$u_{12,4}=\frac{0.5\,x}{\sqrt{(12-4)\cdot 85}}=0.01917x \text{ and } u_{12,9}=\frac{0.5\,x}{\sqrt{(12-9)\cdot 85}}=0.03131\,x,$$

$$u_{13,4}=\frac{0.5\,x}{\sqrt{(13-4)\cdot 85}}=0.01808x \text{ and } u_{13,9}=\frac{0.5\,x}{\sqrt{(13-9)\cdot 85}}=0.02712\,x,$$

$$u_{14,4}=\frac{0.5\,x}{\sqrt{(14-4)\cdot 85}}=0.01715x \text{ and } u_{14,9}=\frac{0.5\,x}{\sqrt{(14-9)\cdot 85}}=0.02425\,x,$$

$$u_{15,4}=\frac{0.5\,x}{\sqrt{(15-4)\cdot 85}}=0.01635x \text{ and } u_{15,9}=\frac{0.5\,x}{\sqrt{(15-9)\cdot 85}}=0.02214\,x,$$

$$u_{16,4}=\frac{0.5\,x}{\sqrt{(16-4)\cdot 85}}=0.01566x \text{ and } u_{16,9}=\frac{0.5\,x}{\sqrt{(16-9)\cdot 85}}=0.02050\,x,$$

The development of the chloride profiles can easily be determined from the equations given above by a spreadsheet, applying an approximation of the complementary error function erfc(u). Here the approximation Equation (3.21) has been applied. The graphs are shown in Figure 2.12 and Figure 2.13.

2.4 Old marine RC structures

Concrete is a kind of material which is characterized by an ongoing change of its microstructure, chemically as well as physically. The greatest change takes place immediately after casting when the hydration of the binders start. However, the micro structural changes will take place even after years with a notable change in properties and characteristics of the concrete. After the development of the final microstructure an ageing process will take place, also resulting in a change of properties and characteristics of the concrete.

The chloride diffusion coefficient of concrete is one of the properties which will be time-dependent. However, when the concrete has been exposed to a chloride laden environment for more than 20–30 years the change in time per year has slowed down, cf. Section 1.2. This means that it is possible to obtain a rough estimate of future chloride ingress and the initiation period by neglecting the time dependency.

Also, the chloride content of the exposed concrete surface C_s is time-dependent. However, concrete submerged in seawater or exposed to the marine splash will reach its (almost) final value after 5–10 years, and it certainly will be a fair estimate to assume that the chloride content of the surface will remain constant after a time-interval of approximately 25 years. On the other hand, concrete which is exposed to a marine atmosphere and traffic splash seems to show a time-dependent chloride surface content for a considerable longer time, cf. Section 1.2.2 and Section 1.2.3.

Estimation of chloride ingress and prediction of the initiation period when the chloride parameters are time-dependent needs extensive calculations. Thus, rough estimates (on the safe side) are justified. Furthermore, the calculations are simple.

In the following such estimates are developed. When applying these estimates it should always be kept in mind that the assumptions made are approximations and they are only valid in special cases.

2.4.1 Fick's second law for constant chloride diffusivity

At an inspection time t_{in} > approximately 25 years the chloride profile of a concrete (submerged in seawater or exposed to the splash zone) is determined. From the chloride profile the chloride diffusion coefficient D_0, the surface chloride content C_0 and the initial chloride content are determined, cf. Section 2.1.3 or 2.1.4. When it is assumed that the chloride parameters remain constant for any time $t \geq t_{\text{in}}$, Fick's second law of diffusion, the boundary and the initial conditions become

$$D = D_0, \qquad \text{for } x \geq 0 \text{ and } t \geq t_{\text{in}} \gg 0, \tag{2.78}$$

$$\frac{\partial C}{\partial t} = D_0 \frac{\partial^2 C}{\partial x^2}, \qquad \text{for } x > 0 \text{ and } t \geq t_{\text{in}} \gg 0, \tag{2.79}$$

$$C(x,t) = C_i \qquad \text{for } x \geq 0 \text{ and } t = 0, \tag{2.80}$$

$$C(x,t) = C_0 \qquad \text{for } x = 0 \text{ and } t \geq t_{\text{in}} \gg 0. \tag{2.81}$$

The solution of Fick's second law of diffusion under these conditions, cf. Equation (3.13), is

$$C = C_i + (C_0 - C_i) \operatorname{erfc}\left(\frac{0.5\, x}{\sqrt{t\, D_0}}\right) \qquad \text{for } t \geq t_{\text{in}}. \tag{2.82}$$

2.4.2 Chloride ingress into old concrete

The future chloride ingress of a reference chloride content C_{re}, cf. Equation (2.12), yields

$$x_{\text{re}} = k_1 \sqrt{t} \qquad \text{for } t \geq t_{\text{in}}, \tag{2.83}$$

where the first year chloride ingress is, cf. Equation (2.13),

$$k_1 = \xi_{\text{re}} \sqrt{D_0} = 2\sqrt{D_0} \times \text{ inv erfc}\left(\frac{C_{\text{re}} - C_i}{C_0 - C_i}\right). \tag{2.84}$$

The following approximation to Equation (2.84) may be used when no table of the inverse complementary error function is available, cf. Summary 2.1.1,

$$k_1 \approx 3.32 \sqrt{D_0} \times \left\{1 - \sqrt{\frac{C_{\text{re}} - C_i}{C_0 - C_i}}\right\}. \tag{2.85}$$

The chloride ingress into concrete calculated in this way will normally be on the safe side. However, the calculation is simple and if the result is satisfactory, no more complicated calculation is needed. This estimate is based upon an approximation of the complementary error function, cf. Summary 2.1.1 and Equation (3.73) with $p = 0$.

2.4.3 Initiation period

With respect to determination of the initiation period of an RC structure it is of interest to know the ingress of the threshold value of chloride in the concrete C_{cr}. By inspection of the concrete and various test methods it is possible to determine the initiation period of time, i.e. the time when the chloride content of the concrete next to the reinforcement reaches the threshold value of chloride in the concrete and starts the corrosion process.

Determination by a chloride profile

It is assumed that a representative chloride profile has been determined from an old marine RC structure in a part of the concrete which has been submerged in seawater or exposed to the splash zone. Furthermore, it is assumed that the age of the structure at the time of inspection t_{in} and the time of the first chloride exposure of the structure are known or could be estimated.

From the chloride profile the chloride parameters D_0, C_0 and C_i are estimated, cf. Section 2.1.3 or 2.1.4. From these parameters the first year chloride ingress k_1 is determined, cf. Equation (2.13),

$$\begin{aligned} k_1 &= 2\sqrt{D_0} \times \text{ inv erfc}\left(\frac{C_{\text{cr}} - C_i}{C_0 - C_i}\right) \\ &= 3.32 \times \sqrt{D_0} \times \left\{1 - \sqrt{\frac{C_{\text{cr}} - C_i}{C_0 - C_i}}\right\}. \end{aligned} \tag{2.86}$$

Thus, the ingress into the concrete of the threshold value of chloride in concrete C_{cr} is given by

$$x_{\text{cr}} = k_1\sqrt{t} \qquad t \geq t_{\text{in}}. \tag{2.87}$$

When the threshold value reaches the position $x_{\text{cr}} = c$, i.e. the concrete cover thickness of the reinforcement, the reinforcement initiates corrosion. Thus, the initiation period t_{cr} is determined from the following equation

$$c = k_1\sqrt{t_{\text{cr}}} \tag{2.88}$$

from which the initiation period of time is calculated

$$t_{\text{cr}} = \left(\frac{c}{k_1}\right)^2 \qquad \text{for } t_{\text{cr}} > t_{\text{in}}. \tag{2.89}$$

Thus, the residual initiation period of time becomes

$$\text{res}\,(t_{\text{cr}}) = \left(\frac{c}{k_1}\right)^2 - t_{\text{in}}. \tag{2.90}$$

The chloride ingress into concrete calculated in this way will normally be on the safe side. However, the calculation is simple and if the result is satisfactory, no more complicated calculation is needed.

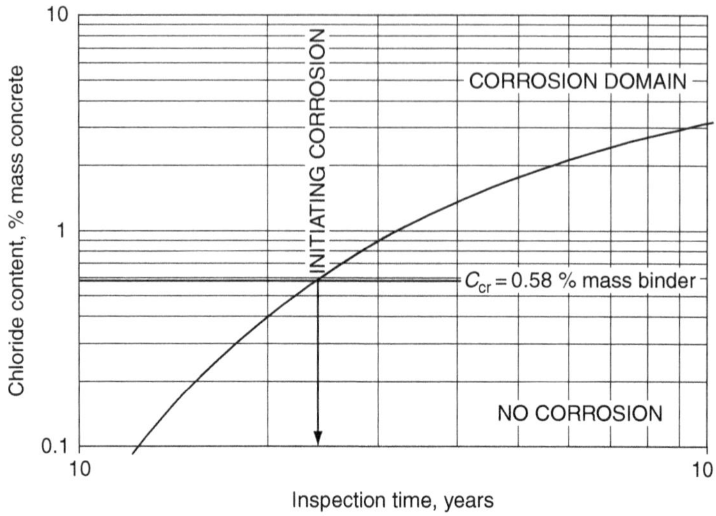

Figure 2.14: A (t, C)-diagram where $C_i = 0$, $C_0 = 6.51$ % mass binder, $D_0 = 15.8$ % mass binder and $C_{cr} = 0.51$ % mass binder, cf. Example 2.4.1. The initiation time is approximately 26 years.

2.4.4 Corrosion domain

The chloride ingress into concrete is represented by the following parameters: The chloride diffusion coefficient $D = D_0$ and the chloride content of the concrete surface $C(0,t) = C_0$. The corrosion of a reinforcing bar is presented by the following parameters: The threshold value of chloride in concrete C_{cr} and the concrete cover c of the reinforcement.

It is convenient to define a diagram where a certain domain represents ongoing corrosion of the reinforcement. In the following such diagrams are presented. When inspecting, testing and determining chloride profiles of a chloride exposed concrete structure it is convenient that these observations can be plotted directly into such diagrams.

(t, C)-diagram

The chloride content $C(x,t)$ at the depth of the reinforcement, $x = c$, and at the inspection time, $t = t_{in}$, is

$$C\left(c, t_{in}\right) = C_i + \left(C_0 - C_i\right) \times \text{erfc}\left(\frac{0.5\, c}{\sqrt{t_{in}\, D_0}}\right). \tag{2.91}$$

The time t_{cr} when the reinforcement initiates corrosion, i.e. at the end of the initiation period of time, is therefore determined by solving the following equation with respect to t_{cr},

$$C_i + (C_0 - C_i) \times \text{erfc}\left(\frac{0.5\, c}{\sqrt{t_{\text{cr}}\, D_0}}\right) = C_{\text{cr}}. \tag{2.92}$$

It is here convenient to define the following functions,

$$C(c,t) = C_i + (C_0 - C_i) \times \text{erfc}\left(\frac{0.5\, c}{\sqrt{t\, D_0}}\right) \quad \text{and} \quad C\,(c, t_{\text{cr}}) = C_{\text{cr}}, \tag{2.93}$$

and plot them in a $(t, C(c,t))$-diagram, cf. Figure 2.14. Now, the domain of corrosion is defined as the set of values of $(t, C(c,t))$ which yield corrosion of the steel reinforcing bar with cover c, i.e. $C(c,t) \geq C_{\text{cr}}$ for all $t > t_{\text{in}}$. Thus, the 'state of corrosion' for a situation $(t_{\text{in}}, C\,(c, t_{\text{in}}))$ is

$$\begin{aligned} C\,(c, t_{\text{in}}) &< C_{\text{cr}} && \text{implies 'no corrosion'}, \\ C\,(c, t_{\text{in}}) &= C_{\text{cr}} && \text{implies 'initiating corrosion'}, \\ C\,(c, t_{\text{in}}) &> C_{\text{cr}} && \text{implies 'ongoing corrosion'}. \end{aligned} \tag{2.94}$$

Testing the chloride content of concrete at the level of the reinforcement is a difficult task. The normal procedure of testing chloride in concrete is to determine the chloride profile of the concrete and find the chloride parameters D_0, C_0 and C_i. From these parameters the chloride content at the depth of the reinforcement may be estimated through a calculation by Equation (2.91). It is, however, more convenient to apply another diagram.

(t,D)-diagram

Now, a domain of corrosion is defined as the set of values of (t, D) which imply corrosion of the steel reinforcing bar with concrete cover c. The set of values of (t, D) which belongs to the border of the domain of corrosion is called $(t_{\text{cr}}, D_{\text{cr}})$, cf. Figure 2.15. Thus, at the time of inspection, $t = t_{\text{in}}$, the 'state of corrosion' for a situation (t_{in}, D_0) is

$$\begin{aligned} D_0 &< D_{\text{cr}} && \text{implies 'no corrosion'}, \\ D_0 &= D_{\text{cr}} && \text{implies 'initiating corrosion'}, \\ D_0 &> D_{\text{cr}} && \text{implies 'ongoing corrosion'}. \end{aligned} \tag{2.95}$$

The ingress x_{cr} into the concrete of the threshold value of chloride in concrete C_{cr} at time of inspection, $t = t_{\text{in}}$, is, cf. Equation (2.12),

$$x_{\text{cr}} = \xi_{\text{cr}} \sqrt{t_{\text{in}}\, D_0}, \tag{2.96}$$

where, cf. Equation (2.11) and (2.85),

$$\xi_{\text{cr}} = 2 \text{ inv erfc}\left(\frac{C_{\text{cr}} - C_i}{C_0 - C_i}\right) = 3.32 \times \left\{1 - \sqrt{\frac{C_{\text{cr}} - C_i}{C_0 - C_i}}\right\}. \tag{2.97}$$

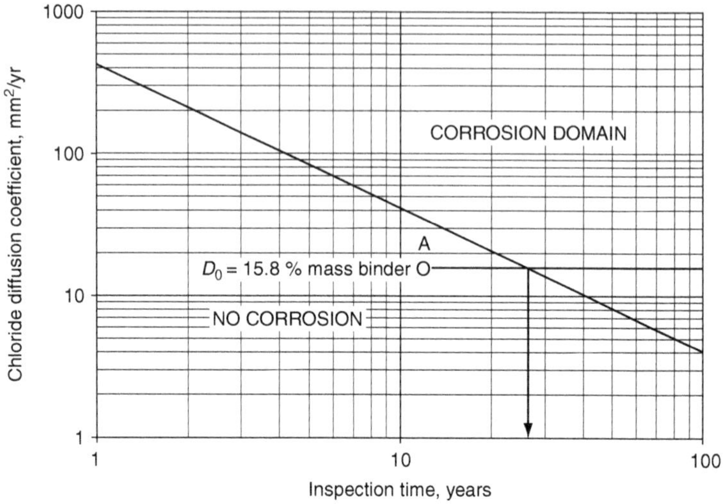

Figure 2.15: A (t, D)-diagram with the same data as in Figure 2.14, cf. Example 2.4.1.

Thus, the value of $D = D_{\text{cr}}$ which at time of inspection, $t = t_{\text{in}}$ leads to corrosion, cf. Equation (2.91), may be determined from the following equation

$$c = \xi_{\text{cr}} \sqrt{t_{\text{in}}\, C_{\text{cr}}}. \tag{2.98}$$

The solution of this equation with respect to D_{cr} is

$$D_{\text{cr}} = \frac{\left(c\,/\xi_{\text{cr}}\right)^2}{t_{\text{in}}}. \tag{2.99}$$

Therefore, the border of the corrosion domain is a hyperbola when plotted in a Cartesian coordinate system. It is seen that the border of the corrosion domain is a straight line when plotted as a logarithmic graph, cf. Figure 2.15.

Example 2.4.1 An area of a concrete bridge exposed to a marine splash environment is inspected and tested. The observations are given in Table 2.5. It is assumed that the initial chloride content of the concrete is $C_i \approx 0$. The cover of the reinforcement is $c = 50$ mm and that the concrete mixture proportions in the area are given in Table 2.6 , cf. Stoltzner (1995) and Frederiksen *et al.* (1997). The threshold value of chloride in concrete exposed to marine splash is estimated in the following way, cf. Section 1.3.1,

Table 2.5: Observations from a marine concrete bridge, cf. Example 2.4.1.

Inspection year	Age years	First chloride exposure, yr	Diffusion coefficient mm^2/yr	Surface chloride % mass binder
1994	12	0.5	15.8	8.51

$$\begin{aligned}
\text{eqv}\{w/c_{\text{cr}}\} &= \frac{W}{PC - 4.7 \times SF - 1.4 \times FA} \\
&= \frac{140}{330 - 4.7 \times 0 - 1.4 \times 40} = 0.51, \\
C_{\text{cr}} &= k_{\text{cr}} \exp(-1.5 \times \ \text{eqv}\{w/c_{\text{cr}}\}) \\
&= 1.25 \times \exp(-1.5 \times 0.51) = 0.58 \text{ \% mass binder.}
\end{aligned}$$

The concrete has an age of $t_{\text{in}} = 12$ years. This is not an age where it is valid to assume a constant chloride diffusivity (it requires an age of at least 25 years). However, in order to achieve a simple estimation (on the safe side), the assumption of constant chloride diffusivity is retained. More relevant conditions by the LIGHTCON model and the HETEK model are shown in Sections 5.1 and 5.2, respectively.

In order to plot the corrosion domain in a (t, D)-diagram the following parameter is determined

$$\xi_{\text{cr}} \approx 3.32 \times \left(1 - \sqrt{\frac{C_{\text{cr}} - C_i}{C_0 - C_i}}\right) = 3.32 \times \left(1 - \sqrt{\frac{0.58 - 0}{8.51 - 0}}\right) = 2.45.$$

Thus, the border of the corrosion domain defines

$$D_{\text{cr}} = \frac{(c\{\xi_{\text{cr}})^2}{t_{\text{in}}} = \frac{(50/2.45)^2}{t_{\text{in}}} = \frac{4.16.5}{t_{\text{in}}} \text{ mm}^2/\text{yr},$$

cf. Figure 2.15, where the present situation is plotted by point A. It is seen that no corrosion takes place at the moment, cf. (2.91 and that the initiation

Table 2.6: Mixture proportions, cf. Example 2.4.1.

Constituents	Contents	Unit
Portland cement	330	kg/m^3 concrete
Fly ash	40	kg/m^3 concrete
Aggregates	1827	kg/m^3 concrete
Water	140	kg/m^3 concrete

period of time is $t_{\mathrm{cr}} \approx 30$ years, if the chloride diffusion coefficient remains constant. Since

$$k_1 = \xi_{\mathrm{cr}} \sqrt{D_0} = 2.45 \times \sqrt{15.8} = 9.74 \text{ mm}/\sqrt{\text{yr}},$$

the initiation period of time becomes

$$t_{\mathrm{cr}} = \left(\frac{c}{k_1}\right)^2 = \left(\frac{50}{9.74}\right)^2 = 26 \text{ years.}$$

This is on the safe side. Prediction of the initiation time by means of the more realistic LIGHTCON model and HETEK model is presented in Sections 5.1 and 5.2 respectively. ◊

2.4.5 Service lifetime

In general the necessary reinforcement is designed in such a way that there is a certain reserve of reinforcement according to the Code of Practice. This means that the initiation period is not identical with the service lifetime (i.e. the time when the reinforcement left from corrosion reaches the requirements in strength and deformation required by the Code of Practice). The actual diameter of the reinforcement is d and the necessary reinforcement is d_{nec} according to the Code of Practice. Thus, if the average corrosion rate of the actual diameter as a result of corrosion is r_{c} the service lifetime t_{LT} yields

$$t_{\mathrm{LT}} = t_{\mathrm{cr}} + \frac{d - d_{\mathrm{nec}}}{r_{\mathrm{c}}}. \tag{2.100}$$

Consequently, the residual lifetime yields

$$\mathrm{res}\,(t_{\mathrm{LT}}) = t_{\mathrm{cr}} + \frac{d - d_{\mathrm{nec}}}{r_{\mathrm{c}}} - t_{\mathrm{in}} = \mathrm{res}\,(t_{\mathrm{cr}}) + \frac{d - d_{\mathrm{nec}}}{r_{\mathrm{c}}}. \tag{2.101}$$

The corrosion rate r_c is time-dependent when the corrosion has started and often the corrosion rate increases exponentially with time, cf. Section 1.3.1. Therefore, the corrosion rate in Equation (2.87) should be the mean value of the corrosion rate during the time interval $\Delta t = t_{\mathrm{LT}} - t_{\mathrm{cr}}$. However, the present knowledge of the corrosion rate of corroding reinforcing bars in concrete is rather limited.

2.4.6 Corrosion multi-probe

When an RC structure is monitored with corrosion multi-probes it is possible to predict the initiation period t_{cr} without any determination of chloride profiles and the threshold value of chloride in concrete, cf. Section 1.4.1.

Observations from a corrosion multi-probe

In case the first anode of the corrosion multi-probe (placed in position x_1) reacts at a time $t_1 >$ approximately 25 years for an RC member submerged into seawater or exposed to the splash zone, it is reasonable to assume that the chloride parameters remain constant in the residual period of the initiation period of time $t_{cr} - t_{in}$. Thus, from Equation (2.6) the first year ingress is determined from

$$x_1 = k_1 \sqrt{t_1}, \qquad \text{i.e.} \qquad k_1 = \frac{x_1}{\sqrt{t_1}}. \tag{2.102}$$

Therefore, when c denotes the reinforcing cover, the initiation time, cf. Equation (2.88), yields

$$t_{cr} = t_1 \times \left(\frac{c}{x_1}\right)^2, \tag{2.103}$$

and in the residual period of initiation period of time

$$\text{res}\,(t_{cr}) = t_1 \times \left(\frac{c}{x_1}\right)^2 - t_{in} = t_{in} \times \left\{\frac{t_1}{t_{in}}\left(\frac{c}{x_1}\right)^2 - 1\right\}. \tag{2.104}$$

The calculated chloride ingress into concrete will be on the safe side. However, the calculation is simple, and if the result is satisfactory, no further complicated calculation may be needed. For further information, cf. Sections 1.3.5, 5.1, 5.2 and 7.1.

Example 2.4.2 The first anode of a multi-probe placed in concrete exposed to the splash zone is reacting at time $t_1 = 25$ yr. The anode has a position of $x_1 = 20$ mm beneath the concrete surface. The time of first exposure is not known. The reinforcement cover is measured to $c = 45$ mm. From the structural design it is known that the diameter of the reinforcement bars achieved is $d_a = 25$ mm, while just the necessary diameter is $d_{nec} = 22.7$ mm. When corrosion takes place it is estimated that the average corrosion rate is $r_c = 50\,\mu$m/yr.

The first year chloride ingress, cf. Equation (2.87), yields

$$k_1 = \frac{x_1}{\sqrt{t_1}} \approx \frac{20}{\sqrt{25}} = 4.0 \text{ mm}/\sqrt{\text{yr}},$$

which involves the initiation period of time, cf. Equation (2.89),

$$t_{cr} = \left(\frac{c}{k_1}\right)^2 = \left(\frac{45}{4.0}\right)^2 = 127 \text{ yr},$$

and the service lifetime, cf. Equation (2.100),

$$t_{LT} = t_{cr} + \frac{d_a - d_{nec}}{r_c} = 127 + \frac{25 - 22.7}{50 \times 10^{-3}} = 173 \text{ yr}.$$

This result has to be followed up by a prediction from the rest of the anodes when they react. By such calculations it can be studied if the assumptions made here are valid, cf. Sections 5.1, 5.2 and 7.1. ◊

Rough estimates by chloride indicators

Before a detailed examination of a chloride profile or waiting for the anode of a corrosion probe to react in order to determine the residual initiation period of time it is suitable to determine the chloride ingress into the concrete by means of a chloride indicator, cf. Section 1.3.1. A chloride indicator does not give an exact measure of the depth of the threshold value of chloride in the concrete, but it gives an estimate suitable for a rough determination.

A representative concrete core, diameter approximately ∅ 75–100 mm, is drilled from the concrete. The core is split along the axis and sprayed by a chloride indication, cf. Section 1.4.1. The part of the broken concrete surface which contains water soluble chloride will keep its colour while the part of the concrete which does not contain water soluble chloride will chance its colour.

The thickness x_{cr} of the chloride containing near-to-surface layer of the concrete is determined (measures). Thus, a rough estimate of the initiation period of time is

$$t_{cr} \approx t_{in} \times \left(\frac{c}{x_{cr}} \right)^2, \tag{2.105}$$

and consequently the residual initiation period of time is

$$\text{res}\,(t_{cr}) \approx t_{in} \times \left(\frac{c}{x_{cr}} \right)^2 - t_{in} = t_{in} \times \left\{ \left(\frac{c}{x_{cr}} \right)^2 - 1 \right\}. \tag{2.106}$$

The chloride ingress into concrete calculated in this way will be on the safe side. The calculation, however, is simple, and if the result is satisfactory, no further complicated calculation may be needed.

Example 2.4.3 Before a detailed examination of a chloride profile from a $t_{in} = 30$ year old marine RC structure with for the purpose of predicting the initiation period of time t_{cr}, a concrete core from a representative position of the concrete from the splash zone is drilled, split and sprayed with a chloride indicator. The concrete cover of the reinforcement is measured to $c = 45$ mm.

The near-to-surface layer of concrete containing the water soluble chloride is measured to $x = 29$ mm. Thus, a rough estimate of the initiation period of time is, cf. Equation (2.106),

$$t_{cr} = t_{in} \left(\frac{c}{x} \right)^2 = 30 \times \left(\frac{45}{29} \right)^2 = 72 \text{ yr.}$$

A rough estimate of the residual initiation period yields $72 - 30 = 42$ yr. ◊

2.4.7 Probabilistic analysis

As long as civil engineers have designed load bearing structures a safety concept has been applied. However, it has not always been a rational safety concept, i.e. it has not taken into account the uncertainties of the various loads and materials. Cornell (1969) has shown the way. It would be sensible to benefit from the experience of the structural designer in this way.

Research, tests and experience show that chloride ingress into a given concrete exposed to a given environment is a stochastic variable. The rebar cover is also a stochastic variable. Therefore, one should have in mind that the necessary cover, nec$\{c\}$, is chosen suitably larger than the chloride ingress at the end of the initial period of time $t = t_{\mathrm{cr}}$. The problem is that the present knowledge is small about the stochastic distribution of the rebar cover c and the depth of the chloride ingress into the concrete x_{cr} at a given time $t = t_{\mathrm{cr}}$. Here, Cornell's method of reliability index makes it possible to obtain some knowledge about the stochastic distribution into account. One should know the expectation (mean value), dispersion (standard deviation or coefficient of variation) and (if correlation exists) also the covariance of the stochastic variables involved. On the other hand, no knowledge about the form of the stochastic distribution is required by Cornell's method.

This is convenient because it may be very long before the distribution of the stochastic variables are known — if ever!

There are two important applications of Cornell's method. The first one is the analysis of the initiation period, and the second one is the design of the concrete cover. Here, the analysis of the initiation time is dealt with, while the design of the cover for reinforcement for new RC structures is presented following the assumptions of the LIGHTCON model and the HETEK model in Sections 5.1 and 5.2, respectively. Karlsson *et al.* (1995), showed how a present code of practice may appear based on the probabilistic calculation of the relation between w/c, service lifetime t_{cr} and the rebar cover c corresponding to a given environment. It is, however, essential that the method is calibrated against structures of an acceptable service lifetime in a similar environment. It is the duty of the society to advise how safe our structures ought to be — and therefore it becomes the task of its Committee for the Code of Practice.

Introduction to the reliability index

The depth of chloride ingress into the concrete x_{cr} of the threshold value of chloride in concrete C_{cr} is not a precise variable, i.e. it is a stochastic variable, cf. Figure 2.16. Therefore, the notation (capital letter) X_{cr} is here used to indicate that it is a stochastic variable.

First we consider a one-dimensional problem, i.e. there is only one stochastic variable X_{cr}. This stochastic variable (i.e. uncertain or random variable) is the chloride ingress of the threshold value of chloride in concrete either now or at a given time, e.g. $t_{\mathrm{cr}} = 100$ years. No corrosion takes place when the cover c of the reinforcing bar is greater than the chloride ingress X_{cr}. Here, the

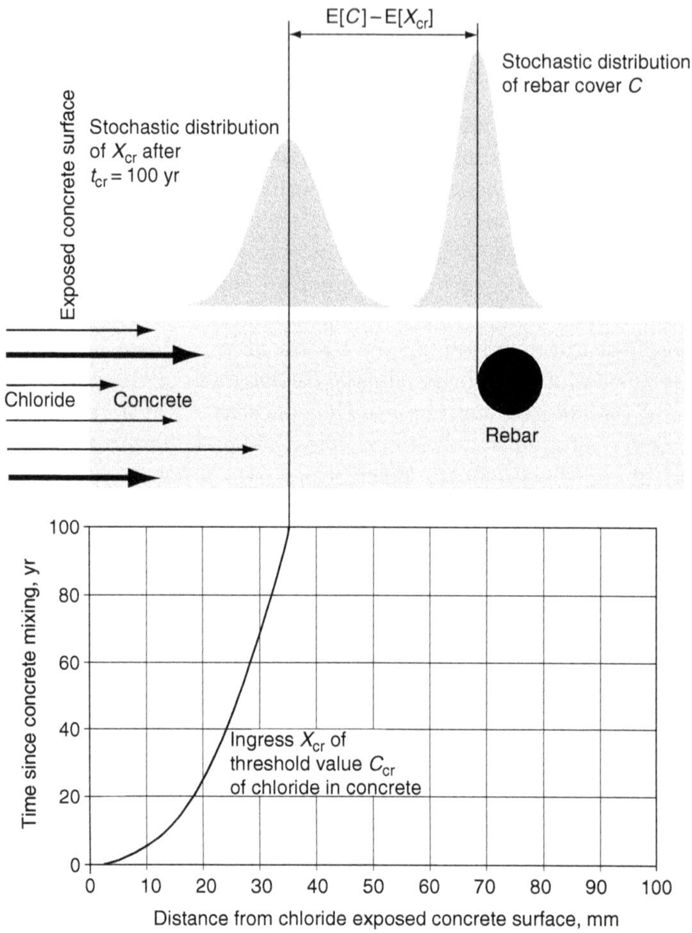

Figure 2.16: The ingress of the threshold value of chloride ${}_{\mathrm{cr}}$ in concrete follows the graph shown. At time $t_{\mathrm{cr}} = 100$ yr the expectation value of the ingress here yields about 35 mm. The rebar cover must be greater than the ingress to prevent corrosion. Since the rebar cover is also a stochastic variable, a suitable distance between the expectation values of the rebar cover $E[C]$ and ingress of the threshold value of chloride in concrete $R\left[X_{\mathrm{cr}}\right]$ must be ensured. Adapted from Mejlbro *et al.* (1999).

cover c is a deterministic parameter (i.e. the standard deviation is $S[c] = 0$). When X_{cr} exceeds the cover c at time t_{cr} corrosion takes place.

The reliability index β is defined as

$$\beta = \frac{c - E\left[X_{\mathrm{cr}}\right]}{S\left[X_{\mathrm{cr}}\right]}, \tag{2.107}$$

where $E\left[X_{\mathrm{cr}}\right]$ and $S\left[X_{\mathrm{cr}}\right]$ are the expectation value and the standard deviation of the stochastic variable X_{cr} respectively.

Table 2.7: Probability of failure p_f and reliability index β in structural design.

Failure type:	Ductile with reserve		Ductile without reserve		Brittle	
Consequence:	p_f	β	p_f	β	p_f	β
Not serious	10^{-3}	3.09	10^{-4}	3.71	10^{-5}	4.26
Serious	10^{-4}	3.71	10^{-5}	4.26	10^{-6}	4.75
Very serious	10^{-5}	4.75	10^{-6}	4.75	10^{-7}	5.20

In other words, the reliability index β times the standard deviation $S[X_{\text{cr}}]$ is equal to what the cover c is greater than the expectation value $E[X_{\text{cr}}]$. It is seen that a high value of β corresponds to a high safety, while a small value of β is obtained for concrete components with a small safety against corrosion.

In order to simplify Equation (2.107) we introduce the non-dimensional stochastic variable

$$X'_{\text{cr}} = \frac{X_{\text{cr}}}{S[X_{\text{cr}}]}, \tag{2.108}$$

and the non-dimensional parameter

$$c' = \frac{c}{S[X_{\text{cr}}]}, \tag{2.109}$$

into Equation 2.107. This yields the simple formula

$$\beta = c' - E[X'_{\text{cr}}]. \tag{2.110}$$

The values of β ought to be specified by the Code of Practice. For structural design one applies the values of β given in Table 2.7, cf. Toft-Christensen *et al.* (1982).

Example 2.4.4 A structural RC member is exposed to a marine environment. The cover of the reinforcing bars is deterministic (i.e. deviation is neglected because of in-situ testing) with a value of $c = 50$ mm. From an inspection and in-situ testing it is estimated that the chloride ingress is a stochastic variable with expectation value of $E[X_{\text{cr}}] = 35$ mm and standard deviation $S[X_{\text{cr}}] = 10$ mm. Thus, the reliability index, Equation (2.107), is given by

$$\beta = \frac{c - E[X_{\text{cr}}]}{S[X_{\text{cr}}]} = \frac{50 - 35}{10} = 1.5. \qquad \Diamond$$

The multi-dimensional problem

Suppose the following situation: Exposed to a given chloride-laden environment the threshold value of chloride in concrete C_{cr} has penetrated to the

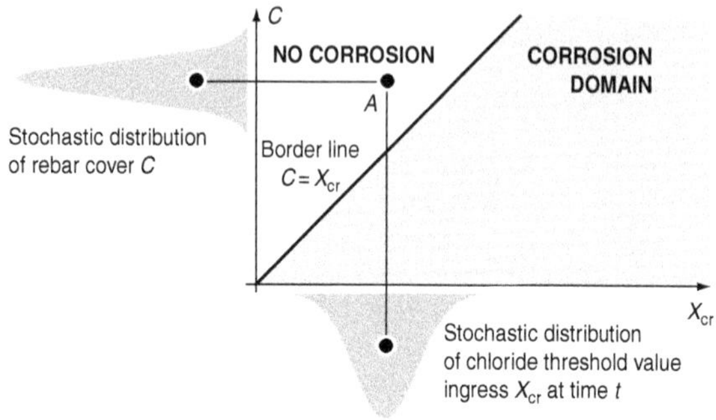

Figure 2.17: The corrosion domain (shaded) shown in a Cartesian coordinate system with the abscissa X_{cr} and the ordinate C. The centroid A of the two-dimensional stochastic distribution lies here in the domain of no corrosion. Adapted from Mejlbro *et al.* (1999).

depth x_{cr}. This is the result of the environment and the properties of the concrete. The depth of ingress x_{cr} is an uncertain variable, i.e. a stochastic variable. Therefore, the notation X_{cr} is used to indicate that it is a stochastic variable. This variable tends to force the steel reinforcement bar to start corroding. It is assumed that the expectation value $E\left[X_{\mathrm{cr}}\right]$ and the standard deviation $S\left[X_{\mathrm{cr}}\right]$ are known at time $t = t_{\mathrm{cr}}$.

The concrete cover c of the reinforcement is a stochastic variable. Therefore, the notation (capital letter) C is used to indicate that it is a stochastic variable. The cover tends to protect the reinforcement from corroding. It is assumed that the expectation value $E[C]$ and the standard deviation $S[C]$ are known and that C and X_{cr} are not correlated.

The cover must be $C > X_{\mathrm{cr}}$ at time t_{cr} as shown in Figure 2.17. Here, the corrosion domain, corresponding to the criterion: "$C \leq X_{\mathrm{cr}}$ implies corrosion", is shown shaded in a Cartesian coordinate system with abscissa C_{cr} and ordinate C. In Figure 2.17 the coordinates are

$$X'_{\mathrm{cr}} = \frac{X_{\mathrm{cr}}}{S\left[X_{\mathrm{cr}}\right]} \quad \text{and} \quad C' = \frac{C}{S[C]}. \tag{2.111}$$

The distance from the centroid $\left(E\left[X'_{\mathrm{cr}}\right], E[C']\right) = (\mu_x, \mu_c)$ of the two dimensional stochastic distribution to the border of the corrosion domain is

$$\beta = \frac{E[C] - E\left[X_{\mathrm{cr}}\right]}{\sqrt{S^2[C] + S^2\left[X_{\mathrm{cr}}\right]}} \tag{2.112}$$

as shown in the following. This distance or parameter is called Cornell's reliability index β.

Example 2.4.5 A structural RC member is exposed to a marine environment. The cover of the reinforcing bars is a stochastic variable with an expectation value of $E[c] = 50$ mm and a standard deviation of $S[c] = 5$ mm.

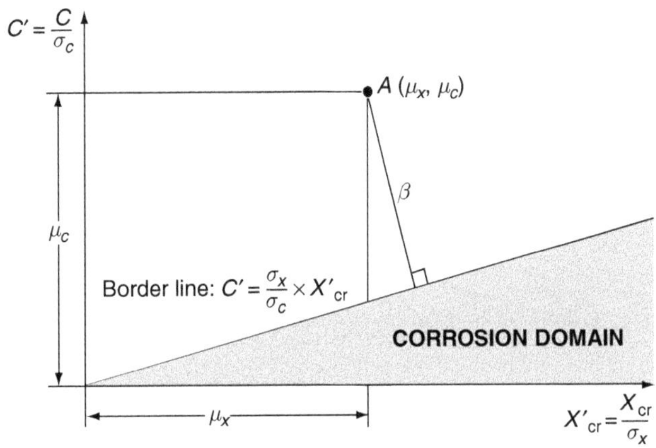

Figure 2.18: This diagram is a transformation of the diagram in Figure 2.17. Here the abscissæ are non-dimensional determined as the threshold value ingress X_{cr} divided by its standard deviation $D[X_{cr}] = \sigma_x$. The ordinates are C divided by its standard deviation $D[C] = \sigma_c$. Adapted from Mejlbro *et al.* (1999).

From inspection and in-situ testing it is observed that the chloride ingress is a stochastic variable with the expectation value of $E[X_{cr}] = 35$ mm and the standard deviation $S[X_{cr}] = 10$ mm. Thus, the reliability index is according to Equation 2.96,

$$\beta = \frac{E[C] - E[X_{cr}]}{\sqrt{S^2[C] + S^2[X_{cr}]}} = \frac{50 - 35}{\sqrt{5^2 + 10^2}} = 1.34,$$

cf. Example 2.4.4. ◊

Geometrical meaning of the reliability index

The corrosion criterion is: "$C \leq X_{cr}$ implies corrosion", and in the Cartesian (X_{cr}, C)-coordinate system, cf. Figure 2.16, the corrosion domain is shown shaded. The border line of the corrosion domain is $C = X_{cr}$. Now, by changing variables

$$X'_{cr} = \frac{X_{cr}}{S[X_{cr}]} \quad \text{and} \quad C' = \frac{C}{S[C]}, \tag{2.113}$$

the equation of the border line with respect to the (X'_{cr}, C')-coordinate system is written

$$C' = \frac{S[X_{cr}]}{S[C]} \times X'_{cr}. \tag{2.114}$$

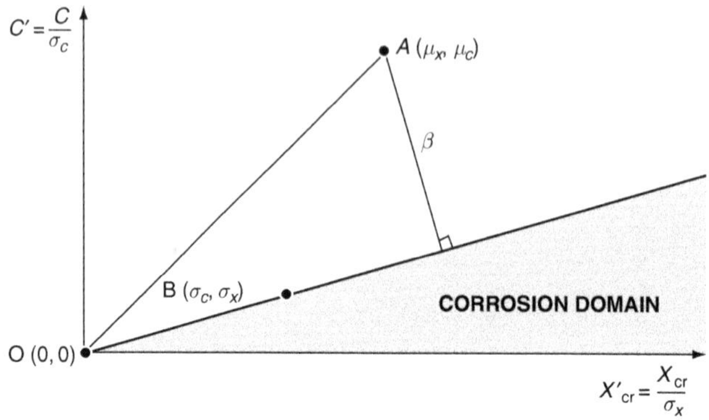

Figure 2.19: The corrosion domain shown in a Cartesian coordinate system with X_{cr}/μ_x as the abscissa and C/μ_c as the ordinate. Point A has the coordinates (μ_x, μ_c) and point B has the coordinates (σ_x, σ_x). Point A has the distance β from the border line of the corrosion domain. Adapted from Mejlbro *et al.* (1999).

This means that the point $(X'_{\mathrm{cr}}, C') = (S[C], S[X_{\mathrm{cr}}])$ are the coordinates of B of the border line, cf. Figure 2.19. The distance from the centroid A: $(X'_{\mathrm{cr}}, C') = (E[X'_{\mathrm{cr}}], E[C'])$ of the stochastic distribution to the border line is called β. This distance is determined by expressing the area of the parallelogram between the vectors $\overline{OA}$ and $\overline{OB}$, cf. Figure 2.20,

$$\text{area}\left\{\overline{OB}, \overline{OA}\right\} = \begin{vmatrix} S[C] & S[X_{\mathrm{cr}}] \\ E[X'_{\mathrm{cr}}] & E[C'] \end{vmatrix} = \beta \times \left|\overline{OB}\right|. \tag{2.115}$$

By calculating the determinant and the length of the vector the distance β is given by

$$\beta = \frac{S[C] \times E[C'] - S[X_{\mathrm{cr}}] \times E[X'_{\mathrm{cr}}]}{\sqrt{S^2[C] + S^2[X_{\mathrm{cr}}]}} = \frac{E[C] - E[X_{\mathrm{cr}}]}{\sqrt{S^2[C] + S^2[X_{\mathrm{cr}}]}}. \tag{2.116}$$

Probability of corrosion

As mentioned before the application of Cornell's reliability index does not assume that the type of the stochastic distribution of the ingress of the threshold value of chloride in concrete and the cover of the rebar are known. However, if they are known it is possible to calculate the probability of corrosion.

Here a normal distribution of X_{cr} and C is considered. When the ingress of the threshold value of chloride in concrete X_{cr} becomes greater than the concrete cover of the reinforcement bar C corrosion of the rebar starts. This means that the probability of corrosion is

$$P[\text{corrosion}] = P[X_{\mathrm{cr}} \geq C] = P[C - X_{\mathrm{cr}} \leq 0], \qquad \text{for } t = t_{\mathrm{cr}}, \tag{2.117}$$

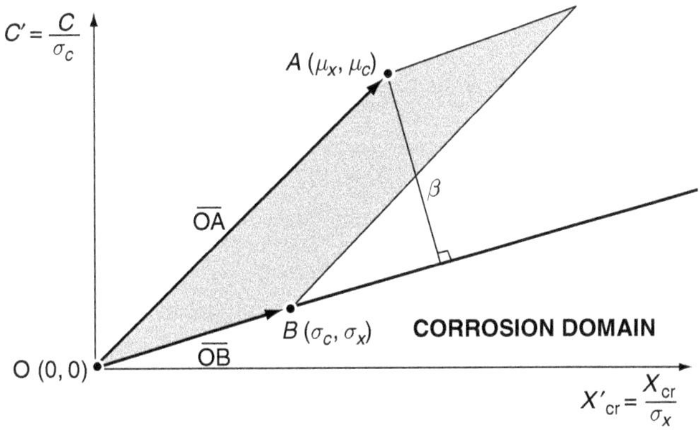

Figure 2.20: The area of the parallelogram defined by the vectors OA and OB is shaded. The area is the product of OA multiplied by β and also the determinant shown in the text. Adapted from Mejlbro *et al.* (1999).

i.e. at the end of the stipulated initiation period t_{cr} where the chloride ingress of the threshold value of chloride in concrete is a stochastic variable X_{cr} and the reinforcement cover C is also a stochastic variable. Assuming that the random variables C and X_{cr} follow the Gaussian distributions $N\left(\mu_c, \sigma_c^2\right)$ and $N\left(\mu_x, \sigma_x^2\right)$ respectively, the safety margin $M = C - X_{\text{cr}}$ is also normally distributed having the expectation value and the standard distribution respectively

$$E[M] = E[C] - E\left[X_{\text{cr}}\right] = \mu_c - \mu_x = \mu_m, \tag{2.118}$$

$$S[M] = \sqrt{\sigma_c^2 + \sigma_x^2} = \sigma_m. \tag{2.119}$$

Thus, the probability of corrosion becomes

$$P[\text{corrosion}] = P[M \leq 0]$$

$$= P\left[\frac{M - \mu_m}{\sigma_m} \leq -\frac{\mu_m}{\sigma_m} = -\frac{\mu_c - \mu_x}{\sqrt{\sigma_c^2 + \sigma_x^2}} = -\beta\right] = \Phi(-\beta). \tag{2.120}$$

It is learned from this that if the ingress X_{cr} of the threshold value or chloride in concrete and the rebar cover C are both stochastic variables and normally distributed, then the probability of corrosion is the standardized normal distribution function of the negative value of Cornell's reliability index β, i.e.

$$P[\text{corrosion}] = \Phi(-\beta).$$

Since $\Phi(-3.09) = 10^{-3}$ it is seen that when Cornell's reliability index is $\beta = 3.09$ the probability of corrosion is 10^{-3} corresponding to 0.1 %.

In the same way it is seen that corresponding to a value of Cornell's reliability index of $\beta = 5.20$ the probability of corrosion is 10^{-7}. Table 2.8 summarizes the probability of corrosion versus Cornell's reliability index when the

Table 2.8: Probability of corrosion versus Cornell's reliability index β. The cover C and the chloride ingress X_{cr} are either normal distributed or log normal distributed.

Reliability index β	Normal distribution	Log normal distribution $V = 5$ %	$V = 10$ %	$V = 20$ %
0	0.50	0.50	0.50	0.50
1	1.59×10^{-1}	1.65×10^{-1}	1.70×10^{-1}	1.81×10^{-1}
2	2.28×10^{-2}	2.83×10^{-2}	3.41×10^{-2}	4.62×10^{-2}
3	1.35×10^{-3}	2.60×10^{-3}	4.30×10^{-3}	9.39×10^{-3}
4	3.17×10^{-5}	13.3×10^{-5}	38.3×10^{-5}	165×10^{-5}
5	2.87×10^{-7}	40.4×10^{-7}	251×10^{-7}	2640×10^{-7}
6	1.00×10^{-9}	77.2×10^{-9}	1300×10^{-9}	40400×10^{-9}

uncertain variables, the cover C and the chloride ingress X_{cr}, are normally distributed (i.e. a Gaussian distribution) and logarithmic normally distributed (a log normal distribution) respectively.

Example 2.4.6 The splash zone of a marine bridge pillar has been inspected. the statistical distribution of the rebar cover C (measured by a cover meter) and the ingress of the threshold value of chloride in concrete X_{cr} (determined by chloride profiles or colorimetric chloride indicator) have been recorded as shown in Table 2.9.

In order to describe the statistical distribution of the ingress of the threshold value of chloride into the concrete in relation to the statistical distribution of the rebar cover by a simple figure Cornell's reliability index β is calculated,

$$\beta = \frac{\mu_c - \mu_x}{\sqrt{\sigma_c^2 + \sigma_x^2}} = \frac{48 - 35}{\sqrt{12^2 + 14^2}} = 0.71.$$

Thus, the probability of corrosion yields, cf. Abramowitz *et al.* (1965),

$$P[\text{corrosion}] = \Phi(-\beta) = \Phi(-0.71) = 23.9\ \%,$$

assuming that C and X_{cr} are normally distributed (i.e. Gaussian distributions).

Table 2.9: Observations from the splash zone of a marine bridge pillar.

Description	Rebar cover C	Chloride ingress X_{cr}	Unit
Expectation value	$\mu_c = 48$	$\mu_x = 35$	mm
Standard deviation	$\sigma_c = 12$	$\sigma_x = 14$	mm
Coeff. of variation	$V_c = 25$	$V_c = 40$	%

Note. *The data given in this example are all chosen for the purpose of illustration. None of the data is found by laboratory tests nor field inspection and testing.* ◊

Characteristic initiation period of time

In Section 2.4.3 the initiation time t_{cr} is treated as a deterministic variable. Now obviously, the initiation time of an RC component is a stochastic variable. This is due to the fact that the chloride front is not a plane parallel to the concrete surface, that the positions of the reinforcement bars are stochastic, and that the threshold value of chloride in concrete is a stochastic variable. This means that the RC component will start corrosion of the reinforcement in spots located at defects and where the cover has a minimum value.

If we assume a constant chloride diffusivity (for an old RC structure), the initiation time is, cf. Equation (2.89),

$$t_{cr} = \left(\frac{c}{k_1}\right)^2 . \tag{2.121}$$

It is possible from this equation to find the stochastic distribution of the initiation time when the distribution of the cover c and the first year chloride ingress k_1 are known. Since the cover of the reinforcement c, the first year chloride ingress k_1, and the initiation time t_{cr} are stochastic variables, the notation C, K_1 and T_{cr} are used instead. It is assumed that the cover C of the reinforcement and the first year chloride ingress K_1 are not correlated.

In analogy with the characteristic strength of the concrete one could define the characteristic value of the initiation time, $K[T_{cr}]$, as the 5 % fractile. In order to be able to find the characteristic value of the initiation time it is necessary to know the statistic distribution of the cover c and the first year chloride ingress k_1. To make it simple (though still realistic) it is here assumed that C and K_1 are log normally distributed. This means that the initiation period of time also becomes log normally distributed. If $E[\ln(T_{cr})]$ and $S[\ln(T_{cr})]$ denote the mean value and the standard deviation of the logarithm of the initiation period of time respectively, the characteristic initiation period of time (i.e. the 5 % fractile), $K[T_{cr}]$, becomes, cf. e.g. Ditlevsen *et al.* (1996),

$$K[T_{cr}] = \exp\left(E[\ln T_{cr}] - 1.65 \times S[\ln T_{cr}]\right). \tag{2.122}$$

The mean value of the logarithm of the initiation period of time (when assuming a constant chloride diffusivity and a log normal distribution) is

$$E[\ln T_{cr}] = \ln\left\{\frac{\left(1+V_k^2\right)\mu_c^2}{\left(1+V_c^2\right)\mu_k^2}\right\}. \tag{2.123}$$

Here the notations $\mu_c = E[C]$, $\sigma_c = S[C]$, and $V_c = \sigma_c/\mu_c$ have been used for the concrete cover of the reinforcing bar, and the following notations $\mu_k = E[K_1]$, $\sigma_k = S[K_1]$, and $V_k = \sigma_k/\mu_k$ for the first year chloride ingress. The standard deviation of the logarithm of the initiation period of time is

$$S[\ln T_{\text{cr}}] = 2\sqrt{\ln\left(\{1+V_c^2\}\{1+V_k^2\}\right)}. \tag{2.124}$$

Thus, inserting Equation (2.123) and (2.124) into Equation (2.122), the characteristic value of the initiation period of time becomes

$$K[T_{\text{cr}}] = \frac{(1+V_k^2)\,\mu_c^2}{(1+V_c^2)\,\mu_k^2} \times \exp\left(-3.3\sqrt{\ln\left(\{1+V_c^2\}\{1+V_k^2\}\right)}\right). \tag{2.125}$$

The expectation value of the initiation period of time is

$$E[T_{\text{cr}}] = \left(\frac{\mu_c}{\mu_k}\right)^2 (1+V_c^2)\,(1+V_k^2)^3. \tag{2.126}$$

The coefficient of variation is given by

$$V[T_{\text{cr}}] = \left(\{1+V_c^2\}\{1+V_k^2\}\right)^4 - 1. \tag{2.127}$$

Thus the standard deviation of the initiation period of time becomes

$$S[T_{\text{cr}}] = E[T_{\text{cr}}] \times V[T_{\text{cr}}], \tag{2.128}$$

where $E[T_{\text{cr}}]$ and $V[T_{\text{cr}}]$ refer respectively to the Equations (2.126) and (2.127) above.

Example 2.4.7 From pretesting of concrete (at $t_{\text{in}} > 25$ years) from a reinforced concrete component exposed to a marine splash environment and an engineering interpretation, the data shown in Table 2.10 is obtained including their standard deviations.

Assuming a constant chloride diffusivity the expectation value of the initiation period of time is, cf. Equation (2.126),

$$E[T_{\text{cr}}] = \left(\frac{50}{4.50}\right)^2 (1+0.10^2)\,(1+0.16^2)^3 = 134.5 \text{ years.}$$

Table 2.10: Data from pretesting and engineering interpretation, cf. Example 2.4.7.

Parameter	Expectation value	Standard deviation	Unit	Coefficient of variation
k_1	$\mu_k = 4.50$	$\sigma_k = 0.72$	$\text{mm}/\sqrt{\text{yr}}$	$V_k = 16$ %
c	$\mu_c = 50$	$\sigma_c = 5.0$	mm	$V_c = 10$ %

The coefficient of variation is given by, cf. Equation (2.127),

$$V^2\left[T_{\text{cr}}\right]=\left\{\left(1+0.10^2\right)\left(1+0.16^2\right)\right\}^4-1=0.151.$$

Thus, the standard deviation of the initiation period of time is, cf. Equation (2.128),

$$S\left[T_{\text{cr}}\right]=134.5\times\sqrt{0.151}=52.3 \text{ years}.$$

Finally, the characteristic value of the initiation time yields, cf. Equation (2.122),

$$\begin{aligned}K\left[T_{\text{cr}}\right] &= \frac{1+0.16^2}{1+0.10^2}\left\{\frac{50}{4.5}\right\}^2\exp\left(-3.3\sqrt{\ln\left(\left\{1+0.10^2\right\}\left\{1+0.16^2\right\}\right)}\right)\\ &= 67.5 \text{ years},\end{aligned}$$

when assuming a constant chloride diffusivity and log normal distributions.

Note. *The data given in this example are all chosen for the purpose of illustration. None of the data is found by laboratory tests nor field inspection and testing.* ◊

Determination of the characteristic value by means of a table

In order to determine the characteristic value (5 % fractile) of the initiation time one has to know the expectation values of the concrete cover and the 'first year chloride ingress' μ_c and μ_k respectively, as well as the coefficients of variation of the concrete cover and the 'first year chloride ingress' V_c and V_k. The relation between the expectation value of the initiation time and μ_c, μ_k, V_c and V_k is, cf. Equation (2.126),

$$E\left[T_{\text{cr}}\right]=\left(\frac{\mu_c}{\mu_k}\right)^2\left(1+V_c^2\right)\left(1+V_k^2\right)^3, \tag{2.129}$$

Table 2.11: The function $g\left(V_c, V_k\right)$, cf. Equation (2.130).

$V_c \setminus V_k$	**.05**	**.06**	**.07**	**.08**	**.09**	**.10**	**.11**	**.12**	**.13**	**.14**	**.15**
.05	.990	.987	.983	.970	.974	.968	.962	.956	.949	.941	.933
.06	.989	.986	.982	.978	.973	.967	.961	.955	.946	.940	.932
.07	.988	.984	.981	.976	.971	.966	.960	.953	.946	.939	.931
.08	.986	.983	.979	.975	.970	.964	.958	.952	.945	.937	.929
.09	.985	.981	.978	.973	.968	.963	.957	.950	.943	.936	.928
.10	.983	.979	.976	.971	.966	.961	.955	.949	.942	.934	.926
.11	.981	.977	.974	.969	.964	.959	.953	.947	.940	.932	.924
.12	.978	.975	.971	.967	.962	.957	.951	.944	.937	.930	.922
.13	.976	.973	.969	.965	.960	.954	.949	.942	.935	.928	.920
.14	.973	.970	.966	.962	.957	.952	.946	.940	.933	.925	.917
.15	.971	.968	.964	.959	.955	.949	.943	.937	.930	.923	.915

Table 2.12: The function $f(V_c, V_k)$, cf. Equation (2.132).

$V_c \setminus V_k$	**.05**	**.06**	**.07**	**.08**	**.09**	**.10**	**.11**	**.12**	**.13**	**.14**	**.15**
.05	.792	.774	.755	.736	.716	.697	.678	.660	.642	.624	.607
.06	.772	.756	.739	.721	.703	.685	.668	.650	.633	.616	.599
.07	.751	.737	.722	.706	.689	.672	.656	.639	.623	.607	.591
.08	.730	.717	.703	.689	.674	.658	.643	.627	.612	.596	.581
.09	.708	.697	.685	.671	.658	.643	.629	.614	.600	.585	.571
.10	.687	.677	.666	.654	.641	.628	.614	.601	.587	.574	.560
.11	.665	.656	.646	.635	.624	.612	.599	.587	.574	.561	.548
.12	.644	.636	.627	.617	.607	.596	.584	.572	.560	.548	.536
.13	.624	.616	.608	.599	.589	.579	.569	.558	.547	.535	.524
.14	.603	.597	.589	.561	.572	.563	.553	.543	.532	.522	.511
.15	.583	.577	.571	.563	.555	.546	.537	.528	.518	.508	.499

or

$$E[T_{\text{cr}}] = \frac{(\mu_c/\mu_k)^2}{g(V_c, V_k)}. \tag{2.130}$$

The relation between the characteristic value of the initiation time and μ_c, μ_k, V_c and V_k is, cf. Equation (2.126),

$$K[T_{\text{cr}}] = \frac{(1+V_k^2)\,\mu_c^2}{(1+V_c^2)\,\mu_k^2} \times \exp\left(-3.3\sqrt{\ln\left((1+V_c^2)(1+V_k^2)\right)}\right), \tag{2.131}$$

or

$$K[T_{\text{cr}}] = f(V_c, V_k) \times \left(\frac{\mu_c}{\mu_k}\right)^2. \tag{2.132}$$

The functions $f(V_c, V_k)$ and $g(V_c, V_k)$ are tabulated in Tables 2.12 and 2.11, respectively.

Example 2.4.8 For a part of a marine RC structure it is estimated that the expectation value of the 'first year chloride ingress' in the submerged zone is $E[k_1] = \mu_c = 5.46$ mm/$\sqrt{yr}$. Assuming that the coefficients of variation are $V_c = 8$ % and $V_k = 14$ %, and that the expectation value of the reinforcement cover is $E[c] = 60$ mm the expectation value and the characteristic value of the initiation period are respectively, cf. Table 2.12 and Table 2.11,

$$E[T_{\text{cr}}] = \frac{(\mu_c/\mu_k)^2}{g(V_c, V_k)} = \frac{(60/5.46)^2}{0.937} = 129 \text{ yr},$$

$$K[T_{\text{cr}}] = f(V_c, V_k) \times \left(\frac{\mu_c}{\mu_k}\right)^2 = 0.596 \times \left(\frac{60}{5.46}\right)^2 = 72 \text{ yr}.$$

The density and the distribution functions of the initiation period of time are shown in Figure 2.21 and Figure 2.22, respectively. This calculation is on the

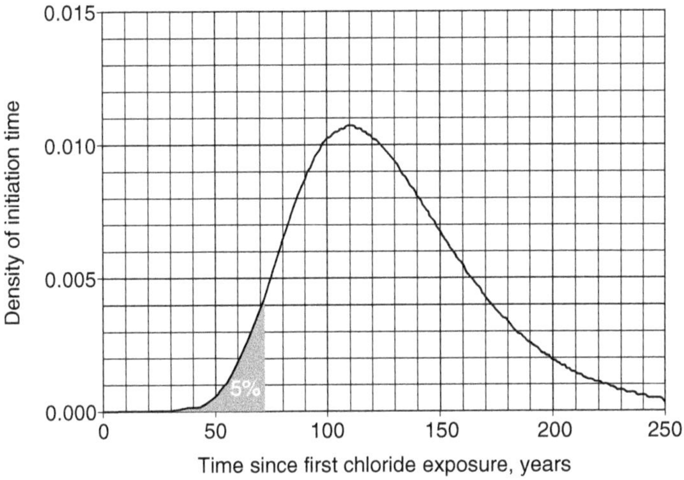

Figure 2.21: Density function of the initiation period of time for an RC component, cf. Example 2.4.8. The lower characteristic value (5 % fractile) is $K\,[T_{\mathrm{cr}}] = 72$ years.

safe side. Prediction of the characteristic value of the initiation period of time by means of the LIGHTCON model and the HETEK model is presented in Section 5.1 and Section 5.2, respectively. The main problem here is that there generally is a lack of knowledge about V_c and V_k. ◊

Summary 2.4 Assuming that the inspection time is $t_{\mathrm{in}} >$ approximately 25 years, the initiation period of time t_{cr} of an RC structure, submerged in seawater or exposed to the splash zone, can be estimated on the safe side from the determination of the depth of chloride ingress x_{cr} corresponding to the threshold value C_{cr} of chloride in concrete. The depth x_{cr} may be determined by means of a chloride profile from the concrete of the structure or estimated from a drilled concrete core which has been sprayed with a chloride indicator. The 'first year chloride ingress' k_1 is

$$k_1 = \frac{x_{\mathrm{cr}}}{\sqrt{t_{\mathrm{in}}}}.$$

Thus, the initiation period of time becomes

$$t_{\mathrm{cr}} = \left(\frac{c}{k_1}\right)^2 = t_{\mathrm{in}}\left(\frac{c}{x_{\mathrm{cr}}}\right)^2,$$

where c is the concrete cover above the reinforcing bar.

The parameters determining the chloride ingress into concrete structures bare all stochastic variables. Thus, the initiation time is also a stochastic

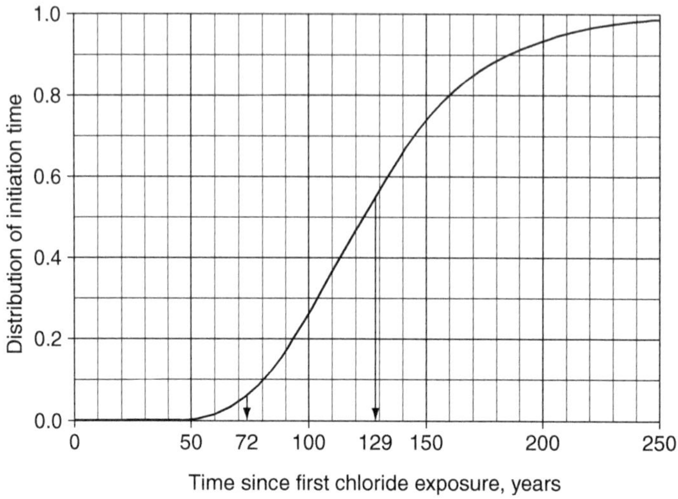

Figure 2.22: Distribution function of the initiation period of time for an RC component, cf. Example 2.4.8. The lower characteristic value (5 % fractile) is $K[T_{\text{cr}}] = 72$ years, and the expectation time is 129 years.

variable. Therefore, the problem of chloride ingress into concrete ought to be treated stochastically. However, the problem is that one does not know much about stochastic distribution of the chloride parameters and the cover of reinforcement. Therefore, one has to make assumptions of the stochastic distributions. For practical use, however, the distribution of the initiation time is not necessary.

It is assumed that the initiation time is greater than 25 years. According to the Theorem of Simplicity it is assumed that the cover of the reinforcement and the 'first year chloride ingress' are log normally distributed. Then the expectation value of the initiation time is

$$E[T_{\text{cr}}] = \frac{(\mu_c/\mu_k)^2}{g(V_c, V_k)}.$$

The characteristic value (the 5 % fractile) of the initiation time is

$$K[T_{\text{cr}}] = f(V_c, V_k) \times \left(\frac{\mu_c}{\mu_k}\right)^2.$$

Here, mu_c and μ_k are the expectation values of the cover of the reinforcement and the 'first year chloride ingress' respectively. Furthermore, V_c and V_k are the coefficients of variation of the same parameters. The functions $f(V_c, V_k)$ and $g(V_c, V_k)$ are tabulated in Tables 2.12 and 2.11.

Chapter 3

Error Function and Related Functions

The purpose of this chapter is to provide the reader with the necessary mathematics for solving Fick's second law in a large number of cases. The assumptions will be calculus as taught at any technical university. The intentions are to give a text which can be used at three levels,

- reference of results or formulæ,
- understanding of the proofs of some key results,
- indications of possible generalizations.

The first point is meant for engineers who in practice just need a model and the procedures behind it. The second one is aiming at students at universities. Paragraphs of this type are always marked by 'Proof' at the beginning and '□' at the end. Finally, the third point clearly refers to research students. Paragraphs of this type will be marked by an asterisk. The end of a remark or an example is marked by the symbol '◊'.

The solution of the basic equation

$$\frac{\partial C}{\partial t} = \frac{\partial^2 C}{\partial x^2}$$

in Fick's second law is in general not a simple matter. Therefore, some hard mathematics are needed. One cannot avoid the gamma function, the error function and the Bessel functions, which therefore are treated in this chapter. There are many books on these functions in the literature, and any one may be used for reference. In particular, we recommend Abramowitz *et al.* (1965), for short reviews of further results and tables, if one is not interested in mathematical proofs. More advanced references are the classical Carslaw *et al.* (1947), and Crank (1975), which are used by many authors for solving

Fick's second law in more general cases. When using Carslaw *et al.* (1947), the reader must be aware that the technical terminology is different, though from a mathematical point of view the heat equation is identical with Fick's second law.

It should be mentioned here that the repeated integrals of $\operatorname{erfc}(u)$ used in Carslaw *et al.* (1947), cf. Section 3.2.7, for generalizing the boundary condition of the heat equation have been further extended by Mejlbro (1996), to the far more flexible family of the $\Psi_p(u)$ functions giving a means to fully solve Fick's second law for constant diffusion coefficient, constant initial condition and any reasonable initial condition in practice (i.e. a nice function in time) for the one dimensional case. The proofs given by Mejlbro (1996), are based on the theory of convergent series. In Section 3.3 we shall give alternative proofs avoiding series as long as possible. However, for numerical reasons we cannot fully exclude series, because they give a means to calculate explicitly the values of the chloride concentration $C(x,t)$ at a given point and time.

3.1 The gamma function

Throughout this subsection the reader is referred to Abramowitz *et al.* (1965), where one should be aware of the fact that they heavily use the more advanced Complex Function Theory, thus considering $1/\Gamma(z)$ as an entire function in the complex plane. This viewpoint is extremely powerful, and many results can be derived by using the more general complex argument $z \in \mathbb{C}$, where these results may be difficult to obtain, when one confines oneself to real arguments $s \in \mathbb{R}$. In the applications, s will always be real. Therefore only the simplest proofs are produced.

3.1.1 Definition and extensions

The basic definition is

Definition 3.1.1 *The gamma function* $\Gamma(s)$ *is for* $s > 0$ *defined by*

$$\Gamma(s) = \int_0^\infty t^{s-1} \exp(-t)\, dt.$$

This definition is easily extended by noting that if $s > 0$, then by a partial integration

$$\Gamma(s+1) = \int_0^\infty t^s e^{-t}\, dt = \left[-t^s e^{-t}\right]_0^\infty + s\int_0^\infty t^{s-1} e^{-t}\, dt = s\Gamma(s),$$

from which we get the functional equations

$$\Gamma(s+1) = s\Gamma(s), \quad \text{i.e.} \quad \Gamma(s) = \frac{1}{s}\Gamma(s+1) \quad \text{for } s > 0. \tag{3.1}$$

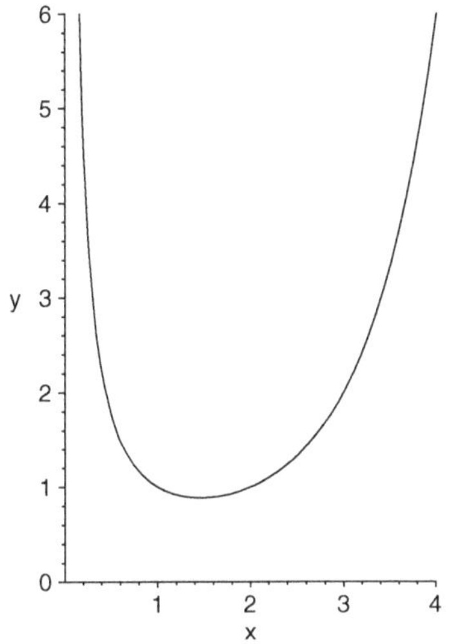

Figure 3.1: The gamma function

The latter equation also makes sense for $-1 < s < 0$, so $\Gamma(s)$ may be extended in this way. By induction, this functional equation can be used to extend $\Gamma(s)$ to all reals s, for which $-s$ is not a nonnegative integer, $-s \notin \mathbb{N}_0$.

When $s = -n$, $n \in \mathbb{N}_0$, one cannot define $\Gamma(-n)$. By continuity we see, however, that the reciprocal $1/\Gamma(-n)$ should be put equal to zero. In fact, by Equation (3.1),

$$\frac{1}{\Gamma(s)} = \frac{s}{\Gamma(s+1)} = \cdots = \frac{s(s+1)\cdots(s+n)}{\Gamma(s+n+1)}, \tag{3.2}$$

so when $s \to -n$ we get $s + n \to 0$ on the right hand side. We may therefore use the convention that

$$\frac{1}{\Gamma(-n)} = 0 \qquad \text{for } n \in \mathbb{N}_0.$$

By a rearrangement of (3.2) we even get

$$\begin{aligned} (s+n)\Gamma(s) &= \frac{\Gamma(s+n+1)}{s(s+1)\cdots(s+n-1)} \\ &\to \frac{(-1)^n}{n!}\Gamma(1) = \frac{(-1)^n}{n!} \qquad \text{for } s \to -n. \end{aligned} \tag{3.3}$$

Remark 3.1.1 Definition 3.1.1 makes sense, if only Re $s > 0$, and it is easily seen that $\Gamma(s)$ becomes an analytic function in this half-plane. By using (3.1) we extend $\Gamma(s)$ analytically to the entire complex plane, save for the points

$$s = -n, \quad n \in \mathbb{N}_0,$$

where it possesses simple poles with residues $(-1)^n/n!$, i.e.

$$\lim_{s\to -n}(s+n)\Gamma(s) = \frac{(-1)^n}{n!}, \qquad n \in \mathbb{N}_0,$$

cf. also (3.3) above. Its reciprocal $1/\Gamma(s)$ is an entire function possessing simple zeros at the points $s=-n$, $n \in \mathbb{N}_0$. Since $1/\Gamma(s) \to 0$ for $s \to +\infty$, real, the numerical procedures for calculating $\Gamma(s)$ are often produced via $1/\Gamma(s)$ instead, when $s > 0$. $\diamondsuit$

3.1.2 Special values

Obviously,

$$\Gamma(1) = \int_0^\infty t^{1-1}e^{-t}\,dt = \left[-e^{-t}\right]_0^\infty = 1,$$

so by iterating the first functional equation of (3.1) we get the important and well-known fact that

$$\Gamma(n+1) = n! \qquad \text{for } n \in \mathbb{N}_0.$$

Another important value of the gamma function is

$$\Gamma\left(\frac{1}{2}\right) = \sqrt{\pi}. \tag{3.4}$$

Since we shall frequently use this value, a proof of (3.4) is given in the following.

PROOF OF (3.4). Perform the change of variable $x = \sqrt{t}$, followed by a transformation into a plane integral. Finally, the plane integral is calculated by means of polar coordinates,

$$\begin{aligned}
\Gamma\left(\frac{1}{2}\right) &= \int_0^\infty t^{-1/2}e^{-t}\,dt = 2\int_0^\infty \exp(-x^2)\,dx \\
&= 2\left\{\int_0^\infty \exp(-x^2)\,dx \cdot \int_0^\infty \exp(-y^2)\,dy\right\}^{1/2} \\
&= 2\left\{\int_0^\infty \int_0^\infty \exp(-(x^2+y^2))\,dxdy\right\}^{1/2} \\
&= 2\left\{\int_0^{\pi/2}\left[\int_0^\infty \exp(-r^2)r\,dr\right]d\varphi\right\}^{1/2} \\
&= 2\left\{\frac{\pi}{2}\left[-\frac{1}{2}\exp(-r^2)\right]_0^\infty\right\}^{1/2} = 2\sqrt{\frac{\pi}{4}} = \sqrt{\pi}. \quad \square
\end{aligned}$$

3.1.3 Important formulæ

We mention without proof the following important formulæ, cf. e.g. Olver (1974).

The reflection formula

$$\Gamma(s)\Gamma(1-s) = -s\Gamma(-s)\Gamma(s) = \frac{\pi}{\sin \pi s} \qquad \text{for } 0 < s < 1.$$

For $s = 1/2$ we find again (3.4), i.e. $\Gamma(1/2) = \sqrt{\pi}$. Combining the reflection formula and the functional equation (3.1) we see that we can find any value of $\Gamma(s)$ if only we know its values in the interval $[0.5\,,\,1]$.

The duplication formula

$$\Gamma(2s) = \frac{1}{2\sqrt{\pi}} \cdot 2^{2s}\Gamma(s)\Gamma\left(s + \frac{1}{2}\right).$$

This formula will be important in Section 3.2.7. It is a special case of

Gauß's multiplication formula

$$\Gamma(ns) = (\sqrt{2\pi})^{1-n} \cdot \frac{n^{ns}}{\sqrt{n}} \prod_{k=0}^{n-1} \Gamma\left(s + \frac{k}{n}\right).$$

3.1.4 Pochhammer's symbol and related symbols

In the mathematical literature one often applies the *Pochhammer's symbol* $(z)_n$ for $z \in \mathbb{C}$ and $n \in \mathbb{N}_0$, defined by

$$(z)_0 = 1 \quad \text{and} \quad (z)_n = z(z+1)\cdots(z+n-1), \qquad \text{for } n \in \mathbb{N},$$

with n factors in ascending order. When $-(z+n) \notin \mathbb{N}_0$, this can also be written

$$(z)_n = \frac{\Gamma(z+n)}{\Gamma(z)}.$$

For our purposes a related symbol $z^{(n)}$ is more convenient. It is defined by

$$z^{(0)} = 1 \quad \text{and} \quad z^{(n)} = z(z-1)\cdots(z-n+1) \qquad \text{for } n \in \mathbb{N},$$

with n factors in descending order. If $z \notin \mathbb{N}$, we may instead write

$$z^{(n)} = \frac{\Gamma(z+1)}{\Gamma(z-n+1)}.$$

The two symbols are related by

$$(z)_n = (z+n-1)^{(n)} \quad \text{and} \quad z^{(n)} = (z-n+1)_n.$$

Note also that

$$n^{(n)} = (1)_n = n!.$$

3.1.5 Approximation formulæ

Not all PCs or pocket calculators have the gamma function in their catalogue of functions. Many useful approximation formulæ exist, cf. e.g. Abramowitz *et al.* (1965). The following two approximation formulæ suffice for our purposes.

Stirling's formula

Let $x > 0$. Then

$$\Gamma(x)=\sqrt{\frac{2\pi}{x}}\left(\frac{x}{e}\right)^x\left\{1+\frac{1}{12x}+\frac{1}{288x^2}-\frac{139}{51840x^3}-\frac{571}{2488320x^4}\right\}+R(x), \tag{3.5}$$

where the absolute value of the error term $|R(x)|$ decreases for increasing $x > 0$.

- For $x = 2$ the relative error of (3.5) without $R(x)$ is $\leq 9 \cdot 10^{-6}$.
- For $x = 3$ the relative error of (3.5) without $R(x)$ is $\leq 3 \cdot 10^{-6}$.
- For $x = 4$ the relative error of (3.5) without $R(x)$ is $\leq 10^{-6}$.
- For $x = 5$ the relative error of (3.5) without $R(x)$ is $\leq 2.5 \cdot 10^{-7}$.

Remark 3.1.2 By using the method of Olver (1974), p. 87 sqq. one may calculate the next terms of (3.5). However, the relative error is already small for $x \geq 3$, so there is no need here to do this. ◊

The second approximation formula gives the values of $\Gamma(x+1)$ for $x \in [0,1]$ by means of a polynomial,

$$\begin{aligned}\Gamma(x+1) &= 1-0.5748646x+0.9512363x^2-0.6998588x^3 \\ &\quad +0.4245549x^4-0.1010678x^5+R(x),\end{aligned} \tag{3.6}$$

where for the error term

$$|R(x)| \leq 5 \cdot 10^{-5} \quad \text{for } x \in [0,1].$$

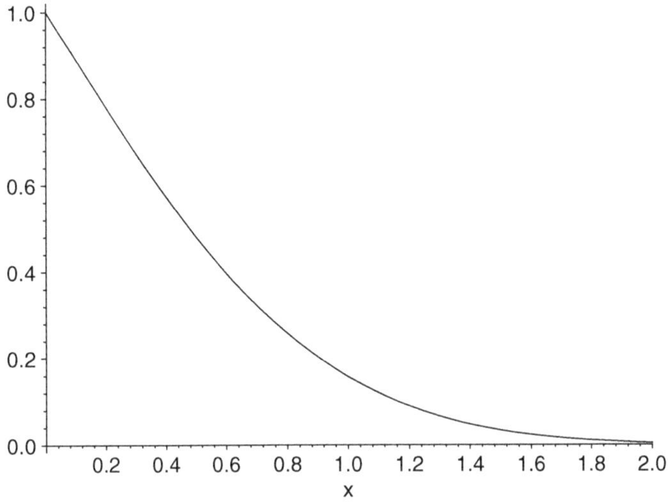

Figure 3.2: The complementary error function.

3.2 The error function and related functions

These functions are central for the classical theory of Fick's second law. Since they also occur in other connections, there is a large literature concerned with them. In order to limit the references we shall only recommend Abramowitz *et al.* (1965), Carslaw *et al.* (1947), and Olver (1974), for further details.

The *error function* is defined by

$$\operatorname{erf}(u) = \frac{2}{\sqrt{\pi}} \int_0^u \exp(-s^2)\, ds, \quad 0 \le u < +\infty, \tag{3.7}$$

and *the complementary error function* is defined by

$$\operatorname{erfc}(u) = \frac{2}{\sqrt{\pi}} \int_u^{\infty} \exp(-s^2)\, ds, \quad 0 \le u < +\infty. \tag{3.8}$$

The name 'error function' was introduced by Gauß, who realized its close connection with the *standardized normal distribution*,

$$\Phi(u) = \frac{1}{\sqrt{2\pi}} \int_{-\infty}^u \exp\left(-\frac{t^2}{2}\right) dt, \quad -\infty < u < +\infty. \tag{3.9}$$

It is easy to show that

$$\operatorname{erf}(u) = 2\Phi(u\sqrt{2}) - 1.$$

For the present theory the relation

$$\operatorname{erfc}(u) = 2\{1 - \Phi(u\sqrt{2})\} \tag{3.10}$$

is more important, because the complementary error function is 'natural' for the simplest nontrivial setup of Fick's second law. The central use of the normal distribution in Statistics means that there exist more tables of $\Phi(u)$ than tables of $\operatorname{erfc}(u)$, so (3.10) can be used to calculate the values of $\operatorname{erfc}(u)$ from the table values of $\Phi(u)$.

A simple statement is the following:

$$0 \leq \operatorname{erf}(u) < 1 \quad \text{and} \quad 0 < \operatorname{erfc}(u) \leq 1 \quad \text{for } 0 \leq u < +\infty. \tag{3.11}$$

PROOF. Formula (3.11) follows immediately from

$$\operatorname{erf}(u) + \operatorname{erfc}(u) = \frac{2}{\sqrt{\pi}} \int_0^\infty \exp(-s^2)\, ds = 1. \quad \square$$

3.2.1 Special values of erfc(u)

Obviously, $\operatorname{erfc}(u)$ is decreasing in u, so it suffices to indicate a few points u, for which $\operatorname{erfc}(u)$ assumes some predetermined values:

$$\operatorname{erfc}(0) = 1, \quad \operatorname{erfc}(0.477) \approx 0.5, \quad \operatorname{erfc}(1.163) \approx 0.1, \quad \operatorname{erfc}(1.821) \approx 0.01.$$

We also notice the estimate

$$\operatorname{erfc}(u+h) \leq \exp(-h(2u+h)) \cdot \operatorname{erfc}(u) \quad \text{for } u, h \geq 0. \tag{3.12}$$

PROOF OF (3.12). By Equation (3.7),

$$\operatorname{erfc}(u) = \frac{2}{\sqrt{\pi}} \int_u^\infty \exp(-s^2)\, ds = \frac{2}{\sqrt{\pi}} \int_0^\infty \exp(-(u+s)^2)\, ds,$$

hence,

$$\begin{aligned}
\operatorname{erfc}(u+h) &= \frac{2}{\sqrt{\pi}} \int_0^\infty \exp(-(u+h+s)^2)\, ds \\
&= \frac{2}{\sqrt{\pi}} \int_0^\infty \exp(-(u+s)^2 - h^2 - 2h(u+s))\, ds \\
&\leq \exp(-h^2 - 2hu) \cdot \frac{2}{\sqrt{\pi}} \int_0^\infty \exp(-(u+s)^2)\, ds \\
&= \exp(-h(2u+h)) \cdot \operatorname{erfc}(u),
\end{aligned}$$

and (3.12) is proved. $\square$

When (3.12) is multiplied by $\exp((u+h)^2)$, we get

$$\exp((u+h)^2)\operatorname{erfc}(u+h) \leq \exp(u^2)\operatorname{erfc}(u) \quad \text{for } u, h \geq 0,$$

so we have proved that the function $\exp(u^2)\operatorname{erfc}(u)$ is decreasing in u.

3.2.2 Connection with Fick's second law

The reason for introducing erfc(u) is its close connection with the solution of Fick's second law in its simplest form, cf. e.g. Carslaw *et al.* (1947). This classical result is so important that we coin it as a theorem.

Theorem 3.2.1 *The solution of*

$$\frac{\partial f}{\partial t}=\frac{\partial^2 f}{\partial x^2} \qquad \textit{for } x,t>0,$$
$$f(0,t)=1, \qquad \textit{for } t>0,$$
$$f(x,0)=0, \qquad \textit{for } x>0,$$

is given by

$$f(x,t)=\mathit{erfc}\left(\frac{x}{\sqrt{4t}}\right). \tag{3.13}$$

PROOF, We shall take the uniqueness of a bounded solution for granted, so we just have to check that (3.13) satisfies the equation and the initial and boundary conditions.

The initial and boundary conditions follow immediately by letting $t\to 0+$, resp. $x\to 0+$ in (3.13), so we only need to check the differential equation.

When (3.13) is differentiated we get

$$\frac{\sqrt{\pi}}{2}\frac{\partial f}{\partial x}=-\frac{1}{\sqrt{4t}}\exp\left(-\frac{x^2}{4t}\right),\quad \frac{\sqrt{\pi}}{2}\frac{\partial^2 f}{\partial x^2}=\frac{2x}{(4t)^{3/2}}\exp\left(-\frac{x^2}{4t}\right),$$

and

$$\frac{\sqrt{\pi}}{2}\frac{\partial f}{\partial t}=\frac{1}{2}\frac{x}{\sqrt{4t^3}}\exp\left(-\frac{x^3}{4t}\right)=\frac{2x}{(4t)^{3/2}}\exp\left(-\frac{x^2}{4t}\right)=\frac{\sqrt{\pi}}{2}\frac{\partial^2 f}{\partial x^2},$$

so $f(x,t)$ satisfies the differential equation as well. □

The physical interpretation of Theorem 3.2.1 is the following. Consider a half-infinite linear bar of concrete, which in its first approximation may be considered as one-dimensional. Place it along the positive X-axis, and assume that $f(x,t)$ represents the chloride concentration at position $x>0$ and time $t>0$. The initial condition means that at time $t=0$ the chloride concentration is $f(x,0)=0$ for $x>0$, and the boundary condition means that at $x=0$ the bar is subjected to an outer pressure from the left of chloride ions. For simplicity this outer pressure is represented by $f(0,t)=1$ for $t>0$, though any other constant may be used. Since the mathematical problem is linear, this just means that the solution should be multiplied by the same constant.

The problem is idealized, because no physical solution can have a jump at (0,0), where the solution (3.13) obviously is discontinuous. Such idealizations are, however, quite useful, because they ease the method of solution, and the error made is only significant in the neighbourhood of the discontinuity, where we for other reasons may expect that the model is not too reliable.

3.2.3 Series expansions of erfc(u)

There are many ways to calculate erfc(u) for given $u \geq 0$. The simplest one is to use a series expansion, which can be found in any book concerned with the error function, cf. e.g. Abramowitz *et al.* (1965), and Carslaw *et al.* (1947). Since

$$\operatorname{erfc}(u) = 1 - \operatorname{erf}(u),$$

it suffices to find a series expansion for erf(u). Using termwise integration we get

$$\begin{aligned}\operatorname{erf}(u) &= \frac{2}{\sqrt{\pi}}\int_0^u \exp(-s^2)\,ds = \frac{2}{\sqrt{\pi}}\int_0^u \sum_{n=0}^{\infty}\frac{(-1)^n}{n!}s^{2n}\,ds \\ &= \frac{2}{\sqrt{\pi}}\sum_{n=0}^{\infty}\frac{(-1)^n}{(2n+1)n!}u^{2n+1},\end{aligned}$$

from which

$$\operatorname{erfc}(u) = 1 - \frac{2}{\sqrt{\pi}}\sum_{n=0}^{\infty}\frac{(-1)^n}{(2n+1)n!}u^{2n+1}. \tag{3.14}$$

In the numerical application of (3.14) only a finite number of terms is summed and one gives an estimate of the error term. We mention two special results.

(a) If $0 \leq u \leq 2$, then

$$\operatorname{erfc}(u) = 1 - \frac{2}{\sqrt{\pi}}\sum_{n=0}^{20}\frac{(-1)^n}{(2n+1)n!}u^{2n+1} + R(u),$$

where

$$|R(u)| < 5\cdot 10^{-8} \quad \text{for } 0 \leq u \leq 2.$$

(b) For the larger interval $0 \leq u \leq 3$ we get

$$\operatorname{erfc}(u) = 1 - \frac{2}{\sqrt{\pi}}\sum_{n=0}^{33}\frac{(-1)^n}{(2n+1)n!}u^{2n+1} + R(u),$$

where

$$|R(u)| < 4\cdot 10^{-8} \quad \text{for } 0 \leq u \leq 3.$$

Although these approximation formulæ are sufficient for our purposes, we shall later give some approximations with fewer terms. The simplest one can already be given here, because it relies on the same idea of a Taylor's expansion as the series (3.14). Without proof we have the following

Variant of Taylor's formula

Let $u_0 \geq 0$, and assume that the value of $\text{erf}(u_0)$ is known, e.g. from a table. If $|p| < 1/200$, then

$$\text{erf}(u_0 + p) = \text{erf}(u_0) + \frac{2p}{\sqrt{\pi}} \exp(-u_0^2) \left\{ 1 - pu_0 + \frac{1}{3} p^2 (2u_0^2 - 1) \right\} + R(u_0, p),$$

where the error term $R(u_0, p)$ satisfies

$$|R(u_0, p)| < 1.2 \cdot 10^{-10}.$$

In the second variant we use a series expansion multiplied by $\exp(-u^2)$,

$$\begin{aligned} \text{erfc}(u) &= 1 - \frac{1}{\sqrt{\pi}} \exp(-u^2) \sum_{n=0}^{\infty} \frac{n!}{(2n+1)!} (2u)^{2n+1} \\ &= 1 - \frac{2}{\sqrt{\pi}} \exp(-u^2) \sum_{n=0}^{\infty} \frac{2^n}{1 \cdot 3 \cdot 5 \cdots (2n+1)} u^{2n+1}. \end{aligned} \tag{3.15}$$

This is closely connected with (3.14), though it requires fewer terms for obtaining the same accuracy. For instance,

$$\text{erfc}(u) = 1 - \frac{2}{\sqrt{\pi}} \exp(-u^2) \sum_{n=0}^{14} \frac{2^n}{1 \cdot 3 \cdot 5 \cdots (2n+1)} u^{2n+1} + R(u),$$

where

$$|R(u)| < 5 \cdot 10^{-8} \qquad \text{for } 0 \leq u \leq 2.$$

PROOF OF (3.15). Define

$$g(u) = \exp(u^2)\text{erf}(u), \qquad g(0) = 0,$$

and notice that

$$g'(u) = 2u \exp(u^2)\text{erf}(u) + \frac{2}{\sqrt{\pi}} \exp(u^2) \exp(-u^2) = \frac{2}{\sqrt{\pi}} + 2ug(u),$$

so

$$g'(u) = \frac{2}{\sqrt{\pi}} + 2ug(u), \qquad g(0) = 0. \tag{3.16}$$

The solution of the differential Equation (3.16) is unique, so we only have to check that

$$\tilde{g}(u) = \frac{2}{\sqrt{\pi}} \sum_{n=0}^{\infty} \frac{2^n}{1 \cdot 3 \cdot 5 \cdots (2n+1)} u^{2n+1}$$

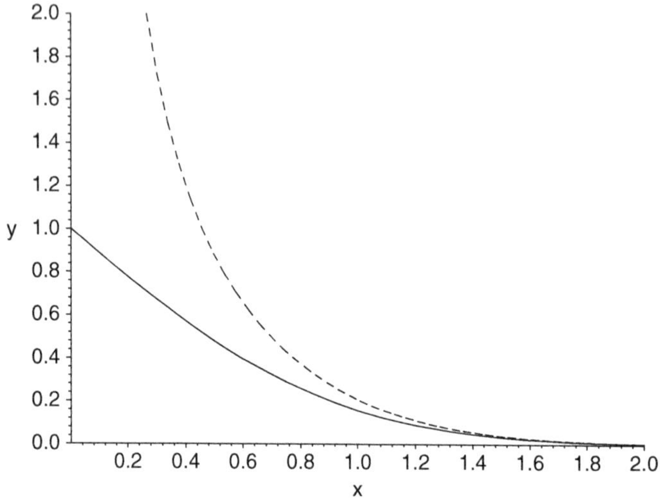

Figure 3.3: Illustration of the estimate in (3.17). The dotted curve is the graph of the function $\frac{1}{\sqrt{\pi}} \cdot \frac{1}{u} \exp\left(-u^2\right)$.

satisfies (3.16). Obviously, $\tilde{g}(0) = 0$. Furthermore, by differentiation and the change $n = m + 1$ of summation variable,

$$\begin{aligned} \tilde{g}'(u) &= \frac{2}{\sqrt{\pi}} \left\{ 1 + \sum_{n=1}^{\infty} \frac{2^n}{1 \cdot 3 \cdot 5 \cdots (2n-1)} u^{2n} \right\} \\ &= \frac{2}{\sqrt{\pi}} + 2u \cdot \frac{2}{\sqrt{\pi}} \sum_{m=0}^{\infty} \frac{2^m}{1 \cdot 3 \cdot 5 \cdots (2m+1)} u^{2m+1} \\ &= \frac{2}{\sqrt{\pi}} + 2u\tilde{g}(u), \end{aligned}$$

so $\tilde{g}(u)$ satisfies (3.16), hence $\tilde{g}(u) = g(u)$ by uniqueness. Finally, (3.15) follows from

$$\operatorname{erfc}(u) = 1 - \operatorname{erf}(u) = 1 - \exp(-u^2)\tilde{g}(u). \qquad \square$$

3.2.4 Estimates

In some cases it suffices to give cruder estimates than those already given in Section 3.2.3. The traditional way of estimating $\operatorname{erfc}(u)$ for large u is folklore,

$$\operatorname{erfc}(u) \le \frac{1}{\sqrt{\pi}} \cdot \frac{1}{u} \exp(-u^2) \qquad \text{for } u > 0. \tag{3.17}$$

PROOF OF (3.2:11). Since $s/u \geq 1$ for $s \geq u$, the estimate follows from

$$\begin{aligned} \operatorname{erfc}(u) &= \frac{2}{\sqrt{\pi}} \int_u^\infty \exp(-s^2)\,ds \leq \frac{2}{\sqrt{\pi}} \int_u^\infty \frac{s}{u} \exp(-s^2)\,ds \\ &= \frac{1}{\sqrt{\pi}} \cdot \frac{1}{u} \exp(-u^2). \qquad \square \end{aligned}$$

Notice, however, that (3.17) is only of interest, when the right hand side is ≤ 1, i.e. we must require that $u \geq 0.4576$. Since u in most cases is somewhat smaller in the present work, this classical estimate will be of little use to us.

Another simple estimate which may be sufficient for giving a rough picture in practical applications is

$$\operatorname{erfc}(u) \approx \left(1 - \frac{u}{\sqrt{3}}\right)^2 \qquad \text{for } 0 \leq u \leq \sqrt{3}.$$

More interesting is the following estimate for $u \geq 0$ which we mention without proof, cf. e.g. Abramowitz *et al.* (1965),

$$\frac{2}{\sqrt{\pi}} \frac{1}{u + \sqrt{u^2 + 2}} \exp(-u^2) < \operatorname{erfc}(u) \leq \frac{2}{\sqrt{\pi}} \frac{1}{u + \sqrt{u^2 + (4/\pi)}} \exp(-u^2) \tag{3.18}$$

Obviously, (3.18) means that one can find a function $c(u)$ for $u \geq 0$, such that $4/\pi \leq c(u) < 2$ and

$$\operatorname{erfc}(u) = \frac{2}{\sqrt{\pi}} \cdot \frac{1}{u + \sqrt{u^2 + c(u)}} \exp(-u^2). \tag{3.19}$$

It can be proved that $c(u)$ is increasing and that

Table 3.1: Table of the function $c(u)$ defined in (3.19).

u	$c(u)$	u	$c(u)$	u	$c(u)$
0.0	1.27324	1.0	1.68621	2.0	1.84750
0.1	1.33261	1.1	1.70999	2.1	1.85697
0.2	1.38743	1.2	1.73158	2.2	1.86566
0.3	1.43781	1.3	1.75120	2.3	1.87365
0.4	1.48396	1.4	1.76904	2.4	1.88100
0.5	1.52610	1.5	1.78528	2.5	1.88779
0.6	1.56451	1.6	1.80008	2.6	1.89405
0.7	1.59945	1.7	1.81358	2.7	1.89984
0.8	1.63120	1.8	1.82591	2.8	1.90520
0.9	1.66003	1.9	1.83718	2.9	1.91017
1.0	1.68621	2.0	1.84750	3.0	1.91479

$$c(0) = \frac{4}{\pi} \qquad \text{and} \qquad \lim_{u\to\infty} c(u) = 2.$$

We get by Equation (3.19) and Table 3.1 a really good estimate of the complementary error function in the domain $u \in [0, 3]$.

Example 3.2.1 We shall find an estimate of erfc(1.234). From Table 3.1 we get by interpolation

$$c(1.234) \approx 1.73825,$$

hence by (3.19),

$$\operatorname{erfc}(1.234) \approx 0.080962.$$

By using (3.14) instead with 21 terms we get for comparison,

$$\operatorname{erfc}(1.234) \approx 0.080961. \qquad \diamond$$

Finally, it should be mentioned that for $u \geq 2$ one may use the simpler estimate, cf. Abramowitz *et al.* (1965) and Equation (3.17),

$$0.5107914 \leq u \exp(u^2)\operatorname{erfc}(u) \leq 0.5641896,$$

i.e.

$$\frac{0.5107914}{u\exp(u^2)} \leq \operatorname{erfc}(u) \leq \frac{0.5641896}{u\exp(u^2)}. \tag{3.20}$$

3.2.5 Approximations

In the dawn of the modern calculating machines there arose a need for rational approximations other than those already mentioned. Some of these are stated below; others can be found in Abramowitz *et al.* (1965).

Whenever $u \geq 0$,

$$\begin{aligned}\operatorname{erfc}(u) &= \frac{0.3480242}{1+0.47047u}\left\{1 - \frac{0.2754975}{1+0.47047u}\left\{1\right.\right.\\ &\left.\left.\quad -\frac{7.7999287}{1+0.47047u}\right\}\right\}\exp\left(-u^2\right) + R(u),\end{aligned} \tag{3.21}$$

where the error term satisfies

$$|R(u)| \leq 2.5 \cdot 10^{-5} \quad \text{for all } u \geq 0.$$

Whenever $u \geq 0$,

$$\begin{aligned}\operatorname{erfc}(u) &= \frac{0.254829592}{1+0.3275911u}\left\{1-\frac{1.116419540}{1+0.3275911u}\left\{1\right.\right.\\ &\quad - \frac{4.996239189}{1+0.3275911u}\left\{1-\frac{1.022328675}{1+0.3275911u}\left\{1\right.\right.\\ &\quad \left.\left.\left.\left.- \frac{0.730415957}{1+0.3275911u}\right\}\right\}\right\}\right\} \exp\left(-u^2\right) + R(u),\end{aligned} \tag{3.22}$$

where the error term satisfies

$$|R(u)| \leq 1.5 \cdot 10^{-7} \qquad \text{for all } u \geq 0.$$

Whenever $u \geq 0$,

$$\begin{aligned}\operatorname{erfc}(u) &= \{1+0.278393u\{1+0.827568u\{1\\ &\quad +0.004219u\{1+80.432922u\}\}\}\}^{-4} + R(u),\end{aligned} \tag{3.23}$$

where the error term satisfies

$$|R(u)| \leq 5 \cdot 10^{-4} \qquad \text{for all } u \geq 0.$$

Whenever $u \geq 0$,

$$\begin{aligned}\operatorname{erfc}(u) &= \{1+0.0705230784u\{1+0.5995218014u\{1\\ &\quad +0.2192644331u\{1+0.0163975896u\{1\\ &\quad +1.8193498901u\{1+0.1557082691u\}\}\}\}\}\}^{-16}\\ &\quad +R(u),\end{aligned} \tag{3.24}$$

where the error term satisfies

$$|R(u)| \leq 3 \cdot 10^{-7} \qquad \text{for all } u \geq 0.$$

All these formulæ are easily implemented on e.g. a pocket calculator.

3.2.6 Inverse of the complementary error function

In some cases we want to solve the equation

$$\operatorname{erfc}(u) = y$$

in u for given $y \in [0,1]$. Since $\operatorname{erfc}(u)$ is continuous and decreasing, this solution exists and is continuous. We denote this inverse function by

$$u = \operatorname{inv\,erfc}(y), \qquad \text{for } y \in [0,1].$$

We have not been able to find an explicit formula for inv erfc(y), and the literature is rather vague on this point. We have therefore developed an iteration procedure, which is easily implemented on e.g. a pocket calculator. The procedure works well for $y \in \,]0,1]$ and $y \gg 3 \cdot 10^{-7}$.

1. Define the functions $f(u)$ and $g(u)$ by

$$\begin{aligned} f(u) &= u(1{+}0.5995485912u(1{+}0.2192546356u(1 \\ &\quad +0.0163975896u(1{+}1.8193498901u(1{+}0.1557082691u))))) \end{aligned}$$

and

$$\begin{aligned} g(u) &= 1{+}1.1990971824u(1{+}0.3288819534u(1{+}0.0218634527u(1 \\ &\quad +2.2741873626u(1{+}0.1868499229u)))). \end{aligned}$$

The numerical error is minimized by the strange factorization.

2. Calculate

$$c^- = 14.1797553749\left(\{y+3\cdot 10^{-7}\}^{-1/16}-1\right),$$

$$c^0 = 14.1797553749\left(y^{-1/16}-1\right),$$

$$c^+ = 14.1797553749\left(\{y-3\cdot 10^{-7}\}^{-1/16}-1\right).$$

The indices have been chosen, such that $c^- < c^0 < c^+$.

3. Solve the three iteration problems for $c = c^-,\, c^0,\, c^+$:

$$u_0 = c, \qquad u_{n+1} = u_n - \frac{f(u_n)-c}{g(u_n)}.$$

Then $u_n^- \to u^-$, $u_n^0 \to u^0$ and $u_n^+ \to u^+$, and

$$u^- < \text{ inv erfc}(y) \approx u^0 < u^+.$$

Example 3.2.2 When $0.001 \le y \le 1$, seven iterations are sufficient for determining inv erfc(y) with a relative error $< 5 \cdot 10^{-5}$. We shall demonstrate this in five cases.

1. For $y = 0.7$ we get after three iterations

$$u_3^- = 0.2724621 < u_3^0 = 0.2724624 < u_3^+ = 0.2724627,$$

which is better than using *regula falsi* on tables.

2. For $y = 0.5$ we get after three iterations

$$u_3^- = 0.4769360 < u_3^0 = 0.4769364 < u_3^+ = 0.4769367.$$

3. For $y = 0.1$ we must use four iterations,

$$u_4^- = 1.1630860 < u_4^0 = 1.1630870 < u_4^+ = 1.1630880.$$

4. For $y = 0.01$ we get after six iterations

$$u_6^- = 1.8213742 < u_6^0 = 1.8213815 < u_6^+ = 1.8213889.$$

5. Finally, we must use seven iterations for $y = 0.001$,

$$u_7^- = 2.3267056 < u_7^0 = 2.3267653 < u_7^+ = 2.3268250.$$

The closer y is to 1, the fewer iterations are needed, and the better are the approximations.

Notice also that in practice one rarely considers $y < 0.001$, so the example covers most of the relevant cases. ◊

3.2.7 Repeated integrals of erfc(u)

The repeated integrals of erfc(u) occur occasionally in the literature concerning more general boundary conditions for Fick's second law, cf. e.g. Carslaw *et al.* (1947), and Abramowitz *et al.* (1965), where more details are given. Since they also represent an important step towards the more general $\Psi_p(u)$ functions, which are dealt with in Section 3.3, a short presentation is given here. They are all special $\Psi_p(u)$ functions, though easier to discuss.

We define by induction

$$\begin{cases} \mathrm{i}^{-1}\mathrm{erfc}(u) = (2/\sqrt{\pi})\exp(-u^2), \quad \mathrm{i}^0\mathrm{erfc}(u) = \mathrm{erfc}(u), \\ \mathrm{i}^n\mathrm{erfc}(u) = \int_u^\infty \mathrm{i}^{n-1}\mathrm{erfc}(t)\,dt, \qquad n \in \mathbb{N}_0, \end{cases} \tag{3.25}$$

and notice that

$$\frac{d}{du}\mathrm{i}^n\mathrm{erfc}(u) = -\mathrm{i}^{n-1}\mathrm{erfc}(u), \qquad n \in \mathbb{N}_0. \tag{3.26}$$

The notation is the classical one. The reader must not be misled to interpret the power of integration 'i^n' as the power 'i^n' of the imaginary unit.

The first result is

$$\mathrm{i}^n\mathrm{erfc}(u) = \frac{2}{\sqrt{\pi}}\int_u^\infty \frac{1}{n!}(t-u)^n\exp(-t^2)\,dt. \tag{3.27}$$

PROOF. When $n = 0$, the formula (3.27) is trivial. Assume that (3.27) holds for some $n \in \mathbb{N}_0$. We shall prove that this implies that (3.27) also holds, when n is replaced by $n + 1$, so (3.27) follows in general by induction.

We get by using (3.25) and the induction hypothesis (3.27),

$$\begin{aligned} \mathrm{i}^{n+1}\mathrm{erfc}(u) &= \int_u^\infty \mathrm{i}^n\mathrm{erfc}(u)\,du \\ &= \frac{2}{\sqrt{\pi}}\int_u^\infty \left\{\int_w^\infty \frac{1}{n!}(t-w)^n \exp(-t^2)\,dt\right\} dw. \end{aligned}$$

This expression is reduced in the following way by interchanging the order of integration,

$$\begin{aligned} \mathrm{i}^{n+1}\mathrm{erfc}(u) &= \frac{1}{\sqrt{\pi}}\int_u^\infty \left\{\int_w^\infty \frac{1}{n!}(t-w)^n \exp(-t^2)\,dt\right\} dw \\ &= \frac{2}{\sqrt{\pi}}\int_u^\infty \left\{\int_w^t \frac{1}{n!}(t-w)^n\,dw\right\} \exp(-t^2)\,dt \\ &= \frac{2}{\sqrt{\pi}}\int_u^\infty \left[-\frac{1}{(n+1)!}(t-w)^{n+1}\right]_{w=u}^t \exp(-t^2)\,dt \\ &= \frac{2}{\sqrt{\pi}}\int_u^\infty \frac{1}{(n+1)!}(t-u)^{n+1}\exp(-t^2)\,dt, \end{aligned}$$

which is precisely (3.27) with n replaced by $n+1$, so (3.27) follows in general by induction. □

An important consequence of (3.27) is

$$\mathrm{i}^n\mathrm{erfc}(0) = \frac{\Gamma\left(\dfrac{n+1}{2}\right)}{\sqrt{\pi}\cdot n!} = \frac{1}{2^n\Gamma\left(\dfrac{n}{2}+1\right)} \qquad \text{for } n \in \mathbb{N}_0. \tag{3.28}$$

For $n=-1$ the first expression does not make sense (∞/∞), but the second one does,

$$\mathrm{i}^{-1}\mathrm{erfc}(0) = \frac{2}{\sqrt{\pi}} \quad \text{and} \quad \frac{1}{2^{-1}\Gamma\left(-\dfrac{1}{2}+1\right)} = \frac{2}{\Gamma\left(\dfrac{1}{2}\right)} = \frac{2}{\sqrt{\pi}}.$$

Proof of (3.28). Let $n \in \mathbb{N}_0$. Then by (3.27) and the change of variable $u=\sqrt{t}$,

$$\begin{aligned} \mathrm{i}^n\mathrm{erfc}(0) &= \frac{2}{\sqrt{\pi}}\int_0^\infty \frac{1}{n!}t^n\exp(-t^2)\,dt \\ &= \frac{1}{\sqrt{\pi}\cdot n!}\int_0^\infty u^{(n-1)/2}e^u\,du = \frac{\Gamma\left(\dfrac{n+1}{2}\right)}{\sqrt{\pi}\cdot n!}, \end{aligned}$$

and we have proved the first equality of (3.28).

Now, by the duplication formula for the gamma function, cf. Section 3.1.3, applied to $n! = \Gamma(n+1)$ with $2s = n+1$,

$$\sqrt{\pi} \cdot n! = \sqrt{\pi}\Gamma(n+1) = \sqrt{\pi} \cdot \frac{1}{\sqrt{\pi}} \cdot 2^n \Gamma\left(\frac{n+1}{2}\right) \Gamma\left(\frac{n}{2}+1\right),$$

hence

$$\mathrm{i}^n\mathrm{erfc}(0) = \frac{\Gamma\left(\frac{n+1}{2}\right)}{\sqrt{\pi} \cdot n!} = \frac{\Gamma\left(\frac{n+1}{2}\right)}{2^n\Gamma\left(\frac{n+1}{2}\right)\Gamma\left(\frac{n}{2}+1\right)} = \frac{1}{2^n\Gamma\left(\frac{n}{2}+1\right)},$$

and we have proved the second equality in (3.28). □

For later reference we define for $n \in \mathbb{N}_0$, using (3.28),

$$\Psi_{n/2}(u) := \frac{\mathrm{i}^n\mathrm{erfc}(u)}{\mathrm{i}^n\mathrm{erfc}(0)} = \frac{2}{\Gamma\left(\frac{n+1}{2}\right)} \int_u^\infty (t-u)^n \exp(-t^2)\, dt, \qquad (3.29)$$

supplied with the trivial modification

$$\Psi_{-1/2}(u) = \frac{\mathrm{i}^{-1}\mathrm{erfc}(u)}{\mathrm{i}^{-1}\mathrm{erfc}(0)} = \exp(-u^2). \qquad (3.30)$$

Obviously, $\Psi_{n/2}(0) = 1$ for $n = -1, 0, 1, 2, \ldots$, and $0 < \Psi_{n/2}(u) < 1$ for $u > 0$.

The following theorem is a key result. It is usually proved by using series expansions. We shall here avoid series in the proof and instead apply (3.27).

Theorem 3.2.2 *The function $y = i^n\,erfc(u)$, $n = -1, 0, 1, 2, \ldots$, is the unique solution of the initial value problem*

$$\frac{d^2y}{du^2} + 2u\frac{dy}{du} - 2ny = 0,$$

$$y(0) = \frac{1}{2^n\Gamma\left(\frac{n}{2}+1\right)}, \qquad y'(0) = -\frac{1}{2^{n-1}\Gamma\left(\frac{n+1}{2}\right)}.$$

PROOF. Let us first consider the special case $n = -1$. The claim is that $y = (2/\sqrt{\pi})\exp(-u^2)$ is the solution of

$$\frac{d^2y}{du^2} + 2u\frac{dy}{du} + 2y = 0, \quad y(0) = \frac{1}{\sqrt{\pi}}, \quad y'(0) = -\frac{4}{\Gamma(0)} = 0.$$

The latter two conditions are trivial. Since

$$\frac{\sqrt{\pi}}{2}\frac{dy}{du} = -2u\exp(-u^2) \quad \text{and} \quad \frac{\sqrt{\pi}}{2}\frac{d^2y}{du^2} = (4u^2-2)\exp(-u^2),$$

it follows that

$$\frac{\sqrt{\pi}}{2}\left\{\frac{d^2y}{du^2}+2u\frac{dy}{du}+2y\right\}=\{(4u^2-2)-4u^2+2\}\exp(-u^2)=0,$$

and the claim is proved for $n=-1$.

Let $n\in\mathbb{N}_0$. Then by (3.28),

$$\mathrm{i}^n\mathrm{erfc}(0)=\frac{1}{2^n\Gamma\left(\frac{n}{2}+1\right)},$$

$$\frac{d}{du}\mathrm{i}^n\mathrm{erfc}(u)_{|u=0}=-\mathrm{i}^{n-1}\mathrm{erfc}(0)=-\frac{1}{2^{n-1}\Gamma\left(\frac{n+1}{2}\right)},$$

and the initial conditions are fulfilled.

It remains to show that $\mathrm{i}^n\mathrm{erfc}(u)$ satisfies the differential equation for $n\in\mathbb{N}_0$.

For $n=0$ this is easy, because

$$y=\mathrm{i}^0\mathrm{erfc}(u)=\frac{2}{\sqrt{\pi}}\int_u^\infty\exp(-t^2)\,dt,$$

so

$$\frac{dy}{du}=-\frac{2}{\sqrt{\pi}}\exp(-u^2)\quad\text{and}\quad\frac{d^2y}{du^2}=\frac{2}{\sqrt{\pi}}\cdot 2u\exp(-u^2),$$

and hence

$$\frac{d^2y}{du^2}+2u\frac{dy}{du}-2\cdot 0\cdot y=\frac{2}{\sqrt{\pi}}\{2u-2u\}\exp(-u^2)=0.$$

For $n\in\mathbb{N}$ we have by (3.27),

$$y=\frac{2}{\sqrt{\pi}}\int_u^\infty\frac{1}{n!}(t-u)^n\exp(-t^2)\,dt, \tag{3.31}$$

thus

$$-2ny=\frac{2}{\sqrt{\pi}}\int_u^\infty(-2)(t-u)\cdot\frac{1}{(n-1)!}(t-u)^{n-1}\exp(-t^2)\,dt. \tag{3.32}$$

When (3.31) is differentiated we get

$$\frac{dy}{du}=-\frac{2}{\sqrt{\pi}}\int_u^\infty\frac{1}{(n-1)!}(t-u)^{n-1}\exp(-t^2)\,dt, \tag{3.33}$$

so

$$2u\frac{dy}{du}=\frac{2}{\sqrt{\pi}}\int_u^\infty\frac{-2u}{(n-1)!}(t-u)^{n-1}\exp(-t^2)\,dt. \tag{3.34}$$

On the other hand, a partial integration of (3.33) gives

$$\begin{aligned}\frac{dy}{du} &= -\frac{2}{\sqrt{\pi}}\left[\frac{1}{n!}(t-u)^n\exp(-t^2)\right]_u^\infty \\ &\quad -\frac{2}{\sqrt{\pi}}\int_u^\infty \frac{1}{n!}(t-u)^n\cdot 2t\exp(-t^2)\,dt \\ &= -\frac{2}{\sqrt{\pi}}\int_u^\infty \frac{1}{n!}(t-u)^n\cdot 2t\exp(-t^2)\,dt. \end{aligned} \tag{3.35}$$

By differentiating (3.35) we finally get

$$\frac{d^2y}{du^2} = \frac{2}{\sqrt{\pi}}\int_u^\infty \frac{2t}{(n-1)!}(t-u)^{n-1}\exp(-t^2)\,dt. \tag{3.36}$$

When (3.36), (3.34) and (3.32) are added we conclude that

$$\begin{aligned}&\frac{d^2y}{du^2} + 2u\frac{dy}{du} - 2ny \\ &\quad = \frac{2}{\sqrt{\pi}}\int_u^\infty \frac{1}{(n-1)!}\{2t-2u-2(t-u)\}(t-u)^{n-1}\exp(-t^2)\,dt = 0,\end{aligned}$$

and the theorem is proved. □

3.2.8 Extension of Fick's second law

Using the repeated integrals of erfc(u) it is easy to prove the following extension of Theorem 3.2.1, cf. Carslaw *et al.* (1947), and more generally Mejlbro (1996).

Theorem 3.2.3 *Let $n \in \mathbb{N}_0$. The solution of*

$$\frac{\partial f}{\partial t} = \frac{\partial^2 f}{\partial x^2} \qquad \textit{for } x, t > 0,$$

$$f(0,t) = t^{n/2} \qquad \textit{for } t > 0,$$

$$f(x,0) = 0 \qquad \textit{for } x > 0,$$

is given by

$$f(x,t) = t^{n/2}\Psi_{n/2}\left(\frac{x}{\sqrt{4t}}\right), \tag{3.37}$$

where

$$\begin{aligned}\Psi_{n/2}(u) &= 2^n\Gamma\left(\frac{n}{2}+1\right) i^n \mathit{erfc}(u) \\ &= \frac{2}{\Gamma\left(\dfrac{n+1}{2}\right)}\int_u^\infty (t-u)^n\exp(-t^2)\,dt.\end{aligned}$$

Remark 3.2.1 Since $\Psi_0(u) = \operatorname{erfc}(u)$, Theorem 3.2.3 reduces to Theorem 3.2.1 when $n = 0$. ◊

PROOF. Let $f(x,t)$ be given by (3.37). From $\Psi_{n/2}(0) = 1$ follows immediately that

$$f(0,t) = t^{n/2},$$

so $f(x,t)$ satisfies the boundary condition.

Since

$$\Psi_{n/2}(u) = \frac{2}{\Gamma\left(\frac{n+1}{2}\right)} \int_0^\infty t^n \exp\left(-(t+u)^2\right)\, dt \to 0, \qquad u \to \infty,$$

we also get

$$f(x,0) = 0,$$

so $f(x,t)$ satisfies the initial condition.

It remains to be shown that $f(x,t)$ satisfies the partial differential equation.

First notice that since $\Psi_{n/2}(u)$ is a constant times $\mathrm{i}^n\operatorname{erfc}(u)$, we get by Theorem 3.2.2 that

$$\Psi''_{n/2}(u) + 2u\Psi'_{n/2}(u) - 2n\Psi_{n/2}(u) = 0. \tag{3.38}$$

Write for short

$$u = \frac{x}{\sqrt{4t}} = \frac{x}{2\sqrt{t}}.$$

Then

$$f(x,t) = t^{n/2}\Psi_{n/2}(u),$$

so by the chain rule,

$$\begin{aligned}
\frac{\partial f}{\partial t} &= \frac{n}{2}t^{(n/2)-1}\Psi_{n/2}(u) + t^{n/2}\frac{\partial u}{\partial t}\Psi'_{n/2}(u) \\
&= \frac{n}{2}t^{(n/2)-1}\Psi_{n/2}(u) - \frac{x}{4}t^{(n-3)/2}\Psi'_{n/2}(u),
\end{aligned}$$

and

$$\frac{\partial f}{\partial x} = t^{n/2}\frac{\partial u}{\partial x}\Psi'_{n/2}(u) = \frac{1}{2}t^{(n-1)/2}\Psi'_{n/2}(u),$$

$$\frac{\partial^2 f}{\partial x^2} = \frac{1}{4}t^{(n-2)/2}\Psi''_{n/2}(u).$$

Hence,

$$\begin{aligned}\frac{\partial^2 f}{\partial x^2}-\frac{\partial f}{\partial t} &= \frac{1}{4}t^{(n-2)/2}\Psi''_{n/2}(u)+\frac{x}{4}t^{(n-3)/2}\Psi'_{n/2}(u)-\frac{n}{2}t^{(n-2)/2}\Psi_{n/2}(u)\\ &= \frac{1}{4}t^{(n-2)/2}\left\{\Psi''_{n/2}(u)+\frac{x}{\sqrt{t}}\Psi'_{n/2}(u)-2n\Psi_{n/2}(u)\right\}\\ &= \frac{1}{4}t^{(n-2)/2}\left\{\Psi''_{n/2}(u)+2u\Psi'_{n/2}(u)-2n\Psi_{n/2}(u)\right\}=0\end{aligned}$$

by (3.38), and the theorem is proved. □

3.2.9 Series expansions for $\mathrm{i}^n\mathrm{erfc}(u)$

We need the series expansions of $\mathrm{i}^n\mathrm{erfc}(u)$ for the numerical calculations, cf. also Abramowitz *et al.* (1965). We shall always use the convention that $1/\Gamma(-m)=0$, when $m\in\mathbb{N}_0$. The following theorem suffices for our purposes.

Theorem 3.2.4 *For $n=-1,\,0,\,1,\,2,\dots,$*

$$i^n\,erfc(u)=\frac{1}{2^n}\sum_{k=0}^{\infty}\frac{(-2u)^k}{k!\Gamma\left(1+\dfrac{n-k}{2}\right)}\qquad \textit{for } |u|<\infty. \tag{3.39}$$

PROOF. When $n=-1$ the terms of odd indices k are zero by the convention above, so (3.39) reduces to

$$\begin{aligned}\frac{1}{2^{-1}}\sum_{k=0}^{\infty}\frac{(-2u)^k}{k!\Gamma\left(1-\dfrac{1+k}{2}\right)} &= \sum_{m=0}^{\infty}\frac{2^{1+2m}(-1)^{2m}u^{2m}}{(2m)!\Gamma\left(\dfrac{1}{2}-m\right)}\\ &= \frac{2}{\sqrt{\pi}}\sum_{m=0}^{\infty}\frac{2^{2m}\Gamma\left(\frac{1}{2}\right)u^{2m}}{(2m)!\Gamma\left(\dfrac{1}{2}-k\right)}.\end{aligned}$$

By (3.1) we get

$$\begin{aligned}\Gamma\left(\frac{1}{2}\right) &= -\frac{1}{2}\Gamma\left(-\frac{1}{2}\right)=\cdots\\ &= \left(-\frac{1}{2}\right)\left(-\frac{3}{2}\right)\left(-\frac{5}{2}\right)\cdots\left(\frac{1}{2}-m\right)\Gamma\left(\frac{1}{2}-m\right)\\ &= (-1)^m\cdot 2^{-m}\cdot 1\cdot 3\cdot 5\cdots(2m-1)\Gamma\left(\frac{1}{2}-m\right)\\ &= (-1)^m\cdot 2^{-m}\cdot\frac{1\cdot 2\cdot 3\cdot 4\cdots(2m-1)\cdot 2m}{(2\cdot 1)(2\cdot 2)\cdots(2\cdot m)}\Gamma\left(\frac{1}{2}-m\right)\\ &= (-1)^m 2^{-2m}\frac{(2m)!}{m!}\Gamma\left(\frac{1}{2}-m\right),\end{aligned}$$

so this series is reduced to

$$\begin{aligned}
&\frac{2}{\sqrt{\pi}}\sum_{m=0}^{\infty}\frac{2^{2m}\Gamma\left(\frac{1}{2}\right)u^{2m}}{(2m)!\Gamma\left(\frac{1}{2}-m\right)} \\
&= \frac{2}{\sqrt{\pi}}\sum_{m=0}^{\infty}\frac{2^{2m}u^{2m}}{(2m)!\Gamma\left(\frac{1}{2}-m\right)}\cdot(-1)^m 2^{-2m}\frac{(2m)!}{m!}\Gamma\left(\frac{1}{2}-m\right) \\
&= \frac{2}{\sqrt{\pi}}\sum_{m=0}^{\infty}\frac{1}{m!}(-u^2)^m = \frac{2}{\sqrt{\pi}}\exp(-u^2),
\end{aligned}$$

and (3.39) holds for $n=-1$.

For $n \in \mathbb{N}_0$, let

$$\varphi_n(u) = \frac{1}{2^n}\sum_{k=0}^{\infty}\frac{(-2u)^k}{k!\Gamma\left(1+\frac{n-k}{2}\right)} = \sum_{k=0}^{\infty}\frac{(-1)^k u^k}{2^{n-k}k!\Gamma\left(1+\frac{n-k}{2}\right)}.$$

This series is clearly convergent for all $u \in \mathbb{C}$, so $\varphi_n(u)$ is an entire function, and it can be differentiated termwise,

$$\varphi_n'(u) = \sum_{k=1}^{\infty}\frac{(-1)^k u^{k-1}}{2^{n-k}(k-1)!\Gamma\left(1+\frac{n-k}{2}\right)} = \sum_{k=0}^{\infty}\frac{(-1)^k k u^{k-1}}{2^{n-k}k!\Gamma\left(1+\frac{n-k}{2}\right)}$$

and

$$\begin{aligned}
\varphi_n''(u) &= \sum_{k=2}^{\infty}\frac{(-1)^k u^{k-2}}{2^{n-k}(k-2)!\Gamma\left(1+\frac{n-k}{2}\right)} = \sum_{k=0}^{\infty}\frac{(-1)^k 4u^k}{2^{n-k}k!\Gamma\left(\frac{n-k}{2}\right)} \\
&= \sum_{k=0}^{\infty}\frac{(-1)^k 2(n-k)u^k}{2^{n-k}k!\Gamma\left(1+\frac{n-k}{2}\right)},
\end{aligned}$$

so

$$\varphi_n''(u) + 2u\varphi_n'(u) - 2n\varphi_n(u)$$

$$= \sum_{k=0}^{\infty} \frac{(-1)^k 2(n-k)u^k}{2^{n-k}k!\Gamma\left(1+\frac{n-k}{2}\right)} + \sum_{k=0}^{\infty} \frac{(-1)^k 2ku^k}{2^{n-k}k!\Gamma\left(1+\frac{n-k}{2}\right)}$$

$$- \sum_{k=0}^{\infty} \frac{(-1)^k 2nu^k}{2^{n-k}k!\Gamma\left(1+\frac{n-k}{2}\right)}$$

$$= \sum_{k=0}^{\infty} \frac{(-1)^k u^k}{2^{n-k}k!\Gamma\left(1+\frac{n-k}{2}\right)} \{2(n-k) + 2k - 2n\} = 0.$$

Since also

$$\varphi_n(0) = \frac{1}{2^n\Gamma\left(1+\frac{n}{2}\right)} \quad \text{and} \quad \varphi_n'(0) = -\frac{1}{2^{n-1}\Gamma\left(\frac{n+1}{2}\right)},$$

we have shown that $\varphi_n(u)$ satisfies the initial value problem of Theorem 3.2.2, so by uniqueness

$$\mathrm{i}^n\mathrm{erfc}(u) = \varphi_n(u) = \frac{1}{2^n} \sum_{k=0}^{\infty} \frac{(-2u)^k}{k!\Gamma\left(1+\frac{n-k}{2}\right)},$$

and the theorem is proved. □

3.3 The $\Psi_p(u)$ functions

The $\Psi_p(u)$ functions were introduced by Mejlbro (1996), in order to solve Fick's second law for time-dependent surface concentration. It was later found that for $p = -\frac{1}{2}, 0, \frac{1}{2}, 1, \frac{3}{2}, \ldots$ they are just a normalized version of the repeated integrals $\mathrm{i}^{2p}\mathrm{erfc}(u)$ of the complementary error function, cf. Section 3.2.7. The new aspect is that $\Psi_p(u)$ is defined for all indices $p \in \mathbb{R}$, so they can be used for interpolation between solutions described by the repeated integrals, thus providing us with a far more flexible model.

Although $\Psi_p(u)$ can be defined for all $p \in \mathbb{R}$, only $p \geq 0$ is relevant for Fick's second law, which will become clear in Section 3.3.2. However, for technical reasons (due to a differentiation) we may also need $\Psi_p(u)$ for $p \geq -0.5$. We shall therefore concentrate on the two cases $p \geq 0$ and $p \geq -0.5$. The reader who is interested in the general case is referred to Mejlbro (1996).

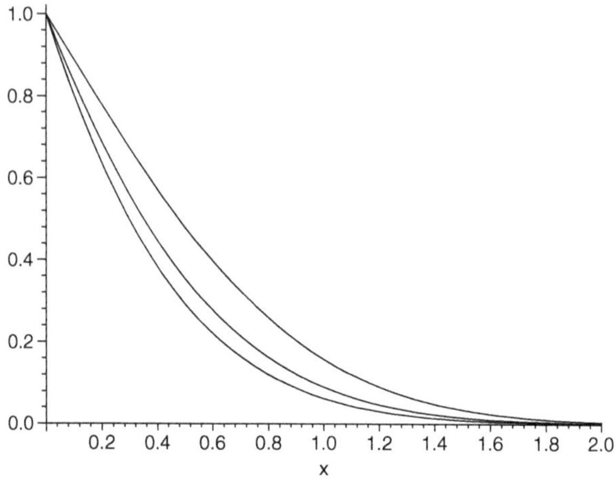

Figure 3.4: The functions $\Psi_p(u)$ for $p = 0.0$, 0.4 and 0.8.

In Mejlbro (1996), all results were derived by using converging series. Here we shall follow a different approach, avoiding series as long as possible and instead use simple properties of the underlying differential equation. For numerical reasons one cannot totally ignore series expansions, so these are also given.

Since these functions cannot be found elsewhere in the literature, we shall allow ourselves to bring the mathematical arguments.

3.3.1 Definition and main theorem

We have already [cf. (3.30) and (3.29)] defined

$$\Psi_{-1/2}(u) = \exp(-u^2), \qquad \Psi_0(u) = \operatorname{erfc}(u),$$

and

$$\Psi_{n/2}(u) = \frac{2}{\Gamma\left(\dfrac{n+1}{2}\right)} \int_u^\infty (w-u)^n \exp(-w^2)\, dw, \qquad n \in \mathbb{N}_0.$$

When n is replaced by $2p$ we see that the latter functions are special cases of the family for $p > -0.5$,

$$\begin{aligned}\Psi_p(u) &= \frac{2}{\Gamma\left(p+\dfrac{1}{2}\right)} \int_u^\infty (w-u)^{2p} \exp(-w^2)\, dw \\ &= \frac{2}{\Gamma\left(p+\dfrac{1}{2}\right)} \int_0^\infty w^{2p} \exp\left(-(w+u)^2\right) dw. \end{aligned} \tag{3.40}$$

Since by Theorem 3.2.2 the function $\Psi_{n/2}(u)$ satisfies

$$\Psi''_{n/2}(u) + 2u\Psi'_{n/2}(u) - 2n\Psi_{n/2}(u) = 0,$$

$$\Psi_{n/2}(0) = 1, \qquad \Psi'_{n/2}(0) = -\frac{2^n\Gamma\left(\frac{n}{2}+1\right)}{2^{n-1}\Gamma\left(\frac{n+1}{2}\right)} = -2\frac{\Gamma\left(\frac{n}{2}+1\right)}{\Gamma\left(\frac{n+1}{2}\right)},$$

it is reasonable to expect that $\Psi_p(u)$ given by (3.40) for $p > -0.5$ satisfies a similar equation with n replaced by $2p$. This is actually true.

Theorem 3.3.1 *Let $p \geq -0.5$. Then $\Psi_p(u)$ is the unique solution of the initial value problem*

$$\Psi''_p(u) + 2u\Psi'_p(u) - 4p\Psi_p(u) = 0,$$

$$\Psi_p(0) = 1, \qquad \Psi'_p(0) = -2\frac{\Gamma(p+1)}{\Gamma\left(p+\frac{1}{2}\right)}.$$

PROOF. For $p = -0.5$ or $2p \in \mathbb{N}_0$, the claim is just Theorem 3.2.2. Let $p > -0.5$ be any other value. Since $2p > -1$, the integrals of (3.40) are convergent, and the definition makes sense.

By the change of variable $u = \sqrt{s}$ and Definition 3.1.1 of the gamma function we get

$$\begin{aligned}\Psi_p(0) &= \frac{2}{\Gamma\left(p+\frac{1}{2}\right)}\int_0^\infty u^{2p}\exp(-u^2)\,du \\ &= \frac{1}{\Gamma\left(p+\frac{1}{2}\right)}\int_0^\infty s^{p-0.5}\exp(-s)ds = 1.\end{aligned}$$

Differentiation of the latter integral of (3.40) gives

$$\Psi'_p(u) = -\frac{4}{\Gamma\left(p+\frac{1}{2}\right)}\int_0^\infty w^{2p}(w+u)\exp\left(-(w+u)^2\right)dw, \tag{3.41}$$

so

$$\begin{aligned}\Psi'_p(0) &= -2\cdot\frac{2}{\Gamma\left(p+\frac{1}{2}\right)}\int_0^\infty w^{2p+1}\exp(-w^2)\,dw \\ &= -\frac{2}{\Gamma\left(p+\frac{1}{2}\right)}\int_0^\infty s^p\exp(-s)\,ds = -2\frac{\Gamma(p+1)}{\Gamma\left(p+\frac{1}{2}\right)},\end{aligned}$$

and the initial conditions of Theorem 3.3.1 are fulfilled.

Differentiation of (3.41) gives

$$\begin{aligned}\Psi_p''(u) &= \frac{4}{\Gamma\left(p+\frac{1}{2}\right)}\int_0^\infty \left\{2(w+u)^2-1\right\} w^{2p}\exp\left(-(w+u)^2\right)\,dw \quad (3.42)\\ &= \frac{8}{\Gamma\left(p+\frac{1}{2}\right)}\int_0^\infty (w+u)^2 w^{2p}\exp\left(-(w+u)^2\right)\,dw-2\Psi_p(u),\end{aligned}$$

hence by (3.42) and (3.41) and a partial integration,

$$\begin{aligned}&\Psi_p''(u)+2u\Psi_p'(u)-4p\Psi_p(u)\\ &= \frac{8}{\Gamma\left(p+\frac{1}{2}\right)}\int_0^\infty \left\{(w+u)^2-u(w+u)\right\} w^{2p}\exp\left(-(w+u)^2\right)\,dw\\ &\quad -(2+4p)\Psi_p(u)\\ &= \frac{8}{\Gamma\left(p+\frac{1}{2}\right)}\int_0^\infty w^{2p+1}\left\{(w+u)\exp\left(-(w+u)^2\right)\right\}\,dw\\ &\quad -(2+4p)\Psi_p(u)\\ &= \frac{4}{\Gamma\left(p+\frac{1}{2}\right)}\left\{\left[-w^{2p+1}\exp\left(-(w+u)^2\right)\right]_0^\infty\right.\\ &\quad \left.+(2p+1)\int_0^\infty w^{2p}\exp\left(-(w+u)^2\right)\,dw\right\}-(2+4p)\Psi_p(u)\\ &= 0+(4p+2)\cdot\frac{2}{\Gamma\left(p+\frac{1}{2}\right)}\int_0^\infty w^{2p}\exp\left(-(w+u)^2\right)\,dw\\ &\quad -(2+4p)\Psi_p(u)\\ &= (4p+2)\Psi_p(u)-(2+4p)\Psi_p(u)=0\end{aligned}$$

by (3.40), and Theorem 3.3.1 is proved. □

3.3.2 Fick's second law for $\Psi_p(u)$

We may expect from the structure of Theorem 3.2.3 that the following more general theorem holds.

Theorem 3.3.2 *Let $p \geq 0$. The solution of*

$$\begin{aligned}\frac{\partial f}{\partial t} &= \frac{\partial^2 f}{\partial x^2} && \text{for } x,t>0,\\ f(0,t) &= t^p && \text{for } t>0,\\ f(x,0) &= 0 && \text{for } x>0,\end{aligned}$$

is given by

$$f(x,t)=t^p\Psi_p\left(\frac{x}{\sqrt{4t}}\right). \quad (3.43)$$

Remark 3.3.1 The proof below carries over for $p > -0.5$ as well, but since $f(0,t) = t^p$ is not bounded for small $t > 0$, when $p < 0$, we do not see any reasonable application as far as chloride ingress is concerned. ◊

PROOF. Let $f(x,t)$ be given by (3.43). Since $\Psi_p(0) = 1$, it is trivial that

$$f(0,t) = t^p \qquad \text{for } t > 0.$$

Since

$$\Psi_p(u) = \frac{2}{\Gamma\left(p+\frac{1}{2}\right)} \int_0^\infty w^{2p} \exp\left(-(w+u)^2\right) dw \to 0 \qquad \text{for } u \to +\infty,$$

it also follows that

$$f(x,t) = t^p \Psi_p\left(\frac{x}{\sqrt{4t}}\right) \to 0 \qquad \text{for } t \to 0+,$$

and we have proved that $f(x,t)$ satisfies the initial and boundary conditions of Theorem 3.3.2.

To ease the notation we put

$$u = \frac{x}{\sqrt{4t}} = \frac{x}{2\sqrt{t}}.$$

Then by the chain rule

$$\frac{\partial f}{\partial t} = pt^{p-1}\Psi_p(u) - \frac{1}{4}t^p \cdot \frac{x}{t\sqrt{t}}\Psi_p'(u) = pt^{p-1}\Psi_p(u) - \frac{1}{2}u\Psi_p'(u),$$

and

$$\frac{\partial f}{\partial x} = \frac{1}{2\sqrt{t}}t^p\Psi_p'(u), \qquad \frac{\partial^2 f}{\partial x^2} = \frac{1}{4}t^{p-1}\Psi_p''(u),$$

so by Theorem 3.3.1,

$$\begin{aligned} \frac{\partial^2 f}{\partial x^2} - \frac{\partial f}{\partial t} &= \frac{1}{4}t^{p-1}\left\{\Psi_p''(u) - 4\left[p\Psi_p(u) - \frac{1}{2}u\Psi_p'(u)\right]\right\} \\ &= \frac{1}{4}t^{p-1}\left\{\Psi_p''(u) + 2u\Psi_p'(u) - 4p\Psi_p(u)\right\} = 0, \end{aligned}$$

and the theorem is proved. □

In pure mathematics one usually expresses the solution of the initial and boundary value problem of Theorem 3.3.2 as a convolution integral. This is important from a theoretical point of view and we shall derive from it a result, which cannot be proved from Theorem 3.3.2. On the other hand, Theorem 3.3.2 is very convenient for practical applications, while Theorem 3.3.3 below is almost impossible to use in explicit calculations. The important link between the two theorems is the uniqueness of the exponentially bounded solution, cf. e.g. Cannon (1984).

Theorem 3.3.3 *Let $p > 0$. The solution of*

$$\frac{\partial f}{\partial t} = \frac{\partial^2 f}{\partial x^2} \qquad \text{for } x, t > 0,$$

$$f(0,t) = t^p \qquad \text{for } t > 0, \qquad\qquad f(x,0) = 0 \qquad \text{for } x > 0,$$

is given by

$$\begin{aligned} f(x,t) &= \int_0^t p(t-\tau)^{p-1}\Psi_0\left(\frac{x}{\sqrt{4\tau}}\right) d\tau \\ &= \int_0^t p\tau^{p-1}\Psi_0\left(\frac{x}{\sqrt{4(t-\tau)}}\right) d\tau. \end{aligned} \tag{3.44}$$

PROOF. By the change of variable $\tau \to t - \tau$ we see that the two integrals of (3.44) are equal.

Let $f(x,t)$ be given by (3.44). Using the fact that

$$0 \le \Psi_0(u) = \operatorname{erfc}(u) \le 1,$$

we get

$$0 \le f(x,t) \le \int_0^t p\tau^{p-1}\, d\tau = t^p, \tag{3.45}$$

so $f(x,0) = 0$, and the initial condition is satisfied.

The boundary condition is trivial

$$f(0,t) = \int_0^t p\tau^{p-1}\, d\tau = t^p.$$

Finally, $\Psi_0(x/\sqrt{4t}) = \operatorname{erfc}(x/\sqrt{4t})$ is by Theorem 3.2.1 a solution of the heat equation, so for any fixed τ

$$\frac{\partial}{\partial t}\Psi_0\left(\frac{x}{\sqrt{4(t-\tau)}}\right) = \frac{\partial^2}{\partial x^2}\Psi_0\left(\frac{x}{\sqrt{4(t-\tau)}}\right).$$

Hence, by differentiating the second integral of (3.44) for $x > 0$ and $t > 0$,

$$\begin{aligned} \frac{\partial f}{\partial t} &= \lim_{\tau \to t-} p\tau^{p-1}\Psi_0\left(\frac{x}{\sqrt{4(t-\tau)}}\right) \\ &\quad + \int_0^t p\tau^{p-1}\frac{\partial}{\partial t}\Psi_0\left(\frac{x}{\sqrt{4(t-\tau)}}\right) d\tau \\ &= 0 + \int_0^t p\tau^{p-1}\frac{\partial^2}{\partial x^2}\Psi_0\left(\frac{x}{\sqrt{4(t-\tau)}}\right) d\tau = \frac{\partial^2 f}{\partial x^2}, \end{aligned}$$

and the theorem is proved. □

By comparing Theorem 3.3.2 and Theorem 3.3.3 and using the uniqueness of the solution we see that

$$\begin{aligned} t^p \Psi_p\left(\frac{x}{\sqrt{4t}}\right) &= \int_0^t p(t-\tau)^{p-1}\Psi_0\left(\frac{x}{\sqrt{4\tau}}\right) d\tau \\ &= t^p \cdot p \int_0^1 (1-w)^{p-1}\Psi_0\left(\frac{x}{\sqrt{4t}}\cdot\frac{1}{\sqrt{w}}\right) dw, \end{aligned}$$

where we have used the change of variable $w = \tau/t$.

Writing $u = x/\sqrt{4t}$ and dividing by t^p we get for $u > 0$ by a partial integration

$$\begin{aligned} \Psi_p(u) &= p\int_0^1 (1-w)^{p-1}\Psi_0\left(\frac{u}{\sqrt{w}}\right) dw \qquad (3.46) \\ &= p\int_0^1 (1-w)^{p-1}\mathrm{erfc}\left(\frac{u}{\sqrt{w}}\right) dw \\ &= \frac{u}{\sqrt{\pi}}\int_0^1 (1-w)^p w^{-3/2}\exp\left(-\frac{u^2}{w}\right) dw. \qquad (3.47) \end{aligned}$$

We can now prove the important

Theorem 3.3.4 *The family of functions $\Psi_p(u)$, $p \geq 0$, is*

1. *decreasing in u for fixed p,*
2. *decreasing in p for fixed u.*

Furthermore,

$$0 \leq \Psi_p(u) \leq 1 \qquad \text{for } p \geq 0 \text{ and } u \geq 0. \qquad (3.48)$$

PROOF. Condition (1) follows from (3.40).
Condition (2) follows from (3.46). Notice that (3.46) relies on Theorem 3.3.3 and apparently cannot be derived from Theorem 3.3.2.

The estimate (3.48) follows from $0 \leq \Psi_0(u/\sqrt{w}) \leq 1$ applied to (3.46). □

Remark 3.3.2 One may wonder, why we solve

$$\frac{\partial f}{\partial t} = \frac{\partial^2 f}{\partial x^2}, \qquad x, t > 0,$$
$$f(0,t) = t^p, \qquad t > 0, \qquad f(x,0) = 0, \qquad x > 0,$$

where t^p is not bounded in t, when $p > 0$. A more reasonable model would be to replace $f(0,t) = t^p$ above by $f(0,t) = \exp(-at)$, $a > 0$, where the latter function is nicely bounded. The reason is a lack of mathematical technique for describing integrals of the form

$$\int_0^t s^{-3/2}\exp\left(as - \frac{x^2}{4s}\right) ds$$

in a simple way, which is inevitable in the case of $f(0,t) = \exp(-at)$. ◊

From Theorem 3.3.2 we get the following useful solution formula. Notice the nice structure. The solution is simply found by adding the factor $\Psi_n(z)$ to each term.

Corollary 3.3.1 *Let $\varphi(t) = \sum_{n=0}^{\infty} a_n t^{b_n}$ be absolutely convergent for $t \geq 0$. The solution of*

$$\frac{\partial f}{\partial t} = \frac{\partial^2 f}{\partial x^2}, \qquad x,\, t > 0,$$

$$f(0,t) = \varphi(t) = \sum_{n=0}^{\infty} a_n t^{b_n}, \qquad t > 0,$$

$$f(x,0) = 0, \qquad x > 0,$$

is given by

$$f(x,t) = \sum_{n=0}^{\infty} a_n t^{b_n} \Psi_{b_n}\left(\frac{x}{\sqrt{4t}}\right). \tag{3.49}$$

Proof. By Theorem 3.3.2 and the principle of superposition we see that (3.49) at least is a formal solution. We shall only prove that (3.49) is convergent. Because $\varphi(t)$ is absolutely convergent for all $t \geq 0$, we get by (3.48),

$$\left|\sum_{n=0}^{\infty} a_n t^{b_n} \Psi_{b_n}\left(\frac{x}{\sqrt{4t}}\right)\right| \leq \sum_{n=0}^{\infty} |a_n| \cdot t^{b_n} \cdot 1 < +\infty,$$

and the corollary is proved. □

Example 3.3.1 We shall consider $\varphi(t) = \exp(-at)$, $a > 0$, already discussed in Remark 3.3.2. A direct approach did not look promising, but since

$$\varphi(t) = \exp(-at) = \sum_{n=0}^{\infty} \frac{1}{n!}(-a)^n t^n$$

is absolutely convergent for all $t \geq 0$, we see by Corollary 3.3.1 that the solution is given by

$$\begin{aligned} f(x,t) &= \sum_{n=0}^{\infty} \frac{1}{n!}(-a)^n t^n \Psi_n\left(\frac{x}{\sqrt{4t}}\right) \\ &= \sum_{n=0}^{\infty} \frac{1}{n!}(-at)^n \Psi_n\left(\frac{x}{\sqrt{4t}}\right). \end{aligned} \tag{3.50}$$

Since $n \in \mathbb{N}_0$, we have by (3.30) and (3.28)

$$\Psi_n(u) = \frac{\mathrm{i}^{2n}\mathrm{erfc}(u)}{\mathrm{i}^{2n}\mathrm{erfc}(0)} = 2^{2n}\Gamma(n+1)\mathrm{i}^{2n}\mathrm{erfc}(u) = 4^n n!\; \mathrm{i}^{2n}\mathrm{erfc}(u),$$

so (3.50) can also be written

$$f(x,t) = \sum_{n=0}^{\infty} (-4at)^n \, \mathrm{i}^{2n}\mathrm{erfc}\left(\frac{x}{\sqrt{4t}}\right). \qquad \diamond$$

We saw that only the repeated integrals of erfc(u) are needed in Example 3.3.1. However, if one instead considers a logistic curve, which again is more natural than an exponential curve, cf. e.g. Swamy *et al.* (1994), then the function

$$\varphi(t) = \exp\left(-at^b\right) = \sum_{n=0}^{\infty} \frac{1}{n!}(-a)^n t^{nb}$$

enters the problem. Obviously, this series is absolutely convergent for all $t \geq 0$, so by superposition the solution is given by

$$f(x,t) = \sum_{n=0}^{\infty} \frac{1}{n!}(-a)^n t^{nb} \Psi_{nb}\left(\frac{x}{\sqrt{4t}}\right),$$

where the functions $\Psi_{nb}(u)$ in general cannot be expressed by means of $\mathrm{i}^m\mathrm{erfc}(u)$, $m \in \mathbb{N}_0$.

The formal generalization of Corollary 3.3.1 to this case is left to the reader. Notice that this result cannot be derived from Theorem 3.3.3.

3.3.3 Hölder's inequality and related results

By using Hölder's inequality we can obtain rather sharp estimates of the considered functions. Since the derivation of Hölder's inequality is simple, it is included here.

Let $p > 1$ and define $q > 1$, such that

$$(p-1)(q-1) = 1, \qquad \text{i.e.} \qquad \frac{1}{p} + \frac{1}{q} = 1. \tag{3.51}$$

One says that (p, q) are *conjugate numbers*.

We begin with a simple geometric result.

Theorem 3.3.5 Young's inequality. *Let (p, q) be conjugate numbers. Then*

$$xy \leq \frac{x^p}{p} + \frac{y^q}{q} \qquad \text{for } x, y \geq 0.$$

PROOF. Due to (3.51) the graphs of the two curves

$$y = x^{p-1} \qquad \text{and} \qquad x = y^{q-1}$$

are congruent. The rectangle $[0, x] \times [0, y]$ is contained in the union of two sets, cf. Figure 3.5,

$$\left\{(\xi,\eta) \;\middle|\; 0 \leq \xi \leq x,\, 0 \leq \eta \leq \xi^{p-1}\right\} \cup \left\{(\xi,\eta) \;\middle|\; 0 \leq \eta \leq y,\, 0 \leq \xi \leq \eta^{q-1}\right\},$$

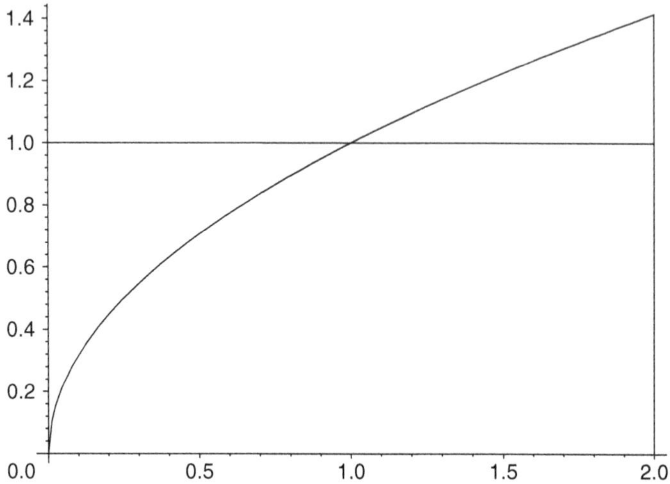

Figure 3.5: Young's inequality for $(p,q) = (\frac{3}{2}, 3)$ and $(x,y) = (2,1)$.

so the area of the rectangle is at most equal to the sum of the areas of this (almost disjoint) splitting, i.e.

$$xy \le \int_0^x \xi^{p-1} d\xi + \int_0^y \eta^{q-1} d\eta = \frac{x^p}{p} + \frac{y^q}{q}. \qquad \square$$

Hölder's inequality is now easily derived from Young's inequality.

PROOF. Due to (3.51) the graphs of the two curves

$$y = x^{p-1} \qquad \text{and} \qquad x = y^{q-1}$$

are congruent. The rectangle $[0,x] \times [0,y]$ is contained in the union of two sets

$$\{(\xi,\eta) \mid 0 \le \xi \le x,\, 0 \le \eta \le \xi^{p-1}\} \\ \cup \{(\xi,\eta) \mid 0 \le \eta \le y,\, 0 \le \xi \le \eta^{q-1}\},$$

so the area of the rectangle is at most equal to the sum of the areas of this (almost disjoint) splitting, i.e.

$$xy \le \int_0^x \xi^{p-1} d\xi + \int_0^y \eta^{q-1} d\eta = \frac{x^p}{p} + \frac{y^q}{q}. \qquad \square$$

Theorem 3.3.6 Hölder's inequality. *Let (p,q) be conjugate numbers, and let $f(t)$ and $g(t)$ be (measurable) functions on an interval I. Define the p- and q-norms of $f(t)$ and $g(t)$ by*

$$\|f\|_p = \left\{\int_I |f(t)|^p \, dt\right\}^{1/p} \quad \textit{and} \quad \|g\|_q = \left\{\int_I |g(t)|^q \, dt\right\}^{1/q}. \tag{3.52}$$

Then

$$\int_I |f(t)g(t)|\,dt \le \|f\|_p\|g\|_q. \tag{3.53}$$

PROOF. If e.g. $\|f\|_p = 0$, then $f(t) = 0$ (almost everywhere) and (3.53) is trivial. We may therefore assume that $\|f\|_p\|g\|_q > 0$. If we put

$$x = |f(t)|/\|f\|_p \quad \text{and} \quad y = |g(t)|/\|g\|_q,$$

then by Young's inequality

$$\frac{|f(t)|}{\|f\|_p}\cdot\frac{|g(t)|}{\|g\|_q} = xy \le \frac{1}{p}\frac{|f(t)|^p}{\|f\|_p^p} + \frac{1}{q}\frac{|g(t)|^q}{\|g\|_q^q},$$

hence by an integration

$$\frac{1}{\|f\|_p\|g\|_q}\int_I |f(t)g(t)|\,dt \le \frac{1}{p}\frac{1}{\|f\|_p^p}\int_I |f(t)|^p dt + \frac{1}{q}\frac{1}{\|g\|_q^q}\int_I |g(t)|^q dt$$
$$= \frac{1}{p}\frac{1}{\|f\|_p^p}\cdot\|f\|_p^p + \frac{1}{q}\frac{1}{\|g\|_q^q}\cdot\|g\|_q^q = \frac{1}{p}+\frac{1}{q} = 1,$$

and the inequality follows. □

Remark 3.3.3 Some mathematical terms like "measurable function" and "almost everywhere" have crept into the formulation. For the applications they are of no importance, because any (piecewise) continuous function is measurable, and the exceptional points, if any, are just the jumps. ◊

Notice that (2, 2) are conjugate numbers. In this case Hölder's inequality reduces to the well-known

Theorem 3.3.7 Cauchy-Schwarz's inequality.

$$\int_I |f(t)g(t)|\,dt \le \left\{\int_I f(t)^2\,dt\right\}^{1/2}\left\{\int_I g(t)^2\,dt\right\}^{1/2}.$$

Remark 3.3.4 In Russian literature one may occasionally find that this estimate is called *(Bohnenblust-Bunjakovski-)Cauchy-Schwarz-(Sobčyk)'s inequality.* Outside Russia Theorem 3.3.3 is called the Cauchy-Schwarz' inequality. ◊

From Hölder's inequality we immediately get the important

Corollary 3.3.2 *Let* $\alpha \in [0,1]$ *be a constant, and let* $f(t)$, $g(t) \ge 0$ *be two nonnegative functions. Then*

$$\int_I f(t)^{1-\alpha}g(t)^\alpha\,dt \le \left\{\int_I f(t)\,dt\right\}^{1-\alpha}\left\{\int_I g(t)\,dt\right\}^{\alpha}.$$

PROOF. When $\alpha = 0$ or $\alpha = 1$ there is nothing to prove. If $0 < \alpha < 1$, put $p = (1-\alpha)^{-1}$ and $q = \alpha^{-1}$, and apply Hölder's inequality. □

Lemma 3.3.1 *Let $a \le b \le c$, where $a < c$. Then for $u > 0$,*

$$u^b \le (1-\alpha)u^a + \alpha u^c, \qquad \text{where } \alpha = \frac{b-a}{c-a}.$$

PROOF. First notice that

$$b = (1-\alpha)a + \alpha c.$$

For $\alpha = 0$ or $\alpha = 1$ there is nothing to prove. If $0 < \alpha < 1$, put $p = (1-\alpha)^{-1}$ and $q = \alpha^{-1}$ and apply Young's inequality,

$$u^b = u^{(1-\alpha)a} \cdot u^{\alpha c} \le (1-\alpha)u^a + \alpha u^c. \qquad \square$$

When Lemma 3.3.1 is applied on the definition (3.40) of the Ψ_p function we get

Corollary 3.3.3 *Let $-1/2 < q \le p \le r$, where $q < r$, and define*

$$\alpha = \frac{p-q}{r-q}, \qquad i.e. \qquad p = (1-\alpha)q + \alpha r.$$

Then

$$\Gamma\left(p+\frac{1}{2}\right)\Psi_p(u) \le (1-\alpha)\Gamma\left(q+\frac{1}{2}\right)\Psi_q(u) + \alpha\Gamma\left(r+\frac{1}{2}\right)\Psi_r(u).$$

PROOF. By Lemma 3.3.1 and (3.40),

$$\begin{aligned}
\Gamma\left(p+\frac{1}{2}\right)\Psi_p(u) &= 2\int_0^\infty w^{2p}\exp\left(-(w+u)^2\right)dw \\
&\le 2\left\{(1-\alpha)\int_0^\infty w^{2q}\exp\left(-(w+u)^2\right)dw\right. \\
&\quad \left. +\,\alpha\int_0^\infty w^{2r}\exp\left(-(w+u)^2\right)dw\right\} \\
&= (1-\alpha)\Gamma\left(q+\frac{1}{2}\right)\Psi_q(u) + \alpha\Gamma\left(r+\frac{1}{2}\right)\Psi_r(u). \qquad \square
\end{aligned}$$

Corollary 3.3.3 implies that the function $\Gamma\left(p+\frac{1}{2}\right)\Psi_p(u)$ is convex in $p > -0.5$ for every fixed $u > 0$.

3.3.4 Differentiation and Taylor expansion of $\Psi_p(u)$

If $p > 0$, then by (3.40),

$$\Psi_p(u) = \frac{2}{\Gamma\left(p+\frac{1}{2}\right)} \int_u^\infty (w-u)^{2p} \exp\left(-w^2\right) dw \tag{3.54}$$

and

$$\Psi_{p-0.5}(u) = \frac{1}{\Gamma(p)} \int_u^\infty (w-u)^{2p-1} \exp\left(-w^2\right) dw. \tag{3.55}$$

These expressions are used to prove

Lemma 3.3.2 *Let $p > 0$. Then*

$$\Psi_p'(u) = -2\frac{\Gamma(p+1)}{\Gamma(p+0.5)}\Psi_{p-0.5}(u).$$

PROOF. When $p = 0$,

$$\Psi_0'(u) = \frac{d}{du}\mathrm{erfc}(u) = -\frac{2}{\sqrt{\pi}} \exp\left(-u^2\right) = -2\frac{\Gamma(1)}{\Gamma(0.5)}\Psi_{-0.5}(u),$$

so the lemma holds for $p = 0$.

When $p > 0$ a differentiation of (3.54) gives

$$\begin{aligned}
\Psi_p'(u) &= \frac{2}{\Gamma(p+0.5)} \left\{ -\left[(w-u)^{2p} \exp\left(-w^2\right)\right]_{w=u} \right. \\
&\quad \left. - \int_u^\infty 2p \cdot (w-u)^{2p-1} \exp\left(-w^2\right) dw \right\} \\
&= -2\frac{2p}{\Gamma(p+0.5)} \int_u^\infty (w-u)^{2p-1} \exp\left(-w^2\right) dw \\
&= -2\frac{p\Gamma(p)}{\Gamma(p+0.5)} \frac{2}{\Gamma(p)} \int_u^\infty (w-u)^{2p-1} \exp\left(-w^2\right) dw \\
&= -2\frac{\Gamma(p+1)}{\Gamma(p+0.5)}\Psi_{p-0.5}(u)
\end{aligned}$$

by (3.55). □

When $p \geq 0$, we have already shown in Theorem 3.3.1 that

$$\Psi_p(0) = 1, \qquad \Psi_p'(0) = -2\frac{\Gamma(p+1)}{\Gamma(p+0.5)} \tag{3.56}$$

and

$$\Psi_p''(u) = 4p\Psi_p(u) - 2u\Psi_p'(u). \tag{3.57}$$

By integrating the result of Lemma 3.3.2 we get from (3.56),

$$\Psi_p(u) = 1 - 2\frac{\Gamma(p+1)}{\Gamma(p+0.5)} \int_0^u \Psi_{p-0.5}(w)\, dw, \tag{3.58}$$

so we have the simple

Corollary 3.3.4 *Let* $p \geq -0.5$. *Then*

$$\int_0^u \Psi_p(w)\, dw = \frac{1}{2}\frac{\Gamma(p+1)}{\Gamma(p+1.5)} \{1 - \Psi_{p+0.5}(u)\} \tag{3.59}$$

and

$$\int_u^\infty \Psi_p(w)\, dw = \frac{1}{2}\frac{\Gamma(p+1)}{\Gamma(p+1.5)} \Psi_{p+0.5}(u). \tag{3.60}$$

PROOF. Replace p by $p+0.5$ in (3.58). Then (3.59) follows by a rearrangement. Since $\Psi_{p+0.5}(u) \to 0$ for $u \to \infty$, we have

$$\int_0^\infty \Psi_p(u)\, du = \frac{1}{2}\frac{\Gamma(p+1)}{\Gamma(p+1.5)},$$

and (3.60) follows by subtracting (3.59) from this equation. □

Theorem 3.3.8 Taylor's formula. *Let* $p \geq 0$, *and let the values of* $\Psi_p(u)$ *and* $\Psi_{p-0.5}(u)$ *be given, e.g. from tables. Then for* $|h| \leq u$

$$\begin{aligned}\Psi_p(u+h) &= \left\{1 + 2ph^2\right\} \Psi_p(u) - 2\frac{\Gamma(p+1)}{\Gamma(p+0.5)} h \{1 - uh\} \Psi_{p-0.5}(u) \\ &\quad + R_p(u;h),\end{aligned} \tag{3.61}$$

where the error term satisfies the estimate

$$\begin{aligned}|R_p(u;h)| &\leq \left\{\frac{4}{3}p(u+|h|) + \frac{2}{3}\left(2p + (u+|h|)^2 + u + |h|\right)\sqrt{p+\frac{1}{2}}\right\} \\ &\quad \times \exp\left(-(u-|h|)^2\right) \cdot |h|^3.\end{aligned} \tag{3.62}$$

PROOF. Use (3.57) to eliminate $\Psi_p''(u)$ in the usual Taylor's formula, followed by an application of Lemma 3.3.2,

$$\begin{aligned}\Psi_p(u+h) &= \Psi_p(u) + \Psi_p'(u)h + \frac{1}{2}\Psi_p''(u) \cdot h^2 + R_p(u;h) \\ &= \left(1 + 2ph^2\right) \Psi_p(u) + \left(h - uh^2\right) \Psi_p'(u) + R_p(u;h) \\ &= \left(1 + 2ph^2\right) \Psi_p(u) - 2\frac{\Gamma(p+1)}{\Gamma(p+0.5)} h(1-uh)\Psi_{p-0.5}(u) \\ &\quad + R_p(u;h),\end{aligned}$$

and (3.61) follows with

$$R_p(u;h) = \frac{1}{3!}h^3\Psi_p^{(3)}(\xi) \tag{3.63}$$

for some ξ between u and $u+h$. In particular, $u - |h| \leq \xi \leq u + |h|$.

Differentiate (3.57) and eliminate $\Psi_p''(u)$ and $\Psi_p'(u)$ as above.

$$\begin{aligned}
\Psi_p^{(3)}(\xi) &= -2\xi\Psi_p''(\xi) + (4p - 2\xi)\Psi_p'(\xi) \\
&= -8p\xi\Psi_p(\xi) + \left(4p - 2\xi + 4\xi^2\right)\Psi_p'(\xi) \\
&= -8p\xi\Psi_p(\xi) - 4\frac{\Gamma(p+1)}{\Gamma(p+0.5)}\left(2\xi^2 - \xi + 2p\right)\Psi_{p-0.5}(\xi).
\end{aligned}$$

It follows from Theorem 3.3.4 that

$$0 \leq \Psi_p(\xi) \leq \Psi_{p-0.5}(\xi) \leq \Psi_{-0.5}(\xi) = \exp\left(-\xi^2\right) \leq \exp\left(-(u - |h|)^2\right),$$

so

$$\left|\Psi_p^{(3)}(\xi)\right| \leq \left(8p\xi + 4\frac{\Gamma(p+1)}{\Gamma(p+0.5)}\left|2\xi^2 - \xi + 2p\right|\right)\exp\left(-(u - |h|)^2\right).$$

By Cauchy-Schwarz's inequality

$$\begin{aligned}
\Gamma(p+1) &= \int_0^\infty s^p e^{-s}\,ds = \int_0^\infty \left\{s^{p/2-1/4}e^{-s/2}\right\}\left\{s^{p/2+1/4}e^{-s/2}\right\}ds \\
&\leq \left\{\int_0^\infty s^{p-1/2}e^{-s}ds\right\}^{1/2}\left\{\int_0^\infty s^{p+1/2}e^{-s}ds\right\}^{1/2} \\
&= \left\{\Gamma\left(p+\frac{1}{2}\right)\Gamma\left(p+\frac{3}{2}\right)\right\}^{1/2} = \left\{\Gamma\left(p+\frac{1}{2}\right)^2\left(p+\frac{1}{2}\right)\right\}^{1/2} \\
&= \Gamma\left(p+\frac{1}{2}\right)\sqrt{p+\frac{1}{2}},
\end{aligned}$$

so

$$\begin{aligned}
|R_p(u;h)| &= \frac{1}{6}|h|^3\left|\Psi_p^{(3)}(\xi)\right| \\
&\leq \frac{2}{3}\left(2p\xi + \sqrt{p+\frac{1}{2}}\cdot\left|2\xi^2 - \xi + 2p\right|\right)\exp\left(-(u-|h|)^2\right)|h|^3 \\
&\leq \frac{2}{3}\left(2p\xi + 2p\sqrt{p+\frac{1}{2}} + \sqrt{p+\frac{1}{2}}\cdot\xi\left|2\xi^2 - 1\right|\right)\exp\left(-(u-|h|)^2\right)|h|^3 \\
&\leq \frac{2}{3}\left\{2p(u+|h|) + \sqrt{p+\frac{1}{2}}\left(2p + u + |h| + (u+|h|)^2\right)\right\} \\
&\quad \times \exp\left(-(u-|h|)^2\right)|h|^3,
\end{aligned}$$

and (3.62) follows. □

Example 3.3.2 When $u = 0$ and $h > 0$ we get

$$\Psi_p(h) = 1 - 2\frac{\Gamma(p+1)}{\Gamma(p+0.5)}h + 2ph^2 + R_p(0;h),$$

cf. also the series expansion in the next section. We cannot use directly the estimate above, because $h > 0$, but from the derived formula

$$\Psi_p^{(3)}(\xi) = -8p\xi\Psi_p(\xi) - 4\frac{\Gamma(p+1)}{\Gamma(p+0.5)}\left(2\xi^2 - \xi + 2p\right)\Psi_{p-0.5}(\xi)$$

and (3.63) it is not hard to show that

$$|R_p(0;h)| \le \frac{2}{3}h^3\left\{2ph + \left(2p + h + 2h^2\right)\sqrt{p + \frac{1}{2}}\right\}. \qquad \Diamond$$

3.3.5 Series expansion of $\Psi_p(u)$

For numerical reasons we need to know the series expansions of $\Psi_p(u)$ and their rate of convergence.

With the usual convention $1/\Gamma(-n) = 0$ for $n \in \mathbb{N}_0$ we define for $p \in \mathbb{R}$, where $-p \notin \mathbb{N}$,

$$\tilde{\Psi}_p(u) = \sum_{n=0}^{\infty} \frac{p^{(n)}}{(2n)!}(2u)^{2n} - \frac{\Gamma(p+1)}{\Gamma(p+\frac{1}{2})}\sum_{n=0}^{\infty}\frac{(p-\frac{1}{2})^{(n)}}{(2n+1)!}(2u)^{2n+1}, \qquad (3.64)$$

where $p^{(n)} = p(p-1)\cdots(p-n+1)$, $n \in \mathbb{N}$, and $p^{(0)} = 1$ were introduced in Section 3.1.4.

The quotient criterion for series shows that (3.64) is convergent for all $u \in \mathbb{C}$, hence $\tilde{\Psi}_p(u)$ is an entire function in $u \in \mathbb{C}$. In particular, termwise differentiation and integration of (3.64) are legal processes.

Since by convention $\Gamma(0.5)/\Gamma(-0.5+0.5) = 0$ we get by choosing $p = -0.5$ in (3.64),

$$\begin{aligned}
\tilde{\Psi}_{-0.5}(u) &= \sum_{n=0}^{\infty}\frac{1}{(2n)!}\left(-\frac{1}{2}\right)^{(n)}(2u)^{2n} \\
&= \sum_{n=0}^{\infty}(-1)^n 2^{-n}\frac{1\cdot 3\cdot 5\cdots(2n-1)}{1\cdot 2\cdot 3\cdots(2n-1)2n}2^{2n}u^{2n} \\
&= \sum_{n=0}^{\infty}\frac{2^n(-u^2)^n}{2\cdot 4\cdot 6\cdots 2n} = \sum_{n=0}^{\infty}\frac{1}{n!}(-u^2)^n = \exp(-u^2) \\
&= \Psi_{-0.5}(u).
\end{aligned}$$

This result suggests that we might expect that $\tilde{\Psi}_p(u) = \Psi_p(u)$ for all $p \ge -0.5$. We have in fact a stronger result.

Theorem 3.3.9 *If $-p \notin \mathbb{N}$, then $\tilde{\Psi}_p(u)$ is the unique solution of*

$$\tilde{\Psi}_p''(u) + 2u\tilde{\Psi}_p' - 4p\tilde{\Psi}_p(u) = 0,$$

$$\tilde{\Psi}_p(0) = 1, \qquad \tilde{\Psi}_p'(0) = -2\frac{\Gamma(p+1)}{\Gamma(p+0.5)}.$$

Combining Theorem 3.3.1 and Theorem 3.3.9 we immediately get

Corollary 3.3.5 *If $p \geq -0.5$, then for all u,*

$$\Psi_p(u) = \sum_{n=0}^{\infty} \frac{p^{(n)}}{(2n)!}(2u)^{2n} - \frac{\Gamma(p+1)}{\Gamma(p+0.5)} \sum_{n=0}^{\infty} \frac{(p-0.5)^{(n)}}{(2n+1)!}(2u)^{2n+1}.$$

PROOF OF THEOREM 3.3.9. In the special case $p = 0$ we get from (3.14)

$$\begin{aligned}
\tilde{\Psi}_0(u) &= \sum_{n=0}^{\infty} \frac{0^{(n)}}{(2n)!}(2u)^{2n} - \frac{\Gamma(1)}{\Gamma(0.5)} \sum_{n=0}^{\infty} \frac{1}{(2n+1)!}\left(-\frac{1}{2}\right)^{(n)} (2u)^{2n+1} \\
&= 1 - \frac{1}{\sqrt{\pi}} \sum_{n=0}^{\infty} \frac{(-1)^n \cdot 2^{-n} \cdot 1 \cdot 3 \cdot 5 \cdots (2n-1)}{1 \cdot 2 \cdot 3 \cdots (2n-1)2n(2n+1)} \cdot 2^{2n+1}u^{2n+1} \\
&= 1 - \frac{1}{\sqrt{\pi}} \sum_{n=0}^{\infty} \frac{(-1)^n}{(2n+1)n!} \cdot u^{2n+1} = \operatorname{erfc}(u) = \Psi_0(u).
\end{aligned}$$

When $-p \notin \mathbb{N}_0$ we obtain by termwise differentiation of (3.64)

$$\begin{aligned}
\tilde{\Psi}_p'(u) &= 2\sum_{n=1}^{\infty} \frac{p^{(n)}}{(2n-1)!}(2u)^{2n-1} \\
&\quad -2\frac{\Gamma(p+1)}{\Gamma(p+0.5)} \sum_{n=0}^{\infty} \frac{1}{(2n)!}\left(p-\frac{1}{2}\right)^{(n)} (2u)^{2n} \\
&= -2\frac{\Gamma(p+1)}{\Gamma(p+0.5)} \sum_{n=0}^{\infty} \frac{1}{(2n)!}\left(p-\frac{1}{2}\right)^{(n)} (2u)^{2n} \\
&\quad +2\sum_{n=0}^{\infty} \frac{p^{(n+1)}}{(2n+1)!}(2u)^{2n+1}
\end{aligned} \tag{3.65}$$

and

$$\begin{aligned}
\tilde{\Psi}_p''(u) &= 4\sum_{n=0}^{\infty} \frac{p^{(n+1)}}{(2n)!}(2u)^{2n} \\
&\quad -4\frac{\Gamma(p+1)}{\Gamma(p+0.5)} \sum_{n=1}^{\infty} \frac{1}{(2n-1)!}\left(p-\frac{1}{2}\right)^{(n)} (2u)^{2n-1} \\
&= 4\sum_{n=0}^{\infty} \frac{p^{(n+1)}}{(2n)!}(2u)^{2n} \\
&\quad -4\frac{\Gamma(p+1)}{\Gamma(p+0.5)} \sum_{n=0}^{\infty} \frac{1}{(2n+1)!}\left(p-\frac{1}{2}\right)^{(n+1)} (2u)^{2n+1}.
\end{aligned} \tag{3.66}$$

It follows from (3.64) and (3.65) that

$$\tilde{\Psi}_p(0) = 1 \qquad \text{and} \qquad \tilde{\Psi}'_p(0) = -2\frac{\Gamma(p+1)}{\Gamma(p+0.5)},$$

so the initial conditions of Theorem 3.3.9 are fulfilled.

By inserting (3.64), (3.65) and (3.66) into the differential equation and isolating the terms on index $n = 0$, we obtain

$$\begin{aligned}
&\tilde{\Psi}''_p(u) + 2u\tilde{\Psi}'_p(u) - 4p\tilde{\Psi}_p(u) \\
&\quad = 4p - 4p + \sum_{n=1}^{\infty} \frac{1}{(2n)!}\left\{4p^{(n+1)} + 2\cdot 2n\cdot p^{(n)} - 4p\cdot p^{(n)}\right\}(2u)^{2n} \\
&\quad - \frac{\Gamma(p+1)}{\Gamma(p+0.5)}\sum_{n=0}^{\infty}\frac{1}{(2n+1)!}\left\{4\left(p-\frac{1}{2}\right)^{(n+1)} + 2(n+1)\left(p-\frac{1}{2}\right)^{(n)}\right. \\
&\quad \left. -4p\cdot\left(p-\frac{1}{2}\right)^{(n)}\right\}(2u)^{2n+1}.
\end{aligned}$$

Now by definition,

$$p^{(n+1)} = p^{(n)}\cdot(p-n) \text{ and } \left(p-\frac{1}{2}\right)^{(n+1)} = \left(p-\frac{1}{2}\right)^{(n)}\cdot\left(p-\frac{1}{2}-n\right),$$

so

$$4p^{(n+1)} + 4n\cdot p^{(n)} - 4p\cdot p^{(n)} = 4p^{(n)}\{(p-n)+n-p\} = 0,$$

and

$$\begin{aligned}
&4\left(p-\frac{1}{2}\right)^{(n+1)} + 4\left(n+\frac{1}{2}\right)\left(p-\frac{1}{2}\right)^{(n)} - 4p\left(p-\frac{1}{2}\right)^{(n)} \\
&\quad = 4\left(p-\frac{1}{2}\right)^{(n)}\left\{\left(p-\frac{1}{2}-n\right)+\left(n+\frac{1}{2}\right)-p\right\} = 0.
\end{aligned}$$

These results imply that both series above disappear, and we have proved that also

$$\tilde{\Psi}''_p(u) + 2u\tilde{\Psi}'_p(u) - 4p\tilde{\Psi}_p(u) = 0. \qquad \square$$

When $-2p \notin \mathbb{N}$ we get from (3.64) and (3.65)

$$
\begin{aligned}
&\Psi_{p-0.5}(u)\\
&= \sum_{n=0}^{\infty} \frac{1}{(2n)!}\left(p-\frac{1}{2}\right)^{(n)}(2u)^{2n} - \frac{\Gamma(p+0.5)}{\Gamma(p)}\sum_{n=0}^{\infty}\frac{(p-1)^{(n)}}{(2n+1)!}(2u)^{2n+1}\\
&= -\frac{1}{2}\frac{\Gamma(p+0.5)}{\Gamma(p+1)}\left\{-2\frac{\Gamma(p+1)}{\Gamma(p+0.5)}\sum_{n=0}^{\infty}\frac{1}{(2n)!}\left(p-\frac{1}{2}\right)^{(n)}(2u)^{2n}\right.\\
&\quad\left. + 2\sum_{n=0}^{\infty}\frac{p^{(n+1)}}{(2n+1)!}(2u)^{2n+1}\right\}\\
&= -\frac{1}{2}\frac{\Gamma(p+0.5)}{\Gamma(p+1)}\Psi_p'(u),
\end{aligned}
$$

so we have proved the first part of

Corollary 3.3.6 *Whenever $p \notin \{-0.5, -1, -1.5, \dots\}$, then*

$$\Psi_p'(u) = -2\frac{\Gamma(p+1)}{\Gamma(p+0.5)}\Psi_{p-0.5}(u). \tag{3.67}$$

Furthermore, if also $p \neq 0$, then

$$\Psi_p''(u) = 4p\Psi_{p-1}(u), \tag{3.68}$$

and we have the recursion formula

$$p\Psi_{p-1}(u) - u\frac{\Gamma(p+1)}{\Gamma(p+0.5)}\Psi_{p-0.5}(u) - p\Psi_p(u) = 0. \tag{3.69}$$

PROOF. Formula (3.67) was proved above. When also $p \neq 0$, then by iteration

$$
\begin{aligned}
\Psi_p''(u) &= -2\frac{\Gamma(p+1)}{\Gamma(p+0.5)}\Psi_{p-0.5}'(u)\\
&= \left(-2\frac{\Gamma(p+1)}{\Gamma(p+0.5)}\right)\left(-2\frac{\Gamma(p+0.5)}{\Gamma(p)}\right)\Psi_{p-1}(u)\\
&= 4\frac{\Gamma(p+1)}{\Gamma(p)}\Psi_{p-1}(u) = 4p\Psi_{p-1}(u).
\end{aligned}
$$

Finally, by using (3.67) and (3.68)

$$
\begin{aligned}
0 &= \frac{1}{4}\left\{\Psi_p''(u) + 2u\Psi_p'(u) - 4p\Psi_p(u)\right\}\\
&= p\Psi_{p-1}(u) - u\frac{\Gamma(p+1)}{\Gamma(p+0.5)}\Psi_{p-0.5}(u) - p\Psi_p(u),
\end{aligned}
$$

and (3.69) is proved. □

We shall in general not be interested in $\Psi_p(u)$ for $p < -0.5$. The reason is the following: The basic problem for Fick's second law is stated in Theorem 3.3.2 with the boundary condition

$$f(0,t) = t^p \qquad \text{for } t > 0.$$

If $p < 0$, this boundary condition would not be bounded in the neighbourhood of $t = 0$, so this indicates that the assumption $p \geq 0$ is essential for the applications of Fick's second law.

Notice, however, that we cannot avoid the derivative $\Psi_p'(u)$, which by Corollary 3.3.6 is a constant $[-2\Gamma(p+1)/\Gamma(p+0.5)]$ times $\Psi_{p-0.5}(u)$, so we are forced to consider $\Psi_q(u)$ for $q \geq -0.5$. No other $\Psi_q(u)$ is needed, because we have

Theorem 3.3.10 *Let $p \geq 0$. For any order n of differentiation there exist unique polynomials $P_n(u)$ and $Q_n(u)$, such that*

$$\Psi_p^{(n)}(u) = P_n(u)\Psi_p(u) + Q_n(u)\Psi_{p-0.5}(u).$$

PROOF. For $n = 1$ we have by (3.67)

$$P_1(u) = 0 \qquad \text{and} \qquad Q_1(u) = -2\frac{\Gamma(p+1)}{\Gamma(p+0.5)}.$$

For $n = 2$ we get from the governing differential equation

$$\Psi_p''(u) = 4p\Psi_p(u) - 2u\Psi_p'(u) = 4p\Psi_p(u) + 4\frac{\Gamma(p+1)}{\Gamma(p+0.5)}u\Psi_{p-0.5}(u),$$

so

$$P_2(u) = 4p \qquad \text{and} \qquad Q_2(u) = 4\frac{\Gamma(p+1)}{\Gamma(p+0.5)}u.$$

Another differentiation of the differential equation gives

$$\Psi_p^{(3)}(u) = (4p-2)\Psi_p'(u) - 2u\Psi_p''(u) - 2\Psi_p(u).$$

Since $\Psi_p'(u)$ and $\Psi_p''(u)$ have the desired structure, so has $\Psi_p^{(3)}(u)$, and the theorem follows by induction. □

Remark 3.3.5 It is possible to derive recursion formulæ for the polynomials $P_n(u)$ and $Q_n(u)$, but since these are not needed here we leave these for the interested reader as an exercise. ◇

3.3.6 Some estimates

We first extend Theorem 3.3.4.

Theorem 3.3.11 *The family of functions* $\Psi_p(u)$, $p \geq 1/2$, $u \geq 0$, *is*

(i) *decreasing in* u *for fixed* p,

(ii) *decreasing in* p *for fixed* u.

PROOF. The first claim follows directly from (3.40) for $p > -1/2$, and it is trivial for $p = -1/2$, where $\Psi_{-0.5}(u) = \exp(-u^2)$.

By a rearrangement of the recursion formula (3.69) we find

$$\Psi_{p-1}(u) = u\frac{\Gamma(p)}{\Gamma(p+0.5)}\Psi_{p-0.5}(u) + \Psi_p(u) \qquad \text{for } p \geq 0.5,$$

or by changing p to $p+1$,

$$\Psi_p(u) = u\frac{\Gamma(p+1)}{\Gamma(p+1.5)}\Psi_{p+0.5}(u) + \Psi_{p+1}(u) \qquad \text{for } p \geq -0.5.$$

Since $p+0.5 \geq 0$ and $p+1 \geq 0$, both $\Psi_{p+0.5}(u)$ and $\Psi_{p+1}(u)$ are decreasing in p by Theorem 3.3.4, and it can be proved that $\Gamma(p+1)/\Gamma(p+1.5)$ is decreasing for $-1/2 \leq p < 0$, and claim (ii) follows. □

In particular,

$$\Psi_p(u) \leq \Psi_{-0.5}(u) = \exp(-u^2) \qquad \text{for } p \geq -0.5. \tag{3.70}$$

Another interesting estimate is obtained when we use Hölder's inequality on (3.46).

Theorem 3.3.12 *Let* $0 < q \leq p \leq r$, $q < r$, *and put*

$$\alpha = \frac{p-q}{r-q} \in [0,1].$$

Then

$$\Psi_p(u) \leq \{\Psi_q(u)\}^{1-\alpha}\{\Psi_r(u)\}^{\alpha} = \Psi_q(u) \cdot \left\{\frac{\Psi_r(u)}{\Psi_q(u)}\right\}^{\alpha}. \tag{3.71}$$

PROOF. Obviously,

$$(1-\alpha)q + \alpha r = p.$$

If $\alpha = 0$ then $p = q$, and if $\alpha = 1$ then $p = q$, and (3.71) is trivial.

Assume that $0 < \alpha < 1$. Then by (3.46) and Hölder's inequality

$$\begin{aligned}
\Psi_p(u) &= \frac{u}{\sqrt{\pi}} \int_0^1 (1-u)^p u^{-3/2} \exp\left(-\frac{u^2}{u}\right) du \\
&= \int_0^1 \left\{ \frac{u}{\sqrt{\pi}} (1-u)^q u^{-3/2} \exp\left(-\frac{u^2}{u}\right) \right\}^{1-\alpha} \\
&\quad \times \left\{ \frac{u}{\sqrt{\pi}} (1-u)^r u^{-3/2} \exp\left(-\frac{u^2}{u}\right) \right\}^{\alpha} du \\
&\le \left\{ \frac{u}{\sqrt{\pi}} \int_0^1 (1-u)^q u^{-3/2} \exp\left(-\frac{u^2}{u}\right) du \right\}^{1-\alpha} \\
&\quad \times \left\{ \frac{u}{\sqrt{\pi}} \int_0^1 (1-u)^r u^{-3/2} \exp\left(-\frac{u^2}{u}\right) du \right\}^{\alpha} \\
&= \{\Psi_q(u)\}^{1-\alpha} \{\Psi_r(u)\}^{\alpha},
\end{aligned}$$

and the theorem is proved. □

Theorem 3.3.12 is actually an approximation of $\Psi_p(u)$ when $\Psi_q(u)$ and $\Psi_r(u)$ are known. The following example illustrates that it is better than the linear approximation.

Example 3.3.3 Choose for simplicity $q = 0$ and $r = 0.5$. The worst case occurs when $p = 0.25$ is the midpoint of the interval $[q, r]$. Derive the values of $\Psi_0(u)$, $\Psi_{0.25}(u)$ and $\Psi_{0.5}(u)$ from the table in Section 8.3, and calculate $\sqrt{\Psi_0(u)\Psi_{0.5}(u)}$ and $\{\Psi_0(u) + \Psi_{0.5}(u)\}/2$ and compare. This gives Table 3.2. Clearly, the relative error is smaller for the approximation given by Theorem 3.3.12 than for the linear approximation. ◊

In some cases rough estimates of a simple type are sufficient for giving an idea of the graph of $\Psi_p(u)$. One such estimate

$$\Psi_0(u) = \operatorname{erfc}(u) \approx \left(1 - \frac{u}{\sqrt{3}}\right)^2 \qquad \text{for } 0 \le u \le \sqrt{3} \tag{3.72}$$

has already been given in Section 3.2.4.

The relative error of (3.72) is $< 20\%$ for $0 \le u \le 1.35$. It is a coincidence that $\sqrt{3}$ can be used in the denominator. The approximation becomes better in the range $0 \le u \le 1.1$, where $\sqrt{3}$ is replaced by 1.66.

More generally one may use

$$\Psi_p(u) \approx \left(1 - \frac{u}{1.66 + 0.3p}\right)^{2+2p} \qquad \text{for } 0 \le u \le 1.1 \text{ and } 0 \le p \le 1, \tag{3.73}$$

which is of the same structure as (3.72). The relative error of (3.73) is $< 10\%$ for all $0 \le u \le 1.1$ and $0 \le p \le 1$, which is sufficient for giving a rough picture of $\Psi_p(u)$, because $\Psi_p(u)$ is small for $u > 1.1$.

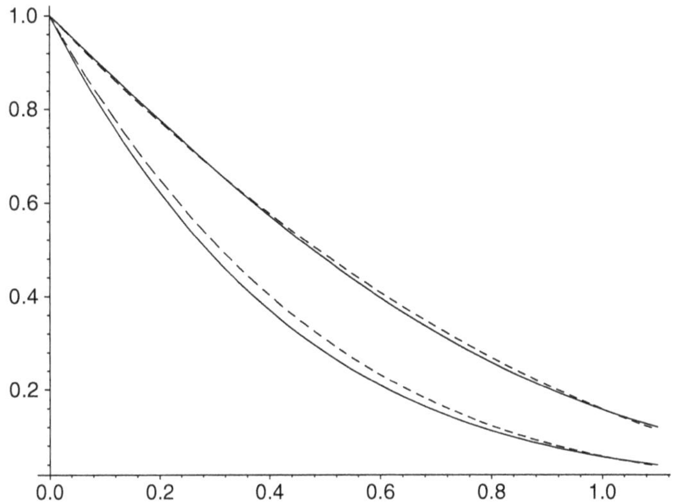

Figure 3.6: Approximations by (3.73) of $\Psi_0(u)$ and $\Psi_1(u)$. The approximations are dotted.

An even better *ad hoc* estimate is the following

$$\Psi_p(u) \approx \left\{1 - 0,4(u^2 - u^4)\right\} \cdot \left(1 - \frac{u}{1,66 + 0,3p}\right)^{2+2p}, \tag{3.74}$$

which gives a good estimate for $0 \le u \le 1.4$. The relative deviations from the right values are shown by drawing the graphs of

$$1 - \frac{1}{\Psi_p(u)} \cdot \left\{1 - 0,4(u^2 - u^4)\right\} \cdot \left(1 - \frac{u}{1,66 + 0,3p}\right)^{2+2p}, \tag{3.75}$$

for various values of p, cf. the Figures 3.7, $\cdots$, 3.17.

Table 3.2: Comparison of $\sqrt{\Psi_0 \Psi_{0.5}}$ and mean $(\Psi_0 + \Psi_{0.5})/2$.

u	Ψ_0	$\Psi_{0.5}$	$\Psi_{0.25}$	$\sqrt{\Psi_0 \Psi_{0.5}}$	Rel. er.	Mean	Rel. er.
0.1	.88754	.83274	.85730	.85970	0.28 %	.86014	3.29 %
0.2	.77730	.68524	.72589	.72982	0.54 %	.73127	0.74 %
0.3	.67137	.55694	.60672	.61148	0.79 %	.61416	1.23 %
0.4	.57161	.44688	.50035	.50541	1.01 %	.50925	1.78 %
0.5	.47950	.35385	.40695	.41191	1.22 %	.41668	2.39 %
1.0	.15730	.08907	.11600	.11837	2.04 %	.12319	6.20 %
1.5	.03390	.01528	.02219	.02276	2.57 %	.02459	10.82 %
2.0	.00468	.00173	.00277	.00285	2.73 %	.00321	15.71 %

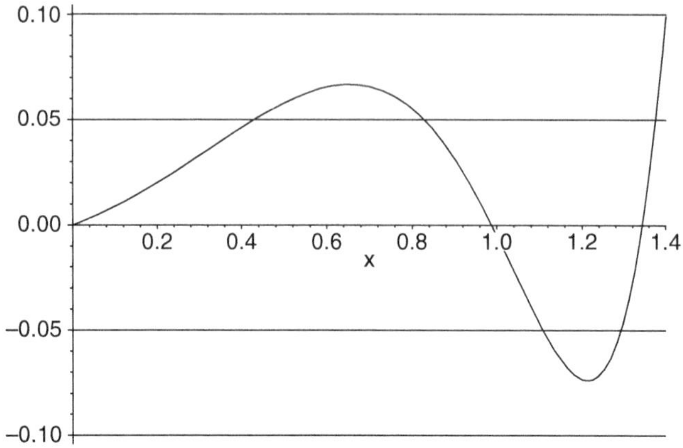

Figure 3.7: The graph of (3.75) for $p = 0$.

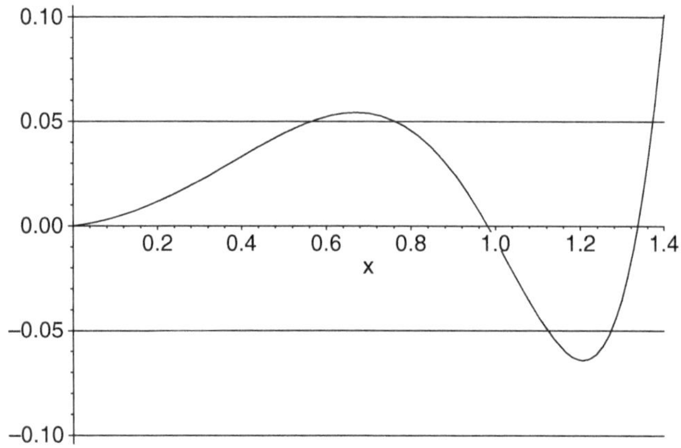

Figure 3.8: The graph of (3.75) for $p = 0.1$.

3.3.7 Polynomial approximation of $\Psi_p(u)$

We have derived the following polynomial approximation of $\Psi_p(u)$, where the absolute error is $< 10^{-5}$ for $0 \le u \le 2$ and $0 \le p \le 1$. For technical reasons u has been replaced by $1 + u$,

$$\Psi_p(1+u) \approx \sum_{n=0}^{9} \psi_n(p) u^n \quad \text{for } -1 \le u \le 1 \text{ and } 0 \le p \le 1, \tag{3.76}$$

where

$$\begin{aligned}\psi_0(p) \quad = \quad & 0.1572\,9775 - 0.2081\,8346p + 0.2136\,1877p^2 \\ & - 0.1997\,1950p^3 + 0.1182\,8390p^4 + 0.1983\,3180p^5 \\ & - 0.9041\,7877p^6 + 1.6611\,0206p^7 - 1.6900\,3807p^8 \\ & + 0.9167\,6313p^9 - 0.2064\,8699p^{10},\end{aligned}$$

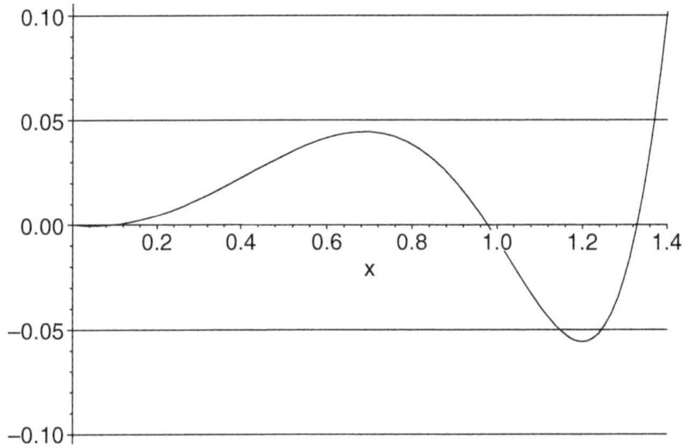

Figure 3.9: The graph of (3.75) for $p = 0.2$.

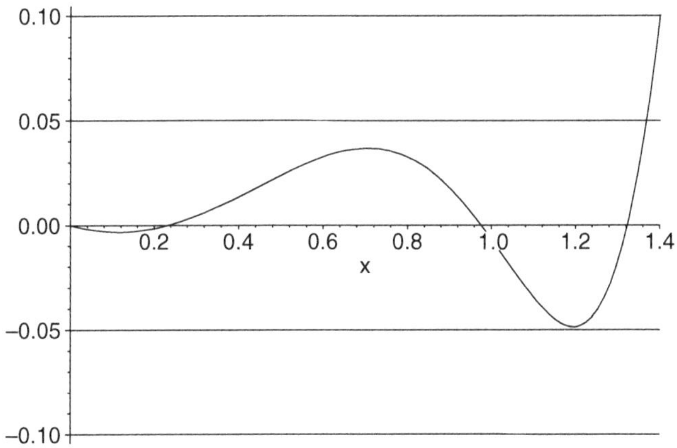

Figure 3.10: The graph of (3.75) for $p = 0.3$.

$$
\begin{aligned}
\psi_1(p) &= -0.4151\,0874 + 0.3757\,4008p - 0.2589\,6448p^2 \\
&\quad - 0.1512\,9193p^3 + 2.1842\,7758p^4 - 8.5111\,2473p^5 \\
&\quad + 19.6590\,9012p^6 - 28.4707\,2172p^7 + 25.2638\,6400p^8 \\
&\quad - 12.5390\,6251p^9 + 2.6622\,8505p^{10},
\end{aligned}
$$

$$
\begin{aligned}
\psi_2(p) &= 0.4151\,8494 + 1.9722\,4214p - 52.9349\,0828p^2 \\
&\quad + 539.2351\,4212p^3 - 2918.5892\,8036p^4 \\
&\quad + 9431.1214\,7934p^5 - 19\,142.5504\,3510p^6 \\
&\quad + 24\,656.7243\,7171p^7 - 19\,580.0513\,2269p^8 \\
&\quad + 8747.1105\,6003p^9 - 1682.1384\,4797p^{10},
\end{aligned}
$$

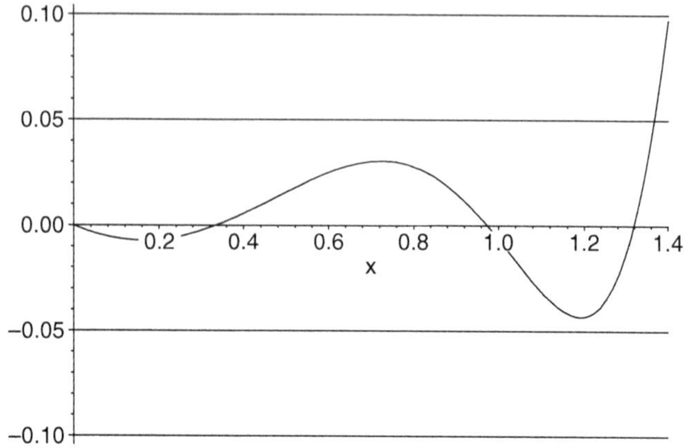

Figure 3.11: The graph of (3.75) for $p = 0.4$.

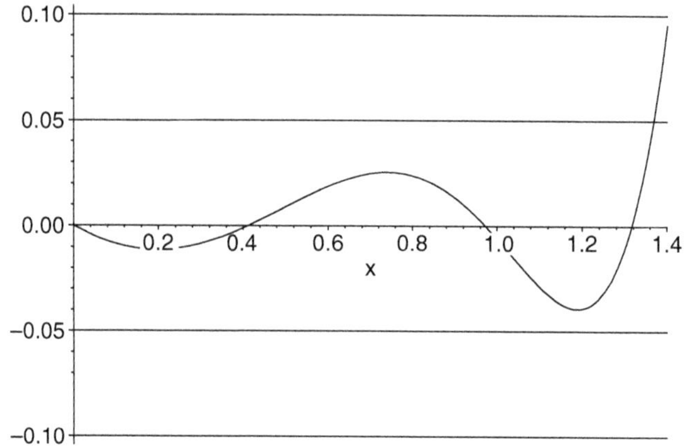

Figure 3.12: The graph of (3.75) for $p = 0.5$.

$$\begin{aligned}\psi_3(p) &= -0.1383\,4376 - 0.3994\,1310p + 1.5117\,9593p^2 \\ &\quad - 12.6797\,2916p^3 + 75.6428\,9914p^4 - 274.1099\,8589p^5 \\ &\quad + 646.0475\,2226p^6 - 952.4213\,8227p^7 + 863.3755\,9534p^8 \\ &\quad - 438.4311\,0672p^9 + 95.3253\,9683p^{10},\end{aligned}$$

$$\begin{aligned}\psi_4(p) &= -0.0698\,2304 + 0.3399\,3674p - 0.1962\,0974p^2 \\ &\quad + 0.1293\,3505p^3 - 0.2600\,0615p^4 + 0.7641\,0925p^5 \\ &\quad - 1.6422\,9259p^6 + 2.2884\,3917p^7 - 1.9663\,3597p^8 \\ &\quad + 0.9462\,0811p^9 - 0.1948\,8536p^{10},\end{aligned}$$

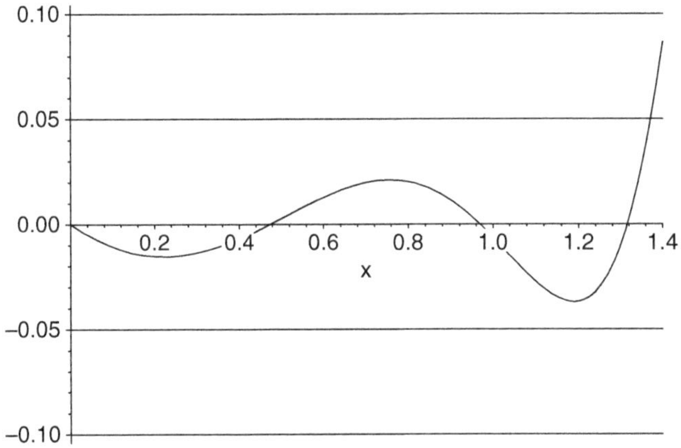

Figure 3.13: The graph of (3.75) for $p = 0.6$.

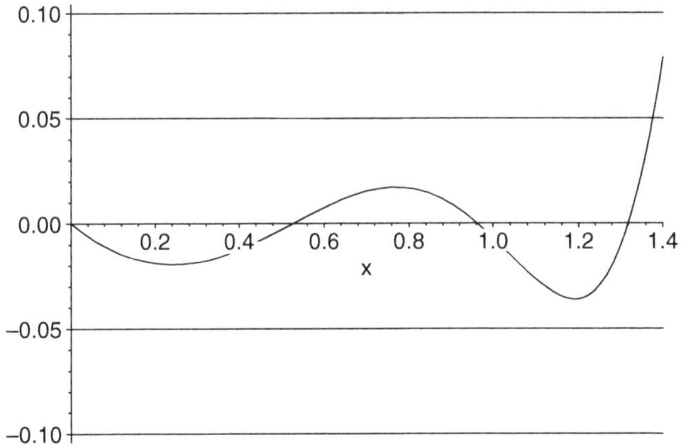

Figure 3.14: The graph of (3.75) for $p = 0.7$.

$$\begin{aligned}
\psi_5(p) &= 0.0690\,4464 - 0.0540\,6877p - 0.1172\,2059p^2 \\
&\quad + 0.0806\,1303p^3 + 0.2941\,6207p^4 - 1.2550\,4537p^5 \\
&\quad + 2.5274\,9074p^6 - 3.0466\,9313p^7 + 2.2144\,1799p^8 \\
&\quad - 0.8946\,2081p^9 + 0.1543\,2099p^{10},
\end{aligned}$$

$$\begin{aligned}
\psi_6(p) &= -0.0027\,3472 - 0.0824\,6215p + 0.1373\,2381p^2 \\
&\quad - 0.1587\,8295p^3 + 0.4344\,4851p^4 - 1.3174\,1204p^5 \\
&\quad + 2.8040\,5740p^6 - 3.8662\,1692p^7 + 3.2957\,6720p^8 \\
&\quad - 1.5771\,6049p^9 + 0.3236\,3316p^{10},
\end{aligned}$$

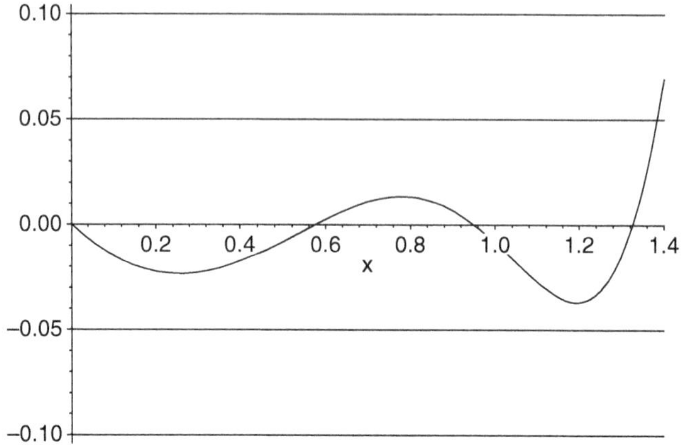

Figure 3.15: The graph of (3.75) for $p = 0.8$.

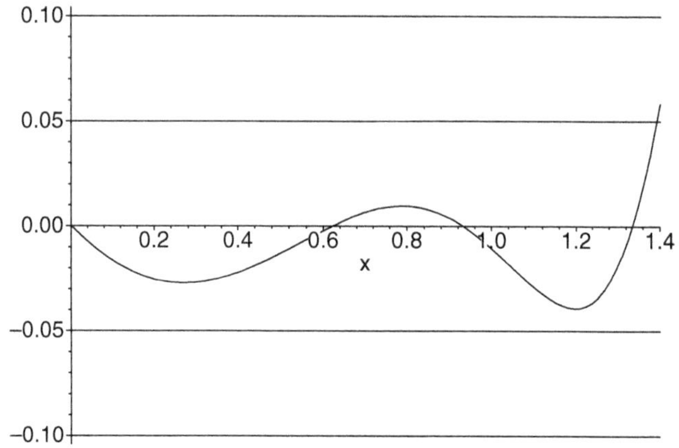

Figure 3.16: The graph of (3.75) for $p = 0.9$.

$$\begin{aligned}
\psi_7(p) &= -0.0148\,4288 + 0.0397\,2439p - 0.0162\,0684p^2 \\
&\quad + 0.0426\,6691p^3 - 0.3032\,0466p^4 + 0.9063\,8519p^5 \\
&\quad - 1.5229\,5556p^6 + 1.4967\,1958p^7 - 0.8158\,7302p^8 \\
&\quad + 0.2045\,8554p^9 - 0.0105\,8201p^{10},
\end{aligned}$$

$$\begin{aligned}
\psi_8(p) &= 0.0024\,1536 + 0.0079\,5870p - 0.0270\,9503p^2 \\
&\quad + 0.0622\,2301p^3 - 0.2373\,3693p^4 + 0.7604\,7778p^5 \\
&\quad - 1.6173\,5556p^6 + 2.2085\,7143p^7 - 1.8634\,9206p^8 \\
&\quad + 0.8835\,9788p^9 - 0.1798\,9418p^{10},
\end{aligned}$$

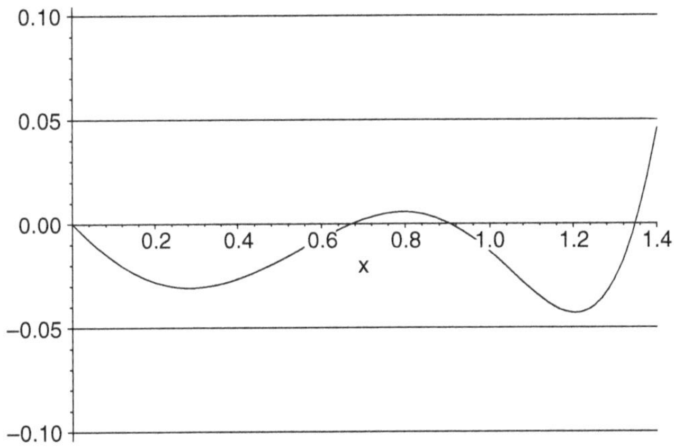

Figure 3.17: The graph of (3.75) for $p = 1$.

$$\begin{aligned}\psi_9(p) &= 0.0015\,8976 - 0.0069\,5879p + 0.0072\,1543p^2 \\ &\quad - 0.0124\,9204p^3 + 0.0467\,0984p^4 - 0.0775\,9259p^5 \\ &\quad - 0.0212\,8889p^6 + 0.2681\,4815p^7 - 0.4000\,0000p^8 \\ &\quad + 0.2574\,9559p^9 - 0.0634\,9206p^{10}.\end{aligned}$$

Once the polynomials $\psi_0(p), \dots, \psi_9(p)$ have been downloaded on the computer it is easy to find the accurate estimates of $\Psi_p(1+u)$. Furthermore, since we are only dealing with polynomials in (3.76) it is easy to use Newton's iteration formula on the problem

$$\Psi_p(1+u) = y, \qquad y \in]\,0,1\,[, \quad y, 1-y \gg 10^{-5} \text{ given,} \tag{3.77}$$

both for fixed p and for fixed u.

1. Let $p \in [0,1]$ and $y \in]\,0,1\,[$ be given, where y, $1-y \gg 10^{-5}$. The solution u of (3.77) is the limit of the recursive sequence

$$u_{m+1} = u_m - \left\{\sum_{n=0}^{9} \psi_n(p)u_m^n - c\right\} \Big/ \left\{\sum_{n=1}^{9} n\psi_n(p)u_m^{n-1}\right\}.$$

   In this case we also write

$$\text{inv}\ \Psi_p(y) = \text{inv}_p\Psi_p(y) = 1 + u = 1 + \lim_{m\to\infty} u_m,$$

   where $\text{inv}_p\Psi_p$ indicates that p is kept fixed.

2. Let $u \in [-1,1]$ and $y \in]\,0,1\,[$ be given, where y, $1-y \gg 10^{-5}$. The solution of (3.77) is the limit of the recursive sequence

$$p_{m+1} = p_m - \left\{\sum_{n=0}^{9} \psi_n(p_m)u^n - c\right\} \Big/ \left\{\sum_{n=0}^{9} \psi_n'(p_m)u^n\right\}.$$

In this case we do not have a reasonable notation for the inverse function. One might use

$$\operatorname{inv}_u \Psi_*(y) = \lim_{m\to\infty} p_m,$$

where $\operatorname{inv}_z \Psi_*$ indicates that u is kept fixed, while $*$ hints that the equation should be solved in p.

In most cases only (1) will be relevant.

3.3.8 Generalized repeated integrals of $\Psi_p(u)$

In Section 3.2.7 we considered the classical repeated integrals of $\operatorname{erfc}(u)$. Since the functions $\Psi_p(u)$, $p \geq -0.5$, are natural extensions of $\operatorname{erfc}(u)$, we may expect similar results for $\Psi_p(u)$, and this is indeed true.

By (3.27),

$$\begin{aligned} \mathrm{i}^n\operatorname{erfc}(u) &= \frac{1}{\sqrt{\pi}}\int_u^{\infty}\frac{(t-u)^n}{n!}\exp(-t^2)\,dt \\ &= \frac{2}{\sqrt{\pi}}\int_u^{\infty}\frac{(t-u)^n}{n!}\Psi_{-0.5}(t)\,dt, \qquad n\in\mathbb{N}_0, \end{aligned}$$

which also can be interpreted as $n+1$ repeated integrals of $\Psi_{-0.5}(t)$. Hence $n+1$ repeated integrals correspond to one single integration of $\Psi_{-0.5}(t)$ times the kernel $(t-u)^n/n!$ (and multiplied by the normalization constant $2/\sqrt{\pi}$), or, when $n+1$ is replaced by n,

$$n \text{ repeated integrals } \sim \int_u^{\infty}\frac{(t-u)^{n-1}}{\Gamma(n)}\cdots dt, \qquad n\in\mathbb{N}. \tag{3.78}$$

The right hand side of (3.78) makes sense for any positive *real* number $n = q > 0$, if only the function represented by the dots is nice (like e.g. the $\Psi_p(u)$ functions). Let us therefore define q *fractional repeated integrals*, $q > 0$, of such a function $f(u)$ by

$$I^q f(u) = \frac{1}{\Gamma(q)}\int_u^{\infty}(t-u)^{q-1}f(t)\,dt, \qquad q>0. \tag{3.79}$$

By means of the beta function it is not hard to prove that

$$I^q\left(I^r f\right)(u) = I^{q+r}f(u), \qquad q,r>0, \tag{3.80}$$

so $q+r$ fractional repeated integrals are the same as first taking r fractional integrals and then apply q fractional integrals on the result. The main ideas in the proof of (3.80) are included in the following example. We leave to the reader the task of deriving (3.80).

Example 3.3.4 We shall prove that two succeeding half integrals actually give one normal integral. We have by (3.79),

$$I^{0.5}f(u) = \frac{1}{\Gamma(0.5)}\int_u^\infty \frac{1}{\sqrt{t-u}}f(t)\,dt = \frac{1}{\sqrt{\pi}}\int_u^\infty \frac{1}{\sqrt{t-u}}f(t)\,dt,$$

so by changing the order of integration,

$$\begin{aligned} I^{0.5}\left(I^{0.5}f\right)(u) &= \frac{1}{\sqrt{\pi}}\int_u^\infty \frac{1}{\sqrt{t-u}}I^{0.5}f(t)\,dt \\ &= \frac{1}{\sqrt{\pi}}\int_u^\infty \frac{1}{\sqrt{t-u}}\left\{\frac{1}{\sqrt{\pi}}\int_t^\infty \frac{1}{\sqrt{v-t}}f(u)\,du\right\}dt \\ &= \frac{1}{\pi}\int_u^\infty f(v)\left\{\int_u^v \frac{1}{\sqrt{(t-u)(v-t)}}dt\right\}dv. \qquad (3.81)\end{aligned}$$

By the change of variables

$$t = \frac{v+u}{2} + \frac{v-u}{2}\tau, \qquad \tau \in [-1,1],$$

in the inner integral we get after some calculus,

$$\frac{1}{\pi}\int_u^v \frac{1}{\sqrt{(t-u)(v-t)}}dt = \frac{1}{\pi}\int_{-1}^1 \frac{1}{\sqrt{1-\tau^2}}d\tau = \frac{\pi}{\pi} = 1,$$

so (3.81) reduces to

$$I^{0.5}\left(I^{0.5}f\right)(u) = \int_u^\infty f(v)\,dv = I^1 f(u),$$

which is (3.80) for $q = r = 0.5$. $\Diamond$

The functions $\Psi_p(u)$ are normalized in the sense that $\Psi_p(0) = 1$. We have in particular by (3.29) for $n \in \mathbb{N}_0$,

$$\begin{aligned} \Psi_{n/2}(u) &= \frac{2}{\Gamma\left(\frac{n+1}{2}\right)}\int_u^\infty (t-u)^n \exp\left(-t^2\right)dt \\ &= \frac{2\Gamma(n+1)}{\Gamma\left(\frac{n+1}{2}\right)}\int_u^\infty \frac{(t-u)^n}{\Gamma(n+1)}\Psi_{-0.5}(t)\,dt \\ &= \frac{n^{n+1}}{\sqrt{\pi}}\Gamma\left(\frac{n}{2}+1\right)I^{n+1}\Psi_{-0.5}(u),\end{aligned}$$

where we have used the duplication formula for the gamma function and (3.79).

The result shows that we in general may expect a normalization factor and that one integration corresponds to an increase of 0.5 in the index of $\Psi_p(u)$.

If $n/2$ is replaced by p in the calculation above we see that the only reasonable extension has the structure

$$\Psi_p(u) = \frac{2^{2p+1}}{\sqrt{\pi}}\Gamma(p+1)I^{2p+1}\Psi_{-0.5}(u) \qquad \text{for } p > -\frac{1}{2}. \tag{3.82}$$

An easy check shows that this is true. By defining I^0 as the identity operator (no integration at all) we see that (3.82) is also true for $p = -1/2$.

A more general result is

Theorem 3.3.13 *Let $p \geq -0.5$ and $q \geq 0$. Then*

$$\begin{aligned} \Psi_{p+q}(u) &= \frac{2^{2q}\Gamma(p+q+1)}{\Gamma(p+1)} I^{2q}\Psi_p(u) \\ &= \frac{2^{2q}\Gamma(p+q+1)}{\Gamma(p+1)\Gamma(2q)} \int_u^{\infty} (t-u)^{2q-1}\Psi_p(t)\,dt, \quad \textit{if also } q > 0. \end{aligned}$$

PROOF. Use (3.82) with p replaced by $p+q$ followed by (3.80),

$$\begin{aligned} \Psi_{p+q}(u) &= \frac{2^{2p+2q+1}}{\sqrt{\pi}}\Gamma(p+q+1)I^{2p+2q+1}\Psi_{-0.5}(u) \\ &= 2^{2q}\Gamma(p+q+1)\frac{2^{2p+1}}{\sqrt{\pi}} I^{2q}\left(I^{2p+1}\Psi_{-0.5}\right)(u) \\ &= \frac{2^{2q}\Gamma(p+q+1)}{\Gamma(p+1)} I^{2q}\Psi_p(u) \\ &= \frac{2^{2q}\Gamma(p+q+1)}{\Gamma(p+1)\Gamma(2q)} \int_u^{\infty} (t-u)^{2q-1}\Psi_p(t)\,dt, \end{aligned}$$

where the latter equality only holds for $q > 0$. □

3.3.9 Connection with hypergeometric functions*

If the reader is familiar with the hypergeometric functions, the approximation formulæ given in the previous subsections can be totally avoided, since we may instead use the following exact formula

$$\begin{aligned} \Psi_p(u) &= \exp\left(-u^2\right) {}_1F_1\left[p+\frac{1}{2};\frac{1}{2},u^2\right] \\ &\quad -2u\cdot\exp\left(-u^2\right)\frac{\Gamma(p+1)}{\Gamma(p+\frac{1}{2})}\,{}_1F_1\left[p+1;\frac{3}{2},u^2\right]. \end{aligned} \tag{3.83}$$

Formula (3.83) has been used by the authors when the tables of Chapter 8 were constructed. We may, however, assume that most of the readers do not know the hypergeometric functions, so we shall not go further into that description, even if we then could save a lot in the previous presentation.

3.3.10 The $\Lambda_p(u)$ functions

The $\Lambda_p(u)$ functions are defined by

$$\Lambda_p(u) := \Psi_p(u)/u^{2p} \qquad \text{for } u > 0. \tag{3.84}$$

They were introduced by Mejlbro (1996). In a sense they are quite superfluous, had it not been for the fact that they are more convenient to use than the $\Psi_p(u)$ functions in problems like: For given x, estimate the time t, for which $C(x,t) = C_{cr}$. Here, C_{cr} can be interpreted as the threshold value of the chloride concentration. For that reason, tables for both $\Psi_p(u)$ and $\Lambda_p(u)$ have been worked out.

Following Mejlbro (1996), we shall in the rest of this subsection describe the technical problems where the $\Lambda_p(u)$ functions are useful. We shall rely on the HETEK model.

Theorem 3.3.14 *The HETEK model. The solution of the initial/boundary value problem*

$$\frac{\partial C}{\partial t} = D(t)\frac{\partial^2 C}{\partial x^2}; \qquad x > 0,\, t > t_{ex},$$

$$C(x, t_{ex}) = C_i, \qquad x \geq 0,$$

$$C(0,t) = C_i + S \cdot \{(t - t_{ex})D_a(t)\}^p, \quad t > t_{ex},\, p > 0 \textit{ const.},$$

is given by

$$\begin{aligned} C_p(x,t) &= C_i + S \cdot \{(t - t_{ex})D_a(t)\}^p \, \Psi_p\left(\frac{x}{\sqrt{4(t-t_{ex})D_a(t)}}\right) \\ &= C_i + S \cdot \left(\frac{x}{2}\right)^{2p} \Lambda_p\left(\frac{x}{\sqrt{4(t-t_{ex})D_a(t)}}\right). \end{aligned}$$

Let $x = c$ be given. We want to solve the equation

$$C_p(c,t) = C_{cr}, \qquad C_{cr} \gg C_i, \tag{3.85}$$

i.e. the critical time t, when the critical level C_{cr} is obtained for $x = c$.

Putting $T = (t - t_{ex})D_a(t)$ and $z = c/\sqrt{4T}$, the Equation (3.85) is written

$$C_{cr} = C_i + S \cdot \left(\frac{c}{2}\right)^{2p} \Lambda_p(z),$$

so by a rearrangement,

$$\Lambda_p(z) = \frac{C_{cr} - C_i}{S} \cdot \left(\frac{2}{c}\right)^{2p}. \tag{3.86}$$

The right hand side of (3.86) is a positive constant, and $\Lambda_p(z)$ is trivially decreasing continuously from $+\infty$ to 0, when z goes from 0 to ∞. Hence, the Equation (3.86) has a unique solution $z > 0$. Once z is found, we get T from

$$T = \left(\frac{c}{2z}\right)^2,$$

so t is finally found by solving

$$(t - t_{ex})D_a(t) = T = \left(\frac{c}{2z}\right)^2. \tag{3.87}$$

The problem with (3.86) is that it is in general difficult to give an expansion of inv $\Lambda_p(\xi)$ for general $p > 0$, so we have to iterate or use tables. Of course, the solution of (3.87) also requires an iteration, once $D_a(t)$ is given. However, both iteration problems are 1-dimensional, so they can be performed on a computer, and in many cases even on a pocket calculator.

This theory is illustrated by the following example from Mejlbro (1996).

Example 3.3.5 Assume the HETEK model of Theorem 3.3.14. Assume furthermore that we have performed some measurements which show that the *mean value chloride diffusion coefficient* $D_a(t)$ [not to be confused with $D(t)$ itself] is approximately given by

$$D_a(t) = D_0 \left(\frac{t_{ex}}{t}\right)^\alpha \qquad \text{for } t \geq t_{ex} \tag{3.88}$$

for some $\alpha > 0$. Notice that the physical dimension of D_0 is length^2/time. The point is that $D_a(t)$ is easier to measure than $D(t)$. Using the assumption (3.88) it follows that

$$\begin{aligned} \int_{t_{ex}}^t D(\tau)\,d\tau &= (t - t_{ex})\left\{\frac{1}{t - t_{ex}}\int_{t_{ex}}^t D(\tau)\,d\tau\right\} \\ &= (t - t_{ex})D_a(t) = D_0(t - t_{ex}) \cdot \left(\frac{t_{ex}}{t}\right)^\alpha. \end{aligned}$$

When we differentiate this equation we get

$$D(t) = D_0\left\{(1-\alpha)\left(\frac{t_{ex}}{t}\right)^\alpha + \alpha\left(\frac{t_{ex}}{t}\right)^{\alpha+1}\right\} \qquad \text{for } t \geq t_{ex},$$

from which we derive the important observation,

$$D_0 = D(t_{ex}) = D_a(t_{ex}).$$

In order to show the principles we choose arbitrarily the following technical data (% by mass concrete):

- Model parameter: $p = 0.2$
- Concrete cover: $c = 50$ mm
- Time of exposure: $t_{ex} = 0.1$ yr
- Critical value: $C_{cr} = 0.1$ %
- Absorption of chloride during the first year: $S = 0.5$ %/yr^p
- Initial content of chloride: $C_i = 0.02\%$
- Water/cement proportion: $w/c = 0.37$
- Exponent: $\alpha = 3(0.55 - w/c) = 3(0.55 - 0.37) = 0.54$
- Diffusion coefficient at time of exposure:

$$D_0 = 0.57 \cdot 10^6 \cdot \exp\left(-\sqrt{\frac{23}{w/c}}\right) \text{ mm}^2/\text{yr} = 215 \text{ mm}^2/\text{yr}.$$

In this case equation (3.86) becomes

$$\Lambda_{0.2}(z) = \frac{C_{cr} - C_i}{S} \cdot \frac{4^{0.2}}{c^{0.4}} = \frac{0.10 - 0.02}{0.5} \cdot \frac{4^{0.2}}{50^{0.4}} = 0.04415.$$

Using the tables of $\Lambda_p(z)$ it is seen that the solution is $z \approx 1.30$, hence

$$T = \left(\frac{c}{2z}\right)^2 = \left(\frac{50}{2 \cdot 1.30}\right)^2 = 369.82.$$

Finally, assuming that $t_{cr} \gg t_{ex}$,

$$(t_{cr} - t_{ex})D_a(t_{cr}) \approx t_{cr} \cdot D_0 \left(\frac{t_{ex}}{t_{cr}}\right)^\alpha = D_0 t_{ex}^\alpha t_{cr}^{1-\alpha} \approx T,$$

i.e.

$$t_{cr} \approx \left\{\frac{T}{D_0 \cdot t_{ex}^\alpha}\right\}^{1/(1-\alpha)} = \left\{\frac{369.82}{215 \cdot 0.1^{0.54}}\right\}^{1/0.46} \approx 48.5 \text{ years},$$

which is the estimated lifetime. We see that an estimated critical time of $t_{cr} \approx 50$ years is quite reasonable.

Note. *The data given in this example are all chosen for the purpose of illustration. None of the data is found by laboratory tests nor field inspection and testing.* ◊

For comparison we also give an example (cf. Mejlbro (1996)) of how the equation $C_p(x,t) = C_{cr}$ is solved in x for given *expected lifetime* t. In this way we can estimate the necessary concrete cover x as a function of p. It follows that the $\Psi_p(u)$ functions are most convenient for this type of problem.

Example 3.3.6 Consider again the HETEK model of Theorem 3.3.14, and let again the mean value chloride diffusion coefficient $D_a(t)$ be modelled by (3.88), i.e.

$$D_a(t) = D_0 \left(\frac{t_{ex}}{t}\right)^{\alpha} \qquad \text{for } t \geq t_{ex}.$$

The boundary value is according to Theorem 3.3.14

$$\begin{aligned} C_p(0,t) &= C_i + S\left\{(t - t_{ex})D_a(t)\right\}^p \\ &\approx C_i + S \cdot D_0^p \cdot t_{ex}^{p\alpha} \cdot t^{(1-\alpha)p} \qquad \text{for } t \gg t_{ex}, \end{aligned}$$

i.e.

$$C_p(0,t) \approx C_i + S_p \cdot t^{(1-\alpha)p}, \qquad \text{where } S_p = S \cdot D_0^p \cdot t_{ex}^{p\alpha}.$$

By Theorem 3.3.14 the approximate solution is

$$C_p(x,t) \approx C_i + S_p t^{(1-\alpha)p} \Psi_p(z), \tag{3.89}$$

where

$$x = \sqrt{4(t - t_{ex})D_a(t)} \cdot z \approx 2\sqrt{D_0 t_{ex}^{\alpha} t^{1-\alpha}} \cdot z. \tag{3.90}$$

Let C_{cr} be the critical level, and let the expected lifetime $t \gg t_{ex}$ be given. We want to estimate the necessary concrete cover x for various values of p.

In the numerical example below we keep the value of S_p fixed. Notice that we are forced to give S_p the strange physical dimension of $\text{length}^{-(1-\alpha)p}$.

Assume the following theoretical, though realistic technical data, where the values of C_{cr}, S_p and C_i are given as % of mass cement:

- Time of exposure: $t_{ex} = 0.1$ yr
- Expected lifetime: $t = 100$ yr
- Critical value: $C_{cr} = 0.8$ %
- Absorption of chloride during the first year: $S_p = 1.5\ \%/\text{yr}^{(1-\alpha)p}$
- Initial content of chloride: $C_i = 0.15$ %
- Exponent: $\alpha = 0.45$
- Diffusion coefficient at time of exposure: $D_0 = 290\ \text{mm}^2/\text{yr}$.

Table 3.3: Concrete cover x as a function of p.

p	$(1-\alpha)p$	$\Psi_p(z)$	z	x mm
0.0	0.000	0.43333	0.55	39.6
0.1	0.055	0.33637	0.64	46.0
0.2	0.110	0.26111	0.71	51.0
0.3	0.165	0.20268	0.78	56.2
0.4	0.220	0.15733	0.84	60.5
0.5	0.275	0.12213	0.90	64.8
0.6	0.330	0.09480	0.95	68.4
0.7	0.385	0.07359	1.00	72.0
0.8	0.440	0.05712	1.05	75.6
0.9	0.495	0.04434	1.09	78.5
1.0	0.550	0.03442	1.14	82.1

Using (3.89) the equation $C_p(x,t) = C_{cr}$ is approximately equivalent to

$$\Psi_p(z) \approx \frac{C_{cr} - C_i}{S_p t^{(1-\alpha)p}} = \frac{0.80 - 0.15}{1.5 \cdot (100^{0.55})^p} = \frac{0.43333}{12.59^p}. \tag{3.91}$$

From (3.90) we get

$$x \approx 2\sqrt{290 \cdot 0.1^{0.45} \cdot 100^{0.55}} \cdot z \text{ mm} \approx 72 \cdot z \text{ mm}.$$

Finally we solve (3.91) by either using tables or the method of Section 3.3.7. In this way we get Table 3.3 of the necessary concrete cover x for various values of p. Notice that the expected thickness of the concrete cover x is quite reasonable for $0.0 \le p \le 1.0$. We need some further measurement to get an estimate of p.

Note. *The data given in this example are all chosen for the purpose of illustration. None of the data is found by laboratory tests nor field inspection and testing.* $\Diamond$

The two examples show that in the HETEK model the problem

$$C_p(x,t) = C_{cr}$$

is solved in t for given x by using the $\Lambda_p(u)$ functions. It is solved in x for given t by using the $\Psi_p(u)$ functions. Both types of problems are relevant, when we are dealing with chloride penetration into concrete.

3.4 Bessel functions

When we try to solve Fick's second law for a circular domain it is natural to change to polar coordinates. The price of this change of coordinate system is that we shall quite often encounter *Bessel's differential equation*, either in the form

$$r^2 \frac{d^2 f}{dr^2} + r \frac{df}{dr} + (r^2 - \nu^2) f(r) = 0, \qquad \nu \in \mathbb{R}, \tag{3.92}$$

or in the equivalent form

$$r \frac{d}{dr} \left\{ r \frac{df}{dr} \right\} = (\nu^2 - r^2) f(r), \qquad \nu \in \mathbb{R}, \tag{3.93}$$

where (3.93) will be quite useful, when we apply partial integration.

The solutions of Bessel's differential equations are called *Bessel functions.* We shall in particular be concerned with Bessel functions of first kind, $J_\nu(r)$, and Bessel functions of second kind, $Y_\nu(r)$.

The theory of Bessel functions was fully developed by the beginning of the last century, cf. e.g. the classical book of Watson (1944). For some reason most universities have only occasionally given courses in Bessel functions during the last 40 years. We shall here present the most important results on Bessel functions, having Fick's second law in our mind all the time. Interested readers are referred to Watson (1944), for a full mathematical treatment of the theory. The important results are of course also listed (without proof) by Abramowitz *et al.* (1965).

3.4.1 Bessel functions

Since the variable r in (3.92) always will be a radius in our applications, we may assume that $r \geq 0$. If furthermore $r > 0$, the theorem of existence and uniqueness of solutions for differential equations of second order tells us that every solution of (3.92) has the structure

$$f(r) = a\ J_\nu(r) + b\ Y_\nu(r), \tag{3.94}$$

where a and b are constants, and where $J_\nu(r)$ and $Y_\nu(r)$ are specific linearly independent solutions, which shall be defined below.

The most primitive idea is to solve (3.92) by a modified series expansion. This works, and by choosing suitable initial conditions we define the *Bessel function of first kind* J_ν as the solution of the form

$$J_\nu(r) = \left(\frac{r}{2}\right)^\nu \sum_{k=0}^{\infty} \frac{(-1)^k}{k!\Gamma(\nu + k + 1)} \left(\frac{r}{2}\right)^{2k}. \tag{3.95}$$

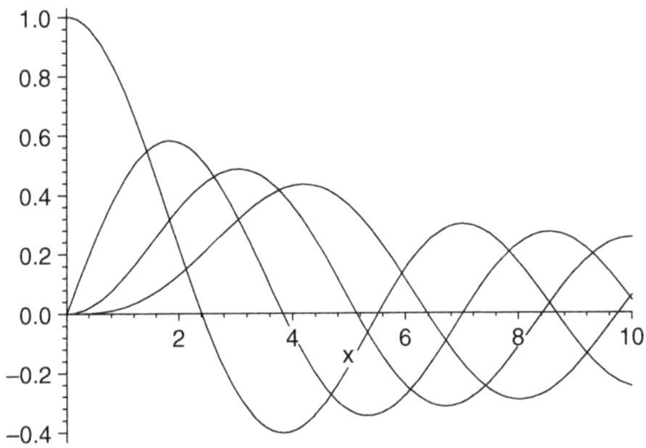

Figure 3.18: The Bessel functions $J_0(x)$, $J_1(x)$, $J_2(x)$, $J_3(x)$.

With the convention $1/\Gamma(-n) = 0$ for $n \in \mathbb{N}_0$ it is easy to check that (3.95) makes sense for $0 < r < \infty$, and even for $r = 0$, if $\nu \geq 0$, and that (3.95) is a solution of (3.92). Obviously, $J_{-\nu}(r)$ is also a solution, so when $J_\nu(r)$ and $J_{-\nu}(r)$ are linearly independent, we might replace (3.94) by

$$f(r) = a' J_\nu(r) + b' J_{-\nu}(r).$$

A century ago one realized that this formally correct formula (at least for $\nu \notin \mathbb{N}_0$) had some drawbacks, so instead one introduced

$$Y_\nu(r) = \frac{J_\nu(r)\cos(\nu\pi) - J_{-\nu}(r)}{\sin(\nu\pi)}, \tag{3.96}$$

when $J_\nu(r)$ and $J_{-\nu}(r)$ are linearly independent, i.e. when ν is not an integer. The function $Y_\nu(r)$ is called a Bessel function of second kind. These are now defined for ν real, but not an integer.

When $\nu = n \in \mathbb{N}_0$ we get from (3.95),

$$J_n(r) = \sum_{k=0}^{\infty} \frac{(-1)^k}{k!(n+k)!} \left(\frac{r}{2}\right)^{2k+n}, \qquad r \geq 0, \tag{3.97}$$

and in particular,

$$J_0(r) = \sum_{k=0}^{\infty} \frac{(-1)^k}{(k!)^2} \left(\frac{r^2}{4}\right)^k, \qquad r \geq 0, \tag{3.98}$$

so

$$J_0(0) = 1, \quad \text{and} \quad J_n(0) = 0 \quad \text{for } n \in \mathbb{N}. \tag{3.99}$$

Furthermore, by the convention $1/\Gamma(-m) = 0$ for $m \in \mathbb{N}_0$ it is easy from (3.97) to derive that

$$J_{-n}(r) = (-1)^n J_n(r) \quad \text{for } n \in \mathbb{N}_0.$$

Hence, we can take the limit $\nu \to n \in \mathbb{N}_0$ and thus define

$$Y_n(r) = \lim_{\nu \to n} Y_\nu(r), \quad \text{for } n \in \mathbb{N}_0,$$

followed by a proof of that $J_n(r)$ and $Y_n(r)$ are linearly independent, and that $Y_n(r)$ is indeed a solution of (3.92).

The functions $Y_\nu(r)$ can be described in an alternative way. This needs a small preparation.

First introduce Euler's constant by

$$\gamma = \lim_{n \to \infty} \left\{ \sum_{k=1}^{n} \frac{1}{k} - \int_1^n \frac{dx}{x} \right\} \approx 0.57722,$$

which is easily given a geometrical interpretation.

Then define a function $\psi : \mathbb{N} \to \mathbb{R}$ by

$$\psi(1) = -\gamma, \quad \psi(n) = -\gamma + \sum_{k=1}^{n-1} \frac{1}{k} \quad \text{for } n = 2, 3, 4, \cdots .$$

Define for $n = 0$, and $r > 0$,

$$Y_0(r) = \frac{2}{\pi} \left\{ \ln\left(\frac{r}{2}\right) + \gamma \right\} J_0(r) + \frac{2}{\pi} \sum_{k=1}^{\infty} \frac{(-1)^{k+1}}{(k!)^2} \left\{ \sum_{j=1}^{k} \frac{1}{j} \right\} \left(\frac{r^2}{4} \right)^k, \tag{3.100}$$

and for $n \in \mathbb{N}$ and $r > 0$,

$$\begin{aligned} Y_n(r) &= -\frac{1}{\pi} \sum_{k=0}^{n-1} \frac{(n-k-1)!}{k!} \left(\frac{r}{2}\right)^{2k-n} + \frac{2}{\pi} \ln\left(\frac{r}{2}\right) J_n(r) \\ &\quad + \frac{1}{\pi} \sum_{k=0}^{\infty} (-1)^{k+1} \frac{\psi(k+1) + \psi(n+k+1)}{k!(n+k)!} \left(\frac{r}{2}\right)^{2k+n} . \end{aligned} \tag{3.101}$$

Then (3.100) and (3.101) are exactly what we get when we take the limit in (3.96). A small check gives that

$$Y_{-n}(r) = (-1)^n Y_n(r), \quad \text{for } n \in \mathbb{N}_0,$$

and that none of the $Y_n(r)$, $n \in \mathbb{N}_0$, are defined for $r = 0$. Returning to (3.96) we see that this holds in general, i.e. all the $Y_\nu(r)$ have a singularity at $r = 0$.

3.4.2 Recurrence formulæ

Using the series expansion (3.92) it is not hard to prove the first column of the following recurrence formulæ,

$$\begin{aligned} J_{\nu-1}(r)+J_{\nu+1}(r)&=\frac{2\nu}{r}J_\nu(r), & Y_{\nu-1}(r)+Y_{\nu+1}(r)&=\frac{2\nu}{r}Y_\nu(r),\\ J_{\nu-1}(r)-J_{\nu+1}(r)&=2J_\nu'(r), & Y_{\nu-1}(r)-Y_{\nu+1}(r)&=2Y_\nu'(r),\\ J_\nu'(r)&=J_{\nu-1}(r)-\frac{\nu}{r}J_\nu(r), & Y_\nu'(r)&=Y_{\nu-1}(r)-\frac{\nu}{r}Y_\nu(r),\\ J_\nu'(r)&=-J_{\nu+1}(r)+\frac{\nu}{r}J_\nu(r). & Y_\nu'(r)&=-Y_{\nu+1}(r)+\frac{\nu}{r}Y_\nu(r). \end{aligned} \tag{3.102}$$

When ν is not an integer, formula (3.96) and the first column of (3.102) give us the second column. The extension to $\nu = n$ an integer follows by taking the limit $\nu \to n$.

We notice in particular that

$$J_0'(r) = -J_1(r) \quad \text{and} \quad Y_0'(r) = -Y_1(r), \tag{3.103}$$

and if z is a zero of $J_\nu(r)$, i.e. $J_\nu(z) = 0$, then

$$J_\nu'(z) = J_{\nu-1}(z) = -J_{\nu+1}(z). \tag{3.104}$$

By the observation (3.104) we anticipate the next section, in which zeros of $J_\nu(r)$, and more generally of $J_\nu(r)Y_\nu(\lambda r) - J_\nu(\lambda r)Y_\nu(r)$ are considered. Why these are important for the solution of Fick's second law on a circular or annular domain will be explained in Chapter 4.

3.4.3 Zeros of $J_\nu(r)$ and $J_\nu(r)Y_\nu(\lambda r) - J_\nu(\lambda r)Y_\nu(r)$

In Chapter 4 we shall need the following orthogonal condition (cf. Watson (1944), or in a slightly different form, cf. Abramowitz *et al.* (1965)),

$$\int_0^1 J_\nu(j_{\nu,m}r)J_{\nu,p}r)r\,dr = \begin{cases} \frac{1}{2}\{J_\nu'(j_{\nu,m})\}^2 & \text{for } p = m,\\ 0 & \text{for } p \neq m, \end{cases}$$

or more generally,

$$\int_0^R J_\nu\left(j_{\nu,m}\frac{r}{R}\right) J_\nu\left(j_{\nu,p}\frac{r}{R}\right) r\,dr = \begin{cases} \frac{1}{2}R^2\left\{J_\nu'\left(j_{\nu,m}\right)\right\}^2 & \text{for } p = m,\\ 0 & \text{for } p \neq m. \end{cases} \tag{3.105}$$

Here, $j_{\nu,m}$ denotes the mth positive zero of $J_\nu(r)$, so the right hand side of (3.105) may be calculated by means of (3.104).

Concerning the zeros of $J_\nu(r)$ this needs an explanation.

Consulting Abramowitz *et al.* (1965), it is seen that the graph of $J_n(r)$, $n \in \mathbb{N}_0$, looks like a damped oscillation. In particular these functions are all bounded,

$$|J_0(r)| \le 1, \qquad \text{and} \qquad |J_n(r)| \le \frac{1}{\sqrt{2}} \qquad \text{for } n \in \mathbb{N},$$

and they tend to 0 for $r \to \infty$. The oscillatory behaviour means that each of them has infinitely many simple positive zeros tending to $+\infty$.

In general, $J_\nu(r)$ has an infinite number of positive zeros, all of which are simple. Notice that $r = 0$ is also a zero, when $\nu > 0$. By (3.95) it is of degree ν, but $r = 0$ will not be important in the applications, because Equation (3.92) here is singular.

When $\nu \ge 0$ we let $j_{\nu,m}$ denote the mth positive zero of $J_\nu(r)$. Then the statements above can be formulated in the following way,

$$J_\nu\left(j_{\nu,m}\right) = 0, \quad J'_\nu\left(j_{\nu,m}\right) \ne 0, \quad j_{\nu,m} \to \infty \quad \text{for } m \to \infty.$$

The importance of the sequences $\{j_{\nu,m}\}$ and $\{J'_\nu\left(j_{\nu,m}\right)\}$ was recognized a century ago. For the same reason one finds tables of these sequences for $\nu = 0, 1, \cdots, 8$ and $m = 1, 2, \cdots, 20$ in Abramowitz *et al.* (1965). These will suffice in many of our applications. Since the sequences occur in quite a variety of physical problems, Abramowitz *et al.* (1965), are unable to also giving hints of applications. Such hints should be given in books on special topics like the present one.

The reader should be aware of the fact that some cases not covered by this book may be solved by copying the methods from Chapter 4 followed by and application of the following formulæ from Abramowitz *et al.* (1965),

- (9.5.12) gives an approximation of $j_{\nu,m}$ for m large,
- (9.5.14) and (9.5.18) give approximations of $j_{\nu,1}$, i.e. the first positive zero, and of $J'_\nu\left(j_{\nu,1}\right)$ for large ν.

The latter approximations are in particular important, because the smallest (i.e. the first) zero in most cases leads to the dominating term of the solution. All this will be explained in Chapter 4.

The theory above applies to a circular domain. When Fick's second law is considered on a pipe we need to generalize to an annular domain. Let $\lambda > 1$ be the ratio between the outer and the inner radius of this annulus. Then it can be shown that the solution of Fick's second law (Chapter 4) depends on the zeros of the function

$$J_\nu(r)Y_\nu(\lambda r) - J_\nu(\lambda r)Y_\nu(r).$$

For given $\nu \geq 0$ and $\lambda > 1$ the theory says that the positive zeros $z_{\nu,\lambda,m}$ are all simple and tend to ∞ for $m \to \infty$.

There is no hint in Watson (1944), and Abramowitz *et al.* (1965), how the $z_{\nu,\lambda,m}$ can be found. Therefore, people would tend to avoid this problem. The following example of a numerical exercise on a pocket calculator shows that the problem of finding the first $z_{\nu,\lambda,m}$ on a computer should be fairly easy and not an unsurmountable obstacle.

Example 3.4.1 We shall here give rough estimates of the first and the second zeros $z_{\nu,\lambda,1}$ and $z_{\nu,\lambda,2}$ for selected values of ν and λ. In the calculations the pocket calculator TI-92 has been used. In order to minimize the error and maximize the speed we choose $\nu = \frac{1}{2}, \frac{3}{2}, \cdots, \frac{13}{2}$ in (3.96), so the denominator is always ± 1, and we need only to calculate $J_\nu(r)$, $J_{-\nu}(r)$, $J_\nu(\lambda r)$ and $J_{-\nu}(\lambda r)$. This is done by arbitrarily truncating (3.95) after fifty terms. The accuracy increases by adding more terms, but at the same time the speed decreases. Therefore, the error may be large in large zeros, so we put a * in the tables, where the values cannot be calculated with only fifty terms. Finally, we choose $\lambda = \frac{11}{10}, \frac{12}{10}, \cdots, 2$. Notice that the pipe is thinnest for $\lambda = \frac{11}{10}$.

The purpose of this numerical example is to show that the first two zeros can be easily calculated, even on a pocket calculator. The missing values for $z_{\nu,\lambda,1}$, when $\lambda = 1.1$, and for $z_{\nu,\lambda,2}$, when $\lambda = 1.1$ and $\lambda = 1.2$, can be found by adding more terms in the programme of $J_\nu(r)$.

The two tables show a lot more. It will be convenient to put $\lambda = 1 + \alpha$, where $\alpha > 0$. Then the tables suggest,

1. When $\nu = 0.5$, the calculated values follow the rule

$$z_{0.5,1+\alpha,1} = \frac{\pi}{\alpha} \quad \text{and} \quad z_{0.5,1+\alpha,2} = \frac{2\pi}{\alpha}.$$

Table 3.4: The first zero $z_{\nu,\lambda,1}$ for $J_\nu(r)Y_\nu(\lambda r) - J_\nu(\lambda r)Y_\nu(r)$.

$z_{\nu,\lambda,1}$	$\nu = 0.5$	$\nu = 1.5$	$\nu = 2.5$	$\nu = 3.5$	$\nu = 4.5$	$\nu = 5.5$	$\nu = 6.5$
$\lambda = 1.1$	*	*	*	*	*	*	*
$\lambda = 1.2$	15.708	15.761	15.866	16.022	16.227	16.480	16.780
$\lambda = 1.3$	10.472	10.544	10.688	10.899	11.175	11.510	11.900
$\lambda = 1.4$	7.8540	7.9429	8.1177	8.3729	8.7013	9.0947	9.5450
$\lambda = 1.5$	6.2832	6.3858	6.5861	6.8754	7.2425	7.6758	8.1650
$\lambda = 1.6$	5.2360	5.3501	5.5711	5.8865	6.2812	6.7406	7.2513
$\lambda = 1.7$	4.4880	4.6118	4.8495	5.1845	5.5982	6.0732	6.5950
$\lambda = 1.8$	3.9270	4.0589	4.3100	4.6594	5.0854	5.5682	6.0921
$\lambda = 1.9$	3.4907	3.6294	3.8909	4.2508	4.6837	5.1686	5.6890
$\lambda = 2.0$	3.1416	3.2860	3.5558	3.9225	4.3584	4.8410	5.3540

Table 3.5: The second zero $z_{\nu,\lambda,2}$ for $J_\nu(r)Y_\nu(\lambda r) - J_\nu(\lambda r)Y_\nu(r)$.

$z_{\nu,\lambda,2}$	$\nu = 0.5$	$\nu = 1.5$	$\nu = 2.5$	$\nu = 3.5$	$\nu = 4.5$	$\nu = 5.5$	$\nu = 6.5$
$\lambda = 1.1$	*	*	*	*	*	*	*
$\lambda = 1.2$	*	*	*	*	*	*	*
$\lambda = 1.3$	20.944	20.982	21.051	21.160	21.306	21.489	21.700
$\lambda = 1.4$	15.708	15.753	15.843	15.977	16.154	16.373	16.630
$\lambda = 1.5$	12.566	12.619	12.724	12.879	13.083	13.334	13.629
$\lambda = 1.6$	10.472	10.531	10.648	10.821	11.048	11.326	11.650
$\lambda = 1.7$	8.9760	9.0405	9.1683	9.3569	9.6029	9.9022	10.250
$\lambda = 1.8$	7.8540	7.9234	8.0606	8.2623	8.5241	8.8411	9.2075
$\lambda = 1.9$	6.9813	7.0550	7.2003	7.4132	7.6884	8.0197	8.4000
$\lambda = 2.0$	6.2832	6.3607	6.5131	6.7356	7.0218	7.3647	7.7566

Our conjecture is that in general $z_{0.5,1+\alpha,m} = \frac{m\pi}{\alpha}, \quad m \in \mathbb{N}$.

2. When $\nu = 0.5$, the calculated values also satisfy

$$z_{0.5,1+\alpha,1} = z_{0.5,1+2\alpha,2},$$

so we conjecture that this equation is true in general. Notice also that

$$z_{\nu,1+\alpha,1} > z_{\nu,1+2\alpha,2} \quad \text{for } \nu > 0.5.$$

3. Both $z_{\nu,\lambda,1}$ and $z_{\nu,\lambda,2}$ are decreasing in λ for fixed ν.

4. Both $z_{\nu,\lambda,1}$ and $z_{\nu,\lambda,2}$ are increasing in ν for fixed λ. Both the absolute and the relative increase in ν become smaller, the smaller λ is.

Unintentionally we have obtained the additional result that Table 3.4 and Table 3.5 may even be used for interpolation. ◊

3.4.4 Bessel functions for $\nu=n+\frac{1}{2}$, $n\in\mathbb{N}_0$

We shall in some special cases consider $J_\nu(r)$ for $\nu = n + 1/2$, $n \in \mathbb{N}_0$. For these special values of ν we have a simpler series expansion for $J_\nu(r)$ than the one derived from (3.95), i.e.

$$J_{n+1/2}(r) = \left(\frac{r}{2}\right)^{n+1/2} \sum_{k=0}^{\infty} \frac{(-1)^k}{k!\,\Gamma(n+k+\frac{1}{2}+1)} \left(\frac{r}{2}\right)^{2k}, \tag{3.106}$$

where $J_{n+1/2}(0) = 0$.

By using the recursion formula (3.1) for $\Gamma(z)$ and the special value $\Gamma(1/2) = \sqrt{\pi}$, cf. (3.4), it follows after a small calculation that

$$\Gamma\left(n+k+\frac{1}{2}+1\right) = \frac{\sqrt{\pi}}{2^{2n+2k+1}} \cdot \frac{(2n+2k+1)!}{(n+k)!}.$$

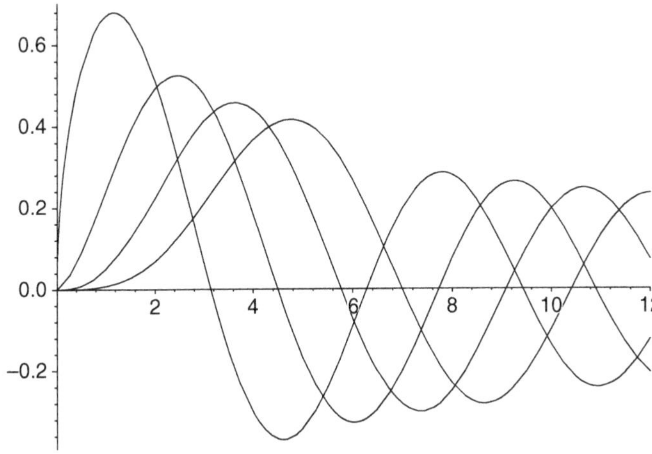

Figure 3.19: The Bessel functions $J_{0.5}(x)$, $J_{1.5}(x)$, $J_{2.5}(x)$, $J_{3.5}(x)$

By using this expression (3.106) is reduced to

$$J_{n+1/2}(r) = \sqrt{\frac{2r}{\pi}} \cdot (2r)^n \sum_{k=0}^{\infty} (-1)^k \cdot \frac{(n+k)!}{k!(2n+2k+1)!} r^{2k}. \tag{3.107}$$

If one does not have a programme for the Bessel functions, (3.107) is more convenient to use in numerical calculations than (3.106). For the lowest relevant index $n = 0$ formula (3.107) becomes particular nice,

$$J_{1/2}(r) = \sqrt{\frac{2}{\pi}} \cdot \frac{\sin r}{\sqrt{r}} \quad \text{for } r > 0. \tag{3.108}$$

Notice that the general series (3.107) is alternating, so an error estimate of any truncated series is given by the absolute value of the first term which is excluded, provided that the truncation takes place at a step, where the absolute values of the terms are decreasing.

It follows from the solution formulæ of Chapter 4 that in practice $0 \leq r < 25$, if the error of the solution should be $< 10^{-5}$. Thus, for any given $n \in \mathbb{N}$ and $r \in [0, 25]$ it is not hard to find the number of terms needed from (3.107) guaranteeing a final error of $< 10^{-5}$, say.

Remark 3.4.1 The upper bound $r < 25$ for r was found when some numerical examples were worked out. This must implicitly have been known in the older literature. For instance, Abramowitz *et al.* (1965), give only tables of the zeros $j_{n+1/2,m}$ smaller than 25 with no hints of that these are sufficient in most cases.

Since $j_{\nu,m}$ and $J_\nu'(j_{\nu,m})$ enter some of the solution formulæ of Chapter 4, we mention here that tables of these numbers can be found in Abramowitz *et al.* (1965), e.g.

1. Table 10.6 for $\nu = n + 1/2$, $n \in \mathbb{N}_0$, and $j_{\nu,s} < 25$,

2. Table 9.5 for $\nu = n$, $n \in \mathbb{N}_0$, and $j_{n,s} < 62$.

For larger zeros apply e.g. Abramowitz *et al.* (1965), formula 9.5.12 sqq. $\Diamond$

3.5 Other useful functions

The special functions considered so far are quite natural when we consider the diffusion equation in general, and most of them have been known for more than a century. When we specialize to chloride ingress into concrete we may expect that the simple geometry of the most often used domains of Fick's second law creates some patterns of the solutions which could be described more easily by introducing some new functions. This principle was implicitly used in Mejlbro (1996), when the $\Psi_p(u)$ functions were introduced.

In this section we list some of these functions. They do not involve much mathematics, so we only define them and sketch a couple of their properties. Tables of them are given at the end of this book. Their convenience will be made clear in Chapter 4.

3.5.1 The function $H(\xi,\tau)$

Using classical Fourier analysis it is easy to expand the constant 1 over the interval $]\,0,1\,[$ as a sinus series,

$$1 = \frac{4}{\pi}\sum_{n=0}^{\infty}\frac{1}{2n+1}\sin((2n+1)\pi\xi), \quad \xi \in]\,0,1\,[. \tag{3.109}$$

Riemann's theorem assures that (3.109) is indeed point-wise convergent with sum 1 for $\xi \in]\,0,1\,[$. The convergence is *not* uniform, and the value of the series is obviously 0 at the end points of the interval.

Define

$$H(\xi,\tau) = \frac{4}{\pi}\sum_{n=0}^{\infty}\frac{1}{2n+1}\exp\left(-(2n+1)^2\pi^2\tau\right)\sin((2n+1)\pi\xi) \tag{3.110}$$

for $0 \leq \xi \leq 1$ and $\tau \geq 0$. Then

$$H(\xi,0) = \begin{cases} 1 & \text{for} \quad 0 < \xi < 1, \\ 0 & \text{for} \quad \xi = 0 \text{ or } \xi = 1, \end{cases} \tag{3.111}$$

$$H(0,\tau) = H(1,\tau) = 0, \tag{3.112}$$

$$\frac{\partial H}{\partial \tau} = \frac{\partial^2 H}{\partial \xi^2} \quad \text{for } 0 < \xi < 1 \text{ and } \tau > 0, \tag{3.113}$$

because (3.110) and all its termwise derivatives are absolutely convergent for $\tau > 0$, hence in particular $H \in C^\infty$.

Now, (3.111)–(3.113) are just defining a special initial/boundary value problem for the diffusion equation, so the solution (3.110) describes a substitute of the so-called fundamental solution.

Remark 3.5.1 Since (3.111) is analogous to the usual *Heaviside function* on the real line,

$$H(t) = \begin{cases} 1 \text{ for } t > 0, \\ \\ 0 \text{ for } t \leq 0, \end{cases}$$

we have found it natural to call it $H(\xi, \tau)$. ◊

Obviously,

$$H(\xi, \tau) = H(1 - \xi, \tau) \text{ for } 0 \leq \xi \leq 1 \text{ and } \tau \geq 0, \tag{3.114}$$

so the tables in Chapter 8 are only given for $0 \leq \xi \leq 0.5$.

Numerical examples show that τ mostly satisfies $0.025 \leq \tau \leq 0.14$, when we consider chloride ingress into concrete represented in the model by a square or a cube. In order to cover rectangles and rectangular boxes as well we have extended the tables to the range $0.0005 \leq \tau \leq 0.25$. On a PC or a pocket calculator one may use the following truncated approximation

$$H(\xi, \tau) \approx \frac{4}{\pi} \sum_{n=0}^{N(\tau)} \frac{1}{2n+1} \exp\left(-(2n+1)^2 \pi^2 \tau\right) \sin((2n+1)\pi\xi) \tag{3.115}$$

of absolute error $< 10^{-7}$. Here the upper bound of the summation $N(\tau)$ is decreasing in τ. Some values of $N(\tau)$ are given in Table 3.6.

Table 3.6: Table of $N(\tau)$ in (3.115).

τ	.0001	.0002	.0003	.0004	.0005	.0006	.0007	.0008	.0009
$N(\tau)$	58	41	34	29	27	24	23	21	20
τ	.001	.002	.003	.004	.005	.007	.009	.010	.015
$N(\tau)$	17	13	10	9	8	7	6	5	4
τ	.03	.04	.07	.20					
$N(\tau)$	3	2	1	0					

3.5.2 The function $H_1(\xi,\tau)$

The function $H_1(\xi,\tau)$ is defined by

$$H_1(\xi,\tau) = 1 - \xi - \frac{2}{\pi}\sum_{n=1}^{\infty}\frac{1}{n}\exp\left(-n^2\pi^2\tau\right)\sin(n\pi\xi), \tag{3.116}$$

for $0 \le \xi \le 1$ and $\tau \ge 0$.

When $\tau \ge \tau_0 > 0$ the series (3.116) and its termwise derivatives are all absolutely convergent, and (3.116) satisfies the diffusion equation (3.113). Furthermore,

$$H_1(0,\tau) = 1 \quad \text{and} \quad H_1(1,\tau) = 0 \quad \text{for all } \tau \ge 0, \tag{3.117}$$

$$H_1(\xi,0) = \begin{cases} 1 & \text{for} \quad \xi = 0, \\ 0 & \text{for} \quad 0 < \xi \le 1, \end{cases} \tag{3.118}$$

$$\lim_{\tau\to\infty} H_1(\xi,\tau) = 1 - \xi. \tag{3.119}$$

The function $H(\xi,\tau)$ can be derived from $H_1(\xi,\tau)$, since we have the relation

$$H_1(\xi,\tau) + H_1(1-\xi,\tau) = 1 - H(\xi,\tau). \tag{3.120}$$

Hence, $H(\xi,\tau)$ is redundant, but since it is interesting in itself and easier to explain we have chosen to introduce both functions. Tables have been worked out for $H_1(\xi,\tau)$ in the range $0.0005 \le \tau \le 0.15$. On a PC or a pocket calculator one may use the following approximation of absolute error $< 10^{-7}$,

$$H_1(\xi,\tau) = 1 - \xi - \frac{1}{\pi}\sum_{n=1}^{N_1(\tau)}\frac{1}{n}\exp\left(-n^2\pi^2\tau\right)\sin(n\pi\xi). \tag{3.121}$$

Here, the upper bound of the summation $N_1(\tau)$ is decreasing in τ. Some values of $N_1(\tau)$ are given in Table 3.7.

Table 3.7: Table of $N_1(\tau)$ in (3.121).

τ	.0001	.0002	.0003	.0004	.0005	.0006	.0007	.0008	.0009
$N(\tau)$	110	82	68	59	53	48	45	42	40
τ	.001	.002	.003	.004	.005	.006	.007	.008	.009
$N(\tau)$	38	27	22	19	17	16	15	14	13
τ	.01	.02	.03	.04	.05	.07	.14	.19	.44
$N(\tau)$	12	9	7	6	5	4	3	2	1

Chapter 4

Fick's Second Law for Constant Diffusion Coefficient

We assume in this chapter that the diffusion coefficient $D > 0$ is either a constant or a function of t alone. When $D(t)$ only depends on time, we can define a new time variable, for which the diffusion coefficient is constant, so it suffices here to assume that D is a constant. The purpose of the chapter is to give a catalogue of solution formulæ for the linear model of Fick's second law in a large number of cases, which often will suffice in a first approximation.

Notice that almost every process in nature is nonlinear, so an application of Fick's second law is in principle 'only' the first linear approximation of the description of chloride ingress into concrete. It is, however, a very important part of the solution, and in some sense the real behaviour can be viewed as perturbations from this linear model. For that reason Fick's second law is central for the theory, though not the whole answer.

In principle, Fick's second law defines a mixed initial/boundary value problem for a parabolic partial differential equation. We ease matters in Section 4.1 by proving that the initial conditions are actually of minor importance, so in the rest of the chapter we simply put the initial conditions equal to zero.

In Section 4.2 we present the mathematical theory needed here. Most readers are acquainted with the error function solution, which was generalized by Mejlbro (1996), to the Ψ_p-solutions. The problem with this approach is that it refers to a half-infinite one-dimensional interval, and it cannot easily be generalized to higher dimensions. This does not fit well into the real world, where the objects are described as bounded three-dimensional domains.

The mathematical theory needed here was developed 150 years ago by the two French mathematicians Sturm and Liouville. A full treatment is given by many authors in the first half of the twentieth century, cf. e.g. Courant *et al.* (1931), and Titchmarsh (1946). Another excellent treatment was given by Morse *et al.* (1953), who more or less wrote the latest work presenting this classical theory in almost full detail. The emphasis in these works, however, was on elliptic and hyperbolic equations, while the parabolic case, containing all sorts of diffusion equations including Fick's second law, was relegated to remarks. This was due to the fact that the mathematical formulæ in the parabolic case are very complicated, involving integrals which could not be calculated reasonably explicit in those days. Therefore one could not give convincing practical examples. Apart for the simplest case of Fourier expansions, and the fully developed theory of finite elements in Numerical Analysis, the general theory was almost forgotten during the past fifty years. The theory remained in textbooks of functional analysis without convincing examples, and it might seem that one had come to a dead end.

The present development of symbolic programmes for PCs, and even pocket calculators, now makes it possible to evaluate the integrals involved, so the theory has come to a revival. This explains why we present the theory in Section 4.2, using rectangular and polar coordinates as examples. In Section 4.3 we continue with a large catalogue of explicit solutions of Fick's second law. The proofs are omitted; they just follow the same pattern as given in Section 4.2, and the reader should be able to produce them all by copying the methods.

In this connection one should also mention Cannon (1984), who gave a very elegant treatment of the one-dimensional heat equation, including nonlinear versions. The book is written so perfectly that without other sources one cannot imagine how this theory can be extended to higher dimensions. Some of the formulæ in Cannon (1984), are quite scaring, if one's main field is not mathematics. For a mathematician, however, this book is really inspiring.

Concerning chloride ingress into concrete we have felt that rectangular boxes and circular columns or pipes are the most important cases. Therefore we have limited ourselves to rectangular, polar and semi-polar coordinate systems. Morse *et al.* (1953), give a full list of 14 coordinate systems, for which the methods of Section 4.2 can be applied, and an additional nonstandard case, in which the coordinates cannot be separated, though Fick's second law can be solved. The first coordinate system on their list, which is not treated here, is the spherical one, which is referring to e.g. a spherical half shell of concrete. The mathematics involved here is very computational, to put it mildly, and we do not feel that this case is in the main stream, as far as chloride penetration is concerned. Therefore we do not treat the spherical coordinates or the even more complicated coordinate systems listed by Morse *et al.* (1953).

Readers who are interested in more complicated problems may also consult Budak *et al.* (1964), in particular Chapter III and Chapter V. This book is a source book on problems with sketched solutions.

The emphasis of this book is laid on Fickean diffusion of chloride into concrete. One may in this connection also consult the classical book of Carslaw *et al.* (1947). This book is formally concerned with the *heat equation*, but since the heat equation and Fick's second law for chloride ingress from a mathematical point of view are identical, the solutions are the same. When reading this book one only has to translate "conduction of heat" into "chloride ingress", and the theory applies as well.

We have not considered non-Fickean or anomalous diffusion occurring mainly in plastics. Nevertheless we would like to mention the excellent book of Crank (1975), on these matters. Combining the theoretical work of Crank (1975), with the mathematical and numerical methods of the present book one will be able to solve quite a few problems. This point of view opens a vast research area which lies far beyond the aim of this book, so we shall not go further into it.

4.1 The general initial/boundary value problem

Let Ω be a bounded domain in $\mathbb{R}^n$, $n = 1, 2, 3$, with a piecewise smooth boundary $\partial\Omega$. Let $\nabla^2 = \partial^2/\partial x_1^2 + \cdots + \partial^2/\partial x_n^2$ denote Laplace's differential operator in the space variables. We assume for simplicity that the diffusion coefficient $D > 0$ is a constant, though the considerations below can be extended to a more general diffusion coefficient.

Under these assumptions the general initial/boundary value problem for Fick's second law over Ω can be written

$$\left\{ \begin{array}{lll} \dfrac{1}{D}\dfrac{\partial C}{\partial t} = \nabla^2 C(\mathbf{x}, t), & \mathbf{x} \in \Omega,\ t > 0, & \text{(Fick's second law)}, \\ C(\mathbf{x}, 0) = f(\mathbf{x}), & \mathbf{x} \in \Omega,\ t = 0, & \text{(initial condition)}, \\ C(\mathbf{x}, t) = g(\mathbf{x}, t), & \mathbf{x} \in \partial\Omega,\ t > 0, & \text{(boundary condition)}. \end{array} \right. \tag{4.1}$$

The purpose of this section is to show that the initial condition is of minor importance for the solution $C(\mathbf{x}, t)$ in the sense that its contribution will die out in time. The trick is easy. Due to the linearity of Fick's second law, (4.1) can be split into a sum of a pure initial value problem over the whole space and a boundary value problem, where the initial condition is put equal to zero. More precisely, the solution of (4.1) can be written

$$C(\mathbf{x}, t) = C_1(\mathbf{x}, t) + C_2(\mathbf{x}, t),$$

where $C_1(\mathbf{x}, t)$ is the solution of the initial value problem in the *full space* $\mathbb{R}^n$,

$$\left\{ \begin{array}{ll} \dfrac{1}{D}\dfrac{\partial C_1}{\partial t} = \nabla^2 C_1(\mathbf{x}, t), & \mathbf{x} \in \Omega^o,\ t > 0, \\ C_1(\mathbf{x}, 0) = \left\{ \begin{array}{l} f(\mathbf{x}), \\ 0, \end{array} \right. & \begin{array}{l} \mathbf{x} \in \Omega^o,\ t = 0, \\ \mathbf{x} \notin \Omega^o,\ t = 0, \end{array} \end{array} \right. \tag{4.2}$$

where Ω^o denotes the interior of Ω, and where $C_2(\mathbf{x}, t)$ is the solution of the special boundary value problem,

$$\begin{cases} \dfrac{1}{D}\dfrac{\partial C_2}{\partial t} = \nabla^2 C_2(\mathbf{x},t), & \mathbf{x} \in \Omega^o,\ t > 0, \\ C_2(\mathbf{x},0) = 0, & \mathbf{x} \in \Omega^o,\ t = 0, \\ C_2(\mathbf{x},t) = g(\mathbf{x},t) - C_1(\mathbf{x},t), & \mathbf{x} \in \partial\Omega,\ t > 0. \end{cases} \tag{4.3}$$

Notice that the solution $C_1(\mathbf{x},t)$ of (4.2) enters the boundary condition of (4.3), so (4.2) must be solved first.

Intuitively, $C_1(\mathbf{x},t) \to 0$ for $t \to \infty$, because initially we only have a finite amount of chloride ions in Ω. By diffusion these ions will spread to a uniform distribution over the whole space for $t \to \infty$, i.e. of density zero. Fortunately, this is one of the cases, where the one-dimensional methods given by e.g. Cannon (1984), or John (1971), can be generalized to higher dimensions. This is due to the fact that the initial value problem can be considered on the full space $\mathbb{R}^n$ in (4.2).

Let us first consider the one-dimensional case. Let $\xi \in \mathbb{R}$ be a parameter and define

$$K(x-\xi,t) = \frac{1}{\sqrt{4\pi t}} \exp\left(-\frac{(x-\xi)^2}{4t}\right), \quad x \in \mathbb{R},\ t > 0.$$

A simple check shows that

$$\frac{\partial K}{\partial t}(x-\xi,t) - \frac{\partial^2 K}{\partial x^2}(x-\xi,t) = 0, \tag{4.4}$$

so $K(x-\xi,t)$ satisfies Fick's second law in one dimension.

By the change of variable $u = (x-\xi)/\sqrt{4t}$ we get

$$\int_{-\infty}^{\infty} K(x-\xi,t)\, d\xi = 1 \quad \text{for } t > 0,$$

hence by continuity also for $t \to 0+$. Our guideline here is

$$\lim_{t\to 0+} K(x-\xi,t) = \lim_{t\to 0+} \frac{1}{\sqrt{4\pi t}} \exp\left(-\frac{(x-\xi)^2}{4t}\right) = \delta_x,$$

where δ_x is Dirac's delta "function" at the point x. Using the theory of generalized functions this idea gives formally,

$$\lim_{t\to 0+} \int_{-\infty}^{\infty} f(\xi)K(x-\xi,t)d\xi = \text{``}\int_{-\infty}^{\infty} f(\xi)K(x-\xi,0)d\xi\text{''} = \delta_x(f) = f(x),$$

when f is continuous.

The theory of generalized functions cannot be assumed here, so although the proof above can be made mathematically correct, we shall here only use it as an *inspiration* for the following definition:

For $t > 0$ we put

$$\begin{aligned} C(x,t) &= \int_{-\infty}^{\infty} f(\xi) K(x-\xi,t)\, d\xi \\ &= \frac{1}{\sqrt{4\pi t}} \int_{-\infty}^{\infty} f(\xi) \exp\left(-\frac{(x-\xi)^2}{4t}\right) d\xi \\ &= \frac{1}{\sqrt{4\pi t}} \int_{-\infty}^{\infty} f(x-\xi) \exp\left(-\frac{\xi^2}{4t}\right) d\xi \\ &= \frac{1}{\sqrt{\pi}} \int_{-\infty}^{\infty} f(x-\sqrt{4t}\cdot\eta) \exp(-\eta^2)\, d\eta. \end{aligned} \tag{4.5}$$

Since x and t only occur as *parameters* in (4.5), we can differentiate (4.5) under the sign of integration with respect to x and t. Using (4.4) we get

$$\frac{\partial C}{\partial t} - \frac{\partial^2 C}{\partial x^2} = \int_{-\infty}^{\infty} f(\xi) \left\{ \frac{\partial K}{\partial t}(x-\xi,t) - \frac{\partial^2 K}{\partial x^2}(x-\xi,t) \right\} d\xi = 0,$$

proving that $C(x,t)$ satisfies Fick's second law. When f is continuous at x, the last equality of (4.5) leads to

$$\begin{aligned} C(x,0) &= \lim_{t\to 0+} \frac{1}{\sqrt{\pi}} \int_{-\infty}^{\infty} f(x-\sqrt{4t}\cdot\eta) \exp(-\eta^2)\, d\eta \\ &= f(x) \cdot \frac{1}{\sqrt{\pi}} \int_{-\infty}^{\infty} \exp(-\eta^2)\, d\eta = f(x). \end{aligned}$$

We conclude that the solution of (4.2) for $n = 1$ is given by

$$C(x,t) = \frac{1}{\sqrt{4\pi t}} \int_{-\infty}^{\infty} f(\xi) \exp\left(-\frac{(x-\xi)^2}{4t}\right) d\xi, \quad t > 0, x \in \mathbb{R}.$$

The total amount of chloride ions is finite,

$$M = \int_{-\infty}^{\infty} f(\xi)\, d\xi < \infty.$$

Hence,

$$C(x,t) \le \frac{1}{\sqrt{4\pi t}} \int_{-\infty}^{\infty} f(\xi)\, d\xi = \frac{M}{\sqrt{4\pi t}} \to 0 \quad \text{for } t \to \infty,$$

and we have proved that the solution of (4.2) dies out in time, when $D = 1$. The general result follows by the chance of variable, $\tau = D \cdot t$.

When we generalize to higher dimensions, just notice that for $n = 2$ the product

$$K(x-\xi,t) \cdot K(y-\eta,t) = \left(\frac{1}{\sqrt{4\pi t}}\right)^2 \exp\left(-\frac{(x-\xi)^2 + (y-\eta)^2}{4t}\right)$$

is a solution of

$$\frac{\partial u}{\partial t} = \frac{\partial^2 u}{\partial x^2} + \frac{\partial^2 u}{\partial y^2}.$$

By copying the approach in the one-dimensional case we see that

$$C_1(x, y, t) = \left(\frac{1}{\sqrt{4\pi t}}\right)^2 \int\int_\Omega f(\xi, \eta) \exp\left(-\frac{(x-\xi)^2 + (y-\eta)^2}{4t}\right) \delta\xi d\eta$$

is indeed a solution of

$$\begin{cases} \dfrac{\partial C_1}{\partial t} = \bigtriangledown^2 C_1 \\ C_1(x, y, 0) = f(x, y) \qquad \text{for } (x, y) \in \Omega. \end{cases}$$

We also have the estimate

$$C_1(x, y, t) \leq \frac{M}{4\pi t} \to 0 \qquad \text{for } t \to \infty,$$

where $M = \int\int_\Omega f(\xi, \eta)\, d\xi d\eta$ is the total amount of chloride ions, so this solution dies out for $t \to \infty$.

The generalization to three dimensions is straightforward.

The conclusion of this section is that the initial conditions in most cases are of minor importance for the solution of Fick's second law. We shall therefore in the rest of this chapter always assume that $C(\mathbf{x}, 0) = 0$, leaving the necessary modifications to the reader in the rare cases, where this cannot be assumed.

4.2 Eigenfunction expansions

The core of the mathematical theory is given in this section. The idea is to choose a coordinate system, such that the bounded domain Ω under consideration in the new variables is described as a rectangle (two dimensions) or as a rectangular box (three dimensions). Then by separating the (new) variables we find the eigenfunctions for $\bigtriangledown^2$ over the new domain, and finally we show that the solution of Fick's second law can be built up by these eigenfunctions.

The theory will be described by examples using either the usual rectangular coordinates in the plane, or the polar coordinates in the plane,

$$x = r\cos\varphi, \qquad y = r\sin\varphi.$$

In the latter case we consider

A a half disc, $\quad r \in]\,0, R\,[,\ \varphi \in]\,0, \pi\,[,$

B a full disc, $\quad r \in [0, R\,[,\ \varphi \in [0, 2\pi\,[\,,$

C a sector of a disc, $\quad r \in]\,0, r\,[,\ \varphi \in]\,0, \alpha\,[,\ 0 < \alpha \leq 2\pi.$

Notice that case A is included in case C for $\alpha = \pi$, while case B is *not* included in case C for $\alpha = 2\pi$. The reason is that $\alpha = 2\pi$ in case C represents a crack along a radius, while case B does not contain a crack. Therefore, the structure of the solution of case B is different from the structure of the solution of case A and case C. Case A has only been chosen, because the solution here is particular nice without too many disturbing strange constants.

4.2.1 Helmholtz's equation

In rectangular coordinates in the plane the gradient of a function $f(x, y)$ is denoted by

$$\bigtriangledown f = \left(\frac{\partial f}{\partial x}, \frac{\partial f}{\partial y}\right), \quad \text{i.e.} \quad \bigtriangledown = \left(\frac{\partial}{\partial x}, \frac{\partial}{\partial y}\right),$$

so $\bigtriangledown$ is a two-dimensional differential operator of first order. By formally taking the dot product we get

$$\bigtriangledown^2 = \bigtriangledown \cdot \bigtriangledown = \frac{\partial^2}{\partial x^2} + \frac{\partial^2}{\partial y^2},$$

which is exactly Laplace's differential operator in the space variables. This explains the notation.

Suppose that we want to find the value of the integral

$$\int\!\!\int_\Omega \bigtriangledown^2 u \cdot v \, dxdy,$$

where we know u, $\bigtriangledown^2 v$ and the boundary values of u and v. Then by two partial integrations (four, if one goes through all the details),

$$\int\!\!\int_\Omega \bigtriangledown^2 u \cdot v \, dx \, dy = \int\!\!\int_\Omega u \cdot \bigtriangledown^2 v \, dx \, dy + \text{ boundary terms.}$$

This reformulation becomes particular simple, when v is an *eigenfunction* of the operator $\bigtriangledown^2$, i.e. v is not identical zero on Ω, though $v(x, y) = 0$ on the boundary, and

$$\bigtriangledown^2 v = \mu v$$

for some constant μ. In technical literature the eigenvalue μ is usually denoted by $-\lambda$. Using this change in terminology we see that we shall solve *Helmholtz's equation*

$$\bigtriangledown^2 v + \lambda v = 0 \quad \text{in } \Omega, \qquad v(x, y) = 0 \quad \text{on the boundary} \tag{4.6}$$

for some constant λ. We shall here omit the simple proof of the fact that if $v \neq 0$ is a solution of (4.6), then the boundary value $v(x, y) = 0$ forces $\lambda > 0$.

4.2.2 Change of coordinates in the operator $\bigtriangledown^2$

For simplicity we shall only consider the two-dimensional case with constant diffusion coefficient $D > 0$. The extension to three dimensions is straightforward.

Let Ω be a bounded domain in the plane. We shall develop a method to solve the boundary value problem

$$\begin{cases} \dfrac{1}{D}\dfrac{\partial C}{\partial t} = \bigtriangledown^2 C(x,y,t), & (x,y) \in \Omega^o \text{ and } t > 0, \\ C(x,y,t) = f(x,y,t), & (x,y) \in \partial\Omega \text{ and } t > 0, \\ C(x,y,0) = 0, & (x,y) \in \Omega^o. \end{cases} \tag{4.7}$$

If e.g. $\Omega = \,]\,0, A\,[\, \times \,]\,0,\, B[$, the space domain is already a rectangle, and we proceed to the next subsection. The only other case of interest for us is when the space domain is changed to a rectangle by using polar coordinates like in the cases A, B and C. There exist other coordinate systems, in which the methods given here can be applied, cf. e.g. Morse *et al.* (1953), but we feel that the rectangular and polar coordinate systems are the most important ones, so these are given a full treatment.

The change of coordinate system also changes the operator $\bigtriangledown^2$. It is well-known, or easily derived, that in polar coordinates,

$$\bigtriangledown^2 \psi(r,\varphi) = \frac{\partial^2 \psi}{\partial r^2} + \frac{1}{r}\frac{\partial \psi}{\partial r} + \frac{1}{r^2}\frac{\partial^2 \psi}{\partial \varphi^2}. \tag{4.8}$$

4.2.3 Separation of the variables

We shall for a while only consider Helmholtz's equation

$$\bigtriangledown^2 \psi(u,v) + \lambda \psi(u,v) = 0,$$

where $\psi(u,v) = 0$ on the boundary of the rectangular parameter domain $[0,A] \times [0,B]$, and where the structure of $\bigtriangledown^2$ depends on the choice of coordinate system. We shall search for solutions of the special structure $\psi(u,v) = f(u)g(v)$, i.e. where the variables are separated. In this case the boundary conditions become very simple,

$$f(0) = f(A) = 0 \qquad \text{and} \qquad g(0) = g(B) = 0.$$

The method will be illustrated by the following examples.

Example 4.2.1 *Rectangular coordinates,* $\Omega = [0,A] \times [0,B]$. In this case Helmholtz's equation is written

$$\frac{\partial^2 \psi}{\partial x^2} + \frac{\partial^2 \psi}{\partial y^2} + \lambda \psi(x,y) = 0.$$

Assuming that $\psi(x, y) = f(x)g(y)$ we get

$$g(y)f''(x) + f(x)g''(y) + \lambda f(x)g(y) = 0.$$

Since also $f(x)g(y) \neq 0$ almost everywhere, we can divide by $f(x)g(y)$, thus obtaining

$$\frac{f''(x)}{f(x)} + \frac{g''(y)}{g(y)} + \lambda = 0.$$

This expression is constant in x and y, so there exist constants c_1 and c_2, such that

$$\frac{f''(x)}{f(x)} = -c_1 \quad \text{and} \quad \frac{g''(y)}{g(y)} = -c_2 \quad \text{and} \quad \lambda = c_1 + c_2.$$

Hence the problem is reduced to the two ordinary second order differential equations

$$\begin{aligned} f''(x) + c_1 f(x) &= 0, \qquad f(0) = f(A) = 0, \\ g''(y) + c_2 g(y) &= 0, \qquad g(0) = g(B) = 0. \end{aligned}$$

By analyzing the complete solutions of the differential equations it is seen that nontrivial solutions only exist, when

$$\begin{aligned} c_{1,m} &= \left(\frac{m\pi}{A}\right)^2, \quad \text{with e.g.} \quad f_m(x) = \sin\left(\frac{m\pi x}{A}\right), \qquad m \in \mathbb{N}, \\ c_{2,m} &= \left(\frac{n\pi}{B}\right)^2, \quad \text{with e.g.} \quad g_n(y) = \sin\left(\frac{n\pi y}{B}\right), \qquad n \in \mathbb{N}, \end{aligned}$$

where we have indexed the infinitely many solutions. This gives the doubly indexed eigenfunctions and eigenvalues of Helmholtz's equation

$$\begin{aligned} \psi_{mn}(x, y) &= \sin\left(\frac{m\pi x}{A}\right) \sin\left(\frac{n\pi y}{B}\right), \\ \lambda_{mn} &= \left(\frac{m\pi}{A}\right)^2 + \left(\frac{n\pi}{B}\right)^2, \qquad m, n \in \mathbb{N}. \end{aligned} \tag{4.9}$$

Obviously, the number of indices is equal to the dimension.

From the classical theory of Fourier series we also know that the system (4.9) of eigenfunctions $\psi_{mn}(x, y)$, m, $n \in \mathbb{N}$, is complete in the following sense: Every continuously differentiable function $f(x, y)$ on $[0, A] \times [0, B]$ has a *unique* expansion

$$\begin{aligned} f(x, y) &= \sum_{m=1}^{\infty} \sum_{n=1}^{\infty} a_{mn} \psi_{mn}(x, y) \\ &= \sum_{m=1}^{\infty} \sum_{n=1}^{\infty} a_{mn} \sin\left(\frac{m\pi x}{A}\right) \sin\left(\frac{n\pi y}{B}\right), \end{aligned} \tag{4.10}$$

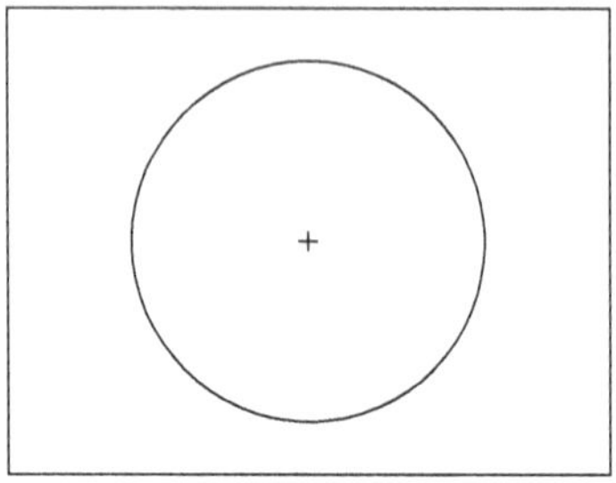

Figure 4.1: Polar coordinates, case A.

where

$$\begin{aligned} a_{mn} &= \frac{2}{A}\cdot\frac{2}{B}\int_0^B\int_0^B f(x,y)\sin\left(\frac{m\pi x}{A}\right)\sin\left(\frac{n\pi y}{B}\right)\,dx\,dy \\ &= \frac{4}{AB}\int\int_\Omega f(x,y)\psi_{mn}(x,y)\,dx\,dy, \end{aligned}$$

and where (4.10) even is true point-wise in the interior of Ω. On the boundary it does not hold in general, because the double series here is trivially zero. $\diamondsuit$

It can be proved that this is true in general for eigenfunction expansions based on Helmholtz's equation. We shall take this for granted in the following. For a proof, cf. e.g. Titchmarsh (1946).

Example 4.2.2 *Polar coordinates*, case A, i.e. the half disc, $\Omega = \{(r,\varphi) \mid 0 \le r \le R, 0 \le \varphi \le \pi\}$. Helmholtz's equation becomes by (4.8),

$$\frac{\partial^2\psi}{\partial r^2} + \frac{1}{r}\frac{\partial\psi}{\partial r} + \frac{1}{r^2}\frac{\partial^2\psi}{\partial\varphi^2} + \lambda\psi(r,\varphi) = 0.$$

If $\psi(r,\varphi) = f(r)g(\varphi)$, this equation is transferred into

$$f''(r)g(\varphi) + \frac{1}{r}f'(r)g(\varphi) + \frac{1}{r^2}f(r)g''(\varphi) + \lambda f(r)g(\varphi) = 0.$$

Division by $f(r)g(\varphi)/r^2$ (defined and $\neq 0$ almost everywhere) gives

$$\left\{r^2\frac{f''(r)}{f(r)} + r\frac{f'(r)}{f(r)} + \lambda r^2\right\} + \frac{g''(\varphi)}{g(\varphi)} = 0.$$

This expression is constant in r and φ, so there is a constant a, such that

$$r^2\frac{f''(r)}{f(r)} + r\frac{f'(r)}{f(r)} + \lambda r^2 = a \quad\text{and}\quad \frac{g''(\varphi)}{g(\varphi)} = -a. \tag{4.11}$$

Adding the boundary conditions to the latter equation,

$$g''(\varphi) + ag(\varphi) = 0, \qquad g(0) = g(\pi) = 0,$$

we see that nontrivial solutions only exist, when $a = n^2$, $n \in \mathbb{N}$, so modulo a constant factor,

$$g_n(\varphi) = \sin n\varphi, \qquad n \in \mathbb{N}. \tag{4.12}$$

For fixed $n \in \mathbb{N}$, put $a = n^2$ in the first equation of (4.11) and add the boundary condition,

$$r^2 \frac{d^2 f}{dr^2} + r\frac{df}{dr} + (\lambda r^2 - n^2) f(r) = 0, \quad f(R) = 0. \tag{4.13}$$

A small argument, which shall not be given here, shows that nontrivial solutions of (4.13) can only exist when $\lambda > 0$. Assuming $\lambda > 0$ we substitute $\rho = \sqrt{\lambda} \cdot r$, by which (4.13) is transferred into the Bessel equation

$$\rho^2 F''(\rho) + \rho F'(\rho) + (\rho^2 - n^2) F(\rho) = 0,$$

where $F(\rho) = F(\sqrt{\lambda} \cdot r) = f(r)$, cf. Section 3.4. Since $F(\rho)$ must be defined for $\rho = 0$, the possible solution $Y_n(\rho)$ is excluded, so

$$f(r) = F(\rho) = J_n(\rho) = J_n(\sqrt{\lambda} \cdot r), \quad n \in \mathbb{N}.$$

Obviously, $F(0) = 0$. The boundary condition gives

$$f(R) = J_n(\sqrt{\lambda} \cdot R) = 0.$$

Hence, λ is determined by the zeros $j_{n,m}$ of $J_n(\rho)$, cf. Section 3.4.3, i.e. (putting indices on λ),

$$\sqrt{\lambda_{mn}} R = j_{n,m}, \qquad \text{or} \qquad \lambda_{mn} = \left(\frac{1}{R} j_{n,m}\right)^2, \quad m, n \in \mathbb{N},$$

so

$$f_{mn}(r) = J_n\left(\sqrt{\lambda_{mn}} r\right) = J_n\left(j_{n,m} \frac{r}{R}\right), \qquad m, n \in \mathbb{N}.$$

Combining this result with (4.2:7) we obtain the doubly indexed set of eigenfuntions,

$$\psi_{mn}(r, \varphi) = J_n\left(j_{n,m} \frac{r}{R}\right) \sin n\varphi, \quad m, n \in \mathbb{N}, \tag{4.14}$$

with the corresponding eigenvalues

$$\lambda_{mn} = \left(\frac{1}{R} j_{n,m}\right)^2, \quad m, n \in \mathbb{N}. \quad \diamond$$

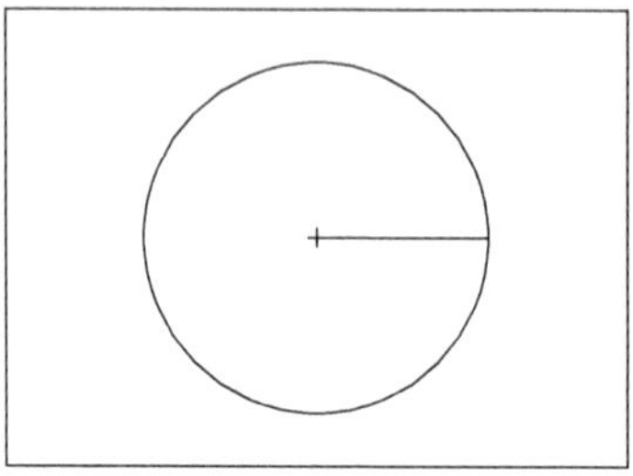

Figure 4.2: Polar coordinates, case B.

Example 4.2.3 *Polar coordinates*, case B, i.e. Ω is a disc, $\Omega = \{(r,\varphi) \mid 0 \leq r \leq R, 0 \leq \varphi \leq 2\pi\}$. By separation of the variables we obtain (4.11) in Example 4.2.2. In the present case, however, we do not have boundary conditions for $g(\varphi)$. Instead we claim that it is periodic,

$$g(0) = g(2\pi).$$

This can be interpreted as if the boundary conditions have been glued together. By solving

$$g''(\varphi) + ag(\varphi) = 0, \quad g(0) = g(2\pi),$$

we get the usual trigonometric system known from the classical theory of Fourier series,

$$g_0(\varphi) = 1, \quad g_{n,1}(\varphi) = \cos n\varphi, \quad g_{n,2}(\varphi) = \sin n\varphi, \quad n \in \mathbb{N}.$$

A straightforward, though tedious calculation shows that (4.13) still holds, even extended to $n = 0$ for $g_0(\varphi)$. By copying Example 4.2.2 we get the three subfamilies of eigenfunctions,

$$\begin{aligned}
\psi_{m0}(r,\varphi) &= J_0\left(j_{0,m}\frac{r}{R}\right), & \lambda_{m0} &= \left(\frac{1}{R}j_{0,m}\right)^2, & m \in \mathbb{N}, n = 0,\\
\psi_{mn1}(r,\varphi) &= J_n\left(j_{n,m}\frac{r}{R}\right)\cos n\varphi, & \lambda_{mn1} &= \left(\frac{1}{R}j_{n,m}\right)^2, & m,n \in \mathbb{N},\\
\psi_{mn2}(r,\varphi) &= J_n\left(j_{n,m}\frac{r}{R}\right)\sin n\varphi, & \lambda_{mn2} &= \left(\frac{1}{R}j_{n,m}\right)^2, & m,n \in \mathbb{N}.
\end{aligned}$$

Notice the notational mess of three subfamilies. This could be avoided by using the complex basis $\exp(in\varphi)$, $n \in \mathbb{Z}$, for the g-functions. We have not done this, because it would cause some other difficulties. The necessity of three subfamilies already indicates the difference between case B and the cases A and C. $\diamondsuit$

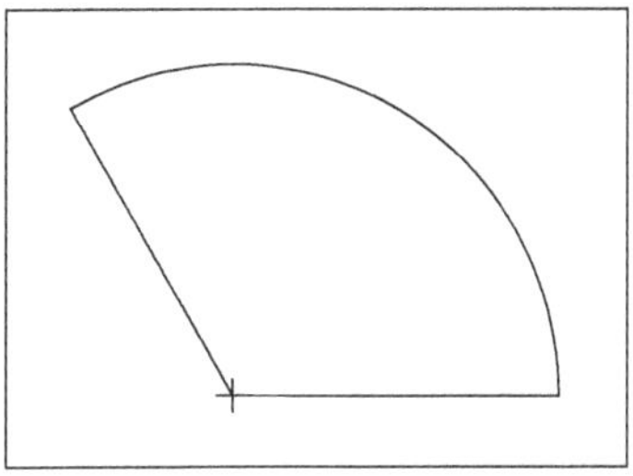

Figure 4.3: Polar coordinates, case C.

Example 4.2.4 *Polar coordinates*, case C, i.e. a sector of a disc $\Omega = \{(r, \varphi) \mid 0 \le r \le R, 0 \le \varphi \le \alpha\}$, where $0 < \alpha \le 2\pi$. Just copy Example 4.2.2 with $n \in \mathbb{N}$ replaced by

$$k_n = \frac{n\pi}{\alpha}, \qquad n \in \mathbb{N}.$$

The nontrivial solutions of Helmholtz's equation

$$\nabla^2 \psi_{mn}(r, \varphi) + \lambda_{mn} \psi_{mn}(r, \varphi) = 0$$

are

$$\psi_{mn}(r, \varphi) = J_{k_n}\left(j_{k_n,m}\frac{r}{R}\right) \sin(k_n \varphi),$$

$$\lambda_{mn} = \left(\frac{1}{R} j_{k_n,m}\right)^2, \quad m, n \in \mathbb{N}. \tag{4.15}$$

We allow $\alpha = 2\pi$, which is *not* the same as case B. In fact, $\alpha = 2\pi$ represents a disc with a crack along a radius, hence not a full disc. The mathematical difference is obvious. For $\alpha = 2\pi$ we get the nice double indexed family (4.15), while we in case B have a more complicated system consisting of three subfamilies. ◊

There are other cases, which should be mentioned here, e.g. polar coordinates for an annulus $\Omega = \{(r, \varphi) \mid a \le r \le b, 0 \le \varphi \le 2\pi\}$. These cases will be dealt with in Section 4.3.

4.2.4 Eigenfunction expansions

Once the eigenfunctions $\psi_{mn}(u, v)$ have been determined we turn to the eigenfunction expansion of a function, i.e. an expansion in a generalized Fourier series. This has already been illustrated in Example 4.2.1, where (4.10) gives an ordinary two dimensional unique expansion of $f(x, y)$ as a Fourier series.

The underlying general mathematical theory is based on Hilbert spaces, which are roughly speaking generalizations of Euclidean spaces to infinite dimensions, preserving as much as possible of the geometrical language. Thus we shall define an inner product on the functions under consideration which formally has the same properties as the inner product in Euclidean spaces. During this process, the functions will play the role of abstract vectors. Once the inner product has been identified we can talk about orthogonal bases, just like we did in the case of Euclidean spaces. Omitting the harder mathematical details of completeness of the basis the main result can be formulated in the following way:

- The total set of eigenfunctions of Helmholtz's equation is exactly the orthogonal basis designed particular for the coordinate system under consideration.

The implication of this observation is that the solution of Fick's second law can be reduced to the solution of a system of linear ordinary differential equations of first order for the coordinate functions of the eigenfunction expansion. This solves the problem in principle, though the practical calculations may be quite large.

We shall start by introducing the *inner product.* Let $x = x(u, v)$, $y = y(u, v)$ be a coordinate transform, which transforms the bounded domain $(x, y) \in \Omega$ in rectangular coordinates into a rectangle $(u, v) \in \tilde{\Omega}$ in the new coordinates. The rectangle is needed for separating the variables, when we solve Helmholtz's equation. The cost is that the area element by this transform is given by

$$dx\, dy = \left| \frac{\partial(x, y)}{\partial(u, v)} \right| du\, dv = w(u, v) du\, dv,$$

where the *weight function* $w(u, v)$ is the absolute value of the Jacobian

$$\frac{\partial(x, y)}{\partial(u, v)} = \left| \begin{array}{cc} \partial x / \partial u & \partial x / \partial v \\ \partial y / \partial u & \partial y / \partial v \end{array} \right| .$$

If we put $F(x, y) = F(x(u, v), y(u, v)) = f(u, v)$ by this transform, then

$$\int\!\!\int_{\Omega} F(x, y)\, dx\, dy = \int\!\!\int_{\tilde{\Omega}} f(u, v) w(u, v)\, du\, dv.$$

All the calculations are carried over to the rectangle $\tilde{\Omega}$, where we have an eigenfunction expansion, at the cost of an extra weight function $w(u, v)$.

From a physical point of view, integrals of the type

$$\int\!\!\int_{\Omega} F(x, y)^2 dx\, dy = \int\!\!\int_{\tilde{\Omega}} f(u, v)^2 w(u, v)\, du\, dv \tag{4.16}$$

may represent an energy, and we only consider systems of finite energy. Hence, we shall only allow functions $f(u, v)$, for which (4.16) is finite. From a mathematical point of view these functions are exactly the elements of the *Hilbert space* $L^2_w(\tilde{\Omega})$, once the *inner product* $\langle\,\cdot,\cdot\,\rangle$ has been defined. This is done by

$$\langle f,g\rangle = \iint_{\tilde{\Omega}} f(u,v)g(u,v)w(u,v)\,du\,dv \tag{4.17}$$

with the corresponding *norm* $\|\cdot\|$ given by

$$\|f\|^2 = \langle f,f\rangle = \iint_{\tilde{\Omega}} f(u,v)^2 w(u,v)\,du\,dv, \qquad \|f\| \geq 0,$$

i.e. the norm is the square root of (4.16), and it is finite in $L^2_w(\tilde{\Omega})$.

We omit the details of proving that (4.17) has the usual properties of an inner product. We just notice that we again find Cauchy-Schwarz's inequality (cf. Section 3.3.3) in the form

$$|\langle f,g\rangle| \leq \|f\|\cdot\|g\|.$$

Lemma 4.2.1 Orthogonality of eigenfunctions. *If $\psi_{mn}(u,v)$ are the set of all the eigenfunctions of ∇^2 in the parameter space, and each λ_{mn} corresponds to just one $\psi_{mn}(u,v)$, i.e. each λ_{mn} is simple, then*

$$\langle\psi_{mn},\psi_{pq}\rangle = \begin{cases} \|\psi_{mn}\|^2 & \text{for } (p,q)=(m,n), \\ 0 & \text{for } (p,q)\neq(m,n). \end{cases}$$

PROOF. Since ψ_{mn} and ψ_{pq} are eigenfunctions, they are zero on the boundary of $\tilde{\Omega}$. The structure of ∇^2 in the (u,v) coordinates matches with the weight function in such a way that by partial integration

$$\langle\nabla^2\psi_{mn},\psi_{pq}\rangle = \langle\psi_{mn},\nabla^2\psi_{pq}\rangle,$$

because the additional boundary terms are all zero for the eigenfunctions.

On the other hand, it follows from Helmholtz's equation that

$$\langle\nabla^2\psi_{mn},\psi_{pq}\rangle = -\lambda_{mn}\langle\psi_{mn},\psi_{pq}\rangle,$$
$$\langle\psi_{mn},\nabla^2\psi_{pq}\rangle = -\lambda_{pq}\langle\psi_{mn},\psi_{pq}\rangle,$$

so

$$(\lambda_{mn}-\lambda_{pq})\langle\psi_{mn},\psi_{pq}\rangle = 0.$$

By assumption, $\lambda_{pq}\neq\lambda_{mn}$ for $(p,q)\neq(m,n)$, so the lemma follows. □

Occasionally we do not have $\lambda_{pq}\neq\lambda_{mn}$ for $(p,q)\neq(m,n)$, so the eigenspace is of higher dimension. One example is already given by Example 4.2.3, where $\lambda_{mn1}=\lambda_{mn2}$. In such cases we shall always give the proper orthogonal system, and omit the details.

The geometric interpretation of Lemma 4.2.1 is that $\{\psi_{mn}\}$ is an orthogonal system, so $e_{mn} = \psi_{mn}/\|\psi_{mn}\|$ must be ortho*normal*. Copying Euclidean geometry we may expect that we have an orthonormal expansion of the form

$$f = \sum_{m,n}\langle f,e_{mn}\rangle e_{mn} = \sum_{m,n}\frac{1}{\|\psi_{mn}\|^2}\langle f,\psi_{mn}\rangle\psi_{mn}, \tag{4.18}$$

or slightly incorrect,

$$f(u,v) = \sum_{m,n} \frac{1}{\|\psi_{mn}\|^2} \langle f, \psi_{mn} \rangle \psi_{mn}(u,v). \tag{4.19}$$

The incorrectness of the point-wise Equation (4.19) is not obvious. It will be explained below.

It can be proved that (4.18) is correct, when ψ_{mn} is the full system of (orthogonal) eigenfunctions and the functions f and ψ_{mn} involved are considered as *vectors* in an infinite dimensional abstract vector space with an inner product. On the other hand, (4.19) indicates a *pointwise* equality, which one cannot expect to be true in general. It is therefore surprising that (4.19) is indeed true in the interior of $\tilde{\Omega}$, when f is of class $C^1 \cap L_w^2$. On the boundary of $\tilde{\Omega}$ the right hand side of (4.19) is zero, so (4.19) is in general wrong on the boundary, though correct in the interior of $\tilde{\Omega}$ in all usual cases involving Fick's second law.

The proof of (4.19) is extremely hard for general L^2 functions, even in the case of ordinary Fourier series, and it was long conjectured that it did not hold. It was proved by Carleson (1966), cf. also Jørsboe *et al.* (1982).

Concerning Fick's second law we may therefore in the interior of $\tilde{\Omega}$ use (4.19) instead of the abstract formula (4.18). We shall later see that Fick's second law also forces the terms on the right hand side of (4.19) to die out rapidly in time the larger one of the indices (m, n) is, as long as we keep some distance from the boundary of $\tilde{\Omega}$. (Something can only go wrong on the boundary.) One may therefore arrive at a good approximation of (4.19) by restricting m, $n \leq N$. In some cases even

$$f(u,v) \approx \frac{1}{\|\psi_{11}\|^2} \langle f, \psi_{11} \rangle \psi_{11}(u,v)$$

may suffice for a first approximation, which makes the numerical calculations easier. The closer we come to the boundary, the more terms are needed for an acceptable approximation, because we here have a generalized Gibbs's phenomenon known from classical Fourier analysis with an overshoot of 8.5 % above the true value. However, for physical reasons one would never rely on this theoretical solution near the boundary, because contributions from unknown sources like the actual structure of the concrete will dominate the theoretical solution. Hence, Gibbs's phenomenon will only occur when the mathematical theory is used beyond its natural physical limitation, and this should of course be avoided.

We conclude from this discussion that once the weight function $w(u, v)$ and the eigenfunctions $\psi_{mn}(u, v)$ of ∇^2 are found, we only need to calculate the squared norms

$$\|\psi_{mn}\|^2 = \int\int_{\tilde{\Omega}} \psi_{mn}(u,v)^2 w(u,v)\, du\, dv, \qquad m, n \in \mathbb{N},$$

and the inner products

$$\langle f, \psi_{mn} \rangle = \int\int_{\tilde{\Omega}} f(u,v)\psi_{mn}(u,v)w(u,v)\, du\, dv, \qquad m,n \in \mathbb{N},$$

in order to get (4.19). This observation reduces the work quite a lot, in particular because one in practice only needs a finite number of terms in (4.19).

Example 4.2.5 *Rectangular coordinates*, cf. Example 4.2.1. We have already found that

$$\psi_{mn}(x,y) = \sin\left(\frac{m\pi x}{A}\right)\sin\left(\frac{n\pi y}{B}\right), \qquad (x,y) \in [0,A] \times [0,B],$$

and the weight function is trivially $w(x,y) = 1$, so the inner product is

$$\langle f, g \rangle = \int_0^B \left\{ \int_0^A f(x,y)g(x,y)\, dx \right\} dy.$$

Let us calculate

$$\langle \nabla^2 \psi_{mn}, \psi_{pq} \rangle = \int_0^B \left\{ \int_0^A \frac{\partial^2 \psi_{mn}}{\partial x^2} \psi_{pq}\, dx \right\} dy + \int_0^A \left\{ \int_0^B \frac{\partial^2 \psi_{mn}}{\partial y^2} \psi_{pq}\, dy \right\} dx.$$

Since ψ_{mn} and ψ_{pq} are zero on the boundary, we get by partial integration of the first inner integral,

$$\int_0^A \frac{\partial^2 \psi_{mn}}{\partial x^2} \psi_{pq}\, dx = \left[\frac{\partial \psi_{mn}}{\partial x} \psi_{pq}(x,y) \right]_0^A - \int_0^A \frac{\partial \psi_{mn}}{\partial x} \frac{\partial \psi_{pq}}{dx}\, dx$$

$$= 0 - \left[\psi_{mn}(x,y) \frac{\partial \psi_{pq}}{\partial x} \right]_0^A + \int_0^A \psi_{mn} \frac{\partial^2 \psi_{pq}}{\partial x^2} dx = \int_0^A \psi_{mn} \frac{\partial^2 \psi_{pq}}{\partial x^2} dx,$$

and similarly for the second inner integral, so we have in this case justified the relation

$$\langle \nabla^2 \psi_{mn}, \psi_{pq} \rangle = \langle \psi_{mn}, \nabla^2 \psi_{pq} \rangle.$$

Furthermore,

$$\|\psi_{mn}\|^2 = \int_0^A \sin^2\left(\frac{m\pi x}{A}\right) dx \int_0^B \sin^2\left(\frac{n\pi y}{B}\right) dy = \frac{A}{2} \cdot \frac{B}{2} = \frac{AB}{4}$$

and

$$\langle f, \psi_{mn} \rangle = \int_0^A \left\{ \int_0^B f(x,y) \sin\left(\frac{n\pi y}{B}\right) dy \right\} \sin\left(\frac{m\pi x}{A}\right) dx,$$

so (4.19) reduces to (4.10) already given in Example 4.2.1. ◊

Example 4.2.6 *Polar coordinates* for a half disc, cf. Example 4.2.2. The eigenfunctions are

$$\psi_{mn}(r,\varphi) = J_n\left(j_{n,m}\frac{r}{R}\right)\sin n\varphi, \quad m,n \in \mathbb{N}, \quad 0 \le r \le R, \quad 0 \le \varphi \le \pi.$$

Since elementary calculus gives $dx\,dy = r\,dr\,d\varphi$, the weight function is $w(r,\varphi) = r$, so the inner product is

$$\langle f, g\rangle = \int_0^\pi \left\{\int_0^R f(r,\varphi)g(r,\varphi)r\,dt\right\}d\varphi$$

The proof of $\langle \nabla^2\psi_{mn}, \psi_{pq}\rangle = \langle \psi_{mn}, \nabla^2\psi_{pq}\rangle$ is left to the reader. It is straightforward, though a little longer than the corresponding proof in Example 4.2.5.

By formula (3.105) we get

$$\begin{aligned}\|\psi_{mn}\|^2 &= \int_0^\pi \left\{\int_0^R J_n\left(j_{n,m}\frac{r}{R}\right)^2 \sin^2(n\varphi) r\,dr\right\}d\varphi \\ &= \int_0^R J_n\left(j_{n,m}\frac{r}{R}\right)^2 r\,dr \cdot \int_0^\pi \sin^2 n\varphi\,d\varphi = \frac{\pi}{4}R^2 J_n'\left(j_{n,m}\right)^2,\end{aligned}$$

so (4.19) gives

$$f(r,\varphi) = \sum_{m=1}^\infty\sum_{n=1}^\infty a_{mn} J_n\left(j_{n,m}\frac{r}{R}\right)\sin n\varphi,$$

where

$$\begin{aligned}a_{mn} &= \frac{1}{\|\psi_{mn}\|^2}\langle f, \psi_{mn}\rangle \\ &= \frac{4}{\pi R^2 J_n'\left(j_{n,m}\right)^2}\int_0^\pi \left\{\int_0^R f(r,\varphi) J_n\left(j_{n,m}\frac{r}{R}\right) r\,dr\right\}\sin n\varphi\,d\varphi. \ \diamond\end{aligned}$$

Example 4.2.7 . *Polar coordinates* for a full disc, cf. Example 4.2.3. The weight function is $w(r,\varphi) = r$ as in Example 4.2.6, and the inner product is

$$\langle f, g\rangle = \int_0^{2\pi} \left\{\int_0^R f(r,\varphi)g(r,\varphi)r\,dr\right\}d\varphi.$$

The eigenfunctions are split into three subfamilies,

$$\begin{aligned}&\psi_{m0}(r,\varphi) = J_0\left(j_{0,m}\tfrac{r}{R}\right), && \|\psi_{m0}\|^2 = \pi R^2 J_0'\left(j_{0,m}\right)^2, \\ &\psi_{mn,1}(r,\varphi) = J_n\left(j_{n,m}\tfrac{r}{R}\right)\cos n\varphi, && \|\psi_{mn,1}\|^2 = \tfrac{\pi}{2}R^2 J_n'\left(j_{n,m}\right)^2, \\ &\psi_{mn,2}(r,\varphi) = J_n\left(j_{n,m}\tfrac{r}{R}\right)\sin n\varphi, && \|\psi_{mn,2}\|^2 = \tfrac{\pi}{2}R^2 J_n'\left(j_{n,m}\right)^2.\end{aligned}$$

Notice that, apart from the missing factor $\frac{1}{2}$ in $\|\psi_{m0}\|^2$, the functions ψ_{m0} have the same structure as $\psi_{mn,1}$. We can therefore define

$$a_{mn} = \frac{2}{\pi R^2 J_n'\left(j_{n,m}\right)^2} \int_0^{2\pi} \left\{ \int_0^R f(r,\varphi) J_n\left(j_{n,m}\frac{r}{R}\right) r\,dr \right\} \cos n\varphi\, d\varphi,$$

$$b_{mn} = \frac{2}{\pi R^2 J_n'\left(j_{n,m}\right)^2} \int_0^{2\pi} \left\{ \int_0^R f(r,\varphi) J_n\left(j_{n,m}\frac{r}{R}\right) r\,dr \right\} \sin n\varphi\, d\varphi,$$

where we also allow $n = 0$ for the a_{mn}. The missing factor $\frac{1}{2}$ creeps in again in the final expansion (compare with ordinary Fourier series),

$$\begin{aligned} f(r,\varphi) &= \frac{1}{2}\sum_{m=1}^{\infty} a_{m0} J_0\left(j_{0,m}\frac{r}{R}\right) + \sum_{m=1}^{\infty}\sum_{n=1}^{\infty} a_{mn} J_n\left(j_{n,m}\frac{r}{R}\right) \cos n\varphi \\ &\quad + \sum_{m=1}^{\infty}\sum_{n=1}^{\infty} b_{mn} J_n\left(j_{n,m}\frac{r}{R}\right) \sin n\varphi. \quad \diamond \end{aligned}$$

Example 4.2.8 *Polar coordinates* for a sector of a disc, cf. Example 4.2.4. Putting $k_n = n\pi/\alpha$, $n \in \mathbb{N}$, we get

$$\psi_{mn}(r,\varphi) = J_{k_n}\left(j_{k_n,m}\frac{r}{R}\right) \sin k_n\varphi, \quad \|\psi_{mn}\|^2 = \frac{\alpha}{4} J_{k_n}'\left(j_{k_n,m}\right)^2.$$

The eigenfunction expansion is

$$f(r,\varphi) = \sum_{m=1}^{\infty}\sum_{n=1}^{\infty} a_{mn} J_{k_n}\left(j_{k_n,m}\frac{r}{R}\right) \sin k_n\varphi,$$

where

$$a_{mn} = \frac{4}{\alpha R^2 J_{k_n}'\left(j_{k_n,m}\right)^2} \int_0^{\alpha} \left\{ \int_0^R f(r,\varphi) J_{k_n}\left(j_{k_n,m}\frac{r}{R}\right) r\,dr \right\} \sin k_n\varphi\, d\varphi. \diamond$$

The Examples 4.2.6–4.2.8 show how useful the Bessel functions are, when polar coordinates are used.

4.2.5 Method of solution of Fick's second law

Let us consider the initial/boundary value problem

$$\begin{cases} \dfrac{1}{D}\dfrac{\partial C}{\partial t} = \nabla^2 C(u,v,t), & (u,v) \in \tilde{\Omega},\ t > 0, \\ C(0,v,t) = f_0(v,t), & C(A,v,t) = f_A(v,t), \\ C(u,0,t) = g_0(u,t), & C(u,B,t) = g_B(u,t), \\ C(u,v,0) = 0, & (u,v) \in \tilde{\Omega}. \end{cases} \tag{4.20}$$

This is Fick's second law transformed from Ω to the new rectangular domain $\tilde{\Omega} =]\,0, A\,[\times]\,0, B\,[$.

Let us assume that the diffusion coefficient either is a constant or only depends on t, i.e. $D = D(t) > 0$.

First consider the time t in (4.20) as a *parameter*. Let $\{\psi_{mn}\}$ be the full system of eigenfunctions associated with ∇^2 on $\tilde{\Omega}$ with corresponding eigenvalues $\{-\lambda_{mn}\}$. Then we have the following two eigenfunction expansions,

$$C(u, v, t) = \sum_{m,n} a_{mn}(t)\psi_{mn}(u, v), \tag{4.21}$$

$$\nabla^2 C(u, v, t) = \sum_{m,n} b_{m,n}(t)\psi_{mn}(u, v), \tag{4.22}$$

where

$$a_{mn}(t) = \frac{1}{\|\psi_{mn}\|^2}\int\!\!\int_{\tilde{\Omega}} C(u, v, t)\psi_{mn}(u, v)w(u, v)\,du\,dv, \tag{4.23}$$

$$b_{mn}(t) = \frac{1}{\|\psi_{mn}\|^2}\int\!\!\int_{\tilde{\Omega}} \nabla^2 C(u, v, t)\psi_{mn}(u, v)w(u, v)\,du\,dv. \tag{4.24}$$

Since the expansion (4.21) is unique, the solution of (4.20) will be complete, if we can derive explicit formulæ for $a_{mn}(t)$ in (4.23), involving only λ_{mn}, ψ_{mn} and the boundary terms f_0, f_A, g_0 and g_B.

Since t only plays the role of a parameter in the expansion (4.21), we may differentiate under the sum, provided that the result again is a convergent double series. Hence, the left hand side of Fick's second law (4.20) is formally

$$\frac{1}{D(t)}\frac{\partial C}{\partial t} = \sum_{m,n} \frac{1}{D(t)} a'_{mn}(t)\psi_{mn}(u, v).$$

The right hand side of Fick's second law is given by (4.22), so these two expansions are equal. From the uniqueness of the expansion we therefore have

$$\frac{1}{D(t)} a'_{mn}(t) = b_{mn}, \qquad a_{mn}(0) = 0 \quad \text{for all } m, n, \tag{4.25}$$

where we have added the initial condition from (4.20).

The trick is that $b_{mn}(t)$ has the structure $\langle \nabla^2 C, \psi_{mn}\rangle / \|\psi_{mn}\|^2$, so by partial integration we can shift the differentiation ∇^2 to ψ_{mn} by adding some boundary term F_{mn}. Thus, in principle,

$$\begin{aligned} b_{mn}(t) &= \frac{1}{\|\psi_{mn}\|^2}\langle \nabla^2 C, \psi_{mn}\rangle = \frac{1}{\|\psi_{mn}\|^2}\langle C, \nabla^2 \psi_{mn}\rangle + F_{mn}(t) \\ &= -\lambda_{mn} \cdot \frac{1}{\|\psi_{mn}\|^2}\langle C, \psi_{mn}\rangle + F_{mn}(t) = -\lambda_{mn} a_{mn}(t) + F_{mn}(t), \end{aligned}$$

where we have used the fact that $\nabla^2\psi_{mn} + \lambda_{mn}\psi_{mn} = 0$ together with (4.23). It will be illustrated in the examples how $F_{mn}(t)$ is derived from the boundary conditions of (4.20).

Using this result, (4.25) is reduced to

$$\frac{1}{D(t)}a'_{mn}(t) + \lambda_{mn}a_{mn}(t) = F_{mn}(t), \quad a_{mn}(0) = 0 \text{ for all } m, n. \tag{4.26}$$

When we multiply by $D(t)\exp\left(\lambda_{mn}\int_0^t D(\tau)\,d\tau\right)$ we find that this fundamental differential equation is transformed into

$$\frac{d}{dt}\left\{a_{mn}(t)\exp\left(\lambda_{mn}\int_0^t D(\tau)d\tau\right)\right\} = D(t)F_{mn}(t)\exp\left(\lambda_{mn}\int_0^t D(\tau)d\tau\right).$$

Since $a_{mn}(0) = 0$, an integration gives the solution formula

$$a_{mn}(t) = \int_0^t D(s)F_{mn}(s)\exp\left(-\lambda_{mn}\int_s^t D(\tau)\,d\tau\right)ds, \tag{4.27}$$

where only the functions $F_{mn}(s)$ remain to be expressed by the boundary conditions from (4.20).

Although (4.27) is a general solution formula which can be used in the following with only minor modifications we choose for simplicity only to consider the case, where $D > 0$ is a constant. Then (4.27) is reduced to the simpler formula

$$a_{mn}(t) = D\int_0^t F_{mn}(\tau)\exp\left(-D\lambda(t-\tau)\right)d\tau, \tag{4.28}$$

which is the convolution of $F_{mn}(t)$ and $\exp(-D\lambda_{mn}t)$, so in calculations in practice Laplace transform methods may be applied.

Summary

We solve (4.20) by first identifying the eigenfunctions ψ_{mn} and the corresponding eigenvalues $-\lambda_{mn}$. Then calculate (4.24), i.e.

$$b_{mn}(t) = \frac{1}{\|\psi_{mn}\|^2}\int\int_{\tilde{\Omega}} \nabla^2 C(u,v,t)\psi_{mn}(u,v)w(u,v)\,du\,dv,$$

by partial integration. In this calculation we need only chase the terms involving boundary functions, because we know that the other terms in the end will add up to $-\lambda_{mn}a_{mn}(t)$. Terms of the latter type can therefore conveniently be indicated by dots, and we save a lot of work. In this way $F_{mn}(t)$ is identified. Put it into (4.27) or (4.28), and we have found the solution of (4.20),

$$C(u,v,t) = \sum_{m,n} a_{mn}(t)\psi_{mn}(u,v).$$

We show by the following examples how this programme is carried out in practice. For convenience we let $D > 0$ be a constant, so we can use (4.18).

Example 4.2.9 *Rectangular coordinates* on $\Omega =]\,0, A\,[\times]\,0, B\,[$. In this case (4.20) is written

$$\frac{1}{D}\frac{\partial C}{\partial t} = \frac{\partial^2 C}{\partial x^2} + \frac{\partial^2 C}{\partial y^2},$$

$$C(0,y,t) = f_0(y,t), \qquad C(A,y,t) = f_A(y,t),$$

$$C(x,0,t) = g_0(x,t), \qquad C(x,B,t) = g_B(y,t),$$

$$C(x,y,0) = 0,$$

where f_0, f_A, g_0, g_B are given functions. From Example 4.2.1 or Example 4.2.5 we have the eigenfunctions

$$\psi_{mn}(x,y) = \sin\left(\frac{m\pi x}{A}\right)\sin\left(\frac{n\pi y}{B}\right), \qquad m,n \in \mathbb{N},$$

and the eigenvalues

$$\lambda_{mn} = \left(\frac{m\pi}{A}\right)^2 + \left(\frac{n\pi}{B}\right)^2, \qquad m,n \in \mathbb{N}.$$

Obviously, $\lambda_{mn} \to \infty$, if either $m \to \infty$ or $n \to \infty$, so terms of (4.23) of at least one large index will only give a small contribution to the solution, which dies out rapidly, when t increases.

We have already found

$$\|\psi_{mn}\|^2 = \frac{A}{2}\cdot\frac{B}{2} = \frac{AB}{4},$$

so

$$\begin{aligned} b_{mn}(t) &= \frac{4}{AB}\int_0^B \left\{\int_0^A \frac{\partial^2 C}{\partial x^2}\sin\left(\frac{m\pi x}{A}\right)dx\right\}\sin\left(\frac{n\pi y}{B}\right)dy \\ &\quad + \frac{4}{AB}\int_0^A \left\{\int_0^B \frac{\partial^2 C}{\partial y^2}\sin\left(\frac{n\pi y}{B}\right)dy\right\}\sin\left(\frac{m\pi x}{A}\right)dx, \quad (4.29) \end{aligned}$$

Partial integration on the first inner integral gives

$$\begin{aligned} \int_0^A \frac{\partial^2 C}{\partial x^2}\sin\left(\frac{m\pi x}{A}\right)dx &= \left[\frac{\partial C}{\partial x}\sin\left(\frac{m\pi x}{A}\right)\right]_0^A - \frac{m\pi}{A}\int_0^A \frac{\partial C}{\partial x}\cos\left(\frac{m\pi x}{A}\right)dx \\ &= 0 - \frac{m\pi}{A}\left[C(x,y,t)\cos\left(\frac{m\pi x}{A}\right)\right]_0^A \\ &\quad - \left(\frac{m\pi}{A}\right)^2\int_0^A C(x,y,t)\sin\left(\frac{m\pi x}{A}\right)dx \\ &= \frac{m\pi}{A}\left\{C(0,y,t) + (-1)^{m+1}C(A,y,t)\right\} + \cdots \\ &= \frac{m\pi}{A}\left\{f_0(y,t) + (-1)^{m+1}f_A(y,t)\right\} + \cdots, \end{aligned}$$

and similarly,

$$\int_0^B \frac{\partial^2 C}{\partial y^2} \sin\left(\frac{n\pi y}{B}\right) dy = \frac{n\pi}{B}\left\{g_0(x,t) + (-1)^{n+1} g_B(x,t)\right\} + \cdots .$$

Since the dots, when integrated, add up to $-\lambda_{mn} a_{mn}(t)$, we get from (4.29),

$$\begin{aligned} b_{mn}(t) &= -\lambda_{mn} a_{mn}(t) \\ &\quad + \frac{4}{AB}\cdot\frac{m\pi}{A}\int_0^B \left\{f_0(y,t) + (-1)^{m+1} f_A(y,t)\right\} \sin\left(\frac{n\pi y}{B}\right) dy \\ &\quad + \frac{4}{AB}\cdot\frac{n\pi}{B}\int_0^A \left\{g_0(x,t) + (-1)^{n+1} g_B(x,t)\right\} \sin\left(\frac{m\pi x}{A}\right) dx. \end{aligned} \tag{4.30}$$

Now, f_0, f_A, g_0 and g_B have ordinary sinus expansions,

$$f_0(y,t) = \sum_{n=1}^{\infty} \alpha_{0,n}(t) \sin\left(\frac{n\pi y}{B}\right), \quad f_A(y,t) = \sum_{n=1}^{\infty} \alpha_{A,n}(t) \sin\left(\frac{n\pi y}{B}\right),$$

$$g_0(x,t) = \sum_{m=1}^{\infty} \beta_{0,m}(t) \sin\left(\frac{m\pi x}{A}\right), \quad g_B(x,t) = \sum_{m=1}^{\infty} \beta_{B,m}(t) \sin\left(\frac{m\pi x}{A}\right),$$

where e.g.

$$\alpha_{0,n}(t) = \frac{2}{B}\int_0^B f_0(y,t) \sin\left(\frac{n\pi y}{B}\right) dy,$$

and similarly for the other coefficients. Using these expressions, (4.30) is reduced to

$$\begin{aligned} b_{mn}(t) &= -\lambda_{mn} a_{mn}(t) + \frac{2m\pi}{A^2}\left\{\alpha_{0,n}(t) + (-1)^{m+1}\alpha_{A,n}(t)\right\} \\ &\quad + \frac{2n\pi}{B^2}\left\{\beta_{0,m}(t) + (-1)^{n+1}\beta_{B,m}(t)\right\}, \end{aligned}$$

so

$$\begin{aligned} F_{mn}(t) &= \frac{2\pi m}{A^2}\left\{\alpha_{0,n}(t) + (-1)^{m+1}\alpha_{A,n}(t)\right\} \\ &\quad + \frac{2\pi n}{B^2}\left\{\beta_{0,m}(t) + (-1)^{n+1}\beta_{B,m}(t)\right\}. \end{aligned}$$

Thus (4.28) is written

$$\begin{aligned} a_{mn}(t) &= \frac{2\pi D m}{A^2}\int_0^t \left\{\alpha_{0,n}(\tau) + (-1)^{m+1}\alpha_{A,n}(\tau)\right\} \exp(-D\lambda_{mn}(t-\tau))\, d\tau \\ &\quad + \frac{2\pi D n}{B^2}\int_0^t \left\{\beta_{0,m}(\tau) + (-1)^{n+1}\beta_{B,m}(\tau)\right\} \exp(-D\lambda_{mn}(t-\tau))\, d\tau, \end{aligned}$$

and the solution is finally given as follows

$$C(x,y,t) = \sum_{m,n=1}^{\infty} a_{mn}(t) \sin\left(\frac{m\pi x}{A}\right) \sin\left(\frac{n\pi y}{B}\right), \qquad \begin{cases} 0 < x < A, \\ 0 < y < B, \end{cases}$$

$$C(0,y,t) = f_0(y,t), \qquad C(A,y,t) = f_A(y,t),$$

$$C(x,0,t) = g_0(x,t), \qquad C(x,B,t) = g_B(y,t),$$

where the boundary conditions are added to the solution, because the series expansion here gives a wrong value.

For computational reasons it is seen that it may be useful immediately to expand the boundary conditions in sinus series,

$$f_0(y,t) = \sum_{n=1}^{\infty} \alpha_{0,n}(t) \sin\left(\frac{n\pi y}{B}\right), \quad f_A(y,t) = \sum_{n=1}^{\infty} \alpha_{A,n}(t) \sin\left(\frac{n\pi y}{B}\right),$$

$$g_0(x,t) = \sum_{m=1}^{\infty} \beta_{0,m}(t) \sin\left(\frac{m\pi x}{A}\right), \quad g_B(x,t) = \sum_{m=1}^{\infty} \beta_{B,m}(t) \sin\left(\frac{m\pi x}{A}\right).$$

In other cases one could change the order of integration, such that the integration with respect to time is performed before the equivalent of (4.30). Anyway we see that even in this case of rectangular coordinates, which is the simplest one, the explicit solution formula becomes quite laborious. This explains why we shall not go into details in Section 4.3 in the following, where we only give the main expressions. ◊

Example 4.2.10 *Polar coordinates* for a half disc. Since Bessel functions cannot be considered familiar for most readers, we shall here give all the details, following the same pattern as in Example 4.2.9. We consider for given functions $f(\varphi,t)$, $g_0(r,t)$, $g_\pi(r,t)$ the problem in polar coordinates,

$$\frac{1}{D}\frac{\partial C}{\partial t} = \frac{\partial^2 C}{\partial r^2} + \frac{1}{r}\frac{\partial C}{\partial r} + \frac{1}{r^2}\frac{\partial^2 C}{\partial \varphi^2},$$

$$C(R,\varphi,t) = f(\varphi,t), \qquad 0 \le \varphi \le \pi,$$

$$C(r,0,t) = g_0(r,t), \qquad C(r,\pi,t) = g_\pi(r,t), \quad 0 < r < R,$$

$$C(r,\varphi,0) = 0.$$

Apparently a condition is missing for $r = 0$, but $r = 0$ corresponds to the singular point $(0,0)$.

The weight function is $w(r,\varphi) = r$, and the eigenfunctions and the eigenvalues are according to Example 4.2.2,

$$\psi_{mn}(r,\varphi) = J_n\left(j_{n,m}\frac{r}{R}\right) \sin n\varphi, \quad \lambda_{mn} = \left(\frac{1}{R} j_{n,m}\right)^2, \qquad m,n \in \mathbb{N},$$

where

$$\|\psi_{mn}\|^2 = \frac{\pi}{4} R^2 J_n'(j_{n,m})^2.$$

The lesson from Example 4.2.9 was that the boundary conditions should be expanded suitably. The right expansions here are

$$\begin{aligned}
f(\varphi,t) &= \sum_{n=1}^{\infty} \alpha_n(t)\sin n\varphi, \\
&\quad \alpha_n(t) = \frac{2}{\pi}\int_0^{\pi} f(\varphi,t)\sin(n\varphi)\,d\varphi, \\
\frac{g_0(r,t)}{r^2} &= \sum_{m=1}^{\infty} \beta_{m,0}^{(n)}(t) J_n\left(j_{n,m}\frac{r}{R}\right), \\
&\quad \beta_{m,0}^{(n)}(t) = \frac{2}{R^2 J_n'(j_{n,m})^2}\int_0^R \frac{g_0(r,t)}{r} J_n\left(j_{n,m}\frac{r}{R}\right) dr, \\
\frac{g_\pi(r,t)}{r^2} &= \sum_{m=1}^{\infty} \beta_{m,\pi}^{(n)}(t) J_n\left(j_{n,m}\frac{r}{R}\right), \\
&\quad \beta_{m,\pi}^{(n)}(t) = \frac{2}{r^2 J_n'(j_{n,m})^2}\int_0^R \frac{g_\pi(r,t)}{r} J_n\left(j_{n,m}\frac{r}{R}\right) dr.
\end{aligned}$$

The reason for expanding $g_0(r,t)/r^2$ and $g_\pi(r,t)/r^2$ instead of $g_0(r,t)$ and $g_\pi(r,t)$ will become clear later. Notice that the integrals defining $\beta_{m,0}^{(n)}(t)$ and $\beta_{m,\pi}^{(n)}(t)$ are well defined, because $J_n(j_{n,m}r/R)$ has a zero of order $n \geq 1$ at $r = 0$.

The coefficients (4.24) are now

$$\begin{aligned}
b_{mn} &= \frac{4}{\pi R^2 J_n'(j_{n,m})^2}\int_0^{\pi}\left(\int_0^R \left\{\frac{\partial^2 C}{\partial r^2} + \frac{1}{r}\frac{\partial C}{\partial r}\right\} J_n\left(j_{n,m}\frac{r}{R}\right) r dr\right)\sin(n\varphi) d\varphi \\
&\quad + \frac{4}{\pi R^2 J_n'(j_{n,m})^2}\int_0^R \left(\int_0^{\pi}\frac{\partial^2 C}{\partial \varphi^2}\sin(n\varphi)\,d\varphi\right)\frac{1}{r^2} J_n\left(j_{n,m}\frac{r}{R}\right) r\,dr.
\end{aligned} \tag{4.31}$$

As before, dots indicate terms which only contribute to the known term

$-\lambda_{mn}a_{mn}(t)$, so we get by partial integration,

$$\begin{aligned}
&\int_0^R \left\{\frac{\partial^2 C}{\partial r^2} + \frac{1}{r}\frac{\partial C}{\partial r}\right\} J_n\left(j_{n,m}\frac{r}{R}\right) r\, dr \\
&= \int_0^R \left\{r\frac{\partial^2 C}{\partial r^2} + 1\cdot\frac{\partial C}{\partial r}\right\} J_n\left(j_{n,m}\frac{r}{R}\right) dr \\
&= \int_0^R \frac{\partial}{\partial r}\left(r\frac{\partial C}{\partial r}\right)\cdot J_n\left(j_{n,m}\frac{r}{R}\right) dr \\
&= \left[r\frac{\partial C}{\partial r}J_n\left(j_{n,m}\frac{r}{R}\right)\right]_0^R - \int_0^R r\frac{\partial C}{\partial r}\frac{d}{dr}\left\{J_n\left(j_{n,m}\frac{r}{R}\right)\right\} dr \\
&= 0 - \left[C(r,\varphi,t)r\cdot\frac{j_{n,m}}{R}J_n'\left(j_{n,m}\frac{r}{R}\right)\right]_0^R \\
&\quad + \int_0^R C(r,\varphi,t)\frac{d}{dr}\left\{r\frac{d}{dr}\left(J_n\left(j_{n,m}\frac{r}{R}\right)\right)\right\} dr \\
&= -j_{n,m}J_n'(j_{n,m})f(\varphi,t) + \cdots,
\end{aligned}$$

and

$$\begin{aligned}
\int_0^\pi \frac{\partial^2 C}{\partial \varphi^2}\sin(n\varphi)\, d\varphi &= \left[\frac{\partial C}{\partial \varphi}\sin(n\varphi)\right]_0^\pi - n\int_0^\pi \frac{\partial C}{\partial \varphi}\cos(n\varphi)\, d\varphi \\
&= 0 - n\left[C(r,\varphi,t)\cos(n\varphi)\right]_0^\pi - n^2\int_0^\pi C(r,\varphi,t)\sin(n\varphi)\, d\varphi \\
&= n\left\{g_0(r,t) + (-1)^{n+1}g_\pi(r,t)\right\} + \cdots,
\end{aligned}$$

so by (4.31),

$$\begin{aligned}
&F_{mn}(t) \\
&= \frac{4}{\pi R^2 J_n'(j_{n,m})^2}\int_0^R n\left\{g_0(r,t)+(-1)^{n+1}g_\pi(r,t)\right\}\frac{1}{r}J_n\left(j_{n,m}\frac{r}{R}\right) dr \\
&\quad + \frac{4}{\pi R^2 J_n'(j_{n,m})^2}\int_0^\pi \left\{-j_{n,m}J_n'(j_{n,m})\right\} f(\varphi,t)\sin(n\varphi)\, d\varphi \\
&= \frac{2n}{\pi}\left\{\beta_{m,0}^{(n)}(t) + (-1)^{n+1}\beta_{m,\pi}^{(n)}(t)\right\} - \frac{2j_{n,m}}{R^2 J_n'(j_{n,m})}\alpha_n(t),
\end{aligned}$$

where we have used the definitions of $\beta_{m,0}^{(n)}(t)$, $\beta_{m,\pi}^{(n)}(t)$ and $\alpha_n(t)$. Then $a_{mn}(t)$ is determined by (4.27) or (4.28), and the solution is

$$C(r,\varphi,t) = \sum_{m=1}^\infty\sum_{n=1}^\infty a_{mn}(t)J_n\left(j_{n,m}\frac{r}{R}\right)\sin(n\varphi). \quad \diamond$$

Example 4.2.11 *Polar coordinates* on the full disc. Fick's second law is here described by

$$\frac{1}{D}\frac{\partial C}{\partial t} = \frac{\partial^2 C}{\partial r^2} + \frac{1}{r}\frac{\partial C}{\partial r} + \frac{1}{r^2}\frac{\partial^2 C}{\partial \varphi^2},$$

$$C(R,\varphi,t) = f(\varphi,t),$$
$$C(R,\varphi,0) = 0,$$

supplied with the periodicity condition

$$C(r,0,t) = C(r,2\pi,t).$$

The derivation of the solution follows the pattern from Example 4.2.10, so we give only the main results.

The weight function is $w(r,\varphi) = r$.

The eigenfunctions and the eigenvalues are

$$\begin{aligned}
\psi_{m0}(r,\varphi) &= J_0\left(j_{0,m}\frac{r}{R}\right), & \lambda_{m0} &= \left(\frac{1}{R}j_{0,m}\right)^2,\\
\psi_{mn,1}(r,\varphi) &= J_n\left(j_{n,m}\frac{r}{R}\right)\cos n\varphi, & \lambda_{mn,1} = \lambda_{mn} &= \left(\frac{1}{R}j_{n,m}\right)^2,\\
\psi_{mn,2}(r,\varphi) &= J_n\left(j_{n,m}\frac{r}{R}\right)\sin n\varphi, & \lambda_{mn,1} = \lambda_{mn} &= \left(\frac{1}{R}j_{n,m}\right)^2,
\end{aligned}$$

and

$$\|\psi_{m0}\|^2 = \pi R^2 J_0'(j_{0,m})^2, \quad \|\psi_{mn,i}\|^2 = \frac{\pi}{2}R^2 J_n'(j_{n,m})^2.$$

Expand the boundary condition in a usual Fourier series,

$$f(\varphi,t) = \frac{1}{2}\alpha_0(t) + \sum_{n=1}^{\infty}\{\alpha_n(t)\cos(n\varphi) + \beta_n(t)\sin(n\varphi)\},$$

where

$$\alpha_n(t) = \frac{1}{\pi}\int_0^{2\pi} f(\varphi,t)\cos(n\varphi)\,d\varphi, \quad \beta_n(t) = \frac{1}{\pi}\int_0^{2\pi} f(\varphi,t)\sin(n\varphi)d\varphi.$$

Then by some calculations,

$$F_{m0}(t) = -\frac{j_{0,m}}{R^2 J_0'(j_{0,m})}\alpha_0(t), \qquad m\in\mathbb{N}, \quad n=0,$$

$$F_{mn1}(t) = -\frac{2j_{n,m}}{R^2 J_n'(j_{n,m})}\alpha_n(t), \quad m\in\mathbb{N}, \quad n\in\mathbb{N},$$

$$F_{mn2}(t) = -\frac{2j_{n,m}}{R^2 J_n'(j_{n,m})}\beta_n(t), \quad m\in\mathbb{N}, \quad n\in\mathbb{N}.$$

By (4.27) or (4.28) we derive $a_{m0}(t)$, $a_{mn1}(t)$ and $a_{mn2}(t)$ from $F_{m0}(t)$, $F_{mn1}(t)$ and $F_{mn2}(t)$, respectively, and the solution is

$$\begin{aligned}
C(r,\varphi,t) &= \sum_{m=1}^{\infty} a_{m0}(t)J_0\left(j_{0,m}\frac{r}{R}\right) \qquad (4.32)\\
&\quad + \sum_{m=1}^{\infty}\sum_{n=1}^{\infty}\{a_{mn1}(t)\cos n\varphi + a_{mn2}(t)\sin n\varphi\}J_n\left(j_{n,m}\frac{r}{R}\right)
\end{aligned}$$

in the interior of the disc, and

$$C(R, \varphi, t) = f(\varphi, t)$$

on the boundary.

Formula (4.32) reduces at the centre $r = 0$ of the disc to

$$C(0, \varphi, t) = \sum_{m=1}^{\infty} a_{m0}(t),$$

because $J_0(0) = 1$, and $J_n(0) = 0$ for $n \in \mathbb{N}$. $\Diamond$

Example 4.2.12 *Polar coordinates* for a sector of the disc, i.e. $0 < r < R$ and $0 < \varphi < \alpha$, where $0 < \alpha \leq 2\pi$. Since Example 4.2.10 is the special case $\alpha = \pi$, we shall be very brief here, leaving the details to the reader. The problem is described by

$$\begin{aligned}
&\frac{1}{D}\frac{\partial C}{\partial t} = \frac{\partial^2 C}{\partial r^2} + \frac{1}{r}\frac{\partial C}{\partial r} + \frac{1}{r^2}\frac{\partial^2 C}{\partial \varphi^2}, \\
&C(R, \varphi, t) = f(\varphi, t), \\
&C(r, 0, t) = g_0(r, t), \qquad C(r, \alpha, t) = g_\alpha(r, t), \\
&C(r, \varphi, 0) = 0.
\end{aligned}$$

The weight function is $w(r, \varphi) = r$. If we let $k_n = n\pi/\alpha$, $n \in \mathbb{N}$, the eigenfunctions, etc., are

$$\psi_{mn}(r, \varphi) = J_{k_n}\left(j_{k_n,m}\frac{r}{R}\right) \sin k_n\varphi, \quad \|\psi_{mn}\|^2 = \frac{\alpha}{4} R^2 J'_{k_n}(j_{k_n,m})^2,$$

$$\lambda_{mn} = \left(\frac{1}{R} j_{k_n,m}\right)^2, \qquad m, n \in \mathbb{N}.$$

When $D > 0$ is constant, we define

$$G_{mn}(r, t) = \int_0^t \exp\left(D\left(j_{k_n,m}/R\right)^2 \tau\right) \{g_0(r, \tau) + (-1)^{n+1} g_\alpha(r, \tau)\}\, d\tau,$$

and

$$H_{mn}(\varphi, t) = \int_0^t \exp\left(D\left(j_{k_n,m}/R\right)^2 \tau\right) f(\varphi, \tau)\, d\tau,$$

so we have changed the order of integration. Then

$$\begin{aligned}
a_{mn}(t) = {} & \exp\left(-D\left(j_{k_n,m}/R\right)^2 t\right) \\
& \cdot \left\{ \frac{2n}{\alpha} \cdot \frac{2}{R^2 J'_{k_n}(j_{k_n,m})^2} \int_0^R \frac{1}{r} G_{mn}(r, t) J_{k_n}\left(j_{k_n,m}\frac{r}{R}\right) dr \right. \\
& \left. - \frac{2 j_{k_n,m}}{R^2 J'_{k_n}(j_{k_n,m})} \cdot \frac{2}{\alpha} \int_0^\alpha H_{mn}(\varphi, t) \sin(k_n\varphi)\, d\varphi \right\},
\end{aligned}$$

and the solution is given by

$$C(r,\varphi,t)=\sum_{m=1}^{\infty}\sum_{n=1}^{\infty}a_{mn}(t)J_{k_n}\left(j_{k_n,m}\frac{r}{R}\right)\sin k_n\varphi,$$

supplied with the boundary conditions. ◊

Example 4.2.13 A crack along a radius of a disc. This is the special case $\alpha=2\pi$ in Example 4.2.12, so $k_n=n/2$, $n\in\mathbb{N}$. Let D be constant, and choose the special boundary conditions

$$C(R,\varphi,t)=1,\quad C(r,0,t)=1,\quad C(r,2\pi,t)=1,\quad C(r,\varphi,0)=0.$$

Using the notation of Example 4.2.12,

$$G_{m,2n}(r,t)=0,$$

$$\begin{aligned}G_{m,2n+1}(r,t)&=2\int_0^t\exp\left(D\left\{\frac{1}{R}j_{n+1/2,m}\right\}^2\tau\right)d\tau\\&=\frac{2}{D}\left\{\frac{R}{j_{n+1/2,m}}\right\}^2\left\{\exp\left(D\left\{j_{n+1/2,m}/R\right\}^2t\right)-1\right\},\end{aligned}$$

$$H_{m,n}(\varphi,t)=\frac{1}{D}\left\{\frac{R}{j_{n/2,m}}\right\}^2\cdot\left\{\exp\left(D\cdot\left\{\frac{1}{R}j_{n/2,m}\right\}^2t\right)-1\right\}.$$

These expressions are all independent of (r,φ), so we just have to calculate

$$\int_0^{2\pi}\sin\left(\frac{n\varphi}{2}\right)d\varphi=\left[-\frac{2}{n}\cos\left(\frac{n\varphi}{2}\right)\right]_0^{2\pi}=\begin{cases}0, & n\text{ even},\\ 2/n, & n\text{ odd},\end{cases}$$

and

$$\int_0^R\frac{1}{r}J_{n+1/2,m}\left(j_{n+1/2,m}\frac{r}{R}\right)dr=\int_0^{j_{n+1/2,m}}\frac{1}{r}J_{n+1/2}(r)dr.\tag{4.33}$$

It follows that $a_{m,2n}(t)=0$, and that

$$\begin{aligned}a_{m,2n+1}(t)=&\frac{1}{D}\left\{\frac{R}{j_{n+1/2,m}}\right\}^2\left\{1-\exp\left(-D\left\{\frac{1}{R}j_{n+1/2,m}\right\}^2t\right)\right\}\\&\cdot\left\{\frac{2(2n+1)}{\pi}\cdot\frac{2}{R^2J'_{n+1/2}\left(j_{n+1/2,m}\right)^2}\int_0^{j_{n+1/2,m}}\frac{1}{r}J_{n+1/2}(r)\,dr\right.\\&\left.-\frac{2j_{n+1/2,m}}{R^2J'_{n+1/2}\left(j_{n+1/2,m}\right)}\cdot\frac{1}{\pi}\cdot\frac{2}{2n+1}\right\},\end{aligned}\tag{4.34}$$

where we only still do not know the value of the integral in (4.33). The solution is given by

$$C(r,\varphi,t)=\sum_{m=1}^{\infty}\sum_{n=0}^{\infty}a_{m,2n+1}(t)J_{n+1/2}\left(j_{n+1/2,m}\frac{r}{R}\right)\sin\left(\frac{2n+1}{2}\varphi\right).\quad\Diamond$$

4.3 A catalogue of solutions of Fick's 2nd law

By using the methods of Section 4.2 we are now able to give solution formulæ for Fick's second law in quite a few cases. The list given below is far from complete. For instance, we do not feel that solid balls or half balls of concrete are that important, as far as ingress of chloride into concrete is concerned. The same can be said about the more exotic coordinate systems, listed by Morse *et al.* (1953), in which an eigenfunction expansion is possible by using the method of separating the variables. Even with these limitations, the catalogue presented here will cover a large number of cases, including rectangular and cylindrical pillars.

Since the section is meant as a catalogue, we shall omit most of the details. In the bounded cases their proofs follow the pattern of Section 4.2, and in the classical unbounded case we either refer to the well known classical theory, or to Mejlbro (1996).

To our knowledge, such a list has never been produced before. We shall in Section 4.3.11 give a plausible reason for why this was never done.

We shall begin with the rectangular cases, which are the simplest ones. Then we shall consider polar coordinates in the plane, followed by the semi-polar coordinates in the space. Finally we sketch the possibilities and obstacles of using spherical coordinates.

4.3.1 Half-infinite interval, 1 dimension

For historical reasons we include this case, although structures in real life are finite. The solution is nice, explaining why it is preferred by so many authors, in particular when it is compared with the more complicated solution of the bounded case in e.g. Section 4.3.2. It is, however, less accurate when applied to finite structures.

The mathematical model is given by

$$\frac{\partial C}{\partial t} = D\frac{\partial^2 C}{\partial x^2}, \qquad x > 0, \quad t > 0,$$

with the *initial condition* $C(x, 0) = 0$ and the *convergence condition*

$$\lim_{x\to\infty} C(x, t) = 0 \qquad \text{for fixed } t > 0.$$

The *boundary condition* is given by

$$C(0, t) = \varphi(t), \qquad t > 0,$$

where $\varphi(t)$ is a given function.

Let us once and for all prove that the case, where the diffusion coefficient only depends on time, can be reduced to the case where $D = D(t) > 0$ is a positive constant. The trick is to change the time variable,

$$T = \int_0^t D(\tau)\, d\tau,$$

because then by the chain rule,

$$\frac{\partial C}{\partial t} = \frac{dT}{dt}\frac{\partial C}{\partial T} = D(t)\frac{\partial C}{\partial T} = D(t)\frac{\partial^2 C}{\partial x^2},$$

so

$$\frac{\partial C}{\partial T} = \frac{\partial^2 C}{\partial x^2},$$

where the new "diffusion coefficient" is just 1. The proof is the same in all dimensions and it does not involve the space variables. We have therefore justified the assumption that $D > 0$ is a constant.

In the present situation we have two cases, which should be mentioned.

(a) *The classical case*, where $\varphi(t) = c$ is a constant. The solution is given by

$$C(x,t) = c \cdot \operatorname{erfc}\left(\frac{x}{\sqrt{4Dt}}\right).$$

(b) *Extensions*, cf. Mejlbro (1996). If $\varphi(t)$ can be written in the form

$$\varphi(t) = \sum_{n=1}^{\infty} a_n \, (D \cdot t)^{p_n}, \tag{4.35}$$

where the series is absolutely convergent for $t \geq 0$, then the solution is given by adding a suitable factor to the series expansion (4.35). We have more precisely,

$$C(x,t) = \sum_{n=1}^{\infty} a_n \, (D \cdot t)^{p_n} \, \Psi_{p_n}\left(\frac{x}{\sqrt{4Dt}}\right), \tag{4.36}$$

where the functions Ψ_p are defined in the Sections 3.3.3 and 3.3.5.

Since $\Psi_0(u) = \operatorname{erfc}(u)$, we see that (a) is contained in (b). The exponents $p_n \geq 0$ of (4.35) need not be integers, so (4.35) is indeed very flexible. Using Weierstraß's approximation theorem it is seen that any *continuous* function $\varphi(t)$ over a closed bounded time interval $[0, T]$ can be *uniformly* approximated over $[0, T]$ by an even *finite* sum (i.e. $a_n = 0$ for $n > N$) of the type (4.35). Since the factors added in (4.36) are bounded by 1 and all satisfy the proper growth conditions, it can be shown that (4.36) also gives a uniform approximation of the solution, when it is truncated in the same way as in (4.35).

4.3.2 Bounded interval, 1 dimension

Let $D > 0$ be a constant. The mathematical model is given by

$$\frac{1}{D}\frac{\partial C}{\partial t} = \frac{\partial^2 C}{\partial x^2}, \qquad 0 < x < A, \quad t > 0,$$

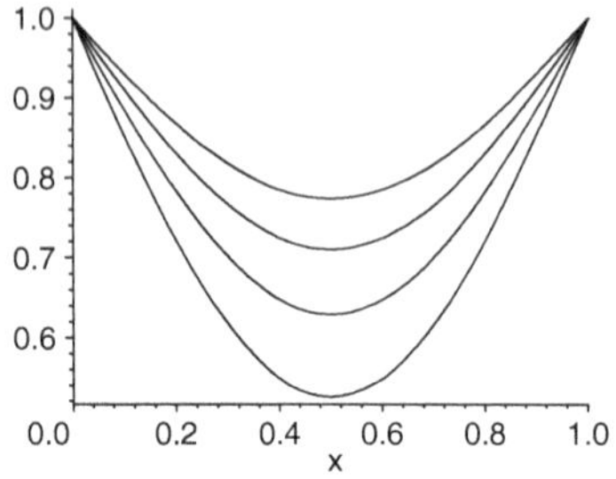

Figure 4.4: The function $C(x,t)$ for a dam with $a = b = 1$ and various values of t. Notice how the curves get closer to the constant solution 1 when t increases.

with the *initial condition* $C(x,0) = 0$ and the two *boundary conditions*,

$$C(0,t) = f(t) \quad \text{and} \quad C(A,t) = g(t).$$

The *weight function* is $w(x) = 1$, and the *eigenfunctions*, etc., are

$$\psi_n(x) = \sin\left(\frac{n\pi x}{A}\right), \quad \|\psi\|^2 = \frac{A}{2}, \quad \lambda_n = \left(\frac{n\pi}{A}\right)^2.$$

The solution is given by

$$C(x,t) = \sum_{n=1}^{\infty} a_n(t) \sin\left(\frac{n\pi x}{A}\right) \qquad \text{for } 0 < x < A \text{ and } t > 0,$$

$$C(0,t) = f(t) \text{ and } C(A,t) = g(t) \quad \text{for } t > 0,$$

where

$$a_n(t) = \frac{2n\pi D}{A^2} \int_0^t \exp\left(-D\left(\frac{n\pi}{A}\right)^2 (t-\tau)\right) \cdot \left\{f(\tau) + (-1)^{n+1} g(\tau)\right\} d\tau.$$

Example 4.3.1 (Cf. also Section 2.2.) A dam of thickness A is separating two reservoirs of water of different (constant) chloride concentration. By choosing the proper coordinate system we can obtain the boundary values

$$C(0,t) = f(t) = a \quad \text{and} \quad C(A,t) = g(t) = b.$$

Using a reduction formula from the classical Fourier theory the solution can be reduced to

$$\begin{aligned} C(x,t) &= a\left(1 - \frac{x}{A}\right) + b \cdot \frac{x}{A} \\ &\quad + \frac{2}{\pi} \sum_{n=1}^{\infty} \frac{(-1)^n b - a}{n} \exp\left(-D\left(\frac{n\pi}{A}\right)^2 t\right) \sin\left(\frac{n\pi x}{A}\right) \\ &= a \cdot H_1\left(\frac{x}{A}, \frac{Dt}{A^2}\right) + b \cdot H_1\left(1 - \frac{x}{A}, \frac{Dt}{A^2}\right), \end{aligned}$$

where $H_1(\xi, \tau)$ is defined in (3.115).

When $t > 0$ is fixed, this solution is even uniformly and absolutely convergent in x, and when $t \to \infty$, the series tends to zero, so

$$\lim_{t\to\infty} C(x,t) = a\left(1 - \frac{x}{A}\right) + b \cdot \frac{x}{A}.$$

which one would expect. ◊

4.3.3 Bounded two-dimensional interval

For constant $D > 0$ the *mathematical model* is given by

$$\frac{1}{D}\frac{\partial C}{\partial t} = \frac{\partial^2 C}{\partial x^2} + \frac{\partial^2 C}{\partial y^2}, \qquad 0 < x < A, \quad 0 < y < B, \quad t > 0,$$

with the *initial condition* $C(x,y,0) = 0$, and the four *boundary conditions* expanded after the relevant eigenfunctions in one dimension,

$$C(x,0,t) = f_0(x,t) = \sum_{m=1}^{\infty} u_{0,m}(t) \sin\left(\frac{m\pi x}{A}\right), \qquad 0 < x < A,$$

$$C(x,B,t) = f_B(x,t) = \sum_{m=1}^{\infty} u_{B,m}(t) \sin\left(\frac{m\pi x}{A}\right), \quad 0 < x < A,$$

$$C(0,y,t) = g_0(y,t) = \sum_{n=1}^{\infty} v_{0,n}(t) \sin\left(\frac{n\pi y}{B}\right), \qquad 0 < y < B,$$

$$C(A,y,t) = g_A(y,t) = \sum_{n=1}^{\infty} v_{A,n}(t) \sin\left(\frac{n\pi y}{B}\right), \qquad 0 < y < B,$$

where

$$u_{0,m}(t) = \frac{2}{A}\int_0^A f_0(x,t) \sin\left(\frac{m\pi x}{A}\right) dx,$$

$$u_{B,m}(t) = \frac{2}{A}\int_0^A f_B(x,t) \sin\left(\frac{m\pi x}{A}\right) dx,$$

$$v_{0,n}(t) = \frac{2}{B}\int_0^B g_0(y,t) \sin\left(\frac{n\pi y}{B}\right) dy,$$

$$v_{A,n}(t) = \frac{2}{B}\int_0^B g_A(y,t) \sin\left(\frac{n\pi y}{B}\right) dy.$$

The *weight function* is $w(x,y) = 1$, and the *eigenfunctions*, etc., are

$$\psi_{mn}(x,y) = \sin\left(\frac{m\pi x}{A}\right) \sin\left(\frac{n\pi y}{B}\right), \qquad \|\psi_{mn}\|^2 = \frac{AB}{4},$$

$$\lambda_{mn} = \left(\frac{m\pi}{A}\right)^2 + \left(\frac{n\pi}{B}\right)^2.$$

In the interior of the domain the solution is given by

$$C(x,y,t) = \sum_{m=1}^{\infty}\sum_{n=1}^{\infty} a_{mn}(t) \sin\left(\frac{m\pi x}{A}\right) \sin\left(\frac{n\pi y}{B}\right),$$

where

$$\begin{aligned} a_{mn}(t) &= \frac{2\pi m D}{A^2} \int_0^t \exp\left(-D\left\{\left(\frac{m\pi}{A}\right)^2 + \left(\frac{n\pi}{B}\right)^2\right\}\tau\right) \\ &\quad \cdot \left\{v_{0,n}(t-\tau) + (-1)^{m+1} v_{A,n}(t-\tau)\right\} d\tau \\ &\quad + \frac{2\pi n D}{B^2} \int_0^t \exp\left(-D\left\{\left(\frac{m\pi}{A}\right)^2 + \left(\frac{n\pi}{B}\right)^2\right\}\tau\right) \\ &\quad \cdot \left\{u_{0,m}(t-\tau) + (-1)^{n+1} u_{B,m}(t-\tau)\right\} d\tau. \end{aligned}$$

Example 4.3.2 Let all the boundary conditions be equal to a constant $c > 0$. From the classical theory of Fourier series it is known that

$$1 = \frac{4}{\pi}\sum_{m=0}^{\infty} \frac{1}{2m+1} \sin\left(\frac{(2m+1)\pi x}{A}\right) \qquad \text{for } 0 < x < A,$$

$$1 = \frac{4}{\pi}\sum_{n=0}^{\infty} \frac{1}{2n+1} \sin\left(\frac{(2n+1)\pi y}{B}\right) \qquad \text{for } 0 < y < B.$$

Using these expansions of 1 it can be shown that for any fixed $t > 0$, the solution is given by the following absolutely and uniformly convergent series

$$\begin{aligned} C(x,y,t) &= c\left\{1 - \frac{16}{\pi^2}\sum_{m=0}^{\infty}\sum_{n=0}^{\infty} \frac{1}{(2m+1)(2n+1)}\right. \\ &\quad \cdot \exp\left(-D\left\{\left(\frac{(2m+1)\pi}{A}\right)^2 + \left(\frac{(2n+1)\pi}{B}\right)^2\right\}t\right) \\ &\quad \left.\cdot \sin\left(\frac{(2m+1)\pi x}{A}\right) \sin\left(\frac{(2n+1)\pi y}{B}\right)\right\} \\ &= c\left\{1 - H\left(\frac{x}{A}, \frac{Dt}{A^2}\right) \cdot H\left(\frac{y}{B}, \frac{Dt}{B^2}\right)\right\}, \end{aligned}$$

where $H(\xi, \tau)$ is defined in (3.110).

If, in particular, $B = A$ and $y = x$ we get the simple expression

$$\begin{aligned} C(x,x,t) &= c\left(1-\left\{\frac{4}{\pi}\sum_{m=0}^{\infty}\frac{1}{2m+1}\exp\left(-D\left\{\frac{(2m+1)\pi}{A}\right\}^2 t\right)\right.\right. \\ &\qquad \left.\left.\cdot\ \sin\left(\frac{(2m+1)\pi x}{A}\right)\right\}^2\right) \\ &= c\left\{1-H\left(\frac{x}{A},\frac{Dt}{A^2}\right)^2\right\}. \end{aligned}$$

Since $C(x,x,t)$ describes $C(x,y,t)$ on a diagonal, we have the first model of $C(x,y,t)$ near a corner. ◊

4.3.4 Bounded three-dimensional interval

This is a trivial generalization of Section 4.3.3. For constant $D > 0$ the *mathematical model* is

$$\frac{1}{D}\frac{\partial C}{\partial t} = \frac{\partial^2 C}{\partial x^2} + \frac{\partial^2 C}{\partial y^2} + \frac{\partial^2 C}{\partial z^2},\ 0 < x < A,\ 0 < y < B,\ 0 < z < C,\ t > 0,$$

with the *initial condition* $C(x,y,z,0) = 0$ and the six *boundary conditions* expanded in suitable two-dimensional Fourier series,

$$C(0,y,z,t) = f_0(y,z,t) = \sum_{n=1}^{\infty}\sum_{p=1}^{\infty}\alpha_{0,n,p}(t)\sin\left(\frac{n\pi y}{B}\right)\sin\left(\frac{p\pi z}{C}\right),$$

$$C(A,y,z,t) = f_A(y,z,t) = \sum_{n=1}^{\infty}\sum_{p=1}^{\infty}\alpha_{A,n,p}(t)\sin\left(\frac{n\pi y}{B}\right)\sin\left(\frac{p\pi z}{C}\right),$$

$$C(x,0,z,t) = g_0(x,z,t) = \sum_{m=1}^{\infty}\sum_{p=1}^{\infty}\beta_{0,m,p}(t)\sin\left(\frac{m\pi x}{A}\right)\sin\left(\frac{p\pi z}{C}\right),$$

$$C(x,B,z,t) = g_B(x,z,t) = \sum_{m=1}^{\infty}\sum_{p=1}^{\infty}\beta_{B,m,p}(t)\sin\left(\frac{m\pi x}{A}\right)\sin\left(\frac{p\pi z}{C}\right),$$

$$C(x,y,0,t) = h_0(x,y,t) = \sum_{m=1}^{\infty}\sum_{n=1}^{\infty}\gamma_{0,m,n}(t)\sin\left(\frac{m\pi x}{A}\right)\sin\left(\frac{n\pi y}{B}\right),$$

$$C(x,y,C,t) = h_C(x,y,t) = \sum_{m=1}^{\infty}\sum_{n=1}^{\infty}\gamma_{C,m,n}(t)\sin\left(\frac{m\pi x}{A}\right)\sin\left(\frac{n\pi y}{B}\right),$$

where

$$\alpha_{0,n,p}(t) = \frac{4}{BC}\int_0^B\int_0^C f_0(y,z,t)\sin\left(\frac{n\pi y}{B}\right)\sin\left(\frac{p\pi z}{C}\right)dzdy,$$

$$\alpha_{A,n,p}(t) = \frac{4}{BC}\int_0^B\int_0^C f_A(y,z,t)\sin\left(\frac{n\pi y}{B}\right)\sin\left(\frac{p\pi z}{C}\right)dzdy,$$

$$\beta_{0,m,p}(t) = \frac{4}{AC}\int_0^A\int_0^C g_0(x,z,t)\sin\left(\frac{m\pi x}{A}\right)\sin\left(\frac{p\pi z}{C}\right)dzdx,$$

$$\beta_{B,m,p}(t) = \frac{4}{AC}\int_0^A\int_0^C g_B(x,z,t)\sin\left(\frac{m\pi x}{A}\right)\sin\left(\frac{p\pi z}{C}\right)dzdx,$$

$$\gamma_{0,m,n}(t) = \frac{4}{AB}\int_0^A\int_0^B h_0(x,y,t)\sin\left(\frac{m\pi x}{A}\right)\sin\left(\frac{n\pi y}{B}\right)dydx,$$

$$\gamma_{C,m,n}(t) = \frac{4}{AB}\int_0^A\int_0^B h_C(x,y,t)\sin\left(\frac{m\pi x}{A}\right)\sin\left(\frac{n\pi y}{B}\right)dydx.$$

The *weight function* is $w(x,y,z) = 1$, and the *eigenfunctions*, etc., are

$$\psi_{mnp}(x,y,z) = \sin\left(\frac{m\pi x}{A}\right)\sin\left(\frac{n\pi y}{B}\right)\sin\left(\frac{p\pi z}{C}\right),$$

$$\|\psi_{mnp}\|^2 = \frac{ABC}{8}, \qquad \lambda_{mnp} = \left(\frac{m\pi}{A}\right)^2 + \left(\frac{n\pi}{B}\right)^2 + \left(\frac{p\pi}{C}\right)^2.$$

The *solution* in the interior of the domain is

$$C(x,y,z,t) = \sum_{m=1}^{\infty}\sum_{n=1}^{\infty}\sum_{p=1}^{\infty} a_{mnp}(t)\sin\left(\frac{m\pi x}{A}\right)\sin\left(\frac{n\pi y}{B}\right)\sin\left(\frac{p\pi z}{C}\right),$$

where

$$\begin{aligned} a_{mnp}(t) &= \frac{2\pi Dm}{A^2}\int_0^t \exp\left(-D\lambda_{mnp}(t-\tau)\right) \\ &\cdot\left\{\alpha_{0,n,p}(\tau) + (-1)^{m+1}\alpha_{A,n,p}(\tau)\right\}d\tau \\ &+\frac{2\pi Dn}{B^2}\int_0^t \exp\left(-D\lambda_{mnp}(t-\tau)\right) \\ &\cdot\left\{\beta_{0,m,p}(\tau) + (-1)^{n+1}\beta_{B,n,p}(\tau)\right\}d\tau \\ &+\frac{2\pi Dp}{C^2}\int_0^t \exp\left(-D\lambda_{mnp}(t-\tau)\right) \\ &\cdot\left\{\gamma_{0,m,n}(\tau) + (-1)^{p+1}\gamma_{C,m,n}(\tau)\right\}d\tau. \end{aligned}$$

One cannot use the series expansion on the boundary. Instead we use the given boundary conditions.

4.3.5 Circle in the plane

This case has already been covered by the Examples 4.2.1, 4.2.4, 4.2.8 and 4.2.12. For constant $D > 0$ the *mathematical model* is given in polar coordinates by

$$\frac{1}{D}\frac{\partial C}{\partial t} = \frac{\partial^2 C}{\partial r^2} + \frac{1}{r}\frac{\partial C}{\partial r} + \frac{1}{r^2}\frac{\partial^2 C}{\partial \varphi^2}, \quad 0 < r < R,\ 0 \le \varphi \le 2\pi,\ t > 0,$$

with the *initial condition* $C(r, \varphi, 0) = 0$, and the *boundary condition*

$$C(R, \varphi, t) = f(\varphi, t) = \frac{1}{2}\alpha_0(t) + \sum_{n=1}^{\infty} \{\alpha_n(t)\cos n\varphi + \beta_n(t)\sin n\varphi\},$$

expanded in an ordinary Fourier series, where

$$\alpha_n(t) = \frac{1}{\pi}\int_0^{2\pi} f(\varphi, t)\cos(n\varphi)\,d\varphi,$$

$$\beta_n(t) = \frac{1}{\pi}\int_0^{2\pi} f(\varphi, t)\sin(n\varphi)\,d\varphi.$$

Furthermore, we have a natural *periodicity condition*,

$$C(r, 0, t) = C(r, 2\pi, t) \quad 0 < r < R,\ t > 0.$$

The *weight function* is $w(r, \varphi) = r$, and the *eigenfunctions*, etc., are

$$\psi_{m0}(r, \varphi) = J_0\left(j_{0,m}\frac{r}{R}\right), \qquad \|\psi_{m0}\|^2 = \pi R^2 J_0'(j_{0,m})^2,$$

$$\psi_{mn1}(r, \varphi) = J_n\left(j_{n,m}\frac{r}{R}\right)\cos n\varphi, \quad \|\psi_{mn1}\|^2 = \frac{\pi}{2}R^2 J_n'(j_{n,m})^2,$$

$$\psi_{mn2}(r, \varphi) = J_n\left(j_{n,m}\frac{r}{R}\right)\sin n\varphi, \quad \|\psi_{mn2}\|^2 = \frac{\pi}{2}R^2 J_n'(j_{n,m})^2,$$

and

$$\lambda_{m0} = \left(\frac{1}{R}j_{0,m}\right)^2, \qquad \lambda_{mni} = \lambda_{mn} = \left(\frac{1}{R}j_{n,m}\right)^2, \quad i = 1, 2.$$

In the interior of the disc the solution is given by

$$\begin{aligned} C(r, \varphi, t) &= \sum_{m=1}^{\infty} a_{m0}(t) J_0\left(j_{n,m}\frac{r}{R}\right) + \sum_{m=1}^{\infty}\sum_{n=1}^{\infty} a_{mn1}(t) J_n\left(j_{n,m}\frac{r}{R}\right)\cos n\varphi \\ &\quad + \sum_{m=1}^{\infty}\sum_{n=1}^{\infty} a_{mn2}(t) J_n\left(j_{n,m}\frac{r}{R}\right)\sin n\varphi, \end{aligned}$$

where

$$a_{m0}(t) = -\frac{D}{R^2}\frac{j_{0,m}}{J_0'(j_{0,m})}\int_0^t \exp(-D\lambda_{m0}(t-\tau))\,\alpha_0(\tau)\,d\tau,$$

$$a_{mn1}(t) = -\frac{2D}{R^2}\frac{j_{n,m}}{J_n'(j_{n,m})}\int_0^t \exp(-D\lambda_{mn}(t-\tau))\,\alpha_n(\tau)\,d\tau,$$

$$a_{mn2}(t) = -\frac{2D}{R^2}\frac{j_{n,m}}{J_n'(j_{n,m})}\int_0^t \exp(-D\lambda_{mn}(t-\tau))\,\beta_n(\tau)\,d\tau.$$

Example 4.3.3 At the centre of the circle the solution reduces to

$$C(0,\varphi,t) = \sum_{m=1}^{\infty} a_{m0}(t).$$

In the special case, where the boundary condition is constant, $f(\varphi,t) = c$, we get

$$\begin{aligned} a_{m0}(t) &= -\frac{2cD}{R^2}\frac{j_{0,m}}{J_0'(j_{0,m})}\int_0^t \exp\left(-D\left(\frac{1}{R}j_{0,m}\right)^2\tau\right)d\tau \\ &= -\frac{2c}{j_{0,m}J_0'(j_{0,m})}\left\{1-\exp\left(-D\left(\frac{1}{R}j_{0,m}\right)^2 t\right)\right\}. \end{aligned}$$

Using some formulæ from Watson (1944), it is possible to reduce the expression for the solution to

$$C(0,\varphi,t) = c\left\{1+\sum_{m=1}^{\infty}\frac{2}{j_{0,m}J_0'(j_{0,m})}\exp\left(-D\left(\frac{1}{R}j_{0,m}\right)^2 t\right)\right\}.$$

It is easy to show that the series is alternating. For $t = 0$ it is conditionally convergent with the correct sum. For $t > 0$ it is even absolutely convergent. Since the terms are alternating and decreasing in absolute value, the error of any truncated sum is smaller than the absolute value of the first rejected term. Thus, it is easy to find a finite approximation with an error smaller than a given $\varepsilon > 0$. ◊

4.3.6 Sector of a circle in the plane

This case has already been covered by the Examples 4.2.1, 4.2.3, 4.2.5, 4.2.7, 4.2.9, 4.2.11 and 4.2.13. For constant $D > 0$ the *mathematical model* is given in polar coordinates by

$$\frac{1}{D}\frac{\partial C}{\partial t} = \frac{\partial^2 C}{\partial r^2} + \frac{1}{r}\frac{\partial C}{\partial r} + \frac{1}{r^2}\frac{\partial^2 C}{\partial \varphi^2}, \qquad 0 < r < R, \quad 0 < \varphi < \alpha,$$

where $0 < \alpha \leq 2\pi$. The *initial condition* is $C(r,\varphi,0) = 0$, and the *boundary conditions* are

$$
\begin{aligned}
C(R,\varphi,t) &= f(\varphi,t), && 0 < \varphi < \alpha, \\
C(r,0,t) &= g_0(r,t), && 0 < r < R, \\
C(r,\alpha,t) &= g_\alpha(r,t), && 0 < r < R.
\end{aligned}
$$

The *weight function* is $w(r,\varphi) = r$, and the *eigenfunctions*, etc., are

$$\psi_{mn}(r,\varphi) = J_{k_n}\left(j_{k_n,m}\frac{r}{R}\right)\sin k_n\varphi, \qquad \lambda_{mn} = \left(j_{k_n,m}/R\right)^2,$$

$$\|\psi_{mn}\|^2 = \frac{\alpha}{4}R^2 J'_{k_n}\left(j_{k_n,m}\right)^2, \qquad k_n = n\pi/\alpha, \qquad m,n \in \mathbb{N}.$$

For constant $D > 0$ we define

$$
\begin{aligned}
G_{mn}(r,t) &= \int_0^t \exp\left(-D\left(j_{k_n,m}/R\right)^2(t-\tau)\right) \\
&\quad \cdot \left\{g_0(r,\tau) + (-1)^{n+1}g_\alpha(r,\tau)\right\}d\tau, \\
H_{mn}(\varphi,t) &= \int_0^t \exp\left(-D\left(j_{k_n,m}/R\right)^2(t-\tau)\right) f(\varphi,\tau)\,d\tau,
\end{aligned}
$$

and

$$
\begin{aligned}
a_{mn}(t) &= \frac{2n}{\alpha}\frac{2}{R^2 J'_{k_n}\left(j_{k_n,m}\right)^2}\int_0^R \frac{1}{r}G_{mn}(r,t)J_{k_n}\left(j_{k_n,m}\frac{r}{R}\right)dr \\
&\quad -\frac{2j_{k_n,m}}{R^2 J'_{k_n}\left(j_{k_n,m}\right)}\cdot\frac{2}{\alpha}\int_0^\alpha H_{mn}(\varphi,t)\sin(k_n\varphi)\,d\varphi.
\end{aligned}
$$

The solution is in the interior of the domain given by

$$C(r,\varphi,t) = \sum_{m=1}^{\infty}\sum_{n=1}^{\infty} a_{mn}(t)J_{k_n}\left(j_{k_n,m}\frac{r}{R}\right)\sin k_n\varphi,$$

together with the given boundary values on the boundary.

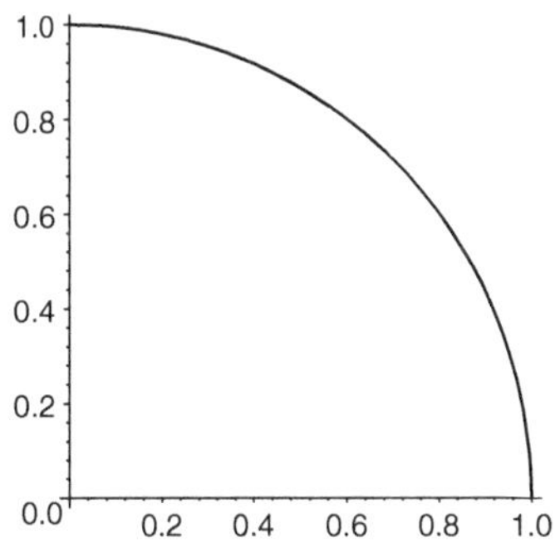

Figure 4.5: Sector of a circle.

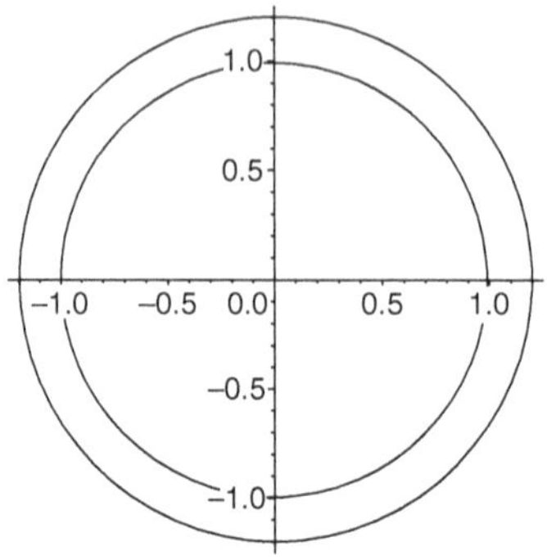

Figure 4.6: The annulus.

4.3.7 Annulus

Again we use polar coordinates. The *mathematical model* is given by

$$\frac{1}{D}\frac{\partial C}{\partial t} = \frac{\partial^2 C}{\partial r^2} + \frac{1}{r}\frac{\partial C}{\partial r} + \frac{1}{r^2}\frac{\partial^2 C}{\partial \varphi^2}, \qquad a < r < b, \quad 0 \le \varphi \le 2\pi,$$

where a is the inner radius, and b the outer radius of the annulus. The *initial condition* is $C(r, \varphi, 0) = 0$, and the two *boundary conditions* are

$$C(a, \varphi, t) = f_a(\varphi, t) \quad \text{and} \quad C(b, \varphi, t) = f_b(\varphi, t).$$

Furthermore, we assume *periodicity*

$$C(r, 0, t) = C(r, 2\pi, t).$$

The *weight function* is $w(r, \varphi) = r$.

When we separate the variables in Helmholtz's equation, we get the general solution of the r-equation,

$$f(r) = c_1 J_n(\sqrt{\lambda} r) + c_2 Y_n(\sqrt{\lambda} r),$$

where $Y_n(\sqrt{\lambda} r)$ makes sense, because $r \ge a > 0$. In order to obtain the proper boundary values for $r = a$ and $r = b$ we are forced to find the zeros of the function

$$J_n(z) Y_n\left(\frac{b}{a} z\right) - J_n\left(\frac{b}{a} z\right) Y_n(z).$$

According to Abramowitz *et al.* (1965), formula 9.5.27, this function has infinitely many positive simple zeros, $z_{n,m} \to \infty$ for $m \to \infty$. When m is small, which is sufficient for our approximate solution, an exercise on a pocket calculator shows that they are not difficult to find, cf. also Example 3.4.1.

Once the $z_{n,m}$ have been found (or the first most important members of this sequence) we define

$$\mu_{n,m} = z_{n,m}/a.$$

Then

$$f_{nm}(r) = Y_n(\mu_{n,m}b)J_n(\mu_{n,m}r) - J_n(\mu_{n,m}b)Y_n(\mu_{n,m}r)$$

obviously implies that $f_{nm}(a) = f_{nm}(b) = 0$, so the boundary conditions are satisfied. By the methods of Section 4.2 we obtain the orthogonal system consisting of the three subfamilies,

$$\psi_{0,m}(r,\varphi) = Y_0(\mu_{0,m}b)J_0(\mu_{0,m}r) - J_0(\mu_{0,m}b)Y_0(\mu_{0,m}r), \quad m \in \mathbb{N}, n = 0,$$

and

$$\psi_{n,m,1}(r,\varphi) = \{Y_n(\mu_{n,m}b)J_n(\mu_{n,m}r) - J_n(\mu_{n,m}b)Y_n(\mu_{n,m}r)\} \cos n\varphi,$$

$$\psi_{n,m,2}(r,\varphi) = \{Y_n(\mu_{n,m}b)J_n(\mu_{n,m}r) - J_n(\mu_{n,m}b)Y_n(\mu_{n,m}r)\} \sin n\varphi,$$

for $m, n \in \mathbb{N}$, where (cf. Abramowitz *et al.* (1965), formula 11.4.2 in the special case $h_1 = 0$ and $h_2 = -1$)

$$\|\psi_{0,m}\|^2 = \pi \left[r^2 \left\{ Y_0(\mu_{0,m}b)J_0'(\mu_{0,m}r) - J_0(\mu_{0,m}b)Y_0'(\mu_{0,m}r) \right\}^2 \right]_{r=a}^{b},$$

$$\|\psi_{n,m,i}\|^2 = \frac{\pi}{2} \left[r^2 \left\{ Y_n(\mu_{n,m}b)J_n'(\mu_{n,m}r) - J_n(\mu_{n,m}b)Y_n'(\mu_{n,m}r) \right\}^2 \right]_{r=a}^{b},$$

for $i = 1, 2$, and

$$\lambda_{n,m} = (\mu_{n,m})^2 = (z_{n,m}/a)^2.$$

Define the constants

$$c_{n,m,a} = \left\{ Y_n\left(\frac{b}{a}z_{n,m}\right) J_n'(z_{n,m}) - J_n\left(\frac{b}{a}z_{n,m}\right) Y_n'(z_{n,m}) \right\} z_{n,m},$$

$$\begin{aligned} c_{n,m,b} &= \left\{ Y_n\left(\frac{b}{a}z_{n,m}\right) J_n'\left(\frac{b}{a}z_{n,m}\right) \right. \\ &\quad \left. - J_n\left(\frac{b}{a}z_{n,m}\right) Y_n'\left(\frac{b}{a}z_{n,m}\right) Y_n'\left(\frac{b}{a}z_{n,m}\right) \right\} \frac{b}{a}z_{n,m}, \end{aligned}$$

and the functions

$$F_{n,m,a}(\varphi,t) = \int_0^t \exp\Big(-D\,(z_{n,m}/a)^2\,(t-\tau)\Big)\, f_a(\varphi,\tau)\, d\tau,$$

$$F_{n,m,b}(\varphi,t) = \int_0^t \exp\Big(-D\,(z_{n,m}/b)^2\,(t-\tau)\Big)\, f_b(\varphi,\tau)\, d\tau.$$

Then the solution is given by

$$\begin{aligned} C(r,\varphi,t) &= \sum_{m=1}^{\infty} a_{0,m}(t)\psi_{0,m}(r,\varphi) + \sum_{m=1}^{\infty}\sum_{n=1}^{\infty} a_{n,m,1}(t)\psi_{n,m,1}(r,\varphi) \\ &\quad + \sum_{m=1}^{\infty}\sum_{n=1}^{\infty} a_{n,m,2}(t)\psi_{n,m,2}(r,\varphi), \end{aligned}$$

where

$$a_{0,m}(t) = \frac{c_{0,m,a}}{\|\psi_{0,m}\|^2} \int_0^{2\pi} F_{0,m,a}(\varphi, t)\, d\varphi - \frac{c_{0,m,b}}{\|\psi_{0,m}\|^2} \int_0^{2\pi} F_{0,m,b}(\varphi, t)\, d\varphi,$$

$$\begin{aligned} a_{n,m,1}(t) &= \frac{c_{n,m,a}}{\|\psi_{n,m,1}\|^2} \int_0^{2\pi} F_{n,m,a}(\varphi, t) \cos(n\varphi) d\varphi \\ &\quad - \frac{c_{n,m,b}}{\|\psi_{n,m,1}\|^2} \int_0^{2\pi} F_{n,m,b}(\varphi, t) \cos(n\varphi) d\varphi, \\ a_{n,m,2}(t) &= \frac{c_{n,m,a}}{\|\psi_{n,m,2}\|^2} \int_0^{2\pi} F_{n,m,a}(\varphi, t) \sin(n\varphi) d\varphi \\ &\quad - \frac{c_{n,m,b}}{\|\psi_{n,m,2}\|^2} \int_0^{2\pi} F_{n,m,b}(\varphi, t) \sin(n\varphi) d\varphi. \end{aligned}$$

4.3.8 Sector of an annulus

The *mathematical model* is in polar coordinates given by

$$\frac{1}{D}\frac{\partial C}{\partial t} = \frac{\partial^2 C}{\partial r^2} + \frac{1}{r}\frac{\partial C}{\partial r} + \frac{1}{r^2}\frac{\partial^2 C}{\partial \varphi^2}, \qquad a < r < b, \quad 0 < \varphi < \alpha,$$

where a is the inner radius, b the outer radius, and α is the angle defining the sector, $0 < \alpha \leq 2\pi$. When $\alpha = 2\pi$, this represents an annulus with a crack. When $\alpha = \pi$ we have one half of an annulus.

The *initial condition* is $C(r, \varphi, 0) = 0$, and the four *boundary conditions* are

$$\begin{array}{lll} C(a, \varphi, t) = f_a(\varphi, t), & C(b, \varphi, t) = f_b(\varphi, t), & 0 < \varphi < \alpha, \\ C(r, 0, t) = g_0(r, t), & C(r, \alpha, t) = g_\alpha(r, t), & a \leq r \leq b. \end{array}$$

The *weight function* is $w(r, \varphi) = r$.

Let $k_n = 2\pi n/\alpha$, and let $z_{k_n,m}$ be the zeros of the function, cf. Section 4.3.7,

$$J_{k_n}(z) Y_{k_n}\left(\frac{b}{a} z\right) - J_{k_n}\left(\frac{b}{a} z\right) Y_{k_n}(z).$$

Define

$$\mu_{k_n,m} = z_{k_n,m}/a.$$

The orthogonal system for this problem is

$$\begin{aligned} \psi_{nm}(r, \varphi) &= \{Y_{k_n}(\mu_{k_n,m} b)\, J_{k_n}(\mu_{k_n,m} r) \\ &\quad - J_{k_n}(\mu_{k_n,m} b)\, Y_{k_n}(\mu_{k_n,m} r)\} \sin k_n \varphi, \end{aligned}$$

where

$$\|\psi_{nm}\|^2 = \frac{\pi}{2}\left[r^2\left\{Y_{k_n}(\mu_{k_n,m}b)\,J'_{k_n}(\mu_{k_n,m}) - J_{k_n}(\mu_{k_n,m})\,Y'_{k_n}(\mu_{k_n,m}r)\right\}^2\right]_{r=a}^{b},$$

and

$$\lambda_{n,m} = (\mu_{k_n,m})^2 = (z_{k_n,m}/a)^2 .$$

Define the constants

$$c_{n,m,a} = \left\{Y_{k_n}\left(\frac{b}{a}z_{k_n,m}\right)J'_{k_n}(z_{k_n,m}) - J'_{k_n}\left(\frac{b}{a}z_{k_n,m}\right)Y'_{k_n}(z_{k_n,m})\right\}z_{k_n,m},$$

$$\begin{aligned} c_{n,m,b} &= \left\{Y_{k_n}\left(\frac{b}{a}z_{k_n,m}\right)J'_{k_n}\left(\frac{b}{a}z_{k_n,m}\right)\right. \\ &\quad \left. - J'_{k_n}\left(\frac{b}{a}z_{k_n,m}\right)Y'_{k_n}\left(\frac{b}{a}z_{k_n,m}\right)\right\}\frac{b}{a}z_{k_n,m}, \end{aligned}$$

and the functions

$$F_{n,m,a}(\varphi,t) = \int_0^t \exp\Big(D\cdot(z_{k_n,m}/a)^2(\tau-t)\Big)\,f_a(\varphi,\tau)\,d\tau,$$

$$F_{n,m,b}(\varphi,t) = \int_0^t \exp\Big(D\cdot(z_{k_n,m}/a)^2(\tau-t)\Big)\,f_b(\varphi,\tau)\,d\tau,$$

$$G_{n,m,0}(r,t) = \int_0^t \exp\Big(D\cdot(z_{k_n,m}/a)^2(\tau-t)\Big)\,g_0(r,\tau)\,d\tau,$$

$$G_{n,m,\alpha}(r,t) = \int_0^t \exp\Big(D\cdot(z_{k_n,m}/a)^2(\tau-t)\Big)\,g_\alpha(r,\tau)\,d\tau.$$

The solution is given by

$$C(r,\varphi,t) = \sum_{m=1}^{\infty}\sum_{n=1}^{\infty} a_{n,m}(t)\psi_{nm}(r,\varphi),$$

where

$$\begin{aligned} a_{nm}(t) &= \frac{c_{n,m,a}}{\|\psi_{nm}\|^2}\int_0^\alpha F_{n,m,a}(\varphi,t)\sin(k_n\varphi)\,d\varphi \\ &\quad -\frac{c_{n,m,b}}{\|\psi_{nm}\|^2}\int_0^\alpha F_{n,m,b}\sin(k_n\varphi)\,d\varphi \\ &\quad +\frac{k_n}{\|\psi_{nm}\|^2}\int_a^b \frac{1}{r}\left\{G_{n,m,0}(r,t) + (-1)^{n+1}G_{n,m,\alpha}(r,t)\right\} \\ &\quad \cdot\left\{Y_{k_n}(\mu_{k_n,m}b)J_{k_n}(\mu_{k_n,m}r) - J_{k_n}(\mu_{k_n,m}b)\,Y_{k_n}(\mu_{k_n,m}r)\right\}dr. \end{aligned}$$

4.3.9 Finite cylindrical column

Using semi-polar coordinates in three dimensions,

$$x = r\cos\varphi, \quad y = r\sin\varphi, \quad z = z,$$

the *mathematical model* is

$$\frac{1}{D}\frac{\partial C}{\partial t} = \frac{\partial^2 C}{\partial r^2} + \frac{1}{r}\frac{\partial C}{\partial r} + \frac{1}{r^2}\frac{\partial^2 C}{\partial \varphi^2} + \frac{\partial^2 C}{\partial z^2}$$

in the open domain

$$0 \le r < R, \quad 0 \le \varphi \le 2\pi, \quad 0 < z < H, \quad t > 0.$$

Here, R is the radius and H is the height of the column.

The *initial condition* is $C(r, \varphi, z, 0) = 0$, and there are three given *boundary conditions*,

$$\begin{array}{ll} C(r,\varphi,0,t) = g_0(r,\varphi,t), & \text{bottom of the column,} \\ C(r,\varphi,H,t) = g_H(r,\varphi,t), & \text{top of the column,} \\ C(R,\varphi,z,t) = f(\varphi,z,t), & \text{cylindric surface.} \end{array}$$

Furthermore, we have the natural *periodicity condition,*

$$C(r, 0, z, t) = C(r, 2\pi, z, t).$$

The *weight function* is $w(r, \varphi, z) = r$, and the *eigenfunctions*, etc., are

$$\psi_{0,m,p}(r,\varphi,z) = J_0\Big(j_{0,m}\frac{r}{R}\Big)\sin\Big(\frac{p\pi z}{H}\Big), \qquad n = 0,\ m, p \in \mathbb{N},$$

$$\psi_{n,m,p,1}(r,\varphi,z) = J_n\Big(j_{n,m}\frac{r}{R}\Big)\cos(n\varphi)\sin\Big(\frac{p\pi z}{H}\Big), \quad n, m, p \in \mathbb{N},$$

$$\psi_{n,m,p,2}(r,\varphi,z) = J_n\Big(j_{n,m}\frac{r}{R}\Big)\sin(n\varphi)\sin\Big(\frac{p\pi z}{H}\Big), \quad n, m, p \in \mathbb{N},$$

where

$$\|\psi_{0,m,p}\|^2 = \frac{\pi H R^2}{2}\left\{J_0'(j_{0,m})\right\}^2,$$

$$\|\psi_{n,m,p,1}\|^2 = \|\psi_{n,m,p,2}\|^2 = \frac{\pi H R^2}{4}\left\{J_n'(j_{n,m})\right\}^2,$$

and

$$\lambda_{nmp} = (j_{n,m}/R)^2 + (p\pi/H)^2, \quad n \in \mathbb{N}_0,\ m \in \mathbb{N},\ p \in \mathbb{N}.$$

Expand the boundary conditions in the following series

$$\begin{aligned} g_0(r,\varphi,t) &= \sum_{m=1}^{\infty} \alpha_{0m}(t) J_0\Big(j_{0,m}\frac{r}{R}\Big) + \sum_{m=1}^{\infty}\sum_{n=1}^{\infty} \alpha_{nm1}(t) J_n\Big(j_{n,m}\frac{r}{R}\Big)\cos n\varphi \\ &\quad + \sum_{m=1}^{\infty}\sum_{n=1}^{\infty} \alpha_{nm2}(t) J_n\Big(j_{n,m}\frac{r}{R}\Big)\sin n\varphi, \end{aligned}$$

$$\begin{aligned} g_H(r,\varphi,t) &= \sum_{m=1}^{\infty} \beta_{0m}(t) J_0\Big(j_{0,m}\frac{r}{R}\Big) + \sum_{m=1}^{\infty}\sum_{n=1}^{\infty} \beta_{nm1}(t) J_n\Big(j_{n,m}\frac{r}{R}\Big)\cos n\varphi \\ &\quad + \sum_{m=1}^{\infty}\sum_{n=1}^{\infty} \beta_{nm2}(t) J_n\Big(j_{n,m}\frac{r}{R}\Big)\sin n\varphi, \end{aligned}$$

$$\begin{aligned} f(\varphi,z,t) &= \sum_{p=1}^{\infty} \gamma_{0p}(t)\sin\Big(\frac{p\pi z}{H}\Big) + \sum_{p=1}^{\infty}\sum_{n=1}^{\infty} \gamma_{np1}(t)\sin\Big(\frac{p\pi z}{H}\Big)\cos n\varphi \\ &\quad + \sum_{p=1}^{\infty}\sum_{n=1}^{\infty} \gamma_{np2}(t)\sin\Big(\frac{p\pi z}{H}\Big)\sin n\varphi, \end{aligned}$$

where

$$\alpha_{0m} = \frac{1}{\pi R^2\{J_0'(j_{0,m})\}^2}\int_0^{2\pi}\left\{\int_0^R g_0(r,\varphi,t)J_0\Big(j_{0,m}\frac{r}{R}\Big)\,r\,dr\right\}d\varphi,$$

$$\begin{aligned} \alpha_{nm1}(t) &= \frac{2}{\pi R^2\{J_n'(j_{n,m})\}^2} \\ &\quad \cdot\int_0^{2\pi}\left\{\int_0^R g_0(r,\varphi,t)J_n\Big(j_{n,m}\frac{r}{R}\Big)\,r\,dr\right\}\cos(n\varphi)\,d\varphi, \end{aligned}$$

$$\begin{aligned} \alpha_{nm2}(t) &= \frac{2}{\pi R^2\{J_n'(j_{n,m})\}^2} \\ &\quad \cdot\int_0^{2\pi}\left\{\int_0^R g_0(r,\varphi,t)J_n\Big(j_{n,m}\frac{r}{R}\Big)\,r\,dr\right\}\sin(n\varphi)\,d\varphi, \end{aligned}$$

and

$$\beta_{0m}(t) = \frac{1}{\pi R^2\{J_0'(j_{0,m})\}^2}\int_0^{2\pi}\left\{\int_0^R g_H(r,\varphi,t)J_0\Big(j_{0,m}\frac{r}{R}\Big)\,r\,dr\right\}d\varphi,$$

$$\begin{aligned} \beta_{nm1}(t) &= \frac{2}{\pi R^2\{J_n'(j_{n,m})\}^2} \\ &\quad \cdot\int_0^{2\pi}\left\{\int_0^R g_H(r,\varphi,t)J_n\Big(j_{n,m}\frac{r}{R}\Big)\,r\,dr\right\}\cos(n\varphi)\,d\varphi, \end{aligned}$$

$$\beta_{nm2}(t) = \frac{2}{\pi R^2\{J_n'(j_{n,m})\}^2} \cdot \int_0^{2\pi}\left\{\int_0^R g_H(r,\varphi,t) J_n\left(j_{n,m}\frac{r}{R}\right) r\,dr\right\}\sin(n\varphi)\,d\varphi,$$

and

$$\gamma_{0p}(t) = \frac{2}{\pi H}\int_0^{2\pi}\left\{\int_0^H f(\varphi,z,t)\sin\left(\frac{p\pi z}{H}\right)dz\right\}d\varphi,$$

$$\gamma_{np1}(t) = \frac{4}{\pi H}\int_0^{2\pi}\left\{\int_0^H f(\varphi,z,t)\sin\left(\frac{p\pi z}{H}\right)dz\right\}\cos(n\varphi)\,d\varphi,$$

$$\gamma_{np2}(t) = \frac{4}{\pi H}\int_0^{2\pi}\left\{\int_0^H f(\varphi,z,t)\sin\left(\frac{p\pi z}{H}\right)dz\right\}\sin(n\varphi)\,d\varphi.$$

Define the functions

$$F_{0mp}(t) = \frac{2p}{H^2}\left\{\alpha_{0m}(t) + (-1)^{p+1}\beta_{0m}(t)\right\} - \frac{j_{0,m}}{\pi R^2 J_0'(j_{0,m})}\gamma_{0p}(t),$$

$$F_{nmpi}(t) = \frac{2p}{H^2}\left\{\alpha_{nmi}(t) + (-1)^{p+1}\beta_{nmi}(t)\right\} - \frac{j_{n,m}}{\pi R^2 J_n'(j_{n,m})}\gamma_{npi}(t),$$

for $i = 1, 2$. The solution of Fick's second law is in the interior of the column given by

$$\begin{aligned} C((r,\varphi,z,t) &= \sum_{m=1}^{\infty}\sum_{p=1}^{\infty} a_{0mp}(t) J_0\left(j_{0,m}\frac{r}{R}\right)\sin\left(\frac{p\pi z}{H}\right) \\ &+ \sum_{m=1}^{\infty}\sum_{n=1}^{\infty}\sum_{p=1}^{\infty} a_{nmp1}(t) J_n\left(j_{n,m}\frac{r}{R}\right)\cos(n\varphi)\sin\left(\frac{p\pi z}{H}\right) \\ &+ \sum_{m=1}^{\infty}\sum_{n=1}^{\infty}\sum_{p=1}^{\infty} a_{nmp2}(t) J_n\left(j_{n,m}\frac{r}{R}\right)\sin(n\varphi)\sin\left(\frac{p\pi z}{H}\right), \end{aligned}$$

where the coefficients are given by $[a_{0mp}(t) = a_{0mp}(t)$ for $i = 1, 2$, when $n = 0]$

$$a_{nmpi}(t) = D\int_0^t \exp\left(-D\left\{\left(\frac{j_{n,m}}{R}\right)^2 + \left(\frac{p\pi}{H}\right)^2\right\}(t-\tau)\right) F_{nmpi}(\tau)\,d\tau.$$

On the boundary we use the given boundary values instead.

4.3.10 Sector of a finite cylindrical column

Using semi-polar coordinates

$$x = r\cos\varphi, \qquad y = r\sin\varphi, \qquad z = z,$$

a sector of a finite circular column is described by

$$\Omega = \{(r, \varphi, z) \mid 0 < r < R,\, 0 < \varphi < v,\, 0 < v < H\},$$

where $0 < v \leq 2\pi$. The cases of greatest interest are $v = 2\pi$ (a column with a vertical crack to the centre line), $v = \pi$ (one half of a column) and $v = \pi/2$ (a quarter of a column).

The *mathematical model* is

$$\frac{1}{D}\frac{\partial C}{\partial t} = \frac{\partial^2 C}{\partial r^2} + \frac{1}{r}\frac{\partial C}{\partial r} + \frac{1}{r^2}\frac{\partial^2 C}{\partial \varphi^2} + \frac{\partial^2 C}{\partial z^2}, \quad (r, \varphi, z) \in \Omega \text{ and } t > 0.$$

The *weight function* is $w(r, \varphi, z) = r$.

Let $k_n = n\pi/v$, $n \in \mathbb{N}$. The eigenfunction method then gives us the following orthogonal basis

$$\psi_{mnp}(r, \varphi, z) = J_{k_n}\left(j_{k_n,m}\frac{r}{R}\right) \sin\left(k_n\varphi\right) \sin\left(\frac{p\pi z}{H}\right), \quad m, n, p \in \mathbb{N},$$

where

$$\|\psi_{mnp}\|^2 = \frac{vH}{8}\left\{RJ'_{k_n}(j_{k_n,m})\right\}^2, \quad m, n, p \in \mathbb{N}.$$

The corresponding λ_{mnp} are

$$\lambda_{mnp} = \left(j_{k_n,m}/R\right)^2 + (p\pi/H)^2, \quad m, n, p \in \mathbb{N}.$$

The *initial condition* is $C(r, \varphi, z, 0) = 0$.

There are five natural *boundary condition.* The structure of the eigenfunctions ψ_{mnp} invites us to the following expansions of these:

(i) *Bottom,*

$$C(r, \varphi, 0, t) = f_0(r, \varphi, t) = \sum_{m=1}^{\infty}\sum_{n=1}^{\infty} \alpha_{mn}(t) J_{k_n}\left(j_{k_n,m}\frac{r}{R}\right) \sin(k_n\varphi),$$

where

$$\alpha_{mn}(t) = \frac{4}{v\{RJ'_{k_n}(j_{k_n,m})\}^2} \cdot \int_0^v \left\{\int_0^R f_0(r, \varphi, t) J_{k_n}\left(j_{k_n,m}\frac{r}{R}\right) r\, dr\right\} \sin(k_n\varphi)\, d\varphi.$$

(ii) *Top,*

$$C(r, \varphi, H, t) = f_H(r, \varphi, t) = \sum_{m=1}^{\infty}\sum_{n=1}^{\infty} \beta_{mn}(t) J_{k_n}\left(j_{k_n,m}\frac{r}{R}\right) \sin(k_n\varphi),$$

where

$$\beta_{mn}(t) = \frac{4}{v\{RJ'_{k_n}(j_{k_n,m})\}^2} \cdot \int_0^v \left\{ \int_0^R f_H(r,\varphi,t) J_{k_n}\left(j_{k_n,m}\frac{r}{R}\right) r\,dr \right\} \sin(k_n\varphi)\,d\varphi.$$

(iii) *First plane surface,*

$$C(r,0,z,t) = g_0(r,z,t) = r^2 \sum_{m=1}^{\infty}\sum_{p=1}^{\infty} \gamma_{mp}(t) J_{k_n}\left(j_{k_n,m}\frac{r}{R}\right) \sin\left(\frac{p\pi z}{H}\right),$$

where

$$\gamma_{mn}(t) = \frac{4}{H\{RJ'_{k_n}(j_{k_n,m})\}^2} \cdot \int_0^H \left\{ \int_0^R g_0(r,z,t) J_{k_n}\left(j_{k_n,m}\frac{r}{R}\right) \frac{dr}{r} \right\} \sin\left(\frac{p\pi z}{H}\right) dz.$$

(iv) *Second plane surface,*

$$C(r,v,z,t) = g_v(r,z,t) = r^2 \sum_{m=1}^{\infty}\sum_{p=1}^{\infty} \delta_{mp}(t) J_{k_n}\left(j_{k_n,m}\frac{r}{R}\right) \sin\left(\frac{p\pi z}{H}\right),$$

where

$$\delta_{mp}(t) = \frac{4}{H\{RJ'_{k_n}(j_{k_n,m})\}^2} \cdot \int_0^H \left\{ \int_0^R g_v(r,z,t) J_{k_n}\left(j_{k_n,m}\frac{r}{R}\right) \frac{dr}{r} \right\} \sin\left(\frac{p\pi z}{H}\right) dz.$$

(v) *Cylindric surface,*

$$C(R,\varphi,z,t) = h(\varphi,z,t) = \sum_{n=1}^{\infty}\sum_{p=1}^{\infty} \varepsilon_{np}(t) \sin(k_n\varphi) \sin\left(\frac{p\pi z}{H}\right),$$

where

$$\varepsilon_{np}(t) = \frac{4}{vH} \int_0^H \left\{ \int_0^v h(\varphi,z,t) \sin(k_n\varphi)\,d\varphi \right\} \sin\left(\frac{p\pi z}{H}\right) dz.$$

Define

$$\begin{aligned} F_{mnp}(t) &= \frac{2p\pi}{H^2}\{\alpha_{mn}(t)+(-1)^{p+1}\beta_{mn}(t)\} \\ &\quad +\frac{2n\pi}{v^2}\{\gamma_{mp}(t)+(-1)^{n+1}\delta_{mp}(t)\} \\ &\quad -2j_{k_n,m}\cdot\left\{R^2J'_{k_n}(j_{k_n,m})\right\}^{-1}\varepsilon_{np}(t), \qquad m,n,p\in\mathbb{N}, \end{aligned}$$

where all the terms are given by the expansion of the boundary conditions. Notice that we expand $g_0(r,z,t)/R^2$ and $g_v(r,z,t)/r^2$ instead of $g_0(r,z,t)$ and $g_v(r,z,t)$.

The solution is in the open set Ω given by

$$C(r,\varphi,z,t)=\sum_{m,n,p=1}^{\infty} a_{mnp}(t)J_{k_n}\left(j_{k_n,m}\frac{r}{R}\right)\sin(k_n\varphi)\sin\left(\frac{p\pi z}{H}\right),$$

where

$$a_{mnp}(t)=D\int_0^t \exp(-D\lambda_{mnp}(t-\tau))F_{mnp}(\tau)\,d\tau.$$

On the boundary of Ω we use the given boundary conditions of the problem.

In practice one only calculates the "lowest" terms of the series. Here, "lowest" is taken in the sense that λ_{mnp} is small, because terms with large λ_{mnp} will only give a negligible contribution as time evolves.

4.3.11 Finite pipe

Using semi-polar coordinates

$$x=r\cos\varphi,\quad y=r\sin\varphi,\quad z=z,$$

a finite pipe is described by

$$\Omega=\{(r,\varphi,z)\mid a<r<b,\ 0\le\varphi\le 2\pi,\ 0<z<H\},$$

where a is the inner radius, b is the outer radius, and H is the height of the pipe.

The *mathematical model*, i.e. Fick's second law, is here

$$\frac{1}{D}\frac{\partial C}{\partial t}=\frac{\partial^2 C}{\partial r^2}+\frac{1}{r}\frac{\partial C}{\partial r}+\frac{1}{r^2}\frac{\partial^2 C}{\partial\varphi^2}+\frac{\partial^2 C}{\partial z^2}\qquad \text{in } \Omega,$$

with the *initial condition* $C(r,\varphi,z,0)=0$.

The *weight function* is $w(r,\varphi,z)=r$.

The present case is obviously an extension of Section 4.3.7 *Annulus*, so we can follow the same procedure, modifying whenever it is necessary.

First we must find the positive zeros $z_{n,m}$ of the function

$$J_n(z)Y_n\left(\frac{b}{a}z\right) - J_n\left(\frac{b}{a}z\right)Y_n(z).$$

By assumption, $0 < a < b$, so we conclude from Abramowitz *et al.* (1965), formula 9.5.27, that the zeros are all simple and that $z_{n,m} \to \infty$ for fixed n and $m \to \infty$. The most important zeros for the solution are the smallest ones. These are easy to find numerically, once programmes for $J_n(z)$ and $Y_n(z)$ have been designed for e.g. a pocket calculator.

Once the $z_{n,m}$ (or a reasonable finite subset of them) have been found, we adjust them for simplicity to

$$\mu_{n,m} = \frac{1}{a}z_{n,m}.$$

Since the z coordinate is orthogonal to the r and φ coordinates, we obtain the orthogonal system here by multiplying the orthogonal system from Section 4.3.7 by the orthogonal system for the z variable over $[0, H]$, i.e. we multiply by $\sin(p\pi z/H)$. Thus the present orthogonal system is the union of the following three subfamilies,

$$\psi_{m0p}(r,\varphi,z) = \{Y_0(\mu_{0,m}b)J_0(\mu_{0,m}r) - J_0(\mu_{0,m}b)Y_0(\mu_{0,m}r)\}\sin\left(\frac{p\pi z}{H}\right),$$

$$\begin{aligned}\psi_{mnp1}(r,\varphi,z) &= \{Y_n(\mu_{n,m}b)J_n(\mu_{n,m}r) - J_n(\mu_{n,m}b)Y_n(\mu_{n,m}r)\}\\ &\quad\cdot\cos(n\varphi)\sin\left(\frac{p\pi z}{H}\right),\end{aligned}$$

$$\begin{aligned}\psi_{mnp2}(r,\varphi,z) &= \{Y_n(\mu_{n,m}b)J_n(\mu_{n,m}r) - J_n(\mu_{n,m}b)Y_n(\mu_{n,m}r)\}\\ &\quad\cdot\sin(n\varphi)\sin\left(\frac{p\pi z}{H}\right),\end{aligned}$$

where $m \in \mathbb{N}$, $n \in \mathbb{N}_0$ and $p \in \mathbb{N}$. The expressions above are too complicated when we later use them in explicit formulæ. To ease matters we introduce the following notation,

$$\mathcal{C}_{n,m}(u) = Y_n(\mu_{n,m}b)J_n(u) - J_n(\mu_{n,m}b)Y_n(u).$$

Then

$$\begin{aligned}\mathcal{C}_{n,m}(\mu_{n,m}r) &= \mathcal{C}_{n,m}\left(z_{n,m}\frac{r}{a}\right)\\ &= Y_n(\mu_{n,m}b)J_n(\mu_{n,m}r) - J_n(\mu_{n,m}b)Y_n(\mu_{n,m}r),\end{aligned}$$

and the orthogonal system can be written in the shorter form

$$\psi_{m0p}(r,\varphi,z) = \mathcal{C}_{0,m}\left(z_{0,m}\frac{r}{a}\right)\sin\left(\frac{p\pi z}{H}\right), \qquad m,p\in\mathbb{N},$$

$$\psi_{mnp1}(r,\varphi,z) = \mathcal{C}_{n,m}\left(z_{n,m}\frac{r}{a}\right)\cos(n\varphi)\sin\left(\frac{p\pi z}{H}\right), \quad m,n,p\in\mathbb{N},$$

$$\psi_{mnp2}(r,\varphi,z) = \mathcal{C}_{n,m}\left(z_{n,m}\frac{r}{a}\right)\sin(n\varphi)\sin\left(\frac{p\pi z}{H}\right), \quad m,n,p\in\mathbb{N}.$$

A simple check of the definitions of $\mathcal{C}_{n,m}(\mu_{n,m}r)$ and $\mu_{n,m}$ shows that

$$\mathcal{C}(\mu_{n,m}a) = \mathcal{C}_{n,m}(\mu_{n,m}b) = 0. \tag{4.37}$$

Hence, all ψ_{mnp} are zero for $r = a$ or $r = b$. By using Abramowitz *et al.* (1965), formula 9.1.27, we get

$$\mathcal{C}'_{n,m}(r) = -\mathcal{C}_{n+1,m}(r) + \frac{n}{r}\mathcal{C}_{n,m}(r),$$

so by Abramowitz *et al.* (1965), formula 11.4.2, and (4.37) above we get

$$\int_a^b \mathcal{C}_{n,m}(\mu_{n,m}r)^2 r\,dr = \frac{1}{2}\left[r^2\mathcal{C}'_{n,m}(\mu_{n,m}r)^2\right]_a^b = \frac{1}{2}\left[r^2\mathcal{C}_{n+1,m}\left(z_{n,m}\frac{r}{a}\right)^2\right]_a^b.$$

By using this result it is easy to derive the expressions

$$\|\psi_{m0p}\|^2 = \frac{\pi H}{2}\left[r^2\mathcal{C}_{1,m}\left(z_{0,m}\frac{r}{a}\right)^2\right]_{r=a}^b, \qquad n=0,\ m,p\in\mathbb{N},$$

$$\|\psi_{mnpi}\|^2 = \frac{\pi H}{4}\left[r^2\mathcal{C}_{n+1,m}\left(z_{n,m}\frac{r}{a}\right)\right]_{r=a}^b, \quad m,n,p\in\mathbb{N}, \quad i=1,2.$$

The corresponding λ_{mnp} are

$$\lambda_{mnp} = \mu_{n,m}^2 + \left(\frac{p\pi}{H}\right)^2 = \frac{1}{a^2}z_{n,m}^2 + \left(\frac{p\pi}{H}\right)^2, \quad m\in\mathbb{N},\ n\in\mathbb{N}_0,\ p\in\mathbb{N},$$

where $\lambda_{mnp1} = \lambda_{mnp2} = \lambda_{mnp}$, when $n\in\mathbb{N}$. When approximating the final solution we see that the only relevant indices (m,n,p) are those, for which λ_{mnp} is small. Given any $C > 0$ we see that $\lambda_{mnp} \le C$ only for a finite number of indices, and the final solution may be truncated by only calculating the contributions from these indices.

The *boundary conditions* are given by four functions supplied with a *periodicity condition* in φ:

(i) *Bottom*, i.e. $z = 0$,

$$C(r,\varphi,0,t) = f_0(r,\varphi,t), \quad a\le r\le b \text{ and } 0\le\varphi\le 2\pi.$$

(ii) *Top*, i.e. $z = H$,

$$C(r, \varphi, H, t) = f_H(r, \varphi, t), \quad a \le r \le b \text{ and } 0 \le \varphi \le 2\pi.$$

(iii) *Inner cylindric surface*, i.e. $r = a$,

$$C(a, \varphi, z, t) = g_a(\varphi, z, t), \quad 0 \le \varphi \le 2\pi \text{ and } 0 \le z \le H.$$

(iv) *Outer cylindric surface*, i.e. $r = b$,

$$C(b, \varphi, z, t) = g_b(\varphi, z, t), \quad 0 \le \varphi \le 2\pi \text{ and } 0 \le z \le H.$$

(v) *Periodicity*,

$$C(b, \varphi, z, t) = C(r, 2\pi, z, t), \quad a \le r \le b \text{ and } 0 \le z \le H.$$

The solution of this initial/boundary value problem for Fick's second law in Ω is given by the following monster,

$$\begin{aligned} C(r, \varphi, z, t) &= \sum_{m=1}^{\infty}\sum_{p=1}^{\infty} a_{m0p}(t)\psi_{m0p}(r, \varphi, z) \\ &\quad + \sum_{m=1}^{\infty}\sum_{n=1}^{\infty}\sum_{p=1}^{\infty} a_{mnp1}(t)\psi_{mnp1}(r, \varphi, z) \\ &\quad + \sum_{m=1}^{\infty}\sum_{n=1}^{\infty}\sum_{p=1}^{\infty} a_{mnp2}(t)\psi_{mnp2}(r, \varphi, z), \end{aligned}$$

together with the boundary conditions; the coefficients $a_{mnpi}(t)$ will be defined in the following. As mentioned above only a finite number of terms are needed for an acceptable approximation of the solution.

The coefficients $a_{mnpi}(t)$ are expressed by the following integrals in terms of some functions $F_{mnpi}(t)$, which are in turn specified by the boundary conditions,

$$a_{m0p}(t) = D\int_0^t \exp(-D \cdot \lambda_{m0p}(t - \tau))\, F_{m0p}(\tau)\, d\tau,$$

$$a_{mnpi}(t) = D\int_0^t \exp(-D \cdot \lambda_{mnp}(t - \tau))\, F_{mnpi}(\tau)\, d\tau, \quad i = 1, 2.$$

At a first glance the following expressions of $F_{mnpi}(t)$ may be quite scaring. However, they only contain known functions, so they can be calculated by using a PC, or just an advanced pocket calculator. The formulæ are

(i) For $n = 0$ we get

$$F_{m0p}(t) = \frac{p\pi}{H\|\psi_{m0p}\|^2}$$

$$\cdot \int_0^{2\pi} \left\{ \int_a^b \{f_0(r,\varphi,t) + (-1)^{p+1} f_H(r,\varphi,t)\} \mathcal{C}_{0,m}(\mu_{0,m} r) r \, dr \right\} d\varphi$$

$$+ \frac{z_{0,m} \mathcal{C}'_{0,m}(z_{0,m})}{\|\psi_{m0p}\|^2} \int_0^{2\pi} \left\{ \int_0^H g_a(\varphi,z,t) \sin\left(\frac{p\pi z}{H}\right) dz \right\} d\varphi$$

$$- \frac{b}{a} \frac{z_{0,m} \mathcal{C}'_{0,m}(\frac{b}{a} z_{0,m})}{\|\psi_{m0p}\|^2} \int_0^{2\pi} \left\{ \int_0^H g_b(\varphi,z,t) \sin\left(\frac{p\pi z}{H}\right) dz \right\} d\varphi.$$

(ii) For $n \in \mathbb{N}$ and $i = 1$ we get

$$F_{mnp1}(t) = \frac{p\pi}{H\|\psi_{mnp1}\|^2}$$

$$\cdot \int_0^{2\pi} \left\{ \int_a^b \{f_0(r,\varphi,t) + (-1)^{p+1} f_H(r,\varphi,t)\} \mathcal{C}_{n,m}(\mu_{n,m} r) r dr \right\} \cos(n\varphi) d\varphi$$

$$+ \frac{z_{n,m} \mathcal{C}'_{n,m}(z_{n,m})}{\|\psi_{mnp1}\|^2} \int_0^{2\pi} \left\{ \int_0^H g_a(\varphi,z,t) \sin\left(\frac{p\pi z}{H}\right) dz \right\} \cos(n\varphi) d\varphi$$

$$- \frac{b}{a} \frac{z_{n,m} \mathcal{C}'_{n,m}(\frac{b}{a} z_{n,m})}{\|\psi_{mnp1}\|^2} \int_0^{2\pi} \left\{ \int_0^H g_b(\varphi,z,t) \sin\left(\frac{p\pi z}{H}\right) dz \right\} \cos(n\varphi) \, d\varphi.$$

(iii) For $n \in \mathbb{N}$ and $i = 2$ we get

$$F_{mnp2}(t) = \frac{p\pi}{H\|\psi_{mnp2}\|^2}$$

$$\cdot \int_0^{2\pi} \left\{ \int_a^b \{f_0(r,\varphi,t) + (-1)^{p+1} f_H(r,\varphi,t)\} \mathcal{C}_{n,m}(\mu_{n,m} r) r dr \right\} \sin(n\varphi) d\varphi$$

$$+ \frac{z_{n,m} \mathcal{C}'_{n,m}(z_{n,m})}{\|\psi_{mnp2}\|^2} \int_0^{2\pi} \left\{ \int_0^H g_a(\varphi,z,t) \sin\left(\frac{p\pi z}{H}\right) dz \right\} \sin(n\varphi) \, d\varphi$$

$$- \frac{b}{a} \frac{z_{n,m} \mathcal{C}'_{n,m}(\frac{b}{a} z_{n,m})}{\|\psi_{mnp2}\|^2} \int_0^{2\pi} \left\{ \int_0^H g_b(\varphi,z,t) \sin\left(\frac{p\pi z}{H}\right) dz \right\} \sin(n\varphi) \, d\varphi.$$

These expressions explain why the applications of the fully developed theory of eigenfunctions was given up 50 years ago. The integrations are so difficult that the old-fashioned way of calculating the coefficients by using paper and pencil would be overwhelming, let alone the estimates of the results. Today one has got the device of symbolic calculations on PCs, so the task has now become possible.

4.3.12 Sector of a finite pipe

The lesson from the previous section is that it should be possible today also to solve the initial/boundary value problem for Fick's second law for a sector of a finite pipe. This is indeed true. The explicit solution formula, however, is extremely complicated, and since this case cannot be said to be in the mainstream, as far as chloride ingress into concrete is concerned, we shall here only sketch the main results, leaving the task of developing the explicit formulæ to the very few readers who are specially interested in this topic. The method is clear. Just copy the approach from Section 4.2. The details are overwhelming.

A sector of a finite pipe is described in semi-polar coordinates by

$$\Omega = \{(r, \varphi, z) \mid a < r < b,\ 0 < \varphi < v,\ 0 < z < H\},$$

where a is the inner radius, b is the outer radius, and H is the height. Furthermore, the sector is defined by an angle v satisfying $0 < v \le 2\pi$, where the special case $v = 2\pi$ represents a pipe with a crack, while $v = \pi$ represents one half of a pipe.

As usual the *mathematical model* (Fick's second law) has the form

$$\frac{1}{D}\frac{\partial C}{\partial t} = \frac{\partial^2 C}{\partial r^2} + \frac{1}{r}\frac{\partial C}{\partial r} + \frac{1}{r^2}\frac{\partial^2 C}{\partial \varphi^2} + \frac{\partial^2 C}{\partial z^2},$$

and we choose the *initial condition* $C(r, \varphi, z, 0) = 0$.

The *weight function* is $w(r, \varphi, z) = r$.

There are six *boundary conditions*,

$$C(a, \varphi, z, t) = f_a(\varphi, z, t), \quad C(b, \varphi, z, t) = f_b(\varphi, z, t),$$

$$C(r, 0, z, t) = g_0(r, z, t), \quad C(r, v, z, t) = g_v(r, z, t),$$

$$C(r, \varphi, 0, t) = h_0(r, \varphi, t), \quad C(r, \varphi, H, t) = h_H(r, \varphi, t).$$

Let $k_n = n\pi/v$, $n \in \mathbb{N}$. Using numerical methods we first find the positive zeros $z_{k_n,m}$, $m \in \mathbb{N}$, for the functions

$$J_{k_n}(z) Y_{k_n}\left(\frac{b}{a} z\right) - J_{k_n}\left(\frac{b}{a} z\right) Y_{k_n}(z), \qquad n \in \mathbb{N}.$$

According to Abramowitz *et al.* (1965), these are all simple, and $z_{n,m} \to \infty$ for $m \to \infty$.

Define for short

$$\mathcal{C}_{k_n,m}(r) = Y_{k_n}\left(z_{k_n,m}\frac{b}{a}\right) J_{k_n}(r) - J_{k_n}\left(z_{k_n.m}\frac{b}{a}\right) Y_{k_n}(r).$$

The orthogonal system is

$$\psi_{mnp}(r, \varphi, z) = \mathcal{C}_{k_n,m}\left(z_{k_n,m}\frac{r}{a}\right) \sin(k_n\varphi) \sin\left(\frac{p\pi z}{H}\right), \quad m, n, p \in \mathbb{N},$$

and the corresponding λ_{mnp} are

$$\lambda_{mnp} = \left(\frac{1}{a} z_{k_n,m}\right)^2 + \left(\frac{p\pi}{H}\right)^2, \qquad m, n, p \in \mathbb{N}.$$

We obtain a fair approximation of the solution by only considering the indices (m, n, p) for which λ_{mnp} is small. Since $\lambda_{mnp} \to \infty$ if just one of the parameters tends to infinity, we shall have $\lambda_{mnp} < C$ for only a finite number of indices (m, n, p).

The squared norms are given by

$$\|\psi_{mnp}\|^2 = \frac{Hv}{8} \left[r^2 C'_{k_n,m} \left(z_{k_n,m} \frac{r}{a} \right)^2 \right]_{r=a}^{b},$$

and the solution in Ω is given as a triple series of the structure

$$C(r, \varphi, z, t) = \sum_{m=1}^{\infty} \sum_{n=1}^{\infty} \sum_{p=1}^{\infty} a_{mnp}(t) \psi_{mnp}(r, \varphi, z),$$

supplied with the boundary conditions.

The coefficients $a_{mnp}(t)$ are given by

$$a_{mnp}(t) = D \int_0^t \exp\left(-D\lambda_{mnp}(t - \tau)\right) F_{mnp}(\tau)\, d\tau,$$

where the functions $F_{mnp}(t)$ can be calculated from the boundary conditions. More explicitly, it is theoretically known that [cf. Section 4.2]

$$\begin{aligned} &\lambda_{mnp} a_{mnp}(t) + F_{mnp}(t) \\ &= \frac{1}{\|\psi_{mnp}\|^2} \int_a^b \left\{ \int_0^v \left\{ \int_0^H \nabla^2 C \cdot \psi_{mnp}\, dz \right\} d\varphi \right\} r dr, \end{aligned} \tag{4.38}$$

where we can use the fact that

$$\begin{aligned} &a_{mnp}(t) \\ &= \frac{1}{\|\psi_{mnp}\|^2} \int_a^b \left\{ \int_0^v \left\{ \int_0^H C(r, \varphi, z, t) \psi_{mnp}(r, \varphi, z)\, dz \right\} d\varphi \right\} r\, dr \end{aligned}$$

when the right hand side of (4.38) is calculated. Such terms can, therefore, be represented by dots, because we already know that they add up to $\lambda_{mnp} a_{mnp}(t)$ in the end.

The technique is therefore: Perform a series of partial integration to reduce

$$\int_a^b \left\{ \int_0^v \left\{ \int_0^H \nabla^2 C \cdot \psi_{mnp} \, dz \right\} d\varphi \right\} r \, dr$$

$$= \int_0^v \left\{ \int_0^H \left\{ \int_a^b \frac{\partial}{\partial r}\left(r \frac{\partial C}{\partial r} \right) \mathcal{C}_{k_n,m}\left(z_{k_n,m} \frac{r}{a} \right) dr \right\} \sin\left(\frac{p\pi z}{H} \right) dz \right\}$$

$$\cdot \sin(k_n \varphi) d\varphi$$

$$+ \int_a^b \left\{ \int_0^H \left\{ \int_0^v \frac{\partial^2 C}{\partial \varphi^2} \sin(k_n \varphi) d\varphi \right\} \sin\left(\frac{p\pi z}{H} \right) dz \right\}$$

$$\cdot \mathcal{C}_{k_n,m}\left(z_{k_n,m} \frac{r}{a} \right) \frac{dr}{r}$$

$$+ \int_a^b \left\{ \int_0^v \left\{ \int_0^H \frac{\partial^2 C}{\partial z^2} \sin\left(\frac{p\pi z}{H} \right) dz \right\} \sin(k_n \varphi) d\varphi \right\}$$

$$\cdot \mathcal{C}_{k_n,m}\left(z_{k_n,m} \frac{r}{a} \right) r dr$$

$$= \| \psi_{mnp} \|^2 \left\{ \lambda_{mnp}(t) + F_{mnp}(t) \right\},$$

by which the function $F_{mnp}(t)$ can be identified.

Chapter 5

Time-Dependent Chloride Diffusivity

Since a mathematical model for the chloride ingress into a marine RC structure was published by Collepardi *et al.* (1970) and (1972), the mathematical model of chloride ingress into marine concrete has been developed in time with observations made in field studies, exposure stations and by laboratory experiments. At present it is known that the chloride transport in concrete free from gross cracks could be modelled by diffusion, i.e. by Fick's laws of diffusion. The physical and boundary conditions have been discussed in the past, but the present knowledge may be as follows

- The achieved chloride diffusion coefficient is decreasing in time as shown by Takewaka *et al.* (1988), and depending on the composition of the concrete and its environment as shown by Frederiksen *et al.* (1997).
- The chloride content of the near-to-surface layer of the exposed concrete surface is increasing in time as shown by Uji *et al.* (1990), and depending on the composition of the concrete and its environment as shown by Frederiksen *et al.* (1997).
- The chloride content of the near-to-surface layer of the exposed concrete surface in marine splash and submerged in seawater may be taken as a constant after approximately 2–5 years as shown by Maage *et al.* (1999).

These conditions may be taken into account by Fick's generalized laws of diffusion as shown by Mejlbro (1996).

On the basis of the conditions given above two mathematical models have been formulated, namely the LIGHTCON model by Maage *et al.* (1995), and the HETEK model by Frederiksen *et al.* (1997). The LIGHTCON model is mainly applicable to concrete in marine splash or submerged in seawater. The HETEK model is applicable to concrete in marine splash, submerged in seawater or exposed to marine atmosphere as well as concrete exposed to de-icing salt.

The LIGHTCON model is described in Section 5.1, and the HETEK model is described in Section 5.2.

5.1 Constant surface chloride content

Chloride ingress into steel reinforced concrete structures, exposed to a chloride-laden environment, e.g. marine environment, depends on the composition, the compactness of the concrete and the chloride intensity of the environment. This section describes the parameters and the formulæ used by the LIGHTCON model when calculating the chloride ingress into the concrete or the initiation period of time either by predicting on the basis of inspection and testing (existing RC structure) or by estimation from the composition of the concrete and its types of marine environment.

5.1.1 LIGHTCON model of chloride ingress

The original LIGHTCON model was developed in 1994–95 by Maage *et al.* (1995). At that time only few observations were available of chloride ingress into concrete versus time. Since then many observations have been reported, from marine bridges, exposure stations and laboratories. Thus, the original LIGHTCON model has been revised by Maage *et al.* (1999). In this chapter the revised LIGHTCON model is based upon the data given in Section 1.2.2 on the 'one year diffusion coefficient'.

The description of the modified LIGHTCON model is given as a step-by-step manual for the prediction of the future chloride ingress, illustrated by numerical examples.

Problem of estimating the chloride ingress into concrete

In order to estimate the future chloride ingress into concrete at least the following information ought to be available

- The age of the concrete structure at the time of inspection t_{in}.
- The age t_{ex} of the concrete structure at the time of first chloride exposure.
- The composition of the concrete, i.e. the type of binder, and the w/c-ratio.
- The environment, i.e. marine atmosphere ATM, marine splash SPL, or submerged in seawater SUB, where the inspected component is situated.

- The chloride profile of the concrete at an inspection time $t_{\text{in}} \gg t_{\text{ex}}$, e.g. by the chloride diffusion coefficient and the chloride content of the concrete surface.

When it is not possible to obtain reliable information from the specification of the concrete structure, the inspection must be supplied with a thin section analysis of the concrete in question.

Testing the concrete by NT Build 443 or by the 'method of inverse cores' may supply the estimation of the future chloride ingress into the concrete.

Assumptions of the LIGHTCON model

The LIGHTCON model for future ingress into concrete is based on the following assumptions

- Chloride C in concrete is here defined as the 'total, acid soluble chloride'.

- Transport of chloride in concrete takes place by diffusion. There is an equilibrium of the mass of ingress of (free) chloride into each element of the concrete, the accumulation of (free and bound) chloride in the element and an ongoing diffusion of (free) chloride in the element towards a neighbouring element, and so on.

- The flow of chloride F is proportional to the gradient of chloride. The factor of proportionality is the achieved chloride diffusion coefficient D_a.

- The achieved chloride diffusion coefficient D_a depends on time, the composition of the concrete and the (local) environment.

- The boundary condition C_s is constant, i.e. independent of time, but it depends on the composition of the concrete and the (local) environment.

- The initial chloride content of the concrete C_i (per unit element of the concrete) is uniformly distributed at time of the first chloride exposure.

- The relations used for the determinative parameters with respect to the environment (ATM, SPL and SUB), the time and the composition of the concrete are documented at the Träslövsläge Marine Exposure Station on the west coast of Sweden (south of Gothenburg) as described by Frederiksen *et al.* (1997).

Mass balance of chloride in an element volume of concrete

The equation of mass balance (equilibrium) for chloride in concrete is expressed by Fick's general law of diffusion,

$$\frac{\partial C}{\partial t} = \frac{\partial}{\partial x}\left(D\,\frac{\partial C}{\partial x}\right), \tag{5.1}$$

where C is the chloride content at the locality x at time t, and D is the chloride diffusion coefficient which may depend on the time t, the locality x, the chloride content C, and the composition and environment of the concrete. It is, however, an assumption of the LIGHTCON model that the chloride diffusion coefficient D is independent of the locality x and the chloride content, though it may depend on time.

Achieved chloride diffusion coefficient

The original version of the LIGHTCON model by Maage *et al.* (1995), implies that the achieved chloride diffusion coefficient is known from testing the concrete at an early age or from a similar marine concrete of the same composition and exposed to a similar marine environment. Furthermore, the original LIGHTCON model did not include any prediction of the achieved chloride diffusion coefficient.

The user of the model was referred to have carried out tests at early ages or to test the concrete according to NT Build 443 at the time of inspection and transform this observation to an early age. However, experience has shown that this has not been valid in practice.

Therefore, it is here shown how it is possible to predict the achieved chloride diffusion coefficient after 1 year of exposure from knowing the eqv$\{w/c\}$ and the type of environment of the concrete in question.

Diffusion coefficient after 1 year of exposure. The study of observations from Träslövsläge Marine Exposure Station by Frederiksen *et al.* (1997), has led to the conclusion that the chloride diffusion coefficient after 1 year of exposure can be expressed by the relation,

$$D_1 = 25,000\, k_D \times \exp\left(-\sqrt{\frac{10}{\text{eqv}\,\{w/c_D\}}}\right) \text{ mm}^2/\text{yr}, \tag{5.2}$$

where the equivalent water/binder-ration (ratio by mass) with reference to diffusion is

$$\text{eqv}\,\{w/c_D\} = \frac{W}{PC + FA + 7 \times DF}. \tag{5.3}$$

Here

W = content of **w**ater in the concrete
PC = content of **p**ortland **c**ement in the concrete
FA = content of **f**ly **a**sh in the concrete
SF = content of **s**ilica **f**ume in the concrete.

Frederiksen *et al.* (1997) define the factor k_D, depending on the environment of concrete in the following way,

$$k_D = \begin{cases} 0.4 \text{ for concrete exposed to marine atmosphere (ATM)} \\ 0.6 \text{ for concrete exposed to marine splash zone (SPL)} \\ 1.0 \text{ for concrete submerged in seawater (SUB).} \end{cases} \tag{5.4}$$

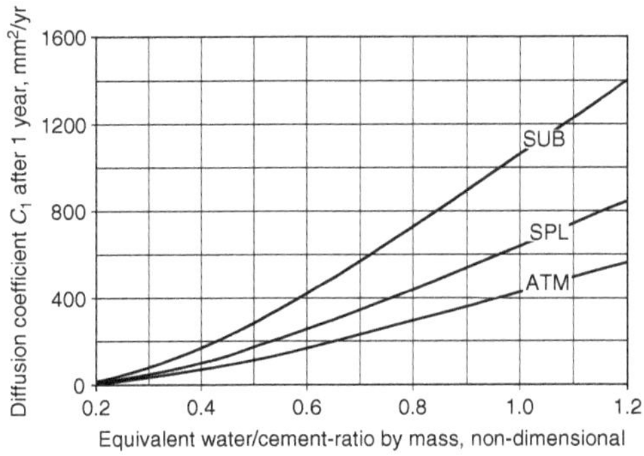

Figure 5.1: The value of D_1 estimated by Equation (5.2) for marine atmosphere (ATM), marine splash (SPL) and submerged in seawater (SUB) respectively.

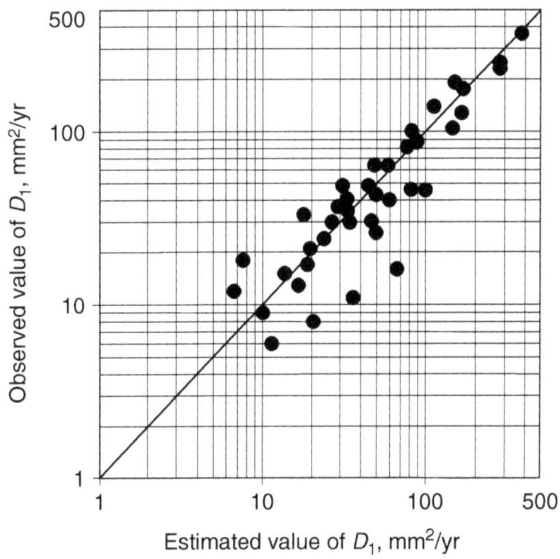

Figure 5.2: The observed value of D_1 at the Träslövsläge Marine Exposure Station, i.e. the chloride diffusion coefficient after one year of exposure, determined by interpolation of measurement versus the values of D_1 determined by Equation (5.2). An acceptable agreement is obtained.

Figure 5.1 shows the relation Equation (5.2), and Figure 5.2 compares the relation with plotted observations from the Träslövsläge Marine Exposure Station.

NT Build 443. When testing the concrete according to NT Build 443 is carried out determining D_{pex} at an exposure of 14 to 28 days, it is possible to write

$$D_1 = \frac{1}{2} k_D \, D_{pex}, \tag{5.5}$$

where k_D is given by (5.4). Equation (5.5) may substitute or supplement Equation (5.2). If, however, the testing is carried out at the time of inspection, a correction is needed as shown by Maage *et al.* (1999).

Method of inverse cores. If a determination of D_1 has been made according to the method of inverse cores, cf. Section 5.2.5, this determination may substitute or supplement Equation (5.2).

Development of the achieved chloride diffusion coefficient. The time dependency of the achieved chloride diffusion coefficient D_a can be expressed mathematically by the power function

$$D_a(t) = D_1 \times \left(\frac{t}{t_1}\right)^{-\alpha} = D_1\, t^{-\alpha}, \tag{5.6}$$

where $t_1 = 1$ yr and t denotes the time (origin equal to the first chloride exposure of the concrete), and D_1 is a factor, which may be explained as the value of the achieved chloride diffusion coefficient at 1 year after the first chloride exposure, and α is an exponent (age parameter), which depends on the composition and environment of the concrete. This is an extension of the findings of Takewaka *et al.* (1988), based upon Maage *et al.* (1995) and Frederiksen *et al.* (1997).

Due to Frederiksen *et al.* (1997) the age parameter α may be estimated by the following relation

$$\alpha = k_\alpha \times (1 - 1.5 \times \ \text{eqv}\{w/c_D\})\,. \tag{5.7}$$

The factor k_α, cf. Frederiksen *et al.* (1997), depends on the environment of the concrete in the following way,

$$k_\alpha = \begin{cases} 1.0 \text{ for concrete exposed to marine atmosphere (ATM)}, \\ 0.1 \text{ for concrete exposed to marine splash zone (SPL)}, \\ 0.6 \text{ for concrete submerged in seawater (SUB)}, \end{cases} \tag{5.8}$$

and the water/cement ratio eqv$\{w/c_D\}$ is determined by Equation (5.3).

The age parameter α. At an inspection, carried out at the time t_{in}, the achieved chloride diffusion coefficient D_{ain} is determined. From the time-dependency of the diffusion coefficient, cf. Equation (5.6), the age parameter α is determined as

$$\alpha = \frac{\ln(D_1/D_{ain})}{\ln(t_{\text{in}}/t_1)}, \tag{5.9}$$

where $t_1 = 1$ year.

Boundary condition

The original version of the LIGHTCON model by Maage *et al.* (1995), implies that the achieved chloride content of the exposed concrete surface predicted from a similar (old) marine concrete of the same composition and exposed to a similar marine environment. The original LIGHTCON model did not include any prediction of the achieved chloride content of the exposed concrete surface. When inspection is carried out at a time $t_{\text{in}} > 2$ years the user of the LIGHTCON model was referred to apply the value of the achieved chloride content of the exposed concrete surface, corrected due to the experience of the user.

Chloride content of the concrete surface. Intensive studies by Mangat *et al.* (1994), Frederiksen *et al.* (1997), and Maage *et al.* (1999), have shown that the achieved chloride content C_{sa} of a chloride exposed concrete surface in a marine environment behaves in the following ways,

- The C_{sa} of a concrete surface which is exposed to a marine atmosphere (ATM) is an increasing function of time and is estimated to increase, in the order of magnitude, by a factor 7 during approximately 100 years. Studies have shown that the rate of C_{sa} seems to be almost a constant through the entire exposure period.

- The C_{sa} of a concrete surface which is exposed to marine splash (SPL) is an increasing function of time and is estimated to increase, in the order of magnitude, by a factor 4.5 during approximately 100 years. However, studies by Maage *et al.* (1999), have shown that the greater part of this increase will take place at approximately the first 2–5 years of exposure.

- The C_{sa} of a concrete surface submerged in seawater (SUB) is an increasing function of time and is estimated to increase, in the order of magnitude, by a factor 1.5 during approximately 100 years. However, studies by Maage *et al.* (1999) and Frederiksen *et al.* (1997), have shown that the greater part of this increase will take place at approximately the first 2–5 years of the exposure.

The conclusion of these findings is

- The LIGHTCON model cannot be applied for concrete exposed to a marine atmosphere without an appreciable deviation from reality.

- The LIGHTCON model may be applied for concrete exposed to marine splash by assuming that C_{sa} is constant for $t \geq 2$ to 5 years.

- The LIGHTCON model may be applied for concrete submerged in seawater by assuming that C_{sa} is constant for $t \geq 2$ to 5 years.

Achieved surface chloride content at the time of inspection. It is here assumed that at the time of inspection, t_{in} lies between 2 and 5 years. The chloride content of the concrete surface $C_{sa} = C_0$ is determined from representative cores of the concrete. This value is used in the calculation of the chloride profiles for $2 \leq t \leq 5$ years.

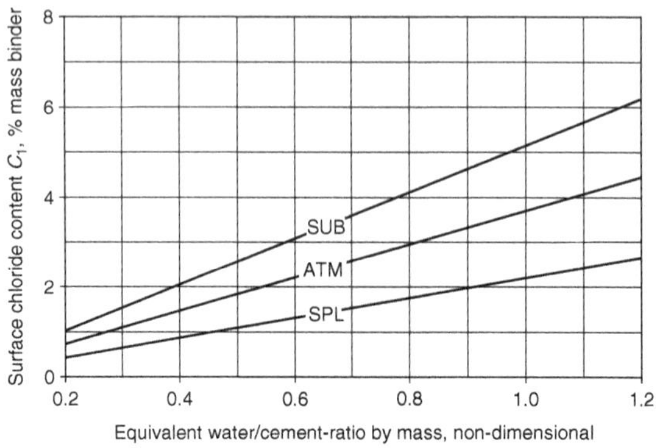

Figure 5.3: The value of C_1 estimated by Equation (5.10) for marine atmosphere (ATM), marine splash (SPL) and submerged in seawater (SUB) respectively.

Estimated achieved surface chloride content for design. There is a need for an estimate of the surface chloride content, when a marine RC structure is under design and the chloride ingress should be predicted. The study of observations from the Träslövsläge Marine Exposure Station by Frederiksen *et al.* (1997), has led to the conclusion that the surface chloride content after 1 year of exposure can be expressed by the relation

$$C_1 = k_b \times \text{ eqv}\left\{w/c_b\right\}. \tag{5.10}$$

Here, $\text{eqv}\{w/c_b\}$ is determined according to

$$\text{eqv}\left\{w/c_b\right\} = \frac{W}{PC + 0.75 \times FA - 1.5 \times SF}. \tag{5.11}$$

The factor k_b defined by Frederiksen *et al.* (1997), depends on the environment of the concrete in the following way,

$$k_b = \begin{cases} 2.20 \text{ for concrete exposed to marine atmosphere (ATM)}, \\ 3.67 \text{ for concrete exposed to marine splach zone (SPL)}, \\ 5.13 \text{ for concrete submerged in seawater (SUB)}. \end{cases} \tag{5.12}$$

Figure 5.3 shows the relation (5.10). Figure 5.4 compares this relation with observations from the Träslövsläge Marine Exposure Station.

Graphs of chloride profiles

After the three parameters of the problem α, D_1 and C_0 have been determined, the ordinates of the chloride profiles $C(x,t)$ at time t years yield

$$C(x,t) = C_i + (C_0 - C_i) \times \text{erfc}\left(\frac{0.5\,x}{\sqrt{t\,D_a}}\right), \tag{5.13}$$

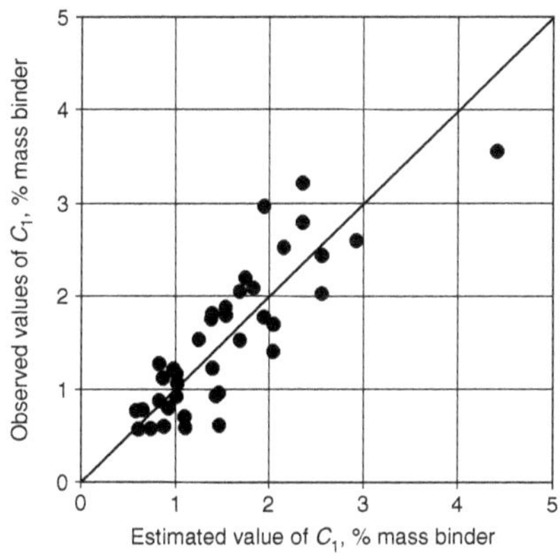

Figure 5.4: The observed value of C_1 at Träslövsläge Marine Exposure Station i.e. the chloride content of the concrete surface, determined by interpolation of measurements versus the values of C_1 determined by Equation (5.10). An acceptable agreement is obtained.

where $C_0 = C_{sa} = C_{\text{sin}}$ and $\text{erfc}(u)$ is the complementary error function. It should be noticed that the achieved chloride diffusion coefficient D_a is not the real diffusion coefficient of the concrete D, but the mean value of the diffusion coefficient in the past, i.e.

$$D_a(t) = \frac{1}{t} \int_0^t D \, dt, \tag{5.14}$$

cf. Section 1.1.3. By inserting D_a from Equation (5.6) into Equation (5.8) the chloride profile becomes

$$C(x,t) = C_i + (C_0 - C_i) \times \text{erfc}\left(\frac{0.5\,x}{\sqrt{t_1\, D_1\, \tau^{1-\alpha}}} \right), \tag{5.15}$$

where $\tau = t/t_1$. Thus, a simple way to calculate the graph of a chloride profile at a specified time $t = t_n$ is as follows: The abscissa u_n and the corresponding ordinates of the complementary error function $\text{erfc}(u)$ are inserted into a spreadsheet, either from a table, cf. Table 8.2, or calculated from an approximation formula, cf. Section 3.1.5. The ordinates of the chloride profile are calculated as

$$C\,(x, t_n) = C_i + (C_0 - C_i) \times \text{erfc}(u_n). \tag{5.16}$$

Finally, the abscissæ of the chloride profile are calculated from

$$u_n = \frac{0.5\,x}{\sqrt{t_n\, D_a}} = \frac{0.5\,x}{\sqrt{t_1\, C_1\, \tau_n^{1-\alpha}}}, \tag{5.17}$$

which yields the abscissæ

$$x_n = 2\,u_n\,\sqrt{t_n\,D_a} = 2\,u_n\,\sqrt{t_1\,D_1\,\tau_n^{1-\alpha}}. \tag{5.18}$$

Example 5.1.1 A structural component of reinforced concrete is exposed to marine splash, inspected and tested. The data of the concrete, inspection, testing and environment are as follows

Concrete

- The binder of the concrete is ordinary Portland cement.
- The water/binder-ratio is $w/c_D = w/c_b = 0.48$ by mass.
- The chloride content in unit '% mass binder' is 6.11 times the chloride content in unit '% mass concrete'.
- The concrete was exposed to chloride for the first time at 0.027 yr.

Inspection

- The concrete has the age of $t_{\text{in}} = 18$ years at the time of inspection.
- $D_{ain} = 32.9$ mm^2/yr.
- $C_{sin} = 0.333$ % mass concrete $= 0.333 \times 6.11 = 2.035$ % mass binder.
- $C_i = 0.008$ % mass concrete $= 0.008 \times 6.11 = 0.049$ % mass binder.

Environment

- Marine splash zone (SPL).

The calculation of the chloride profiles at the time $t_2 = 2$ yr, $t_{10} = 10$ yr and $t_{50} = 50$ yr is carried out as follows,

Step 1. The achieved chloride diffusion coefficient after 1 year of exposure is estimated to

$$D_1 = 0.6 \times 25{,}000 \times \exp\left(-\sqrt{\frac{10}{0.48}}\right) = 156.2 \text{ mm}^2/\text{yr}, \quad \text{cf. (5.2).}$$

Step 2. The age parameter is estimated to

$$\alpha = \frac{\ln(156.2/32.9)}{\ln(18/1)} = 0.539, \qquad \text{cf. (5.9).}$$

Step 3. The abscissæ u_n and the corresponding ordinates of the complementary error function $\text{erfc}(u_n)$ are inserted into the spreadsheet, either from a table or calculated from an approximation formula, cf. Section 3.1.5.

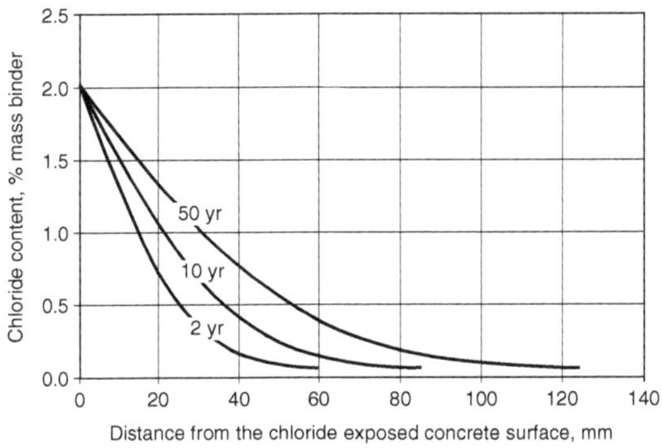

Figure 5.5: The chloride profiles after 2, 10 and 50 years of exposure as determined in Example 5.1.1. The LIGHTCON model was assumed to be valid for the chloride ingress into the concrete.

Step 4. The ordinates of the chloride profiles are calculated according to the following formulæ

$$C(x, 2\text{ yr}) = 0.049 + (2.035 - 0.049) \times \operatorname{erfc}(u_2), \qquad \text{cf. (5.12)},$$

$$C(x, 10\text{ yr}) = 0.049 + (2.035 - 0.049) \times \operatorname{erfc}(u_{10}), \qquad \text{cf. (5.12)},$$

and

$$C(x, 50\text{ yr}) = 0.049 + (2.035 - 0.049) \times \operatorname{erfc}(u_{50}), \qquad \text{cf. (5.12)}.$$

Step 5. The achieved chloride diffusion coefficients at time 2, 10 and 50 years are estimated to, cf. (5.6),

$$D_a(2\text{ yr}) = 156.2 \times 2^{-0.539} = 107.5\text{ mm}^2/\text{yr},$$
$$D_a(10\text{ yr}) = 156.2 \times 10^{-0.539} = 44.3\text{ mm}^2/\text{yr},$$
$$D_a(50\text{ yr}) = 156.2 \times 50^{-0.539} = 19.0\text{ mm}^2/\text{yr}.$$

Step 6. The abscissæ of the chloride profiles are calculated according to the following formulæ, cf. (5.13)

$$x_2 = 2u_2 \times \sqrt{2 \times 107.5} = 29.33 \times u_2,$$
$$x_{10} = 2u_{10} \times \sqrt{10 \times 44.3} = 42.10 \times u_{10},$$
$$x_{50} = 2u_{50} \times \sqrt{50 \times 19.0} = 61.64 \times u_{50},$$

where u is the abscissa of the complementary error function erfc(u), cf. Step 4.

Step 7. The chloride profiles are plotted, cf. Figure 5.5. ◊

5.1.2 Chloride ingress into concrete

The location x_{re} of a reference chloride content C_{re}, cf. Equation (5.8), may be determined by solving the equation of the LIGHTCON model

$$C_{re} = C_i - (C_0 - C_i)\ \mathrm{erfc}\left(\frac{0.5\,x_{re}}{\sqrt{t\,D_a}}\right) \qquad \text{for } t > 0 \text{ and } x_{re} > 0, \quad (5.19)$$

where the achieved chloride diffusion coefficient obeys Equation (5.6), i.e.

$$D_a(t) = D_1 \times \left(\frac{t}{t_1}\right)^{-\alpha} = D_1\,\tau^{-\alpha}. \quad (5.20)$$

By inserting Equation (5.15) into Equation (5.14) and solving with respect to x_{re} we get the following relation

$$x_{re} = k_1\,\sqrt{\tau^{1-\alpha}} \qquad \text{for } \tau = \frac{t}{t_1} > 0, \quad (5.21)$$

where the first year chloride ingress is

$$k_1 = 2\,\sqrt{t_1\,D_1} \times\ \mathrm{inv\ erfc}\left(\frac{C_{re} - C_i}{C_0 - C_i}\right). \quad (5.22)$$

The following estimate may be used in lack of a table of the inverse complementary error function. This estimate is based upon an approximation of the complementary error function, cf. Equation (3.73),

$$k_1 \approx 3.32\,\sqrt{t_1\,D_1} \times \left(1 - \sqrt{\frac{C_{re} - C_i}{C_0 - C_i}}\right). \quad (5.23)$$

Example 5.1.2 A structural component of reinforced concrete shall be exposed to marine splash through 20 years. The concrete should be made of normal Portland cement with af w/cD-ratio $= w/c_b$-ratio of 0.35 by mass. The ingress of the reference value of chloride content $C_{re} = 0.2$ % by mass binder during 20 years shall be estimated. It is assumed that the surface chloride content is a constant equal to $C_0 = 3.0$ % mass binder and that the initial chloride content of the concrete is $C_i = 0$.

Step 1. The achieved chloride diffusion coefficient after one year of exposure is estimated to

$$D_1 = 0.6 \times 25,000 \times \exp\left(-\sqrt{\frac{10}{0.35}}\right) = 71.6\ \mathrm{mm^2/yr}, \quad \text{cf. (5.2).}$$

Step 2. The first year chloride ingress is estimated to

$$k_1 \approx 3.32\,\sqrt{1 \times 71.6} \times \left(1 - \sqrt{\frac{0.2 - 0}{3.0 - 0}}\right) = 20.8\ \mathrm{mm}, \quad \text{cf. (5.18).}$$

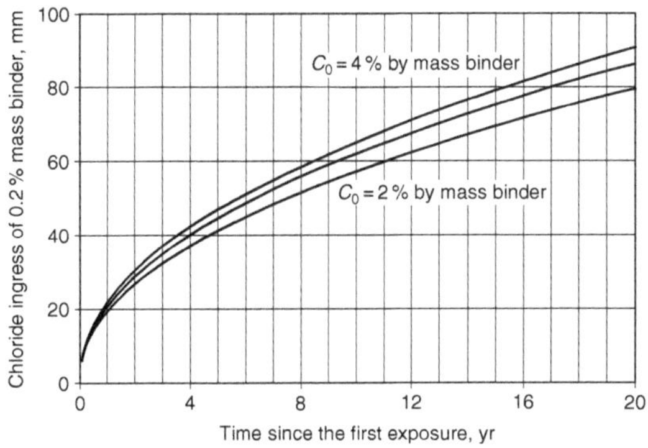

Figure 5.6: Ingress into a concrete of a reference value of chloride content $C_{re} = 0.2$ % by mass binder assuming a constant surface chloride content of $C_0 = 2$, 3 and 4 % by mass binder, cf. Example 5.1.2. It is seen that the chloride ingress depends on the value of C_0 without being too sensitive.

Step 3. The age parameter is estimated to

$$\alpha = 0.1 \times (1 - 1.5 \times 0.35) = 0.048,$$

corresponding to $1 - \alpha = 0.952$, cf. (5.7).

Step 4. Thus, after 20 years of exposure the chloride ingress of 0.2 % by mass binder will reach the depth of

$$x_{re}(20 \text{ yr}) = 20.8 \times \sqrt{20^{0.952}} = 87 \text{ mm}, \qquad \text{cf. (5.18).}$$

Figure 5.6 shows the ingress of $x_{re} = 0.2$ % mass binder when constant surface chloride contents are chosen equal to 2, 3 and 4 % mass binder respectively. It is seen that the chloride ingress into the concrete depends on the value of C_0 without being too sensitive. ◊

5.1.3 Initiation period of time

Concerning the determination of the initiation period of an RC structure it is of interest to know the ingress of the threshold value of chloride in the concrete C_{cr}. By inspection of the concrete and various test methods it is possible to determine the initiation period of time, i.e. the time when the chloride content of the concrete next to the reinforcement reaches the threshold value of chloride in the concrete and starts the corrosion process.

The threshold value of chloride in concrete is estimated by

$$C_{cr} = k_{cr} \times \exp\left(-1.5 \times \text{ eqv}\left\{w/c_{cr}\right\}\right), \tag{5.24}$$

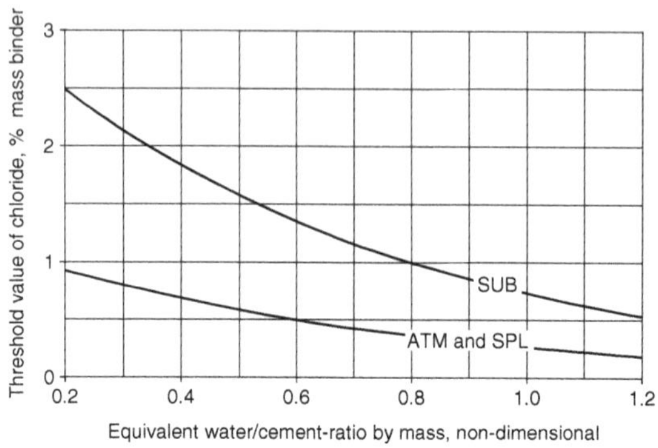

Figure 5.7: The threshold value of chloride in concrete (here defined as the minimum total acid soluble chloride content of the concrete which initiates corrosion of the rebars in the concrete) versus the equivalent water/cement-ratio by mass and various exposure regimes, cf. Frederiksen *et al.* (1997).

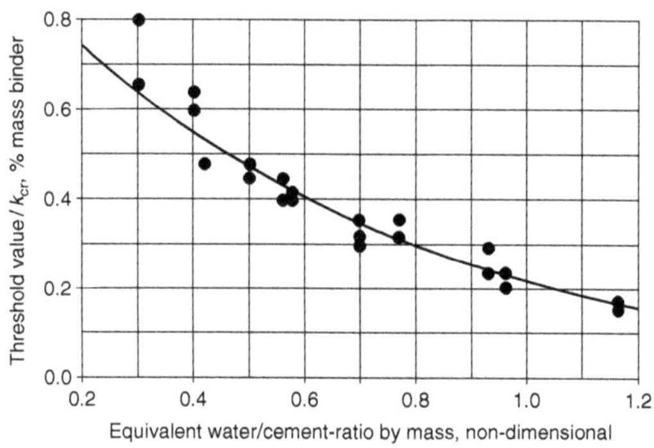

Figure 5.8: The threshold value of chloride in concrete for various environments versus the equivalent water/cement-ratio, cf. Equation (5.25), compared with observations from Träslövsläge Marine Exposure Station, cf. Frederiksen *et al.* (1997).

where the equivalent w/c-ratio here is

$$\text{eqv}\{w/c_{cr}\} = \frac{W}{PC - 1.4 \times FA - 4.7 \times SF}, \tag{5.25}$$

compare with Equation (5.3). The factor k_{cr} defined by Frederiksen *et al.* (1997), depends on the environment of the concrete in the following way,

$$k_{cr} = \begin{cases} 1.25 \text{ for concrete exposed to marine atmosphere (ATM)}, \\ 1.25 \text{ for concrete exposed to marine splash (SPL)}, \\ 3.35 \text{ for concrete submerged in seawater (SUB)}. \end{cases} \tag{5.26}$$

Figure 5.7 shows the relation (5.24), and Figure 5.8 compares this relation with plotted observations from the Träslövsläge Marine Exposure Station.

Determination by a chloride profile

It is assumed that a representative chloride profile has been determined from a marine RC structure in a part of the concrete which has been submerged in seawater or exposed to the splash zone. Furthermore, it is assumed that the age of the structure at time of inspection, t_{in} and the time of first chloride exposure of the structure are known or could be estimated.

From the chloride profile the chloride parameters D_{ain}, C_0 and C_i are estimated, cf. Section 2.1.3 or 2.1.4. From these parameters the first year chloride ingress k_1 is determined, cf. (5.22) or (5.23),

$$k_1 = 2\sqrt{t_1\, D_1} \times \text{ inv erfc}\left(\frac{C_{cr} - C_i}{C_0 - C_i}\right) \approx 3.32 \times \sqrt{t_1\, D_1}\left(1 - \sqrt{\frac{C_{re} - C_i}{C_0 - C_i}}\right). \tag{5.27}$$

Thus, the ingress into the concrete of the threshold value of chloride in concrete C_{cr} yields

$$x_{cr} = k_1 \sqrt{\left(\frac{t}{t_1}\right)^{1-\alpha}} \qquad \text{for } t \geq t_{\text{in}}. \tag{5.28}$$

When the threshold value reaches the position $x_{cr} = c$, i.e. the concrete cover thickness of the reinforcement, the reinforcement initiates corrosion. Thus, the initiation period t_{cr} is determined from the following equation

$$c = k_1 \sqrt{\left(\frac{t_{cr}}{t_1}\right)^{1-\alpha}} \qquad \text{for } t_{cr} \geq t_{\text{in}}, \tag{5.29}$$

from which the initiation period is determined by

$$t_{cr} = t_1 \times \left(\frac{c}{k_1}\right)^{\frac{2}{1-\alpha}} \qquad \text{for } t_{cr} > t_{\text{in}}. \tag{5.30}$$

The age parameter α may be determined in two different ways,

- Direct estimation of α by Equation (5.7).
- Indirect estimation of D_1 by Equation (5.2) and α by Equation (5.9).

In the last case the initiation period of time is given by

$$t_{cr} = t_{\text{in}} \times \left(\frac{c/\xi_{cr}}{\sqrt{t_{\text{in}}\, D_{in}}}\right)^{\frac{2}{1-\alpha}} \qquad \text{for } t_{cr} > t_{\text{in}}. \tag{5.31}$$

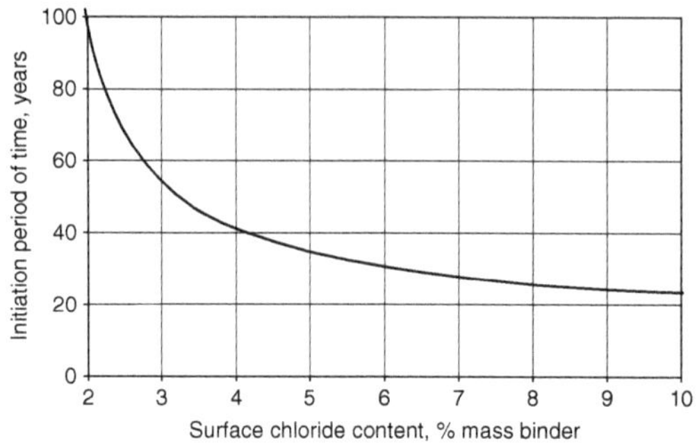

Figure 5.9: The initiation period of time versus the surface chloride content of the concrete given in Example 5.1.3. The LIGHTCON model was assumed to be valid for the chloride ingress into the concrete.

Notice that the initiation period of time found by Equation (5.30) and Equation (5.31) may not be equal. Gross deviation may occur. Therefore, several observations of chloride ingress at various times will increase the accuracy of the prediction of the initiation period of time, cf. Example 5.2.4.

Example 5.1.3 A marine concrete bridge is under design. The initiation time of an area of the concrete which is exposed to a marine splash environment shall be estimated. The cover of the reinforcement is 75 mm. The following mixture proportion is planned,

- Sulphate resistant Portland cement: $PC = 400$ kg/m^3 concrete,
- Silica fume: $SF = 20$ kg/m^3 concrete,
- Mixing water: $W = 165$ kg/m^3 concrete.

It is assumed that $C_i \approx 0$. Thus, the prediction of the initiation period of time is carried out in the following way:

Step 1. The equivalent water/binder ratio with respect to diffusion is

$$\text{eqv}\{w/c_D\} = \frac{165}{400 + 7 \times 20} = 0.306, \qquad \text{cf. (5.3).}$$

Step 2. The achieved diffusion coefficient after 1 year of exposure is

$$D_1 = 0.6 \times 25,000 \times \exp\left(-\sqrt{\frac{10}{0.306}}\right) = 49.4 \text{ mm}^2/\text{yr}, \quad \text{cf. (5.2).}$$

Step 3. Calculation of the age parameter gives

$$\alpha = 0.1 \times (1 - 1.5 \times 0.306) = 0.0541, \qquad \text{cf. (5.7).}$$

Step 4. The equivalent water/binder ratio with respect to corrosion is

$$\text{eqv}\,\{w/b_{cr}\} = \frac{165}{400 - 4.7 \times 20} = 0.539, \qquad \text{cf. (5.20).}$$

Step 5. The threshold value of chloride in the concrete is

$$C_{cr} = 1.25 \times \exp(-1.5 \times 0.306) = 0.790\ \%\ \text{mass binder, cf. (5.19).}$$

Step 6. Assuming that the surface chloride content is constant and equal to C_0, the first year chloride ingress is given by, cf. (5.18),

$$k_1 \approx 3.32\sqrt{1 \times 49.4} \times \left(1 - \sqrt{\frac{0.790 - 0}{C_0 - 0}}\right) = 23.33 \times \left(1 - \sqrt{\frac{0.79}{C_0}}\right).$$

Step 7. Finally, the initiation period of time is given by

$$t_{cr} = 1 \times \left(\frac{75}{k_1}\right)^{2.114}, \qquad \text{cf. (5.25),}$$

where the first year chloride ingress k_1 depends on C_0, the surface chloride content, cf. Step 6. Assuming that $C_0 = 8$ % mass binder, the initiation time becomes $t_{cr} = 26$ years, while $C_0 = 5$ % mass binder gives $t_{cr} = 35$ years. One must therefore be very careful in estimating the surface chloride content. Figure 5.9 shows the graph of t_{cr} versus C_0. There are in practice various techniques to minimize the surface chloride content, e.g. surface treatment, choice of binders and form liner (controlled permeability formwork). ◊

5.1.4 Corrosion domain

In order to get an overview of the corrosion situation it is convenient to draw the corrosion domain, cf. Section 2.4.3.

(t, D)-diagram

A domain of corrosion is e.g. defined as the set of values of (t, D), for which the steel reinforcing bar with concrete cover c corrodes. The values of (t, D) which belong to the border of the domain of corrosion are called (t_{cr}, D_{cr}), cf. Figure 5.10. Thus, at the time of inspection, $t = t_{\text{in}}$, the 'state of corrosion' for a situation (t_{in}, D_0) is

$$\begin{aligned} &D_0 < D_{cr} \text{ implies 'no corrosion'},\\ &D_0 = D_{cr} \text{ implies 'initiating corrosion'},\\ &D_0 > D_{cr} \text{ implies 'ongoing corrosion'}. \end{aligned} \qquad (5.32)$$

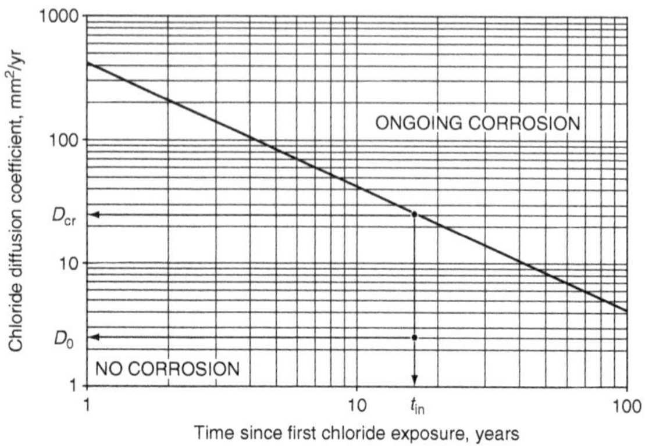

Figure 5.10: An example of a corrosion domain. At time of inspection, t_{in}, the chloride diffusion coefficient is D_0. When $D_0 < D_{cr}$ it implies 'no corrosion', where D_{cr} denotes the border of the corrosion domain. When $D_0 = D_{cr}$, it implies 'initiating corrosion', cf. Equation (5.32).

The condition of initiation of the reinforcement with a cover c may be found by solving the following equation, cf. Equation (5.10),

$$C_{cr} = C_i + (C_0 - C_i) \times \operatorname{erfc}\left(\frac{0.5\,c}{\sqrt{t_{\text{in}}\,D_a}}\right). \tag{5.33}$$

Thus, the value of $D_a = D_{cr}$ which at time of inspection, t_{in} leads to corrosion may be determined as

$$D_{cr} = \frac{(c/\xi_{cr})^2}{t_{\text{in}}}, \tag{5.34}$$

where

$$\xi_{cr} = 2 \text{ inv erfc}\left(\frac{C_{cr} - C_i}{C_0 - C_i}\right) \approx 3.32 \times \left(1 - \sqrt{\frac{C_{cr} - C_i}{C_0 - C_i}}\right). \tag{5.35}$$

Therefore, the border of the corrosion domain, Equation (5.34), is a straight line when plotted as a logarithmic graph, cf. Figure 5.10.

Example 5.1.4 An area of a concrete bridge exposed to a marine splash environment is inspected and tested. The observations are listed in Table 5.1. It is assumed that the initial chloride content of the concrete is $C_i \approx 0$. The cover of the reinforcement is $c = 50$ mm and the concrete mixture proportions in the area are as stated in Table 5.2, cf. Stoltzner (1995) and Frederiksen *et al.* (1997).

The threshold value of chloride in the concrete exposed to marine splash is estimated in the following way, cf. Equation (5.24) and Section 1.3.1. First

Table 5.1: Observations from a marine concrete bridge, cf. Example 5.1.4.

Inspection year	Age years	Diffusion coefficient mm^2/yr	Surface chloride % mass binder
1994	12	15.8	8.51

we determine the water/binder ratio with respect to corrosion,

$$\begin{aligned} \text{eqv}\{w/c_{cr}\} &= \frac{W}{PC - 4.7 \times SF - 1.4 \times FA} \\ &= \frac{140}{330 - 4.7 \times 0 - 1.4 \times 40} = 0.51. \end{aligned}$$

Thus the threshold value of chloride in concrete becomes

$$\begin{aligned} C_{cr} &= k_{cr} \exp\left(-1.5 \times \ \text{eqv}\{w/c_{cr}\}\right) \\ &= 1.25 \times \exp(-1.5 \times 0.51) = 0.58 \ \% \text{ mass binder.} \end{aligned}$$

The concrete has an age of $t_{\text{in}} = 12$ years at inspection, cf. Table 5.1.

In order to plot the corrosion domain in a (t, D)-diagram the following parameter is determined

$$\xi_{cr} = 3.32 \times \left(1 - \sqrt{\frac{C_{cr} - C_i}{C_0 - C_i}}\right) = 3.32 \times \left(1 - \sqrt{\frac{0.58 - 0}{8.51 - 0}}\right) = 2.45.$$

Thus, the border of the corrosion domain becomes, cf. Figure 5.11,

$$D_{cr} = \frac{(c/\xi_{cr})^2}{t_{\text{in}}} = \frac{(50/2.45)^2}{t_{\text{in}}} = \frac{416.5}{t_{\text{in}}} \ \text{mm}^2/\text{yr}.$$

By plotting the point A: $(t_{\text{in}}, D_{in}) = \left(12 \text{ yr}, 15.8 \text{ mm}^2/\text{yr}\right)$, it is observed that no corrosion takes place at the moment, cf. Figure 5.11. In order to predict the future chloride ingress the movement of point A in time should be estimated. This can be done by determination of the age parameter α, either directly by

Table 5.2: Mixture proportions, cf. Example 5.1.4.

Constituents	Contents	Unit
Portland cement	330	kg/m^3 concrete
Fly ash	40	kg/m^3 concrete
aggregates	1827	kg/m^3 concrete
Water	140	kg/m^3 concrete

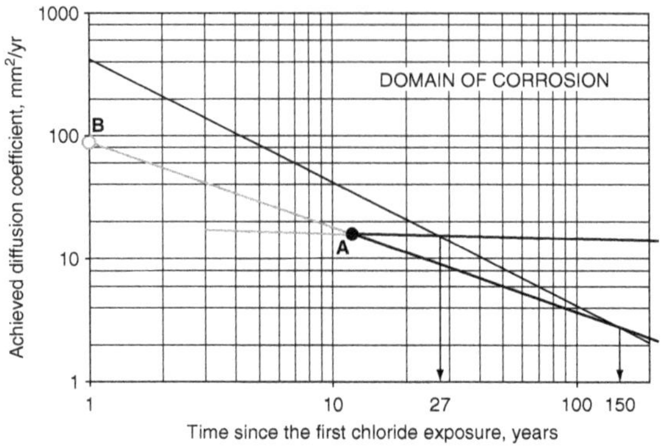

Figure 5.11: Corrosion domain of an inspection area of an RC bridge, cf. Example 5.1.4. Point A represents the inspection. From the data given in the example it is predicted that the initiation period of time lies in the interval between 27 years and 150 years.

Equation (5.7), or indirectly by Equation (5.2) and Equation (5.9). This will not result in an unambiguous value of the initiation period of time. Thus further data are necessary, though the methods are illustrated. Example 5.2.10 illustrates the methods when more data are available.

The equivalent water/binder ratio with respect to diffusion in concrete is determined, cf. Equation (5.3),

$$\text{eqv}\{w/b_D\} = \frac{W}{PC + FA + 7 \times SF} = \frac{140}{330 + 40} = 0.378.$$

Thus, the achieved chloride diffusion coefficient after 1 year of exposure is, cf. Equation (5.2) and point B in Figure 5.11,

$$D_1 = 25,000 \times 0.6 \times \exp\left(-\sqrt{\frac{10}{0.378}}\right) = 87.56 \text{ mm}^2/\text{yr}.$$

Furthermore, the age parameter is, cf. Equation (5.9),

$$\alpha = \frac{\ln(D_1/D_{ain})}{\ln(t_{\text{in}}/t_1)} = \frac{\ln(87.56/15.8)}{\ln(12/1)} = 0.689.$$

Since

$$k_1 = \xi_{cr}\sqrt{t_1 D_1} = 2.45 \times \sqrt{1 \times 87.56} = 22.93 \text{ mm},$$

cf. Equation (5.27), the initiation period of time is, cf. Equation (5.30),

$$t_{cr} = \left(\frac{c}{k_1}\right)^{\frac{2}{1-\alpha}} = \left(\frac{50}{22.93}\right)^{6.43} = 150 \text{ years}.$$

The age parameter has the following value by a direct estimation

$$\alpha = 0.1 \times (1 - 1.5 \times \text{ eqv}\{w/b_D\}) = 0.1 \times (1 - 1.5 \times 0.378) = 0.0433,$$

cf. Equation (5.7). Thus, the initiation period of time under these assumptions is, cf. Equation (5.31),

$$t_{cr} = t_{\text{in}} \left(\frac{c/\xi_{cr}}{\sqrt{t_{\text{in}}\, D_{in}}} \right)^{\frac{2}{1-\alpha}} = 12 \times \left(\frac{50/2.45}{\sqrt{12 \times 15.8}} \right)^{2.091} = 27 \text{ years.}$$

Therefore, it may be concluded that the initiation period of time lies in the interval between 27 years and 150 years, cf. Figure 5.11. This is not a satisfactory result, and more observations are needed in order to obtain a more accurate result. ◊

5.1.5 Corrosion multiprobe

As seen above the determination of one chloride profile does not seem to be sufficient in order to determine the initiation period of time. Chloride profiles at various times are needed. By monitoring an RC structure with corrosion multiprobes it is possible to predict the initiation period of time without any determination of chloride profiles, cf. Section 1.4.1.

Observations from a corrosion multiprobe

It is assumed that the corrosion multiprobe has n anodes: 1, 2, 3, ... , n, placed in the near-by-the-surface layer at the following distances from the exposed concrete,

$$x_I,\, x_{II},\, x_{III},\, \ldots\, ,\, x_N. \tag{5.36}$$

An anode will react when the chloride content of concrete in contact with the anode reaches the threshold value of chloride in concrete. The anodes are assumed to react at times

$$t_I,\, t_{II},\, t_{III},\, \ldots\, ,\, t_N. \tag{5.37}$$

The ingress into the concrete of the threshold value of chloride in concrete obeys the following relation

$$x_{cr} = k_1 \sqrt{\left(\frac{t}{t_1}\right)^{1-\alpha}} = k_1\, \tau^{0.5(1-\alpha)}, \tag{5.38}$$

where $t_1 = 1$ year and, cf. Equation (5.27),

$$\begin{aligned} k_1 &= 2\sqrt{t_1\, D_1} \times \text{ inv erfc} \left(\frac{C_{cr} - C_i}{C_0 - C_i} \right) \\ &\approx 3.32 \times \sqrt{t_1\, D_1} \times \left(1 - \sqrt{\frac{C_{cr} - C_i}{C_0 - C_i}} \right). \end{aligned} \tag{5.39}$$

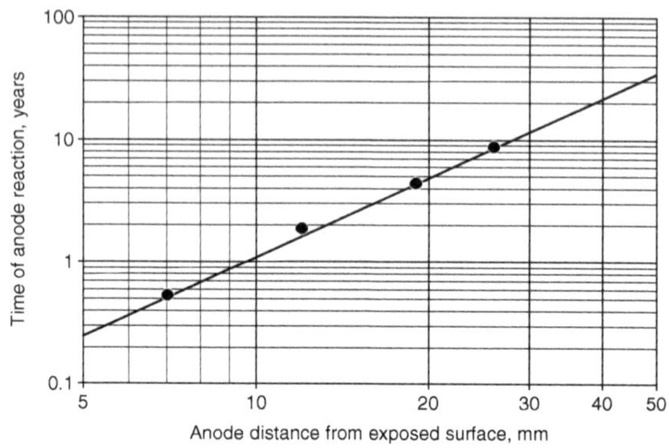

Figure 5.12: The reaction time for anodes of a CorroWatch multiprobe in concrete versus the distance of the anode from the chloride exposed surface of the concrete. It is predicted from the regression line that the reinforcing bar (having a cover of 50 mm) will initiate corrosion approximately 35 years after casting, cf. Example 5.1.5.

Equation (5.38) may also be written as

$$\ln x_{cr} = \ln k_1 + 0.5\,(1 - \alpha)\,\ln \tau. \tag{5.40}$$

The logarithmic graph of Equation (5.39) is a straight line if the threshold value of chloride in concrete C_{cr} is constant, i.e. independent of time and location. This can be utilized by the interpretation of the observations. In principle just two anodes are sufficient, but due to the deviation of the observations it is common to use four to twenty anodes. In this way k_1 and α may be determined by a linear regression analysis.

If c denotes the cover of the reinforcement, the initiation period of time may be determined from the following equation, cf. Equation (5.39),

$$c = k_1 \left(\frac{t_{cr}}{t_1} \right)^{\frac{1-\alpha}{2}}. \tag{5.41}$$

By solving Equation (5.39) with respect to the initiation period of time one gets

$$t_{\mathrm{cr}} = t_1 \left(\frac{c}{k_1} \right)^{\frac{2}{1-\alpha}}. \tag{5.42}$$

Example 5.1.5 A corrosion macrocell, the multiprobe CorroWatch from the company FORCE with $n = 4$ anodes is cast into concrete exposed to marine splash. The positions of the anodes and their times of reaction are shown in Table 5.3.

The concrete cover of the reinforcement is $c = 50$ mm and there is no special surface treatment of the concrete.

Table 5.3: Observations from a CorroWatch Multiprobe.

Anode n, nos	Position x, mm	Reaction t, yr	$\tau = t/t_1$ non-dimensional
1	7	0.53	0.53
2	13	1.89	1.89
3	19	4.42	4.42
4	26	8.81	8.81

In this case the reaction time of the first anode is $t_I < 2$ years. Therefore, the LIGHTCON model is barely valid, but in order to see the agreement with the model the regression analysis is carried out, cf. Example 5.2.11 where the HETEK model is applied. From the regression analysis it is seen that

$$\ln x = A + B \times \ln \tau = 2.2512 + 0.46631 \times \ln \tau,$$

which means that $\ln k_1 = 2.2512$ and $0.5(1 - \alpha) = 0.46631$, cf. Figure 5.12. From this we get $k_1 = 9.4991$ and $\alpha = 0.06738$. Thus, the initiation period of time is

$$t_{cr} = t_1 \left\{ \frac{c}{k_1} \right\}^{\frac{2}{1-\alpha}} = 1 \times \left\{ \frac{50}{9.4991} \right\}^{\frac{1}{0.46631}} = 35 \text{ years.} \quad \diamondsuit$$

5.1.6 Chloride ingress into prismatic RC components

The description above of chloride ingress into concrete according to the LIGHTCON model assumes that the concrete massive is semi-infinite. Thus, structural components like concrete walls and columns with square or quadratic cross-sections cannot under these assumptions be treated with the LIGHTCON model. In the following the chloride ingress into prismatic RC components is described.

Chloride ingress into walls from opposite sides

Section 2.2.1 gives the solution of chloride ingress into concrete walls from opposite sides under the assumptions that the surface chloride content C_0 and the diffusion coefficient D_0 remain constant. The solution assuming the LIGHTCON model derives from the solution assuming a constant diffusion coefficient D_0 by replacing the constant diffusion coefficient with the time-dependent achieved diffusion coefficient D_a, all other parameters being equal, cf. Poulsen (1993) and Mejlbro (1996).

The solution of Fick's second law of diffusion under these conditions, cf. Equation (2.49), is

$$C\,(x, t_{\text{in}}) = C_i + (C_{00} - C_i) \times H_0(\xi, \tau) + (C_{0d} - C_i) \times H_0(1 - \xi, \tau), \quad (5.43)$$

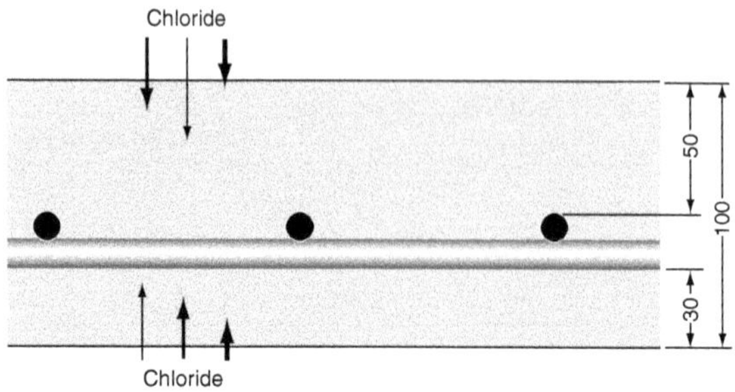

Figure 5.13: Geometry of a reinforced concrete wall submerged in seawater. The wall is reinforced with a net of reinforcing bars, each having a diameter of $\varnothing = 10$ mm, cf. Example 5.1.6. Measures shown are in mm. the Chloride penetrates from both sides into the concrete wall. The chloride content of concrete in contact with reinforcement versus time is shown in Figure 5.14.

where the function $H_0(\xi, \tau)$ is defined by

$$H_0(\xi, \tau) = 1 - \xi - \frac{2}{\pi} \sum_{n=1}^{\infty} \frac{1}{n} \exp\left(-n^2 \pi^2 \tau\right) \sin(n \pi \xi). \tag{5.44}$$

It is tabulated in Table 8.8. The parameters ξ and τ are given by

$$\xi = \frac{x}{d} \qquad \text{and} \qquad \tau = \frac{t\, D_a}{d^2} \tag{5.45}$$

respectively, where t is the time since the first chloride exposure and D_a is the achieved diffusion coefficient, defined as

$$D_a(t) = \frac{1}{t} \int_0^t D(u)\, du, \tag{5.46}$$

cf. Section 1.1.3. Here, the time-dependency of the achieved diffusion coefficient is expressed mathematically by the power function

$$D_a(t) = D_1 \times \left(\frac{t}{t_1}\right)^{-\alpha} = D_1\, t^{-\alpha}, \tag{5.47}$$

where $t_1 = 1$ yr, cf. Equation (5.6).

In general, the chloride content of the surfaces, C_{00} and C_{0d}, are different. However, in the special case where $C_{00} = C_{0d} = C_0$, the chloride profile is given by

$$C(x, t) \tag{5.48}$$

$$= \quad C_i + (C_0 - C_i) \times \left\{ 1 - \frac{4}{\pi} \sum_{n=0}^{\infty} \frac{1}{2n+1} \exp\left(-(2n+1)^2 \pi^2 \tau\right) \sin((2n+1)\pi\xi) \right\}.$$

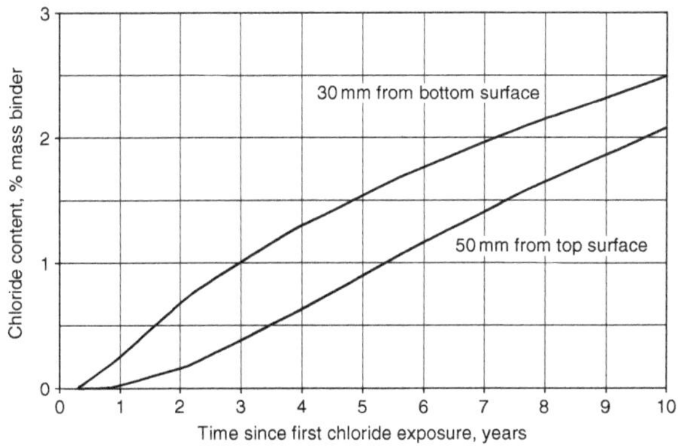

Figure 5.14: The chloride content of concrete in contact with the reinforcing bars versus time since the first chloride exposure. The concrete is submerged in seawater, cf. Example 5.1.6. The threshold value of chloride in the concrete is 2.0 % mass binder. Thus, the initiation period of time is approximately 7 years, and the corrosion starts at the bottom side of the reinforcing bars.

Example 5.1.6 A wall of an RC structure shall be submerged in seawater. The wall has the thickness of $d = 100$ mm. The wall is reinforced with a net of $\varnothing = 10$ mm reinforcing bars, having a concrete cover of $c_1 = 30$ mm and $c_2 = 50$ mm from the bottom and the top surface respectively, cf. Figure 5.13. The following conditions of the concrete are assumed:

- Type of binder: normal Portland cement,
- $w/b_D = w/b_b = w/b_{cr} = 0.35$ by mass,
- $C_i \approx 0$,
- $C_{0n} = 5.50$ % mass binder,
- $C_{0d} = 8.50$ % mass binder.

It is assumed that the concrete obeys the LIGHTCON model. The initiation time shall be predicted.

Step 1. The achieved chloride diffusion coefficient after one year of exposure is estimated to the following value

$$D_1 = 1.0 \times 25,000 \times \exp\left(-\sqrt{\frac{10}{0.35}}\right) = 119 \text{ mm}^2/\text{yr}, \qquad \text{cf. (5.2).}$$

Step 2. The age parameter is estimated to

$$\alpha = 0.6 \times (1 - 1.5 \times 0.35) = 0.285, \qquad \text{cf. (5.7).}$$

Table 5.4: Determination of chloride ingress into a wall specified in Example 5.1.6.

τ non-dimensional	t years	$H_0(0.3, \tau)$	$H_0(0.7, \tau)$	$C(30, t)$
0.005	0.308	0.00270	0.00000	0.015
0.010	0.813	0.03389	0.00000	0.186
0.020	2.142	0.13361	0.00047	0.739
0.030	3.777	0.22067	0.00427	1.250
0.040	5.648	0.28884	0.01332	1.702
0.050	7.717	0.34278	0.02682	2.113
0.060	9.958	0.38646	0.04313	2.482

Step 3. The threshold value of chloride in concrete, which is submerged in seawater, is

$$C_{cr} = 3.35 \times \exp(-1.5 \times 0.35) = 2\ \%\ \text{mass binder}, \qquad \text{cf. (5.24)}.$$

Step 4. The chloride content of the concrete at the distance $x = 30$ mm from the bottom surface, cf. Equation (5.43), is

$$C(30, t) = C_i + (C_{00} - C_i) \times H_0(\xi, \tau) + (C_{0d} - C_i) \times H_0(1 - \xi, \tau),$$

or

$$C(30, t) = 5.5 \times H_0(0.3, \tau) + 8.5 \times H_0(0.7, \tau),$$

where $\xi = x/d = 30/100 = 0.30$ and

$$\tau = \frac{t\,D}{d^2} = \frac{t_1 D_1}{d^2}\left(\frac{t}{t_1}\right)^{1-\alpha} = \frac{1 \times 119}{100^2} \times \left(\frac{t}{1}\right)^{0.715} = 0.0119 \times t^{0.715}.$$

Calculation of the increase of the chloride content is carried out by means of the spreadsheet shown in Table 5.4. The chloride increase is shown in Figure 5.14.

Step 5. The chloride content of the concrete at the distance $x = 50$ mm from the top surface, cf. Equation 5.43, is

$$\begin{aligned} C(50, t) &= C_i + (C_{00} - C_i) \times H_0(\xi, \tau) + (C_{0d} - C_i) \times H_0(1 - \xi, \tau) \\ &= 5.5 \times H_0(0.5, \tau) + 8.5 \times H_0(0.5, \tau) = 14.0 \times H_0(0, 5, \tau). \end{aligned}$$

Calculation of the increase of the chloride content is carried out by means of the spread sheet shown in Table 5.5. The chloride increase is shown in Figure 5.14.

It is seen from the graph that the initiation period of time is $t_{cr} \approx 7$ years and that the reinforcing bars start the corrosion from the bottom side. $\diamond$

Table 5.5: Determination of chloride ingress into a wall specified in Example 5.1.6.

τ non-dimensional	t years	$H_0(0.5,\tau)$	$C(30,t)$
0.005	0.308	0.00000	0.0000
0.010	0.813	0.00041	0.006
0.020	2.142	0.01242	0.174
0.030	3.777	0.04123	0.557
0.040	5.648	0.07710	1.079
0.050	7.717	0.11384	1.594
0.060	9.958	0.14890	2.085

5.1.7 Chloride ingress from de-icing salt

It is characteristic for the effect on concrete by using de-icing salt, that the exposure from chloride is not constant. This means that the surface chloride content of concrete, C_s, does not remain a constant. It is an assumption of the LIGHTCON model that C_s remains constant. It is, however, possible to adopt a simple model based upon the LIGHTCON model and obtain reasonable results. It is assumed that C_s is a periodic function of time corresponding to the progress of the de-icing period of time and the summer season.

In the following it is shown by a simple method how it is possible to predict the chloride ingress into concrete from de-icing salt based upon the simple error-function solution. Later, cf. Section 5.2.3, it is shown how it is possible to model the boundary condition by means of a set of piecewise linear functions versus time by the HETEK model. It requires, however, that there exist observations concerning the boundary condition versus time for similar RC structures.

Equations of the problem

Fick's generalized (second) law of diffusion is given by the following partial differential equation

$$\frac{\partial C}{\partial t} = \frac{\partial}{\partial x}\left\{D\frac{\partial C}{\partial x}\right\}. \tag{5.49}$$

Here, D denotes the true diffusion coefficient of the concrete. In general D may depend on time t, location x, chloride content C and temperature T. Due to seasonal variation, the temperature depends on time. Hence, we may assume that D is a function in t alone, i.e. $D = D(t)$. Then Equation (5.49) is transformed into the following simpler form

$$\frac{\partial C}{\partial t} = D_a(t) \times \frac{\partial^2 C}{\partial x^2}, \tag{5.50}$$

where $D_a(t)$ is the achieved chloride diffusion coefficient

$$D_a(t) = \frac{1}{t - t_{\text{ex}}} \int_{t=t_{\text{ex}}}^{t} D(u)\, du, \tag{5.51}$$

cf. Poulsen (1993).

Physical condition. According to observations of Sandberg (1996), Takewaka *et al.* (1988), Maage *et al.* (1995), and others the physical condition may be described by the relation

$$D_a(t) = D_{aex} \left\{ \frac{t}{t_{\text{ex}}} \right\}^{-\alpha} \qquad \text{for } t \geq t_{\text{ex}}. \tag{5.52}$$

Here, D_{aex} is the value of the chloride diffusion coefficient at a given time of reference, e.g. the time of the first chloride exposure t_{ex}. Another reference time may be used if convenient. It is assumed that de-iced concrete structures behave like marine concrete structures, e.g. like concrete exposed to marine atmosphere.

Initial condition. It is assumed that the initial (original) chloride content of the concrete C_i is uniformly distributed throughout the concrete. For convenience the initial chloride content is put to zero in the calculations. Later on the result (the chloride profiles) may be adjusted by adding the initial chloride C_i to the derived solution (if C_i is known).

Boundary condition. The chloride content of the surface layer of the concrete depends on time $C(0,t) = C_s$, i.e. the periods of de-icing, the composition of the concrete, the environment (wet or dry local road environment). If the graph of C_s is known, it is possible to approximate this graph by a set of piecewise linear functions versus time. In this way the problem may be solved, cf. Mejlbro *et al.* (2000). However, the lack of available observations do not allow such detailed information on the boundary conditions. Thus, a simple boundary condition may be used like a step function until more detailed observations are available.

Step function as a boundary condition

Periods of summer and winter. It is assumed that the concrete is cast during the end of a summer period and that the concrete is exposed to de-icing salt containing chloride during the following winter period. It is assumed that the chloride content of the concrete surface is increasing momentary to the value C_s. The de-icing is assumed to remain constant during the winter season and it is assumed that the chloride content of the concrete surface remains a constant equal to C_s until de-icing stops. When de-icing stops it is assumed that the chloride content of the concrete surface decreases momentarily to zero. During the summer seasons there is no chloride exposure of the concrete. It is finally assumed that the periods of de-icing are repeated periodically.

Chloride content of the concrete surface. It is assumed that the boundary condition is given in the following way, cf. Figure 5.15,

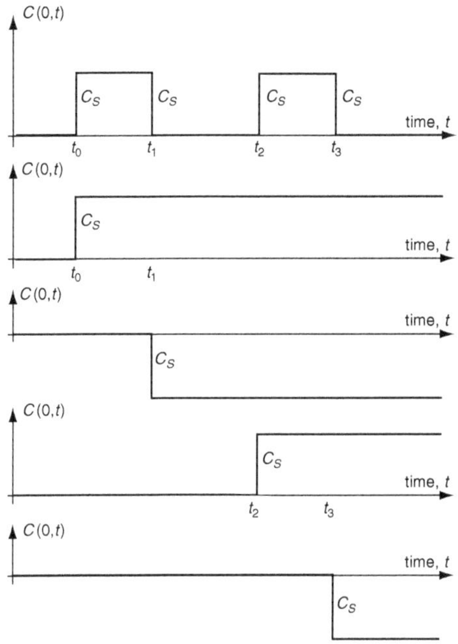

Figure 5.15: Simplified boundary condition for de-iced concrete, when knowledge is limited. The chloride profiles may be calculated by superposition of a number of error-function solutions as shown in the text. When more information is available, a more advanced method of calculation can be applied, cf. the HETEK model.

- The first period of de-icing starts at time $t = t_0$ and will last until $t = t_1$. During this period the chloride content of the concrete surface remains constant $C = C_s$.

- The first summer season starts at time $t = t_1$ and lasts until $t = t_2$. during this period the chloride content of the concrete surface is $C = C_s = 0$.

- The exposure is repeated periodically, so that $t_2 = t_0 + 12$ months and $t_3 = t_1 + 12$ months, etc.

In order to model the chloride content of the concrete surface it is convenient to use Heaviside's function

$$H(t - t_j) = \begin{cases} 0 & \text{for } t < t_j, \\ 1 & \text{for } t \geq t_j. \end{cases} \tag{5.53}$$

By applying this function it is possible to write the chloride content of the concrete surface in the following way, cf. Figure 5.15,

$$C(0,t) = 0, \qquad 0 \le t < t_0, \tag{5.54}$$
$$C(0,t) = C_s \times H(t-t_0) \qquad t_0 \le t < t_1 \tag{5.55}$$
$$C(0,t) = C_s \times \{H(t-t_0) - H(t-t_1)\} \qquad t_1 \le t < t_2 \tag{5.56}$$
$$C(0,t) = C_s \times \{H(t-t_0) - H(t-t_1) + H(t-t_2)\} \qquad t_2 \le t < t_3, \tag{5.57}$$
$$\cdots \quad \text{etc.}$$

It is seen that the Equations (5.54)–(5.57) may be written in the following short form,

$$C(0,t) = C_s \times \sum_{j=0}^{n} (-1)^j \, H(t - t_j), \qquad t_n \le t < t_{n+1}, \quad n \in \mathbb{N}. \tag{5.58}$$

Chloride profiles of the concrete

In order to determine the chloride profiles the following arguments u_{t,t_j} are defined as

$$u_{t,t_j} = \frac{0.5\,x}{\sqrt{(t - t_j) \times D_a(t)}} \qquad t_j \le t < t_{j+1}. \tag{5.59}$$

The argument u_{t,t_j} belongs to the part of the boundary condition $H(t - t_j)$, cf. Figure 5.15. It contributes to the chloride profile with the amount

$$C(x,t) = C_s \times \operatorname{erfc}\left(u_{t,t_j}\right), \qquad t_j \le t < t_{j+1}. \tag{5.60}$$

Hence, the chloride profiles $C(x,t)$ for the boundary conditions (5.54–5.57) are found in the following way by superposition,

$$C(x,t) = 0 \qquad 0 \le t < t_0 \tag{5.61}$$
$$C(x,t) = C_s \times \operatorname{erfc}\left(u_{t,t_0}\right) \qquad t_0 \le t < t_1 \tag{5.62}$$
$$C(x,t) = C_s \times \{\operatorname{erfc}\left(u_{t,t_0}\right) - \operatorname{erfc}\left(u_{t,t_1}\right)\} \qquad t_1 \le t < t_2 \tag{5.63}$$
$$C(x,t) = C_s \times \{\operatorname{erfc}\left(u_{t,t_0}\right) - \operatorname{erfc}\left(u_{t,t_1}\right) + \operatorname{erfc}\left(u_{t,t_2}\right)\}, \quad t_2 \le t < t_3 \tag{5.64}$$
$$\cdots \text{ etc.}$$

Example 5.1.7 A motorway bridge has been cast and is ready for use on the First of August. The initial chloride content of the concrete is $C_i = 0$. De-icing starts on the First of December and continues until the First of May. The bridge is not exposed to chloride during the period from the First of May until the First of December. Then the chloride exposure of the concrete bridge continues periodically.

The chloride content of the concrete surface is assumed to vary in the following simple way,

- $0 \leq t < 4$ months : $C_s = 0$ % mass binder,
- $4 \leq t < 9$ months : $C_s = 6$ % mass binder,
- $9 \leq t < 16$ months : $C_s = 0$ % mass binder,
- $16 \leq t < 21$ months : $C_s = 6$ % mass binder,
- $21 \leq t < 28$ months : $C_s = 0$ % mass binder,
- etc.

An inspection section of the concrete surface of the bridge shall be tested and the development of the chloride profiles determined. It is assumed that the achieved chloride diffusion coefficient of the concrete can be estimated from Takewaka's equation, cf. (5.52), with $D_{aex} = 85$ mm^2/month and the age parameter $\alpha = 0.45$. Thus, the chloride ingress is predicted from the boundary conditions and the physical condition mentioned above with the following achieved chloride diffusion coefficient for $t \geq 4$ months,

$$D_{at} = D_{aex} \times \left(\frac{t}{t_{\text{ex}}}\right)^{-\alpha} = 85 \times \left(\frac{t}{4}\right)^{-0.45} = 158.62 \times t^{-0.45} \text{ mm}^2/\text{month}.$$

Note. *The data given in this example are all chosen for the purpose of illustration. None of the data is found by laboratory tests nor field inspection and testing.*

The predicted chloride profiles are determined month by month in the following way.

Period 0: First summer season, i.e. $0 \leq t < 4$ *months*
The concrete of the motorway bridge is cast at time $t = 0$. During the summer season $0 \leq t < 4$ months the concrete is not exposed to chloride and the chloride profiles are all zero, i.e.

$$C(x,t) = 0, \qquad 0 \leq t < 4 \text{ months}.$$

Period 1: First winter season, i.e. $4 \leq t < 9$ *months*
The de-icing starts at the First of December, i.e. at time $t = 4$ months.

Step 1. The achieved chloride diffusion coefficients have the following values, calculated month by month,

- $t = 5$ months: $D_a(5) = 158.62 \times 5^{-0.45} = 76.88$ mm^2/month,
- $t = 6$ months: $D_a(6) = 158.62 \times 6^{-0.45} = 70.83$ mm^2/month,
- $t = 7$ months: $D_a(7) = 158.62 \times 7^{-0.45} = 66.08$ mm^2/month,

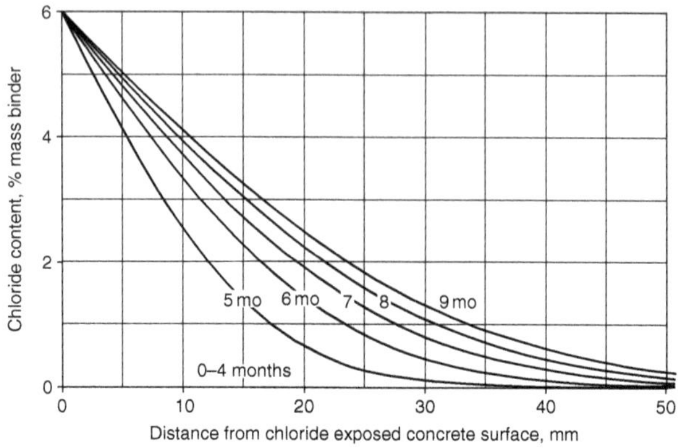

Figure 5.16: Chloride profiles of the concrete month by month from de-icing salt during the period $0 \le t \le 9$ months, cf. Example 5.1.7. The surface chloride content of the concrete increases momentarily from zero to 6 % mass binder when de-icing starts.

- $t = 8$ months: $D_a(8) = 158.62 \times 8^{-0.45} = 62.23$ mm^2/month,
- $t = 9$ months: $D_a(9) = 158.62 \times 9^{-0.45} = 59.01$ mm^2/month.

Step 2. During period 1, i.e. $4 \le t < 9$ months, it is assumed that the chloride content of the concrete surface is constantly equal to

$$C(0,t) = C_s = b_1 = 6.0 \text{ \% mass binder.}$$

Step 3. During period 1, i.e. $4 \le t < 9$ months, the chloride profiles of the concrete are given by

- $t = 5$ months, $C(x,t) = 6.0 \times \text{erfc}\,(u_{5,4})$, where

$$u_{5,4} = \frac{0.5\,x}{\sqrt{(5-4) \times 76.88}} = 0.05702\,x.$$

- $t = 6$ months, $C(x,t) = 6.0 \times \text{erfc}\,(u_{6,4})$, where

$$u_{6,4} = \frac{0.5\,x}{\sqrt{(6-4) \times 70.83}} = 0.04201\,x.$$

- $t = 7$ months, $C(x,t) = 6.0 \times \text{erfc}\,(u_{7,4})$, where

$$u_{7,4} = \frac{0.5\,x}{\sqrt{(7-4) \times 66.08}} = 0.03551\,x.$$

- $t = 8$ months, $C(x,t) = 6.0 \times \text{erfc}\,(u_{8,4})$, where

$$u_{8,4} = \frac{0.5\,x}{\sqrt{(8-4) \times 62.23}} = 0.03169\,x.$$

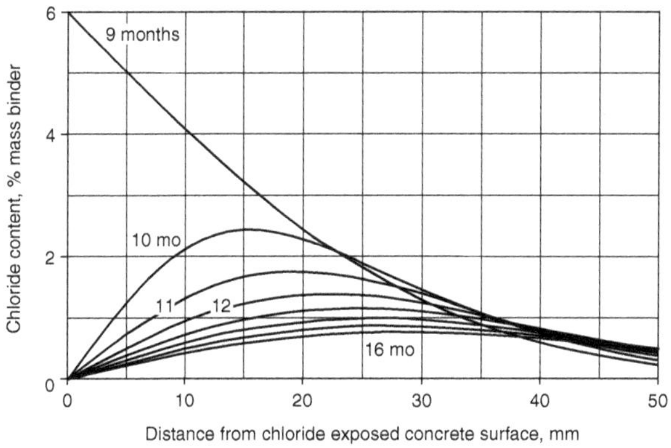

Figure 5.17: Chloride profiles of the concrete month by month from de-icing salt during the period $9 \leq t \leq 16$ months, cf. Example 5.1.7. The surface chloride content of the concrete decreases momentarily from 6 % mass binder to zero when de-icing stops.

- $t = 9$ months, $C(x, t) = 6.0 \times \text{erfc}\,(u_{9,4})$, where

$$u_{9,4} = \frac{0.5\,x}{\sqrt{(9-4) \times 59.01}} = 0.02911\,x.$$

Period 2: Second summer season, i.e. $9 \leq t < 16$ *months.*
The de-icing stops when summer season stars, i.e. at time $t = 9$ months. Then the chloride of the concrete surface is leached out momentarily so that the chloride content of the concrete surface is $C_s = 0$ during the period $9 \leq t < 16$ months.

Step 4. The achieved chloride diffusion coefficient takes on the following values, determined month by month

- $t = 10$ months: $D_a(10) = 158.62 \times 10^{-0.45} = 56.28$ mm^2/month,
- $t = 11$ months: $D_a(11) = 158.62 \times 11^{-0.45} = 53.92$ mm^2/month,
- $t = 12$ months: $D_a(12) = 158.62 \times 12^{-0.45} = 51.85$ mm^2/month,
- $t = 13$ months: $D_a(13) = 158.62 \times 13^{-0.45} = 50.01$ mm^2/month,
- $t = 14$ months: $D_a(14) = 158.62 \times 14^{-0.45} = 48.37$ mm^2/month,
- $t = 15$ months: $D_a(15) = 158.62 \times 15^{-0.45} = 46.89$ mm^2/month,
- $t = 16$ months: $D_a(16) = 158.62 \times 16^{-0.45} = 45.55$ mm^2/month,

Step 5. During the period 2, i.e. $9 \leq t < 16$ months, it is assumed that the chloride content of the concrete surface remains constant equal to

$$C(0, t) = C_s = b_2 = 0.0 \text{ \% mass binder.}$$

Step 6. During the period 2, i.e. $9 \leq t < 16$ months, the chloride profiles of the concrete are determined in the following way, month by month

- $t = 10$ months: $C(x,t) = 6.0 \times \text{erfc}(u_{10,4}) - 6.0 \times \text{erfc}(u_{10,9})$,
- $t = 11$ months: $C(x,t) = 6.0 \times \text{erfc}(u_{11,4}) - 6.0 \times \text{erfc}(u_{11,9})$,
- $t = 12$ months: $C(x,t) = 6.0 \times \text{erfc}(u_{12,4}) - 6.0 \times \text{erfc}(u_{12,9})$,
- $t = 13$ months: $C(x,t) = 6.0 \times \text{erfc}(u_{13,4}) - 6.0 \times \text{erfc}(u_{13,9})$,
- $t = 14$ months: $C(x,t) = 6.0 \times \text{erfc}(u_{14,4}) - 6.0 \times \text{erfc}(u_{14,9})$,
- $t = 15$ months: $C(x,t) = 6.0 \times \text{erfc}(u_{15,4}) - 6.0 \times \text{erfc}(u_{15,9})$,
- $t = 16$ months: $C(x,t) = 6.0 \times \text{erfc}(u_{16,4}) - 6.0 \times \text{erfc}(u_{16,9})$,

where

$$u_{10,4} = \frac{0.5\,x}{\sqrt{(10-4) \times 56.28}} = 0.02721\,x,$$

$$u_{10,9} = \frac{0.5\,x}{\sqrt{(10-9) \times 56.28}} = 0.06665\,x,$$

$$u_{11,4} = \frac{0.5\,x}{\sqrt{(11-4) \times 53.92}} = 0.02574\,x,$$

$$u_{11,9} = \frac{0.5\,x}{\sqrt{(11-9) \times 53.92}} = 0.04815\,x,$$

$$u_{12,4} = \frac{0.5\,x}{\sqrt{(12-4) \times 51.85}} = 0.02455\,x,$$

$$u_{12,9} = \frac{0.5\,x}{\sqrt{(12-9) \times 51.85}} = 0.04009\,x,$$

$$u_{13,4} = \frac{0.5\,x}{\sqrt{(13-4) \times 50.01}} = 0.02357\,x,$$

$$u_{13,9} = \frac{0.5\,x}{\sqrt{(13-9) \times 50.01}} = 0.03535\,x,$$

$$u_{14,4} = \frac{0.5\,x}{\sqrt{(14-4) \times 48.37}} = 0.02273\,x,$$

$$u_{14,9} = \frac{0.5\,x}{\sqrt{(14-9) \times 48.37}} = 0.03215\,x,$$

$$u_{15,4} = \frac{0.5\,x}{\sqrt{(15-4) \times 46.89}} = 0.02202\,x,$$

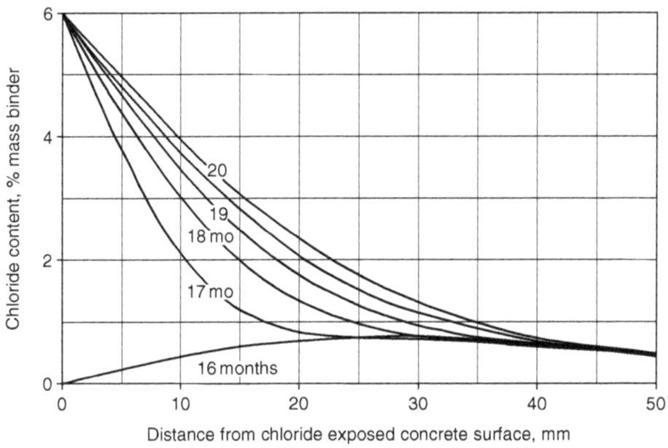

Figure 5.18: Chloride profiles of the concrete month by month from de-icing salt during the period $16 \leq t \leq 20$ months, cf. Example 5.1.7. The surface chloride content of the concrete increases momentarily from zero to 6 % mass binder when de-icing starts.

$$u_{15,9} = \frac{0.5\,x}{\sqrt{(15-9) \times 46.89}} = 0.02981\,x,$$
$$u_{16,4} = \frac{0.5\,x}{\sqrt{(16-4) \times 45.55}} = 0.02139\,x,$$
$$u_{16,9} = \frac{0.5\,x}{\sqrt{(16-9) \times 45.55}} = 0.02800\,x.$$

Period 3: Second winter season, i.e. $16 \leq t < 21$ *months.*
The de-icing starts again on the First of December, i.e. at time $t = 16$ months. Then the chloride content of the concrete of the concrete surface increases momentarily to the value $C_s = 6.0$ % mass binder.

Step 7. The achieved chloride diffusion coefficient takes on the following values, determined month by month,

- $t = 17$ months: $D_a(17) = 158.62 \times 17^{-0.45} = 44.33$ mm^2/month,
- $t = 18$ months: $D_a(18) = 158.62 \times 18^{-0.45} = 43.20$ mm^2/month,
- $t = 19$ months: $D_a(19) = 158.62 \times 19^{-0.45} = 42.16$ mm^2/month,
- $t = 20$ months: $D_a(20) = 158.62 \times 20^{-0.45} = 41.20$ mm^2/month,
- $t = 21$ months: $D_a(21) = 158.62 \times 21^{-0.45} = 40.31$ mm^2/month,

Step 8. During the period 3, i.e. $16 \leq t < 21$ months, the chloride content of the concrete surface is assumed to remain constant equal to

$$C(0,t)C_s = b_3 = 6.0 \text{ \% mass binder}.$$

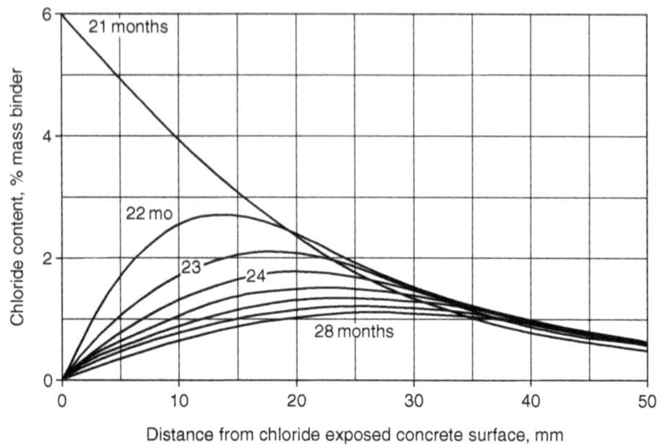

Figure 5.19: Chloride profiles of the concrete month by month from de-icing salt during the period $20 \le t \le 28$ months, cf. Example 5.1.7. The surface chloride content of the concrete decreases momentarily from 6 % mass binder to zero when de-icing stops.

Step 9. During period 3, i.e. $16 \le t < 21$ months, the chloride profiles of the concrete are month by month,

- $t = 17$ months: $C(x,t) = 6.0\times\{\mathrm{erfc}(u_{17,4}) - \mathrm{erfc}(u_{17,9}) + \mathrm{erfc}(u_{17,16})\}$,
- $t = 18$ months: $C(x,t) = 6.0\times\{\mathrm{erfc}(u_{18,4}) - \mathrm{erfc}(u_{18,9}) + \mathrm{erfc}(u_{18,16})\}$,
- $t = 19$ months: $C(x,t) = 6.0\times\{\mathrm{erfc}(u_{19,4}) - \mathrm{erfc}(u_{19,9}) + \mathrm{erfc}(u_{19,16})\}$,
- $t = 20$ months: $C(x,t) = 6.0\times\{\mathrm{erfc}(u_{20,4}) - \mathrm{erfc}(u_{20,9}) + \mathrm{erfc}(u_{20,16})\}$,
- $t = 21$ months: $C(x,t) = 6.0\times\{\mathrm{erfc}(u_{21,4}) - \mathrm{erfc}(u_{21,9}) + \mathrm{erfc}(u_{21,16})\}$,

where

$$u_{t,t_j} = \frac{0.5\,x}{\sqrt{(t-t_j)\times D_a(t)}}.$$

The development of the chloride profiles is calculated by means of the following approximation of the complementary error function

$$\mathrm{erfc}(y) \approx \frac{0.3480242}{1+0.47047u}\left(1 - \frac{0.2754975}{1+0.47047u}\left(1 - \frac{7.7999287}{1+0.47047u}\right)\right)\exp\left(-u^2\right).$$

This approximation of the complementary error function has an error $R(u)$ which satisfies the estimate $|R(u)| \le 2.5\cdot 10^{-5}$ for all $u \ge 0$. One may instead use interpolations in tables. Which method to use depends on the aim of the project and the calculator available. ◊

Discussion and conclusion. The calculation method gives for a minimum of estimated parameters a distribution of the chloride in the concrete depending

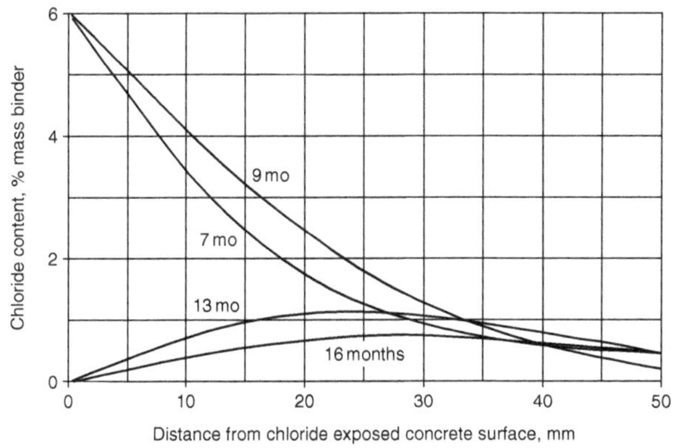

Figure 5.20: Chloride profiles in the middle of the de-icing period and in the middle of the summer season, cf. Example 5.1.7.

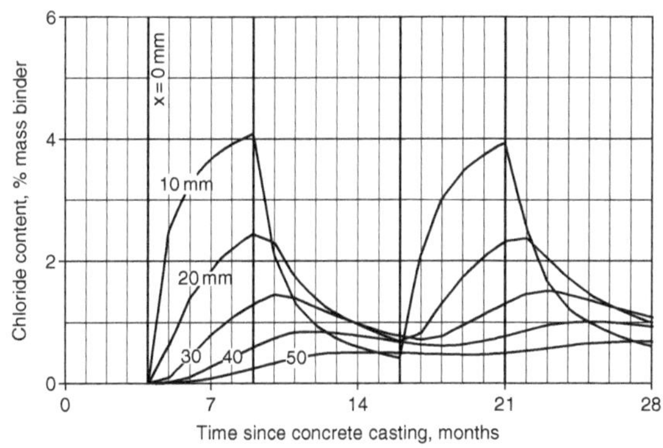

Figure 5.21: The chloride content of the concrete versus the depths below the chloride exposed concrete surface and time since concrete casting, cf. Example 5.1.7. Notice that the influence of the de-icing decreases with the depth below the surface ant that there is a slight increase in the chloride content in the depth of e.g. 40 mm.

on location and time, cf. Figure 5.16 to Figure 5.21. If de-icing remains constant (i.e. the whole year) the chloride content of the concrete will increase monotonously in time. If de-icing is carried out periodically, the chloride content of the concrete will sometimes increase and sometimes decrease, cf. Figure 5.21. This leads to the conclusion, that the chloride content of the concrete at a certain period of de-icing has an upper limit.

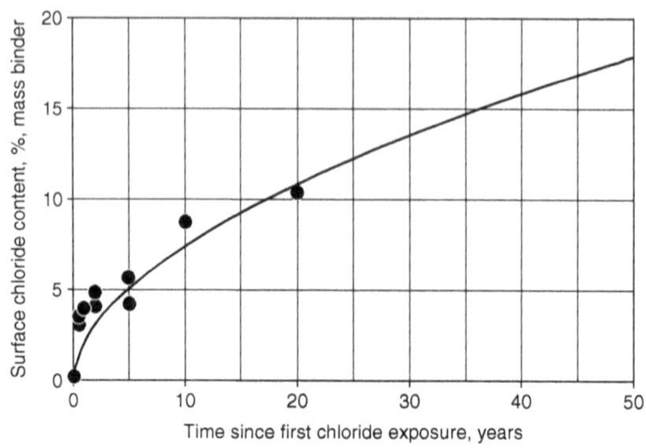

Figure 5.22: An example of the development of the chloride content of a concrete surface exposed to a marine environment according to Swamy *et al.* (1995). The shaded curve shown is a result of a regression analysis. When applying the LIGHTCON model and trying to predict the chloride ingress into the concrete e.g. during 50 years a constant surface chloride content of 10 % mass binder is estimated.

5.2 Time-dependent surface chloride content

Chloride ingress into steel reinforced concrete structures, exposed to a chloride-laden environment, e.g. marine environment, depends on the composition, the compactness of the concrete and the chloride intensity of the environment. This section describes the parameters and the formulæ used by the HETEK model when we are going to predict the future chloride ingress into the concrete on the basis of an inspection and testing (a chloride profile). The HETEK model was developed by Frederiksen *et al.* (1997). At that time about three years of observations of chloride ingress into concrete versus time were available from the Träslövsläge Marine Exposure Station, situated south of Gothenburg on the west coast of Sweden. The HETEK model was developed in such a way that it was able to cover these observations. This is here named the 'HETEK model, Version 1997'. Since then five years of observations have been reported from the Träslövsläge Marine Exposure Station and the HETEK model Version 1997 has been updated by Frederiksen *et al.* (2000).

The description of the HETEK model is given as a step-by-step manual for the prediction of the future chloride ingress, illustrated by numerical examples. These examples correspond to the former examples, applying the LIGHTCON model, cf. Section 5.1.

5.2.1 Extension of the LIGHTCON model

Before dealing with the HETEK model it is here shown that it is possible to extend the LIGHTCON model to cover the time dependence of the surface

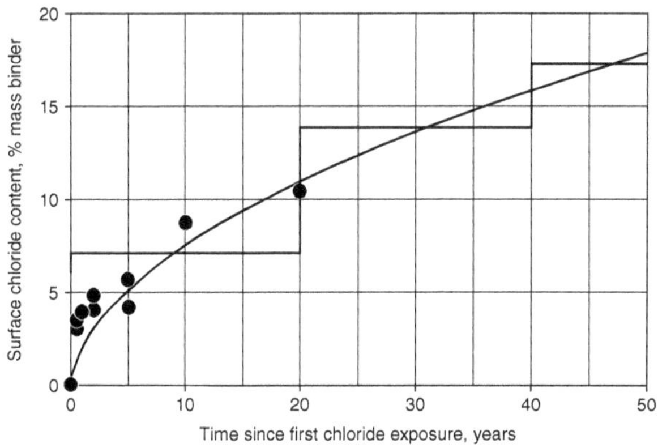

Figure 5.23: The estimate in Figure 5.22 is rather subjective. A better result is assumed by approximating the surface chloride content by a step function as shown here.

chloride content of the concrete surface.

The surface chloride content of the exposed concrete surface is time-dependent, cf. Figure 5.22. This is ignored by the LIGHTCON model.

In fact, it is usually estimated which value of a constant surface chloride content will result in the same chloride ingress into the concrete as the time-dependent surface chloride content after a certain period of time, cf. Figure 5.22. Although this estimate is subjective and apparently will not lead to an unambiguous result, it can be shown that the governing equation of Fick's second law will diminish the error made by such an approximation as the time increases. The method can be simplified by approximating the time-dependent chloride content versus time by a step function, cf. Figure 5.23.

Step function as a boundary condition

It is assumed that the surface chloride content of the exposed concrete surface versus time is known. This boundary condition is estimated by means of a step function in the following way, cf. Figure 5.24.

- The first period starts at time $t = 0$ and lasts until $t = t_1$. During this period the chloride content of the concrete surface remains constant at the value $C = C_{s1}$.

- The second period starts at time $t = t_1$ and lasts until $t = t_2$. During this period the chloride content of the concrete surface remains constant at the value $C = C_{s1} + C_{s2}$.

- The third period starts at time $t = t_2$, and so on, cf. Figure 5.25.

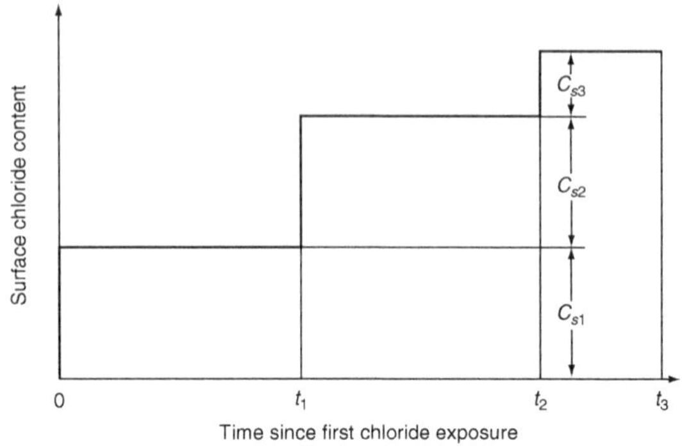

Figure 5.24: The time-dependent surface chloride content of the concrete is approximated by means of a step function, cf. Figure 5.23.

In order to model the chloride content of the exposed concrete surface it is convenient to use Heaviside's function

$$H(t - t_j) = \begin{cases} 0 & \text{for } t < t_j, \\ 1 & \text{for } t \geq t_j. \end{cases} \tag{5.65}$$

Using this function it is possible to write the chloride content of the exposed concrete surface in the following way, cf. Figure 5.25,

$$C(0,t) = C_{s1} \times H(t), \tag{5.66}$$

$$C(0,t) = C_{s1} \times H(t) + C_{s2} \times H(t - t_1) \qquad t_1 \leq t < t_2, \tag{5.67}$$

$$C(0,t) = C_{s1} \times H(t) + C_{s2} \times H(t - t_1) + C_{s3} \times H(t - t_2) \quad t_2 \leq t < t_3, \tag{5.68}$$

etc. It is seen that the Equations (5.66)–(5.68) may be written in a short form as

$$C(0,t) = \sum_{j=1}^{n} C_{sj} \times H(t - t_{j-1}), \qquad t_n \leq t < t_{n+1}. \tag{5.69}$$

Chloride profiles of the concrete

In order to determine the chloride profiles the following arguments u_{t,t_j} are defined as

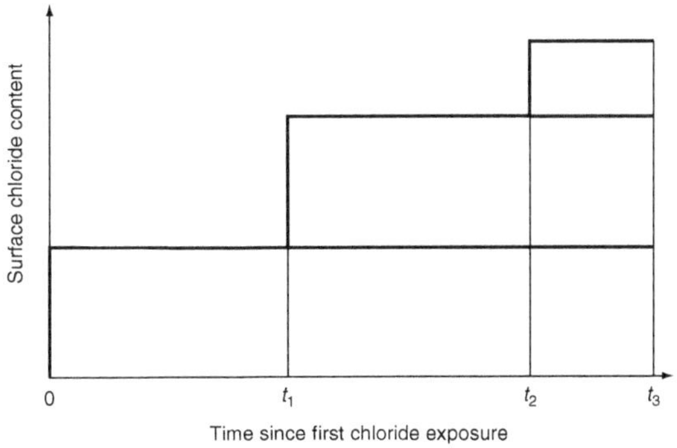

Figure 5.25: The time-dependent surface chloride content of the concrete is approximated by means of the sum of three Heaviside's functions. When the surface chloride content is a constant the LIGHTCON model is valid. Then, the chloride profiles are found by superposition

$$u_{t,t_j} = \frac{0.5\,x}{\sqrt{(t-t_j)\times D_a(t)}}, \qquad t_j \le t < t_{j+1}. \tag{5.70}$$

The argument u_{t,t_j} belongs to the boundary condition $H(t-t_j)$, cf. Figure 5.25, and it contributes to the chloride profile with the amount

$$C(x,t) = C_{sj} \times \operatorname{erfc}\left(u_{t,t_j}\right). \tag{5.71}$$

Hence, the chloride profile for the boundary conditions (5.66)–(5.68) are found by superposition in the following way,

$$0 < t < t_1: \qquad C(x,t) = C_{s0} \times \operatorname{erfc}\left(u_{t,t_0}\right), \tag{5.72}$$

$$t_1 < t < t_2: \qquad C(x,t) = C_{s0} \times \operatorname{erfc}\left(u_{t,t_0}\right) + C_{s1} \times \operatorname{erfc}\left(u_{t,t_1}\right), \tag{5.73}$$

$$t_2 < t < t_3: \qquad C(x,t) = \sum_{j=0}^{2} C_{sj} \times \operatorname{erfc}\left(u_{t,t_j}\right), \tag{5.74}$$

etc., where

$$u_{t,t_j} = \frac{0.5\,x}{\sqrt{(t-t_j)\times D_a(t)}}. \tag{5.75}$$

The corresponding abscissæ of the chloride profiles are

$$x_t = 2\,u_{t,t_j} \times \sqrt{(t-t_j)\times D_a(t)}. \tag{5.76}$$

Table 5.6: Determination of chloride profiles specified in Example 5.2.1.

u	erfc(u)	x_{10}	x_{30}	x_{50}	$C(x,t)$
-	-	mm	mm	mm	% mass binder
0.00	1.0000	0.0	0.0	0.0	10.0000
0.10	0.8875	5.1	8.6	10.9	8.8752
0.20	0.7773	10.1	17.1	21.8	7.7730
0.30	0.6714	15.2	25.7	32.7	6.7139
0.40	0.5715	20.3	34.2	43.6	5.7163
0.50	0.4795	25.4	42.8	54.6	4.7951
0.60	0.3961	30.4	51.3	65.5	3.9614
0.70	0.3222	35.5	59.9	76.4	3.2218
0.80	0.2579	40.6	68.4	87.3	2.5788
0.90	0.2031	45.6	77.0	98.2	2.0307
1.00	0.1573	50.7	85.5	109.1	1.5728
1.25	0.0771	63.4	106.9	136.4	0.7711
1.50	0.0339	76.1	128.3	163.7	0.3391
1.75	0.0133	88.7	149.6	190.9	0.1335
2.00	0.0047	101.4	171.0	218.2	0.0469
2.25	0.0015	114.1	192.4	245.5	0.0147
2.50	0.0004	126.8	213.8	272.8	0.0041
2.75	0.0001	139.4	235.1	300.0	0.0010
3.00	0.0000	152.1	256.5	327.3	0.0002

Example 5.2.1 A structural component of reinforced concrete has been exposed to marine splash for 20 years. During this period of time the concrete has been inspected and tested. The chloride content of the concrete surface has varied with time as shown in Figure 5.22. The shaded curve is a power function estimating the theoretical relation between time and surface chloride content. The equivalent water/binder ratio is $\text{eqv}\{w/b_D\} = \text{eqv}\{w/b_b\} = 0.35$ by mass.

In order to make the calculation simple, the surface chloride content versus time is approximated by a constant value $C_0 = 10$ % mass binder, cf. Figure 5.22. The initial chloride content of the concrete is here assumed to be $C_i = 0$. The approximation $C_0 = 10$ % mass binder is rather subjective and it is therefore not an unambiguous estimate. Various estimates are treated in Examples 5.2.2–5.2.6. The calculation of the chloride profile at time $t_{10} = 10$ yr, $t_{30} = 30$ yr, and $t_{50} = 50$ yr is carried out as follows:

Step 1. The chloride diffusion coefficient after 1 year of exposure is estimated to

$$D_1 = 0.6 \times 25,000 \times \exp\left(-\sqrt{\frac{10}{0.35}}\right) = 71.6 \text{ mm}^2/\text{yr}, \quad \text{cf. (5.2).}$$

The measurement of any chloride diffusion coefficient during the period of the first 20 years may be used instead.

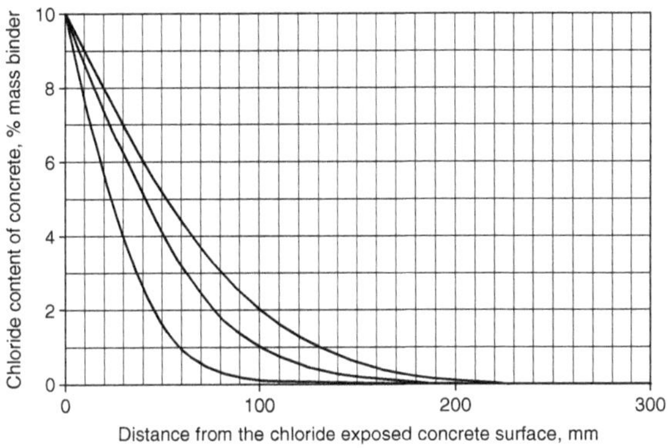

Figure 5.26: The chloride profiles after 10, 30 and 50 years of chloride exposure as described and determined in Example 5.2.1. The chloride content of the concrete surface is time-dependent as given in Figure 5.22. This relation was approximated by a constant value $C_0 = 10$ % by mass binder, and the LIGHTCON model was assumed to be valid for the chloride ingress into the concrete, cf. Figure 5.22.

Step 2. The age parameter is estimated to

$$\alpha = 0.1 \times (1 - 1.5 \times 0.35) = 0.0475, \qquad \text{cf. (5.7).}$$

Step 3. Thus, the chloride diffusion coefficients are given by

$$D_a(10 \text{ yr}) = 71.6 \times 10^{-0.0475} = 64.2 \text{ mm}^2/\text{yr} \qquad \text{cf. (5.6),}$$

$$D_u(30 \text{ yr}) = 71.6 \times 30^{-0.0475} = 60.9 \text{ mm}^2/\text{yr} \qquad \text{cf. (5.6),}$$

$$D_a(50 \text{ yr}) = 71.6 \times 50^{-0.0475} = 59.5 \text{ mm}^2/\text{yr} \qquad \text{cf. (5.6).}$$

Step 4. The abscissæ u and the corresponding ordinates of the complementary error function erfc(u) are inserted into a spreadsheet, either from a table or calculated from an approximation formula, cf. Section 3.2.5.

Step 5. The ordinates of the chloride profile are calculated according to the following formulæ

$$C(x, 10 \text{ yr}) = 0 + (10.0 - 0) \times \text{erfc}\,(u_{10}) = 10.0 \times \text{erfc}\,(u_{10}) \text{ cf. (5.13),}$$

$$C(x, 30 \text{ yr}) = 0 + (10.0 - 0) \times \text{erfc}\,(u_{30}) = 10.0 \times \text{erfc}\,(u_{30}) \text{ cf. (5.13),}$$

$$C(x, 50 \text{ yr}) = 0 + (10.0 - 0) \times \text{erfc}\,(u_{50}) = 10.0 \times \text{erfc}\,(u_{50}) \text{ cf. (5.13),}$$

where

$$u_n = \frac{0.5\,x}{\sqrt{t_n \times D_a\,(t_n)}} \qquad \text{cf. (5.17).}$$

Step 6. The abscissæ of the chloride profiles are calculated according to the following formulæ

$$x_{10} = 2\,u_{10} \times \sqrt{10 \times 64.2} = 50.7 \times u_{10}, \qquad \text{cf. (5.18),}$$

$$x_{30} = 2\,u_{30} \times \sqrt{30 \times 60.9} = 85.5 \times u_{30}, \qquad \text{cf. (5.18),}$$

$$x_{50} = 2\,u_{50} \times \sqrt{50 \times 59.5} = 109.1 \times u_{50}, \qquad \text{cf. (5.18),}$$

where u_n is the abscissa of the complementary error function, cf. **Step 4.**

Step 7. The chloride profiles are plotted, cf. Figure 5.26. ◊

Example 5.2.2 A structural component of reinforced concrete has been exposed to marine splash for 20 years. During this period of time the concrete has been inspected and tested. The chloride content of the concrete surface has varied in time as shown in Figure 5.23. The shaded curve is a power function estimating the theoretical relation between time and surface chloride content. The equivalent water/binder ratio is 0.35 by mass.

In order to make the calculation simple the surface chloride content versus time is approximated by a step function, cf. Figure 5.23,

$$\{C_{s1},\, C_{s2},\, C_{s3}\} = \{7.0,\, 5.0,\, 4.0\}\ \%\ \text{mass binder.}$$

The initial chloride content of the concrete is here assumed to be $C_i = 0$. The approximation by this step function is still rather subjective (but not as subjective as in Example 5.2.1). Various estimates are treated by Examples 5.2.1–5.2.6.

The calculation of the chloride profile at the time $t_{10} = 10$ yr, $t_{30} = 30$ yr and $t_{50} = 50$ yr is carried out as follows:

Step 1. The chloride diffusion coefficient after 1 year of exposure is

$$D_1 = 0.6 \times 25,000 \times \exp\left(-\sqrt{\frac{10}{0.35}}\right) = 71.6\ \text{mm}^2/\text{yr}, \qquad \text{cf. (5.2).}$$

The measurement of any chloride diffusion coefficient during the period of the first 20 years may be used instead.

Step 2. The age parameter is estimated to

$$\alpha = 0.1 \times (1 - 1.5 \times 0.35) = 0.0475, \qquad \text{cf. (5.7).}$$

Table 5.7: Chloride ingress into concrete after 50 years as specified by Example 5.2.2

$u_{50,0}$	erfc $(u_{50,0})$	$u_{50,20}$	erfc $(u_{50,20})$	$u_{50,40}$	erfc $(u_{50,40})$	x mm	$C(x,50)$ % mass binder
0.0000	1.0000	0.0000	1.0000	0.0000	1.0000	0	16.000
0.0498	0.9439	0.0590	0.9335	0.1025	0.8847	5	14.814
0.0995	0.8881	0.1180	0.8674	0.2050	0.7719	10	13.641
0.1493	0.8328	0.1770	0.8023	0.3075	0.6637	15	12.496
0.1990	0.7784	0.2360	0.7386	0.4100	0.5621	20	11.390
0.2488	0.7250	0.2950	0.6766	0.5125	0.4686	25	10.332
0.2985	0.6729	0.3540	0.6167	0.6150	0.3844	30	9.332
0.3980	0.5736	0.4720	0.5045	0.8200	0.2462	40	7.522
0.4975	0.4817	0.5900	0.4041	1.0250	0.1472	50	5.981
0.5970	0.3985	0.7080	0.3167	1.2300	0.0820	60	4.701
0.6965	0.3246	0.8260	0.2427	1.4350	0.0424	70	3.656
0.7960	0.2603	0.9440	0.1819	1.6400	0.0204	80	2.813
0.8955	0.2053	1.0620	0.1331	1.8450	0.0091	90	2.139
0.9950	0.1594	1.1800	0.0952	2.0500	0.0038	100	1.606
1.2438	0.0786	1.4750	0.0370	2.5625	0.0003	125	0.736
1.4925	0.0348	1.7700	0.0123	3.0750	0.0000	150	0.305
1.7413	0.0138	2.0650	0.0035	3.5875	0.0000	175	0.114
1.9900	0.0049	2.3600	0.0008	4.1000	0.0000	200	0.039
2.2388	0.0016	2.6550	0.0002	4.6125	0.0000	225	0.012
2.4875	0.0004	2.9500	0.0000	5.1250	0.0000	250	0.003

Step 3. Thus, the chloride diffusion coefficients are found to be

$$D_a(10\text{ yr}) = 71.6 \times 10^{-0.0475} = 64.2\text{ mm}^2/\text{yr} \qquad \text{cf. (5.6)},$$

$$D_a(30\text{ yr}) = 71.6 \times 30^{-0.0475} = 60.9\text{ mm}^2/\text{yr} \qquad \text{cf. (5.6)},$$

$$D_a(50\text{ yr}) = 71.6 \times 50^{-0.0475} = 59.5\text{ mm}^2/\text{yr} \qquad \text{cf. (5.6)}.$$

Step 4. The abscissæ x of the chloride profiles $C(x,t)$ for the wanted interval are inserted into the spreadsheet.

Step 5. The ordinates of the chloride profile are calculated according to the following formulæ:

$$C(x, 10\text{ yr}) = 7.0 \times \text{erfc}\,(u_{10,0}) \text{ for } 0 < t < 20 \qquad \text{cf. (5.72)}.$$

$$\begin{aligned} C(x, 30\text{ yr}) &= 7.0 \times \text{erfc}\,(u_{30,0}) + 5.0 \times \text{erfc}\,(u_{30,20}) \\ &\quad \text{for } 20 < t < 40, \qquad \text{cf. (5.73)}. \\ C(x, 50\text{ yr}) &= 7.0 \times \text{erfc}\,(u_{50,0}) + 5.0 \times \text{erfc}\,(u_{50,20}) \\ &\quad +4.0 \times \text{erfc}\,(u_{50,40}), \text{ for } 40<t<50, \qquad \text{cf. (5.74)}, \end{aligned}$$

where

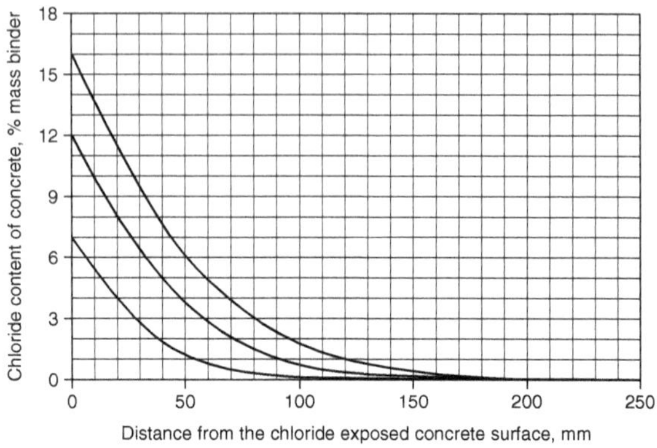

Figure 5.27: The chloride profiles after 10, 30 and 50 years of chloride exposure as described in Example 5.2.2. The chloride content of the concrete surface is time-dependent as given in Figure 5.23. This relation was approximated by a step function as shown, and the LIGHTCON model was assumed to be valid for the chloride ingress into the concrete, cf. Figure 5.23.

$$u_{10,0} = \frac{0.5\,x}{\sqrt{(10-0)\times 64.2}} = 0.0197\,x, \qquad \text{cf. (5.75)},$$

$$u_{30,0} = \frac{0.5\,x}{\sqrt{(30-0)\times 60.9}} = 0.0117\,x, \qquad \text{cf. (5.75)},$$

$$u_{30,20} = \frac{0.5\,x}{\sqrt{(30-20)\times 60.9}} = 0.0203\,x, \qquad \text{cf. (5.75)},$$

$$u_{50,0} = \frac{0.5\,x}{\sqrt{(50-0)\times 59.5}} = 0.00995\,x, \qquad \text{cf. (5.75)},$$

$$u_{50,20} = \frac{0.5\,x}{\sqrt{(50-20)\times 59.5}} = 0.0118\,x, \qquad \text{cf. (5.75)},$$

$$u_{50,40} = \frac{0.5\,x}{\sqrt{(50-40)\times 59.5}} = 0.0205\,x, \qquad \text{cf. (5.75)}.$$

Step 6. The abscissæ of the chloride profiles are calculated according to the following formulæ,

$$x_{10} = 2\,u_{10} \times \sqrt{10 \times 64.2} = 25.3 \times u_{10}, \qquad \text{cf. (5.76)},$$

$$x_{30} = 2\,u_{30} \times \sqrt{30 \times 60.9} = 42.7 \times u_{30}, \qquad \text{cf. (5.76)},$$

$$x_{50} = 2\,u_{50} \times \sqrt{50 \times 59.5} = 54.5 \times u_{50}, \qquad \text{cf. (5.76)},$$

where u_n is the abscissa of the complementary error function, cf. **Step 4**.

Step 7. The chloride profiles are plotted, cf. Figure 5.27. ◊

5.2.2 The HETEK model of chloride ingress into concrete

A complete description and documentation of the HETEK model Version 1997 is given by Frederiksen *et al.* (1997). The entire series of HETEK reports are recommended when a detailed study of the model is wanted. There, a more detailed description is given than in this short section. Here, the necessary formulæ for the estimation of future chloride ingress are presented for a concrete with types of binders like Portland cement, silica fume and fly ash and having a known w/c ratio as well as (sometimes) a chloride profile observed at an inspection test – and no further information.

The original HETEK model assumes that the chloride content of the exposed concrete surface versus time obeys a power function which will be explained later. It is, however, possible to approximate the surface chloride content versus time with a set of piecewise linear functions versus time. It is in this way possible to treat an arbitrary relation, cf. Section 5.2.3.

The problem of estimating the chloride ingress into concrete

In order to estimate the future chloride ingress into concrete it is necessary to have certain information on the concrete exposed. The general cases of information are:

- *Existing marine RC structure.* The structure has been inspected in a period of time so that it has been possible to establish a relation of the surface chloride content versus time and a relation of the achieved diffusion coefficient versus time. The future chloride profiles shall be predicted, cf. Example 5.2.3.

- *Existing marine RC structure.* The structure is inspected twice during the lifetime and the future chloride profiles shall be predicted, cf. Example 5.2.5.

- *Marine RC structure under design.* No information exists on the concrete, but the concrete composition is assumed. The future chloride profiles shall be predicted, cf. Example 5.2.4.

In practice, combinations of these cases will often take place.

When it is not possible to obtain reliable information on the concrete. i.e. the water/cement ratio from the specification of the concrete structure, the inspection must be supplied with a thin section analysis of the concrete in question in order to determine the concrete composition.

Testing the concrete by NT Build 443 and/or by the 'method of inverse cores' (as shown in Section 5.2.5) may improve the estimation of the future chloride ingress into the concrete.

Assumptions of the HETEK model

The HETEK model for future chloride ingress into concrete is based on the following assumptions:

- Chloride C in concrete is defined as the 'total, acid soluble chloride'.

- Transport of chloride in concrete takes place by diffusion. There is an equilibrium of the mass of ingress of (free) chloride into each element of the concrete, the accumulation of (free and bound) chloride in the element and an ongoing diffusion of (free) chloride in the element towards a neighbouring element, and so on.

- The flow of chloride F is proportional to the gradient of chloride $\partial C/\partial x$. The factor of proportionality is the achieved chloride diffusion coefficient D_a of the concrete.

- The achieved chloride diffusion coefficient D_a depends (may decrease) on the time t, the composition of the concrete and the environment.

- The boundary condition C_s depends (may increase) on time t, the composition of the concrete and the environment.

- The initial chloride content of the concrete C_i (per unit element of the concrete) is uniformly distributed at exposure time $t = 0$.

- The relations used for the determinative parameters with respect to the environment (ATM, SPL and SUB), the time and the composition of the concrete. These are documented at the Träslövsläge Marine Exposure Station on the west coast of Sweden (south of Gothenburg), as described by Frederiksen *et al.* (1997).

Mass balance of chloride in a volume element of concrete

The equation of mass balance (equilibrium) for chloride in concrete is expressed by Fick's general law of diffusion:

$$\frac{\partial C}{\partial t} = \frac{\partial}{\partial x}\left(D\,\frac{\partial C}{\partial x}\right), \tag{5.77}$$

where C is the chloride content at the locality x at time t, and D is the chloride diffusion coefficient which may depend on time t, the locality x, the chloride content C, and the composition of the concrete and the environment. Notice, however, that it is an assumption of the HETEK model that the chloride diffusion coefficient D is independent of the locality x and the chloride content.

Achieved chloride diffusion coefficient

Diffusion coefficient after 1 year of exposure. The study of observations from the Träslövsläge Marine Exposure Station has led to the conclusion that the chloride diffusion coefficient after 1 year of exposure with suitable accuracy may be expressed as given by Equation (5.2).

Development of the achieved chloride diffusion coefficient. The time dependence of the achieved chloride diffusion coefficient D_a can be expressed mathematically by the power function given by Equation (5.6).

The age parameter. At the inspection of the chloride ingress into the concrete, carried out at the time $t_{\text{in}} \gg$ approximately 5 year the achieved chloride diffusion coefficient D_{ain} is determined. From the time dependence of the diffusion coefficient the age parameter α is determined by Equation (5.9).

Boundary condition

Chloride content of the concrete surface after 1 year of exposure. The HETEK model assumes that the achieved chloride content of the exposed concrete surface after 1 year of exposure is given by

$$C_1 = k_b \times \text{ eqv}\left\{w/b_b\right\} \text{ \% mass binder.} \tag{5.78}$$

The equivalent water/binder-ratio (by mass) with respect to binding is

$$\text{eqv}\left\{w/b_b\right\} = \frac{W}{PC + 0.75 \times FA - 1.5 \times SF}. \tag{5.79}$$

Here

- W = content of mixing water of the concrete,
- PC = content of Portland cement of the concrete,
- FA = content of fly ash of the concrete,
- SF = content of silica fume of the concrete.

The factor k_b depends on the environment of the concrete in the following way,

$$k_b = \begin{cases} 2.20 \text{ for concrete exposed to marine atmosphere (ATM)} \\ 3.67 \text{ for concrete exposed to marine splash zone (SPL)} \\ 5.13 \text{ for concrete submerged in seawater (SUB).} \end{cases} \tag{5.80}$$

Figure 5.28 shows the relation Equation (5.78) with plotted observations from the Träslövsläge Marine Exposure Station.

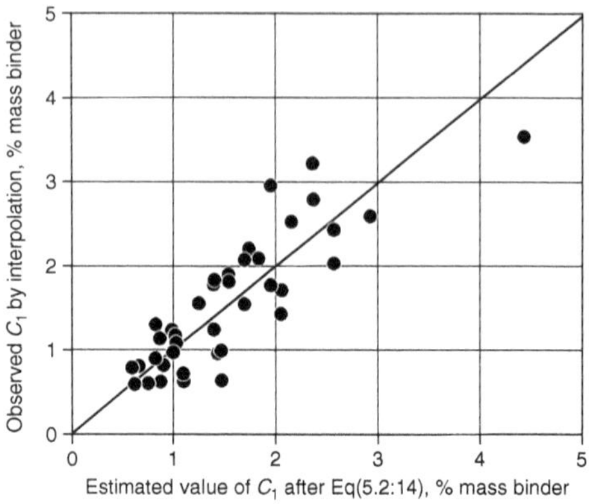

Figure 5.28: The observed value of C_1, i.e. the chloride content of the concrete surface after one year of exposure to marine environment, determined by interpolation of measurements, plotted versus the values of C_1 determined by Equation (5.78). An acceptable agreement is obtained.

Method of inverse cores. A determination of C_1 can be made according to the method of inverse cores, cf. Poulsen *et al.* (1998)).

Development of the chloride content of the concrete surface. If the achieved chloride content of the exposed concrete surface C_{sa} versus time obeys the following relation, it has been shown by Mejlbro (1996) that the solution of Fick's second law of diffusion is simple, cf. also Frederiksen *et al* (1997),

$$\begin{aligned} C_{sa} &= C_i + S \times (t\, D_a)^p = C_i + S \times \left\{ t\, D_1\, (t/t_1)^{-\alpha} \right\}^p \\ &= C_i + S_1 \times (t/t_1)^{(1-\alpha)p}, \end{aligned} \tag{5.81}$$

cf. Equation (5.6), page 246. Here $t_1 = 1$ yr, and $S_1 = S \times D_1^p$, and C_i is the initial (uniformly distributed) chloride content of the concrete. The exponents α and p and the factor S_1 depend on the composition of the concrete and the environment. The geometrical meaning is shown on Figure 5.29.

Surface chloride content after 100 years of exposure. It has been possible to determine an expression for the exponent p by studying the chloride ingress into concrete specimens at the Träslövsläge Marine Exposure Station. However, for mnemonic reasons we define a surface chloride content after 100 years of exposure

$$C_{100} = k_{100} \times C_1. \tag{5.82}$$

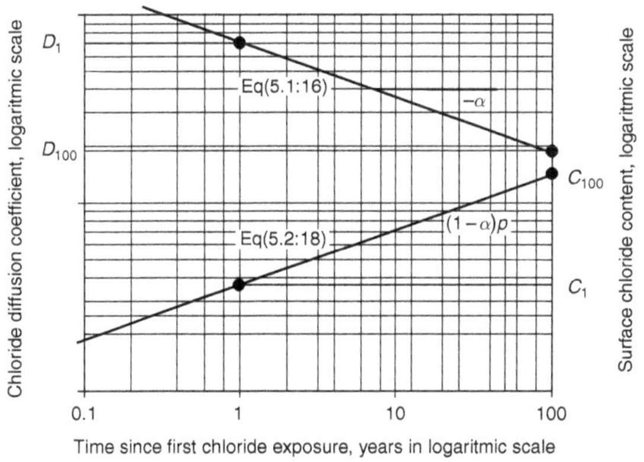

Figure 5.29: The geometrical meaning of the diffusion parameters of the HETEK model of chloride ingress into concrete.

The factor k_{100} depends on the environment of the concrete,

$$k_{100} = \begin{cases} 7.0 \text{ for concrete exposed to marine atmosphere (ATM)}, \\ 4.5 \text{ for concrete exposed to marine splash zone (SPL)}, \\ 1.5 \text{ for concrete submerged in seawater (SUB)}. \end{cases} \tag{5.83}$$

The chloride content of the concrete surface. From the estimation of C_1 and C_{100} it is possible to calculate the exponent p and the factor S_1 in the following way, cf. Figure 5.29,

$$p = \frac{\log_{10}(C_{100} - C_i) - \log_{10}(C_1 - C_i)}{2 \times (1 - \alpha)}, \tag{5.84}$$

and

$$S_1 = C_1 - C_i. \tag{5.85}$$

By inspecting and testing (determination of a representative chloride profile) at the time of inspection $t_{\text{in}} \gg t_{\text{ex}}$ (unit: years), the achieved chloride content of the exposed concrete surface C_{sin} and the initial chloride content of the concrete C_i are determined. These observations are more reliable than the estimations given above. Thus, the exponent p becomes, cf. Figure 5.29,

$$p = \frac{\ln(C_{sin} - C_i) - \ln(C_1 - C_i)}{(1 - \alpha)\ln t_{\text{in}}}. \tag{5.86}$$

Plotting the chloride profiles

As described above the four determining parameters α, D_1, p and S_1 of the mathematical expression of the HETEK model describe unambiguously the chloride profiles and their developments versus time. Thus, from the following equation according to Mejlbro (1996), the equation of the chloride profile can be written, cf. also Equation (3.43),

$$\begin{aligned} C &= C_i + (C_{sa} - C_i) \times \Phi_p\left(\frac{0.5\,x}{\sqrt{t\,D_a}}\right) \\ &= C_i + S_1 \times \left\{\frac{t}{t_1}\right\}^{(1-\alpha)p} \Psi_p\left(\frac{0.5\,x}{\sqrt{t_1 D_1 \times (t/t_1)^{1-\alpha}}}\right), \end{aligned} \tag{5.87}$$

where $t_1 = 1$ yr. The chloride profiles can be calculated at any time wanted.

$\Psi_p(u)$*-functions.* The functions $\Psi_p(u)$ are defined in Section 3.3.1 and tabulated in Section 8.3. Furthermore, polynomial approximations are available, cf. Section 3.3.7, as well as simple estimates, cf. Equation (3.72). The application of these tables or approximations is as follows: When α, D_1, p and S_1 are given or determined by testing, the chloride profiles of the concrete at a given time (age) t_n is easily determined by use of a spreadsheet. First the abscissæ and the ordinates of the functions u and $\Psi_p(u)$ respectively are inserted into the spreadsheet. If p is not a multiple of 0.05, an interpolation (linear or parabolic) may be used. However, the nearest value for p to the multiple of 0.05 may be used, since the problem is not sensitive to small changes of the parameter p.

At time $t = t_n$ the chloride content $C = C(x_n, t_n)$ at location x_n is calculated from

$$C = C_i + S_1\,\tau_n^{(1-\alpha)p}\,\Psi_p(u), \tag{5.88}$$

where

$$\tau_n = \frac{t_n}{t_1}, \tag{5.89}$$

and $t_1 = 1$ yr. Then the corresponding abscissæ x_n are determined from

$$\begin{aligned} x_n &= u \times \sqrt{4\,t_n D_a} = u \times \sqrt{4 t_n D_1 \cdot (t_n/t_1)^{-\alpha}} \\ &= u \times \sqrt{4 t_n^{1-\alpha} t_1^{\alpha} D_1}, \end{aligned} \tag{5.90}$$

where $t_1 = 1$ year.

When $p = 0$, 0.5 and 1, the functions $\Psi_p(u)$ become fairly simple,

$$p = 0.0 \text{ implies that } \Psi_0(u) = \operatorname{erfc}(u). \tag{5.91}$$

$$p = 0.5 \text{ implies that } \Psi_{0.5}(u) = \exp\left(-u^2\right) - \sqrt{\pi}\,u\,\operatorname{erfc}(u). \tag{5.92}$$

$$p = 1.0 \text{ implies that } \Psi_1(u) = \left(1 + 2u^2\right)\operatorname{erfc}(u) - \frac{2u}{\sqrt{\pi}}\exp\left(-u^2\right). \tag{5.93}$$

Table 5.8: Chloride ingress into concrete specified by Example 5.2.3.

u	$\Psi_{0.6}(u)$	x_{10}	$C(x_{10}, 10)$	x_{30}	$C(x_{30}, 10)$	x_{50}	$C(x, 50)$
-	-	mm	% MB	mm	% MB	mm	% MB
0.0	1.00000	0.0	7.450	0.0	13.630	0.0	18.060
0.1	0.82402	5.1	6.139	8.6	11.231	10.9	14.882
0.2	0.67107	10.1	4.999	17.1	9.147	21.8	12.120
0.3	0.53987	15.2	4.022	25.7	7.358	32.7	9.750
0.4	0.42887	20.3	3.195	34.2	5.845	43.6	7.745
0.5	0.33626	25.4	2.505	42.8	4.583	54.6	6.073
0.6	0.26012	30.4	1.938	51.3	3.545	65.5	4.698
0.7	0.19846	35.5	1.479	59.9	2.705	76.4	3.584
0.8	0.14927	40.6	1.112	68.4	2.035	87.3	2.696
0.9	0.11066	45.6	0.824	77.0	1.508	98.2	1.999
1.0	0.08082	50.7	0.602	85.5	1.102	109.1	1.460
1.1	0.05814	55.8	0.433	94.1	0.792	120.0	1.050
1.2	0.04119	60.8	0.307	102.6	0.561	130.9	0.744
1.3	0.02872	65.9	0.214	111.2	0.391	141.8	0.519
1.4	0.01971	71.0	0.147	119.7	0.269	152.7	0.356
1.5	0.01331	76.1	0.099	128.3	0.181	163.7	0.240
1.6	0.00884	81.1	0.066	136.8	0.120	174.6	0.160
1.7	0.00578	86.2	0.043	145.4	0.079	185.5	0.104
1.8	0.00371	91.3	0.028	153.9	0.051	196.4	0.067
1.9	0.00234	96.3	0.017	162.5	0.032	207.3	0.042
2.0	0.00145	101.4	0.011	171.0	0.020	218.2	0.026

Example 5.2.3 A structural component of reinforced concrete has been exposed to marine splash for 20 years. During this period of time the concrete has been inspected and tested. The chloride content of the concrete surface has varied in time as shown in Figure 5.22. The shaded curve is a power function estimating the theoretical relation between time and surface chloride content. The equivalent water/binder-ratio with reference to diffusion is $\text{eqv}(w/c_D) = 0.35$ by mass.

The initial chloride content of the concrete is here assumed to be $C_i = 0$. It is assumed that the boundary condition obeys the HETEK model, i.e.

$$C_{sa} = S_1\, t^{(1-\alpha)p} = 2.1 \times t^{0.55}.$$

Here, $(1-\alpha)p = 0.55$ (non-dimensional), and $S_1 = 2.1$ % mass binder are not found by regression analysis of the measurements, but it is desirable to know the consequence of such a boundary condition. Other estimates of the boundary condition are treated in Examples 5.2.1, 5.2.2 and 5.2.4.

The calculation of the chloride profile at the time $t_{10} = 10$ yr, $t_{30} = 30$ yr and $t_{50} = 50$ yr is carried out as follows:

Step 1. The chloride diffusion coefficient after 1 year of exposure is

$$D_1 = 0.6 \times 25,000 \times \exp\left(-\sqrt{\frac{10}{0.35}}\right) = 71.6 \text{ mm}^2/\text{yr}, \quad \text{cf. (5.2)}.$$

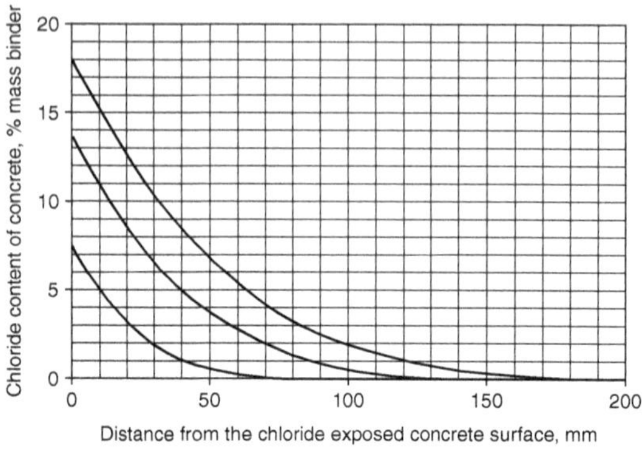

Figure 5.30: The chloride profiles after 10, 30 and 50 years of chloride exposure as described and determined in Example 5.2.3. The chloride content of the concrete surface is time dependent as given in Figure 5.23. The HETEK model was assumed to be valid for the chloride ingress into the concrete.

The measurement of any chloride diffusion coefficient during the period of the first 20 years may be used instead.

Step 2. The age parameter is estimated to

$$\alpha = 0.1 \times (1 - 1.5 \times 0.35) = 0.0475, \qquad \text{cf. (5.7).}$$

Step 3. The chloride diffusion coefficients become

$$D_a(10 \text{ yr}) = 71.6 \times 10^{-0.0475} = 64.2 \text{ mm}^2/\text{yr}, \qquad \text{cf. (5.6),}$$

$$D_a(30 \text{ yr}) = 71.6 \times 30^{-0.0475} = 60.9 \text{ mm}^2/\text{yr}, \qquad \text{cf. (5.6),}$$

$$D_a(50 \text{ yr}) = 71.6 \times 50^{-0.0475} = 59.5 \text{ mm}^2/\text{yr}, \qquad \text{cf. (5.6).}$$

Step 4. The surface chloride contents are

$$C_s(10 \text{ yr}) = 2.1 \times 10^{0.55} = 7.45 \text{ \% mass binder}, \qquad \text{cf. (5.81),}$$

$$C_s(30 \text{ yr}) = 2.1 \times 30^{0.55} = 13.63 \text{ \% mass binder}, \qquad \text{cf. (5.81),}$$

$$C_s(50 \text{ yr}) = 2.1 \times 50^{0.55} = 18.06 \text{ \% mass binder}, \qquad \text{cf. (5.81).}$$

Step 5. Since $(1 - \alpha)p = 0.55$ it is assumed that $p \approx 0.6$. Thus, the abscissæ u and the corresponding ordinates of the function $\Psi_{0.6}(u)$ are inserted into the spreadsheet, either from Table 8.3 or calculated from an approximation formula, cf. Section 3.2.4.

Step 6. The ordinates of the chloride profiles are, since $C_i = 0$, calculated according to the following formulæ, cf. Equation (5.87),

$$C(x, 10 \text{ yr}) = 0 + (7.45 - 0) \times \Psi_{0.6}(u) = 7.45 \times \Psi_{0.6}(u),$$

$$C(x, 30 \text{ yr}) = 0 + (13.63 - 0) \times \Psi_{0.6}(u) = 13.63 \times \Psi_{0.6}(u),$$
$$C(x, 50 \text{ yr}) = 0 + (18.06 - 0) \times \Psi_{0.6}(u) = 18.06 \times \Psi_{0.6}(u),$$

Step 7. The corresponding abscissæ are, cf. Equation (5.90),

$$x_{10} = u \times \sqrt{4t_{10} \times D_a(10 \text{ yr})} = u \times \sqrt{4 \times 10 \times 64.2} = 50.7 \times u,$$
$$x_{30} = u \times \sqrt{4t_{30} \times D_a(30 \text{ yr})} = u \times \sqrt{4 \times 30 \times 60.9} = 85.5 \times u,$$
$$x_{50} = u \times \sqrt{4t_{50} \times D_a(50 \text{ yr})} = u \times \sqrt{4 \times 50 \times 59.9} = 109.1 \times u.$$

Step 8. Finally, the chloride profiles are plotted, cf. Figure 5.30. ◊

Example 5.2.4 A structural component of reinforced concrete has been exposed to marine environment for 10 years. The concrete has been inspected and tested at 1 year and at 10 years after the first chloride exposure. The following observations were made:

$C_{s1} = 1.69$ % mass binder after 1 year of exposure,
$C_{sin} = C_{s10} = 2.93$ % mass binder after 10 years of exposure,
$D_{a1} = 91.8$ mm^2/yr after 1 year of exposure,
$D_{ain} = D_{a10} = 58.0$ mm^2/yr after 10 years of exposure.

The initial chloride content of the concrete was measured to $C_i = 0$.

The calculation of the chloride profile at the time $t_{20} = 20$ yr, and at $t_{50} = 50$ yr is carried out as follows:

Step 1. The age parameter is

$$\alpha = \frac{\ln D_1 - \ln D_{ain}}{\ln (t_{10}/t_1)} = \frac{\ln(91.8/58.0)}{\ln(10/1)} = 0.20, \qquad \text{cf. (5.9).}$$

Step 2. The exponent p is

$$p = \frac{\ln C_{sin} - \ln C_1}{(1-\alpha) \ln 10} = \frac{\ln(2.93/1.69)}{0.80 \times \ln 10} = 0.30, \qquad \text{cf. (5.86).}$$

Step 3. The factor S_1 is

$$S_1 = C_1 - C_i = 1.69 - 0 = 1.69 \text{ \% mass binder,} \qquad \text{cf. (5.85).}$$

Step 4. The achieved chloride diffusion coefficients become, cf. (5.6),

$$D_a(20 \text{ yr}) = D_1 \times \left(\frac{t_{20}}{t_1}\right)^{-\alpha} = 91.8 \times \left(\frac{20}{1}\right)^{-0.20} = 50.42 \text{ mm}^2/\text{yr},$$

$$D_a(50 \text{ yr}) = D_1 \times \left(\frac{t_{50}}{t_1}\right)^{-\alpha} = 91.8 \times \left(\frac{50}{1}\right)^{-0.20} = 41.98 \text{ mm}^2/\text{yr}.$$

Table 5.9: Chloride ingress into concrete as specified by Example 5.2.4.

u	$\Psi_{0.3}(u_1)$	x_{20}	$C(x, 20\text{ yr})$	x_{50}	$C(x, 50\text{yr})$
-	-	mm	% MB	mm	% MB
0.0	1.00000	0.0	3.610	0.0	4.750
0.1	0.85202	6.4	3.076	9.2	4.047
0.2	0.71706	12.7	2.589	18.3	3.406
0.3	0.59580	19.1	2.151	27.5	2.830
0.4	0.48851	25.4	1.764	36.7	2.320
0.5	0.39508	31.8	1.426	45.8	1.877
0.6	0.31503	38.1	1.137	55.0	1.496
0.7	0.24757	44.5	0.894	64.1	1.176
0.8	0.19168	50.8	0.692	73.3	0.910
0.9	0.14614	57.2	0.528	82.5	0.694
1.0	0.10973	63.5	0.396	91.6	0.521
1.1	0.08108	69.9	0.293	100.8	0.385
1.2	0.05895	76.2	0.213	110.0	0.280
1.3	0.04217	82.6	0.152	119.1	0.200
1.4	0.02967	88.9	0.107	128.3	0.141
1.5	0.02052	95.3	0.074	137.4	0.097
1.6	0.01396	101.6	0.050	146.6	0.066
1.7	0.00933	108.0	0.034	155.8	0.044
1.8	0.00613	114.3	0.022	164.9	0.029
1.9	0.00396	120.7	0.014	174.1	0.019
2.0	0.00251	127.0	0.009	183.3	0.012

Step 5. Since $C_i = 0$, the surface chloride content after 20 and 50 years are, cf. (5.81),

$$C_s(20\text{ yr}) = 1.69 \times (20/1)^{(1-0.20)\times 0.30} = 3.47\ \%\text{ mass binder},$$

$$C_s(50\text{ yr}) = 1.69 \times (50/1)^{(1-0.20)\times 0.30} = 4.32\ \%\text{ mass binder}.$$

Step 6. Since $C_i = 0$, the ordinates of the chloride profiles are calculated according to the following formulæ

$$C(x, 20\text{ yr}) = 0 + (3.47 - 0) \times \Psi_{0.30}(u) = 3.47 \times \Psi_{0.30}(u),$$

$$C(x, 50\text{ yr}) = 0 + (4.32 - 0) \times \Psi_{0.30}(u) = 4.32 \times \Psi_{0.30}(u).$$

Step 7. The corresponding abscissæ are, cf. Equation (5.90):

$$x_{20} = u \times \sqrt{4t_{20} \times D_a(20\text{ yr})} = u \times \sqrt{4 \times 20 \times 50.42} = 63.51 \times u,$$

$$x_{50} = u \times \sqrt{4t_{50} \times D_a(50\text{ yr})} = u \times \sqrt{4 \times 50 \times 41.98} = 91.63 \times u.$$

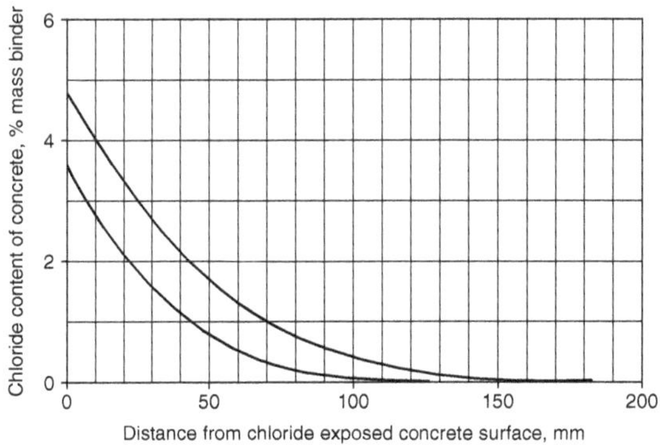

Figure 5.31: The chloride profiles after 20 and 50 years of chloride exposure as described and determined in Example 5.2.4. The chloride diffusion coefficients and the surface chloride contents were determined after 1 yr and 10 yr of exposure. It was assumed that the HETEK model was valid for the chloride ingress into the concrete.

Step 8. Finally, the chloride profiles are plotted, cf. Figure 5.31. $\diamond$

Example 5.2.5 In many cases there will be no information on the surface chloride content and the chloride diffusion coefficient of the concrete. This is e.g. the case when a new marine RC structure is under design. In such cases the surface chloride content of the concrete versus time has to be estimated. It is here assumed that a marine RC component has to be cast of concrete having the following composition:

- Portland cement: 330 kg/m^3 of concrete,
- Fly ash: 40 kg/m^3 of concrete,
- Silica fume: 20 kg/m^3 of concrete,
- Mixing water: 133 kg/m^3 of concrete,

with a sufficient amount of admixtures, e.g. air-entraining and plasticising admixtures (it is similar to the concrete of the pillars of the Great Belt Link), cf. Nielsen *et al.* (1994).

The structure has to be submerged in seawater. It is assumed that the initial chloride content is $C_i = 0$. The equivalent water/binder-ratio of the concrete with reference to diffusion is, cf. Equation (5.3),

$$\text{eqv}\{w/b_D\} = \frac{133}{330 + 40 + 7 \times 20} = 0.261.$$

The equivalent water/binder-ratio of the concrete with reference to binding is, cf. Equation (5.79),

$$\text{eqv}\{w/b_b\} = \frac{133}{330 + 0.75 \times 40 - 1.5 \times 20} = 0.403.$$

Table 5.10: Chloride ingress into concrete as specified by Example 5.2.5.

u	$\Psi_{0.30}$	$\Psi_{0.35}$	$\Psi_{0.327}$	x_{10} mm	$C(x,10)$ % MB	x_{30} mm	$C(x,30)$ % MB	x_{50} mm	$C(x,50)$ % MB
0.0	1.00000	1.00000	1.00000	0.0	3.140	0.0	4.500	0.0	5.320
0.1	0.85202	0.84694	0.84928	2.3	2.667	3.3	3.822	3.8	4.518
0.2	0.71706	0.70861	0.71250	4.6	2.237	6.5	3.206	7.7	3.790
0.3	0.59580	0.58541	0.59019	6.9	1.853	9.8	2.656	11.5	3.140
0.4	0.48851	0.47730	0.48246	9.2	1.515	13.1	2.171	15.4	2.567
0.5	0.39508	0.38390	0.38904	11.5	1.222	16.3	1.751	19.2	2.070
0.6	0.31503	0.30448	0.30933	13.8	0.971	19.6	1.392	23.0	1.646
0.7	0.24757	0.23804	0.24242	16.1	0.761	22.9	1.091	26.9	1.290
0.8	0.19168	0.18336	0.18719	18.4	0.588	26.1	0.842	30.7	0.996
0.9	0.14614	0.13913	0.14235	20.7	0.447	29.4	0.641	34.6	0.757
1.0	0.10973	0.10395	0.10661	23.0	0.335	32.7	0.480	38.4	0.567
1.1	0.08108	0.07645	0.07858	25.3	0.247	35.9	0.354	42.2	0.418
1.2	0.05895	0.05534	0.05700	27.6	0.179	39.2	0.257	46.1	0.303
1.3	0.04217	0.03941	0.04068	30.0	0.128	42.5	0.183	49.9	0.216
1.4	0.02967	0.02760	0.02855	32.3	0.090	45.7	0.128	53.7	0.152
1.5	0.02052	0.01902	0.01971	34.6	0.062	49.0	0.089	57.6	0.105
1.6	0.01396	0.01288	0.01338	36.9	0.042	52.3	0.060	61.4	0.071
1.7	0.00933	0.00858	0.00893	39.2	0.028	55.5	0.040	65.3	0.047
1.8	0.00613	0.00561	0.00585	41.5	0.018	58.8	0.026	69.1	0.031
1.9	0.00396	0.00361	0.00377	43.8	0.012	62.1	0.017	72.9	0.020
2.0	0.00251	0.00228	0.00239	46.1	0.007	65.3	0.011	76.8	0.013

The calculation of the chloride profile at the time $t_{10} = 10$ yr, $t_{30} = 30$ yr and $t_{50} = 50$ yr is carried out as follows:

Step 1. The chloride diffusion coefficient after 1 year of exposure is estimated to

$$D_1 = 1.0 \times 25,000 \times \exp\left(-\sqrt{\frac{10}{0.261}}\right) = 51.3 \text{ mm}^2/\text{yr}, \quad \text{cf. (5.2).}$$

Step 2. The age parameter is estimated to

$$\alpha = 0.6 \times (1 - 1.5 \times 0.261) = 0.365 \qquad \text{cf. (5.7).}$$

Step 3. The chloride diffusion coefficients are

$$D_a(10 \text{ yr}) = 51.3 \times 10^{-0.365} = 22.1 \text{ mm}^2/\text{yr} \qquad \text{cf. (5.6),}$$

$$D_a(30 \text{ yr}) = 51.3 \times 30^{-0.365} = 14.8 \text{ mm}^2/\text{yr} \qquad \text{cf. (5.6),}$$

$$D_a(50 \text{ yr}) = 51.3 \times 50^{-0.365} = 12.3 \text{ mm}^2/\text{yr} \qquad \text{cf. (5.6).}$$

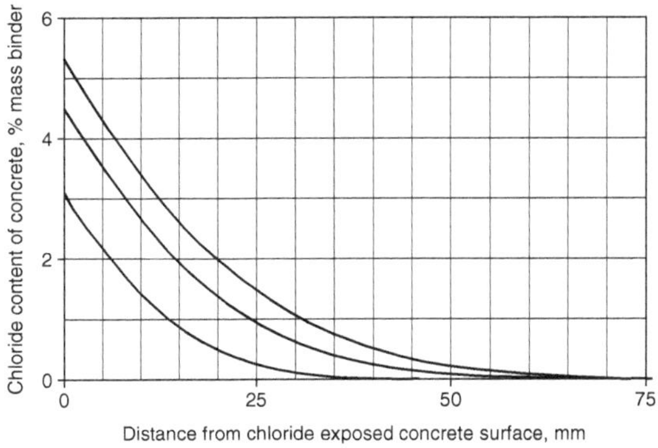

Figure 5.32: The chloride profiles after 10, 20 and 50 years of chloride exposure as described and determined in Example 5.2.5. The values of the function $\Psi_{0.327}(u)$ are found by linear interpolation between $\Psi_{0.30}(u)$ and $\Psi_{0.35}(u)$ from Table 8.3. It was assumed that the HETEK model was valid for the chloride ingress into the concrete.

Step 4. The surface chloride contents after 1 yr and 100 yr are

$$C_1 = 5.13 \times 0.403 = 2.07 \text{ \% mass binder}, \qquad \text{cf. (5.78)},$$

$$C_{100} = 1.5 \times 2.07 = 3.10 \text{ \% mass binder}, \qquad \text{cf. (5.82)}.$$

Step 5. The exponent p is given by

$$p = \frac{\log_{10}(3.10 - 0) - \log_{10}(2.07 - 0)}{2 \times (1 - 0.365)} = 0.138, \qquad \text{cf. (5.84)}.$$

Step 6. The factor S_1 becomes

$$S_1 = 2.07 - 0 = 2.07 \text{ \% by mass binder}, \qquad \text{cf. (5.85)}.$$

Step 7. The surface chloride contents are, cf. (5.81)

$$C_s(10 \text{ yr}) = 2.07 \times 10^{(1-0.365)\times 0.138} = 2.53 \text{ \% mass binder},$$

$$C_s(30 \text{ yr}) = 2.07 \times 30^{(1-0.365)\times 0.138} = 2.79 \text{ \% mass binder},$$

$$C_s(50 \text{ yr}) = 2.07 \times 50^{(1-0.365)\times 0.138} = 2.92 \text{ \% mass binder}.$$

Step 8. The abscissæ u and the corresponding ordinates of the function $\Psi_{0.138}(u)$ by interpolation are inserted into the spreadsheet, either from Table 8.3 or calculated from an approximation formula, cf. Section 3.2.5.

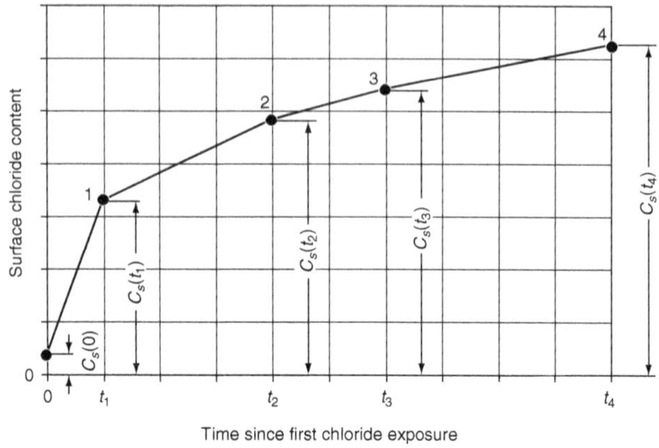

Figure 5.33: An example of an imaginary graph 0-1-2-3-4 of the surface chloride content versus time. The relation is approximated by a set of piecewise linear functions versus time. This procedure is called a linearization of the HETEK model.

Step 9. Since $C_i = 0$, the ordinates of the chloride profiles are calculated according to the following formulæ, cf. (5.87):

$$C(x, 10 \text{ yr}) = 0 + (2.53 - 0) \times \Psi_{0.138}(u) = 2.53 \times \Psi_{0.138}(u),$$

$$C(x, 30 \text{ yr}) = 0 + (2.79 - 0) \times \Psi_{0.138}(u) = 2.79 \times \Psi_{0.138}(u),$$

$$C(x, 50 \text{ yr}) = 0 + (2.92 - 0) \times \Psi_{0.138}(u) = 2.92 \times \Psi_{0.138}(u),$$

with the corresponding abscissæ:

$$x_{10} = u \times \sqrt{4 \times 10 \times 22.1} = 29.7 \times u, \qquad \text{cf. (5.90)},$$

$$x_{30} = u \times \sqrt{4 \times 30 \times 14.8} = 42.1 \times u, \qquad \text{cf. (5.90)},$$

$$x_{50} = u \times \sqrt{4 \times 50 \times 12.3} = 49.6 \times u, \qquad \text{cf. (5.90)}.$$

Step 10. Finally, the chloride profiles are plotted, cf. Figure 5.32. ◊

5.2.3 Linearization of the HETEK model

In cases where the surface chloride content versus time is not well-known it may be convenient to approximate this relation with a set of piecewise linear functions versus time, cf. Mejlbro *et al.* (2000). Also, in cases where the surface chloride content versus time is established by measurements, it may be convenient to linearize the problem as explained in this section. In Figure 5.33 the surface chloride content versus time is given by measurements. A set of piecewise linear functions may be as shown.

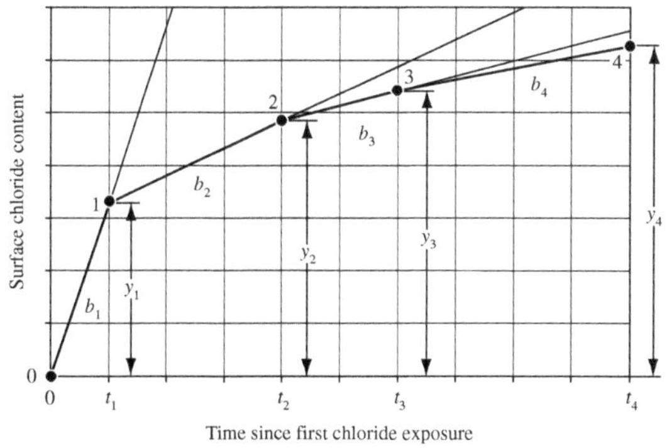

Figure 5.34: A set of piecewise linear functions 0-1-2-3-4 which approximate the surface chloride content versus time. The analytical model of this boundary condition is found to be as described by Equations (5.95)–(5.98), where b_j is given by the relations Equation (5.100) and Equation (5.101).

Chloride content of surface

As mentioned before, the $\Psi_p(u)$ functions are relatively simple when $p = 0$ and $p = 1$, cf. Equation (5.90) and Equation (5.91),

- $p = 0$ implies that

$$\Psi_0(u) = \text{erfc}(u), \tag{5.94}$$

- $p = 1$ implies that

$$\Psi_1(u) = \left(1 + 2u^2\right) \text{erfc}(u) - \frac{2u}{\sqrt{\pi}} \exp\left(-u^2\right). \tag{5.95}$$

The set of piecewise linear functions which describe the chloride content of the concrete surface (i.e. the boundary condition) is given by the set of ordinates, cf. Figure 5.34,

$$y_0, y_1, y_2, \ldots, y_j, \tag{5.96}$$

where $y_0 = 0$, cf. Figure 5.34. In order to determine an expression for the chloride content of the concrete surface it is convenient to define a set of constants

$$b_1, b_2, b_3, \ldots, \tag{5.97}$$

so that the variation of the chloride content of the exposed concrete surface can be written as

$$C_s = C(0, t) = 0, \qquad \text{for } t = t_0 = 0, \tag{5.98}$$

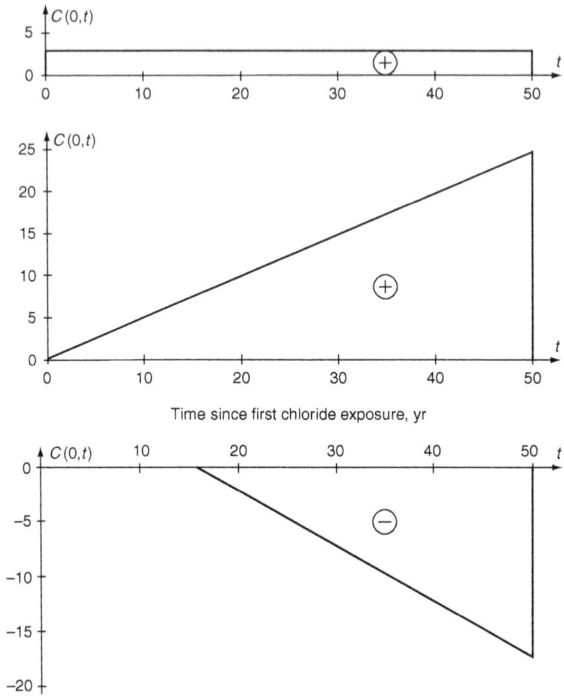

Figure 5.35: The boundary condition (the surface chloride content versus time) shown in Figure 5.33 is split into three functions in such a way that the sum of the three functions equals the surface chloride content. Thus, the chloride profiles may be calculated by superposition. When the boundary condition is a constant, the solution is proportional to the function $\Psi_0(u) = \text{erfc}(u)$, cf. Equation (5.81). When the boundary condition is a straight line, the solution is proportional to $\Psi_1(u)$, cf. Equation (5.91).
The calculations of the profiles at 10 yr, 30 yr and 50 yr are carried out in Example 5.2.4.

$$C_s = C(0,t) = b_1\left(t - t_0\right) = b_1 t, \qquad \text{for } 0 \le t \le t_1, \tag{5.99}$$

$$C_s = C(0,t) = b_1\left(t - t_0\right) + b_2\left(t - t_1\right), \qquad \text{for } t_1 \le t \le t_2, \tag{5.100}$$

$$C_s = C(0,t) = b_1\left(t - t_0\right) + b_2\left(t - t_1\right) + b_3\left(t - t_2\right), \qquad \text{for } t_2 \le t \le t_3, \tag{5.101}$$

$$C_s = C(0,t) = \sum_{j=1}^{n}\left\{b_j\left(t - t_{j-1}\right)\right\}, \qquad \text{for } t_n \le t < t_{n+1}. \tag{5.102}$$

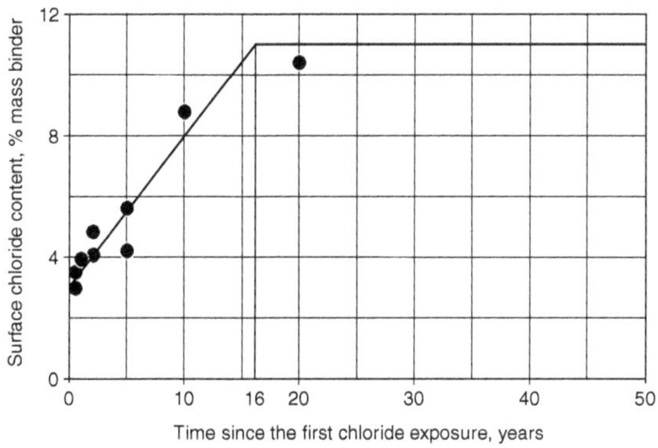

Figure 5.36: The information by the measurements in the period $0 \leq t \leq 20$ yr gives an uncertain information for the period $20 < t \leq 50$ yr. Example 5.2.6 studies the influence of a maximum surface chloride content of 11 % mass binder. The linearization of the HETEK model (as shown by the shaded piecewise linear function) is assumed to be valid for the chloride ingress into the concrete.

The chloride content C_s of the concrete surface can be expressed by means of the ordinates $y_0, y_1, y_2, \ldots$ in the following way. It is seen that

$$b_1 = \frac{y_1 - y_0}{t_1 - t_0} = \frac{y_1}{t_1}, \tag{5.103}$$

and we have for $n \geq 1$,

$$\sum_{j=1}^{n} b_j = \frac{y_n - y_{n-1}}{t_n - t_{n-1}}. \tag{5.104}$$

This means that the constants b_n, $n \geq 2$, are given by

$$b_n = \sum_{j=1}^{n} b_j - \sum_{j=1}^{n-1} b_j = \frac{y_n - y_{n-1}}{t_n - t_{n-1}} - \frac{y_{n-1} - y_{n-2}}{t_{n-1} - t_{n-2}}. \tag{5.105}$$

Equation (5.102) is then proved by induction.

Chloride profiles of the concrete

By using Equation (5.81) and Equation (5.102) we can now establish the equations of chloride ingress into the concrete. The initial chloride content at time $t = 0$ is fixed to $C_i = 0$. If $C_i \neq 0$, then the value of C_i is added to the chloride profiles. At first the chloride diffusion coefficient is given the value $D = 1$. Later it is possible by a suitable substitution to insert the wanted physical condition, cf. Equation (5.77) and Equation (5.14). Thus, Fick's

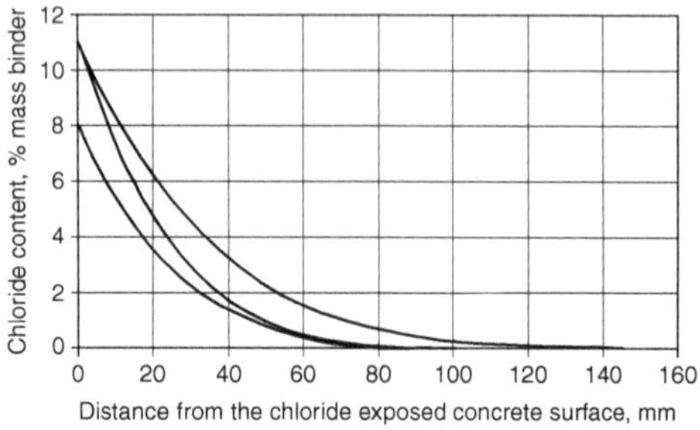

Figure 5.37: The chloride profiles after 10, 30 and 50 years of chloride exposure as described and determined in Example 5.2.6. The chloride content of the concrete surface is time-dependent as given in Figure 5.36. The linearization of the HETEK model shown in Figure 5.36 was assumed to be valid for the chloride ingress.

second law of diffusion and its corresponding initial and boundary conditions are given by

$$\frac{\partial C}{\partial t} = \frac{\partial^2 C}{\partial x^2}, \qquad \text{for } x \geq 0 \text{ and } t \geq 0, \tag{5.106}$$

$$C_s = C(0,t) = \sum_{j=1}^{n} b_j \times (t - t_{j-1}), \qquad \text{for } t_n \leq t \leq t_{n+1}, \tag{5.107}$$

$$C(x,0) = C_i = 0, \qquad \text{for } x \geq 0. \tag{5.108}$$

As previously shown the solution may be found by superposition with the boundary conditions

$$C_s = C(0,t) = \begin{cases} 0, & \text{for } t \leq t_{j-1}, \\ b_j \times (t - t_j), & \text{for } t \geq t_{j-1}. \end{cases} \tag{5.109}$$

If we write for short

$$u_j = \frac{0.5\,x}{\sqrt{t - t_{j-1}}}, \qquad \text{for } x \geq 0 \text{ and } t_{j-1} \leq t \leq t_j, \tag{5.110}$$

then the solution of (5.106)–(5.108) with respect to Equation (5.109) can be summarized in the following way (for details, cf. Example 5.2.6):

$$C_1(x,t) = b_1 \times (t - t_0) \times \Psi_1(u_1), \qquad \text{for } t_0 \leq t \leq t_1, \tag{5.111}$$

$$C_2(x,t) = C_1(x,t) + b_2 \times (t - t_1) \times \Psi_1(u_2), \quad \text{for } t_1 \leq t \leq t_2, \tag{5.112}$$

$$C_3(x,t) = C_2(x,t) + b_3 \times (t - t_2) \times \Psi_1(u_3), \quad \text{for } t_2 \leq t \leq t_3, \tag{5.113}$$

$$C_n(x,t) = \sum_{j=1}^{n} b_j \times (t - t_{j-1}) \times \Psi_1(u_j), \qquad \text{for } t_{n-1} \leq t \leq t_n, \quad (5.114)$$

or

$$C_n(x,t) = C_{n-1}(x,t) + b_n \times (t - t_{n-1}) \times \Psi_1(u_n), \qquad \text{for } t_{n-1} \leq t \leq t_n. \quad (5.115)$$

Example 5.2.6 A structural component of reinforced concrete has been submerged in seawater for 20 years. During this period of time the concrete has been inspected and tested. The chloride content of the concrete surface has varied with time as shown in Figure 5.36. The piecewise linear function on Figure 5.36, given by

$$C_s = \begin{cases} 3 + 0.5 \times (t - t_0) = 3 + 0.5\,t & \text{for } t_0 = 0 \leq t \leq t_{16}, \\ 3 + 0.5 \times (t - t_0) - 0.5 \times (t - t_{16}), & \text{for } t_{16} \leq t \leq t_{50}, \end{cases}$$

where $t_0 = 0$, $t_{16} = 16$ yr and $t_{50} = 50$ yr, represents the best estimate of the surface chloride content versus time from 0 to 50 years. In order to have $C(0,0) = 0$ we define the chloride exposure as shown in Figure 5.35. Thus, the chloride profiles are given by

$$C_1(x,t) = 3.0 \times \Psi_0(u_0) + 0.5\,t \times \Psi_1(u_0), \qquad \text{for } t_0 = 0 \leq t \leq t_{16},$$

$$C_2(x,t) = C_1(x,t) - 0.5(t - 16) \times \Psi_1(u_{16}), \qquad \text{for } t_{16} \leq t \leq t_{50},$$

where the arguments are

$$u_0 = \frac{0.5\,x}{\sqrt{(t-0) \times D_a(t)}}, \qquad \text{for } 0 \leq t \leq 16 \text{ yr},$$

$$u_{16} = \frac{0.5\,x}{\sqrt{(t-16) \times D_a(t)}}, \qquad \text{for } 16 \leq t \leq 50\text{yr}.$$

We shall under these circumstances estimate the chloride ingress into the concrete at 10, 30 and 50 years from the chloride exposure.

The calculation of the chloride profile at the time $t_{10} = 10$ yr, $t_{30} = 30$ yr and $y_{50} = 50$ yr is carried out in the following way:

Step 1. The chloride diffusion coefficient after 1 year of exposure is

$$D_1 \approx 1.0 \times 25{,}000 \times \exp\left(-\sqrt{\frac{10}{0.35}}\right) = 119.3 \text{ mm}^2/\text{yr}, \quad \text{cf. (5.2)}.$$

The measurement of any chloride diffusion coefficient during the period of the first 20 years may be used instead.

Table 5.11: Chloride ingress into concrete specified by Example 5.2.6.

u	$\Psi_0(u)$	$\Psi_1(u)$	x_{10}	$C(x, 10\text{ yr})$	x_{30}	$C(x, 30\text{ yr})$	x_{50}	$C(x, 50\text{ yr})$
-	-	-	mm	% MB	mm	% MB	mm	% MB
0.0	1.0000	1.0000	0.0	8.000	0.0	11.000	0.0	11.000
0.1	0.8875	0.7936	5.0	6.630	5.0	9.011	7.3	9.011
0.2	0.7773	0.6227	10.0	5.445	10.1	7.313	14.6	7.313
0.3	0.6714	0.4828	14.9	4.428	15.1	5.877	21.9	5.877
0.4	0.5716	0.3699	19.9	3.564	20.2	4.674	29.2	4.674
0.5	0.4795	0.2799	24.9	2.838	25.2	3.677	36.5	3.677
0.6	0.3961	0.2090	29.9	2.234	30.2	2.861	43.7	2.861
0.7	0.3222	0.1541	34.9	1.737	35.3	2.199	51.0	2.199
0.8	0.2579	0.1120	39.8	1.334	40.3	1.670	58.3	1.670
0.9	0.2031	0.0803	44.8	1.011	45.4	1.252	65.6	1.252
1.0	0.1573	0.0568	49.8	0.756	50.4	0.926	72.9	0.926
1.1	0.1198	0.0396	54.8	0.557	55.4	0.676	80.2	0.676
1.2	0.0897	0.0272	59.8	0.405	60.5	0.486	87.5	0.486
1.3	0.0660	0.0184	64.7	0.290	65.5	0.345	94.8	0.345
1.4	0.0477	0.0122	69.7	0.204	70.6	0.241	102.1	0.241
1.5	0.0339	0.0080	74.7	0.142	75.6	0.166	109.4	0.166
1.6	0.0237	0.0052	79.7	0.097	80.6	0.112	116.6	0.112
1.7	0.0162	0.0033	84.7	0.065	85.7	0.075	123.9	0.075
1.8	0.0109	0.0021	89.6	0.043	90.7	0.049	131.2	0.049
1.9	0.0072	0.0013	94.6	0.028	95.8	0.032	138.5	0.032
2.0	0.0047	0.0008	99.6	0.018	100.8	0.020	145.8	0.020

Step 2. The age parameter is estimated to

$$\alpha = 0.6 \times (1 - 1.5 \times 0.35) = 0.285, \qquad \text{cf. (5.7).}$$

Step 3. The achieved chloride diffusion coefficients are

$$D_a(10\text{ yr}) = 119.3 \times 10^{-0.285} = 61.9\text{ mm}^2/\text{yr}, \qquad \text{cf. (5.6),}$$

$$D_a(30\text{ yr}) = 119.3 \times 30^{-0.285} = 45.3\text{ mm}^2/\text{yr}, \qquad \text{cf. (5.6),}$$

$$D_a(50\text{ yr}) = 119.3 \times 50^{-0.285} = 39.1\text{ mm}^2/\text{yr}, \qquad \text{cf. (5.6).}$$

Step 4. The abscissæ u and the corresponding ordinates of the functions $\Psi_0(u)$ and $\Psi_1(u)$ are inserted into the spreadsheet, either from Table 8.3 or calculated from an approximation formula, cf. Section 3.2.5.

Step 5. Since $C_i = 0$, the ordinates of the chloride profiles are calculated according to the following formulæ, cf. Table 5.9,

$$\begin{aligned} C(x, 10\text{ yr}) &= 3.0 \times \Psi_0(u_0) + 0.5 \times 10 \times \Psi_1(u_0) \\ &= 3.0 \times \Psi_0(u_0) + 5 \times \Psi_1(u_0), \end{aligned} \qquad \text{cf. (5.85).}$$

$$\begin{aligned} C(x, 30\text{ yr}) &= 3.0 \times \Psi_0(u_{16}) - 0.5 \times (30 - 16) \times \Psi_1(u_{16}) \\ &= 3.0 \times \Psi_0(u_{16}) + 8.0 \times \Psi_1(u_{16}), \end{aligned} \qquad \text{cf. (5.85),}$$

$$
\begin{aligned}
&C(x, 50\text{ yr}) \\
&= 3.0 \times \Psi_0(u_{16}) - 0.5 \times 50 \times \Psi_1(u_{16}) - 17.0 \times \Psi_1(u_{16}) \\
&= 3.0 \times \Psi_0(u_{16}) + 8.0 \times \Psi_1(u_{16}), \qquad \text{cf. (5.85).}
\end{aligned}
$$

Step 6. The corresponding abscissæ are

$$
\begin{aligned}
x_{10} &= u \times \sqrt{4 \times (10 - 0) \times D_a(10\text{ yr})} \\
&= u \times \sqrt{4 \times 10 \times 37.15} = 38.55 \times u, \qquad \text{cf. (5.90),}
\end{aligned}
$$

$$
\begin{aligned}
x_{30} &= u \times \sqrt{4 \times (30 - 16) \times D_a(30\text{ yr})} \\
&= u \times \sqrt{4 \times 14 \times 27.16} = 39.00 \times u, \qquad \text{cf. (5.90),}
\end{aligned}
$$

$$
\begin{aligned}
x_{50} &= u \times \sqrt{4 \times (50 - 16) \times D_a(50\text{ yr})} \\
&= u \times \sqrt{4 \times 34 \times 39.1} = 72.9 \times u, \qquad \text{cf. (5.90).}
\end{aligned}
$$

Step 7. Finally, the chloride profiles are plotted, cf. Figure 5.37. ◊

Discussion and conclusion. Examples 5.2.1 to 5.2.6 are predictions of the chloride profiles under various conditions: From pure estimation to pure interpolation and extrapolation. As expected the predictions vary with the assumptions made. ◊

5.2.4 HETEK model by approximations

The solution of Fick's second law of diffusion becomes rather complicated, when we model the chloride ingress into concrete by accounting for the time-dependent boundary condition, i.e. the change versus time of chloride content of the near-to-surface layer of the concrete under consideration.

The solution of chloride ingress into concrete with a time-dependent chloride content at the surface has been solved theoretically, and numerical solutions are presented in Section 5.2.2. The exact solution of this case may however be troublesome, especially when an extended application is needed, e.g. when taking the probability approach into account. For these cases and others we present a simple approximation to the solution of Fick's second law of diffusion with time-dependent boundary condition.

For concrete exposed to marine environment (marine atmosphere ATM, marine splash SPL, and submerged in seawater SUB) this section describes the simple calculation (by a pocket calculator) of the predicted chloride ingress (described by the chloride profiles). The method is illustrated by a numerical example.

Approximation of the $\Psi_p(u)$ functions

A simple approximation of the Ψ_p functions has been proposed by Mejlbro, cf. Equation (3.73),

$$\Psi_p(u) \approx \left(1 - \frac{u}{1.66 + 0.3p}\right)^{2+2p} . \tag{5.116}$$

This approximative formula is valid for $0 \le p \le 1$ and $0 \le u \le 1.1$ with a relative error of less that 10 %. The present section describes the most common calculations in connection with chloride ingress into concrete exposed to marine environment.

If a more accurate approximation is wanted, the following formula may be used:

$$\Psi_p(u) \approx \left\{1 - 0.4\,u^2\left(1 - u^2\right)\right\} \times \left(1 - \frac{u}{1.66 + 0.3p}\right)^{2+2p} . \tag{5.117}$$

This approximative formula is valid for $0 \le p \le 1$ and $0 \le u \le 1.35$ with a relative error of less that 10 %.

Plotting the chloride profiles

When the governing parameters are measured, estimated or calculated, it is convenient to plot the graphs of the chloride profiles in the following way: The arguments are chosen in the interval $0 \le u \le 1.1$ e.g. with subintervals, e.g. $\Delta u = 0.1$. Accordingly, the ordinates and the abscissæ of the chloride profiles are calculated by the following formulæ:

$$C(x,t) = C_i + S_p\, t^p \times \left(1 - \frac{u}{1.66 + 0.3p}\right)^{2+2p} , \tag{5.118}$$

$$x = u \times \sqrt{4 \times t \times D_a(t)} = u \times \sqrt{4 \times D_1 \times t^{1-\alpha}}, \tag{5.119}$$

since

$$\tau = \frac{t}{t_1} = t \quad \text{and} \quad D_a(t) = D_1 \times \left(\frac{t}{t_1}\right)^{-\alpha} = D_1 \times t^{-\alpha}, \tag{5.120}$$

and $t_1 = 1$ yr. It is seen that only simple calculations are needed and that they can be carried out by means of a simple pocket calculator or a simple spreadsheet. The application of the approximation of the Ψ_p functions is illustrated by resolving the former Example 5.2.4.

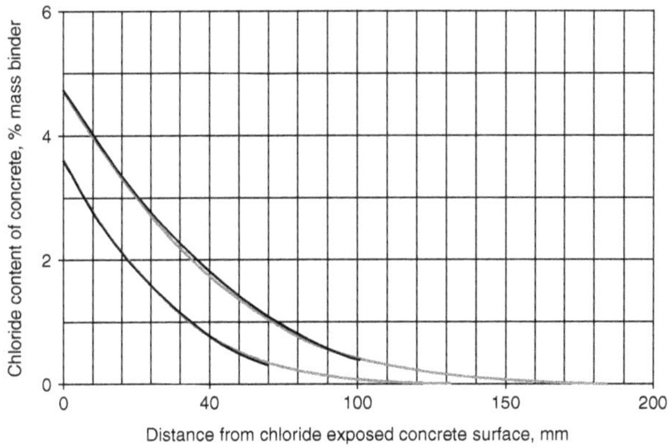

Figure 5.38: The Chloride profiles after 20 and 50 years of chloride exposure as described and determined in Example 5.2.7 and Example 5.2.4. The shaded curves represent the solution by the HETEK model, cf. Example 5.2.4, and the solid curves represent the HETEK model by approximation, cf. Example 5.2.7.

Example 5.2.7 We found in Example 5.2.4, page 295, the following governing parameters by measurements and estimation:

$C_1 = 1.47$ % mass binder
$C_{10} = 2.93$ % mass binder
$C_i = 0$
$D_1 = 91.8\ \text{mm}^2/\text{yr}$
$D_{10} = 58.0\ \text{mm}^2/\text{yr}$
$\alpha = 0.20$ non-dimensional
$p = 0.30$ non-dimensional
$S_p = 1.47$ % mass binder

By using the simple Equation (5.118) we get the chloride profile after 20 yr of exposure,

$$\begin{aligned} C(x, 20\text{ yr}) &= C_i + S_p\,\tau^p \times \Psi_p(u) \\ &\approx 0 + 1.47 \times 20^{0.30} \times \left(1 - \frac{u}{1.66 + 0.3 \times 0.30}\right)^{2+2\times 0.30} \\ &= 3.61 \times \left(1 - \frac{u}{1.75}\right)^{2.6} \%\text{ mass binder.} \end{aligned}$$

We get in a similar way the chloride profile after 50 yr of exposure:

$$C(x, 50\text{ yr}) \approx 4.75 \times \left(1 - \frac{u}{1.75}\right)^{2.6} \%\text{ mass binder,}$$

where the corresponding abscissæ are, cf. (5.119),

$$x_{20} = u \times \sqrt{4 \times 91.8 \times 20^{1-0.20}} = 63.51 \times u\text{ mm},$$

Table 5.12: Chloride ingress into concrete as specified by Example 5.2.7.

u	x_{20}	$C(x, 20 \text{ yr})$	x_{50}	$C(x, 50 \text{ yr})$
-	mm	% mass binder	mm	% mass binder
0.00	0.0	3.610	0.0	4.750
0.05	3.2	3.348	4.6	4.405
0.10	6.4	3.098	9.2	4.076
0.15	9.5	2.860	13.7	3.763
0.20	12.7	2.633	18.3	3.465
0.25	15.9	2.418	22.9	3.182
0.30	19.1	2.214	27.5	2.913
0.35	22.2	2.021	32.1	2.659
0.40	25.4	1.839	36.7	2.419
0.45	28.6	1.667	41.2	2.193
0.50	31.8	1.505	45.8	1.980
0.55	34.9	1.354	50.4	1.781
0.60	38.1	1.212	55.0	1.594
0.65	41.3	1.080	59.6	1.420
0.70	44.5	0.957	64.1	1.259
0.75	47.6	0.843	68.7	1.109
0.80	50.8	0.737	73.3	0.970
0.85	54.0	0.641	77.9	0.843
0.90	57.2	0.552	82.5	0.727
0.95	60.3	0.472	87.0	0.621
1.00	63.5	0.399	91.6	0.525
1.05	66.7	0.333	96.2	0.439
1.10	69.9	0.275	100.8	0.362

$$x_{50} = u \times \sqrt{4 \times 91.8 \times 50^{1-0.20}} = 91.63 \times u \text{ mm}.$$

It is seen that the chloride profiles are presented by very simple expressions, which are easy to use by simple spreadsheets or simple pocket calculators.

The chloride profiles are calculated by means of the spreadsheet in Table 5.10 and shown in Figure 5.38 together with the profiles calculated in Example 5.2.4. We see that the deviations are insignificant. ◊

5.2.5 Method of inverse cores

This section describes a procedure for the determination of penetration parameters of concrete structures exposed to chloride in order to estimate the chloride ingress into concrete versus time.

The mathematical model behind the procedure allows for time-dependency of both the surface boundary condition and the chloride diffusion coefficient. the mathematical model is the complete solution to 'the general Fick's second law of diffusion' by Mejlbro (1996). The HETEK model in Section 5.2.2 is an extension of that method.

A typical problem is that the chloride profile of a marine RC member is known at the inspection time, but its further development is not known. In order to find the unknown parameters it is necessary to know a chloride profile of the same RC member and its environment at another time. To overcome this, an approximation is proposed: The mature but unaffected concrete situated deeper than the measurable part of the chloride profile can be exposed to chloride at exactly the same position of the RC member. By doing this for one year an estimate of the chloride profile is obtained. Exactly one year is chosen in order to take the variability of the year into account.

It can be argued that this is not the same as if the concrete were immature ('green') – and that is true. On the other hand, it can also be argued that it is the best possible approximation if no other information exists. A correction factor may be looked for in the near future.

This section shows how to handle the mathematics of the HETEK model, when it is applied to the 'Method of Inverse Cores'.

It is assumed that the reader of this section is familiar with the HETEK model of determination of chloride ingress into concrete, cf. Section 5.2.2 and Section 5.2.3.

Method of test

The procedure is as follows: At least three cores ⌀75 mm are drilled from the structure. These cores must be placed so that they represent the local environment in question, i.e. they must have the same horizontal position and a distance of 200 mm between the centres of the cores. Reinforcement bars in cores should be avoided.

After drilling the cores they are rinsed in fresh water, dried and wrapped in plastic sheets in order to prevent evaporation. The cores are sent to a laboratory for inspection and testing. The chloride profiles of the near-to-surface layer of the chloride exposed concrete are determined by means of the test method NT Build 208 (1996b), cf. Section 1.4, or a similar test method. The diffusion coefficient D_{ain} and the calculated surface chloride concentration C_{sin} are determined by regression analysis, cf. Section 2.1.4.

Then the cores are painted with a chloride tight epoxy or polyurethane membrane in order to protect the cores from chloride ingress through the surface. The painted cores are cut approximately 50 mm from the virgin part and placed into the original holes so that the virgin part of the cores are exposed to the chloride laden environment for $t_1 = 1$ yr. The 50 mm of the virgin parts of the cores are tested according to the test method NT Build 443 (1996a), 'Accelerated Chloride Penetration'. Here, the potential chloride diffusion coefficient D_{pex} and the calculated surface chloride concentration C_{sp} should be determined.

After 1 year of exposure the achieved chloride profiles of the inverse cores placed in the holes are determined according to the test method NT Build 208

(1996b), cf. Section 1.4. Here, the achieved chloride diffusion coefficient D_{a1} and the calculated concrete surface concentration C_{s1} should be determined.

Formulæ for parameters of chloride ingress

When D_{ain} and C_{sin} have been determined after inspection and testing the concrete, and D_{a1} and C_{s1} have been determined according to the method of inverse cores, the following parameters are calculated:

$$\alpha = \frac{\ln D_{a1} - \ln D_{ain}}{\ln t_{\text{in}} - \ln t_1}, \tag{5.121}$$

$$p = \frac{\ln (C_{sin} - C_i) - \ln (C_{s1} - C_i)}{\ln t_{\text{in}} - \ln t_1}, \tag{5.122}$$

$$S_p = \frac{C_{sin} - C_i}{t_{\text{in}}^p} = \frac{S_{s1} - C_i}{t_1^p} = C_{s1} - C_i, \tag{5.123}$$

where t_{in} is the time at which the inspection of the structure is carried out, and $t_1 = 1$ yr. Thus, the chloride profile developed at time t is

$$C(x,t) = C_i + S_p t^p \times \Psi_p(u), \tag{5.124}$$

where

$$x = u \times \sqrt{4t \times D_1 \times t^{-\alpha}} = u \times \sqrt{4D_1 \times t^{1-\alpha}}. \tag{5.125}$$

5.2.6 Corrosion

The main purpose of the prediction of chloride ingress into reinforced concrete is to estimate where and when the reinforcement initiates corrosion. In connection with the LIGHTCON model prediction of corrosion has been dealt with in the following way assuming the assumptions of the LIGHTCON model (i.e. time-dependent chloride diffusion coefficient and a constant surface chloride content):

- A model of the initiation period of time is determined, cf. Section 5.1.3.
- Corrosion domain is determined, cf. Section 5.1.4.
- A model of a corrosion multiprobe is developed, cf. Section 5.1.5.

When the HETEK model is assumed instead of the LIGHTCON model, it will result in slight, though important changes. The present section deals with this subject.

Threshold value of chloride in concrete

The threshold value of chloride in concrete depends on the composition of the concrete and the environment. Frederiksen *et al.* (1997) have proposed the following simple model,

$$C_{cr} = k_{cr} \times \exp\left(-1.5 \times \ \text{eqv}\left\{w/c_{cr}\right\}\right), \tag{5.126}$$

where eqv$\{w/c_{cr}\}$ is the equivalent water/binder ratio with respect to corrosion, cf. Equation (5.25), and k_{cr} is the factor defined by Equation (5.26). The model is compared with the observations in Figure 5.8.

Initiation period of time

Assuming the HETEK model, the initiation period of time t_{cr} is determined by the following equation,

$$C_{cr} = C_i + S_p \tau^p \times \Psi_p\left(u_{cr}\right), \tag{5.127}$$

where $t_1 = 1$ yr and

$$\tau = \frac{t_{\text{cr}}}{t_1} = t_{\text{cr}} \quad \text{and} \quad u_{cr} = \frac{c}{\sqrt{t_{\text{cr}} \times D_a\left(t_{\text{cr}}\right)}}, \tag{5.128}$$

and c is the thickness of the concrete cover of the reinforcement bar. It is seen that the determination of t_{cr} is not an easy job. However, by introducing the $\Lambda_p(u)$ function, cf. Equation (3.84),

$$\Lambda_p(u) = \frac{\Psi_p(u)}{u^{2p}} \quad \text{or} \quad \Psi_p\left(u_{cr}\right) = u_{cr}^{2p} \times \Lambda_p\left(u_{cr}\right), \tag{5.129}$$

we get the following more convenient equation

$$C_{cr} = C_i + S_p \times (0.5)^{2p} \times \Lambda_p\left(u_{cr}\right). \tag{5.130}$$

The solution of Equation (5.130) with respect to u_{cr} is given by

$$u_{cr} = \ \text{inv}\ \Lambda_p\left(\nu_{cr}\right), \tag{5.131}$$

where inv Λ_p is the inverse function of Λ_p and

$$\nu_{cr} = \frac{C_{cr} - C_i}{S_p \times (0.5\,c)^{2p}}. \tag{5.132}$$

When we solve with respect to t_{cr}, the initiation period of time is given by

$$t_{\text{cr}} = t_1 \times \left\{\frac{0.5\,c}{\sqrt{t_1 D_1} \times \ \text{inv}\ \Lambda_p\left(\nu_{cr}\right)}\right\}^{\frac{2}{1-\alpha}}. \tag{5.133}$$

Notice that at the end of the initiation period of time the reinforcement will start corrosion, though it will take several years before the formation of rust will cause the concrete cover to spall, cf. Section 1.3.1.

Example 5.2.8 A structural component of reinforced concrete has been exposed to marine splash for 10 years. The concrete has been inspected and tested at 1 year and at 10 years after the first chloride exposure. The following observations were made, cf. Example 5.2.4:

- $C_{s1} = 1.47$ % mass binder after 1 year of exposure
- $C_{sin} = C_{s10} = 2.93$ % mass binder after 10 years of exposure
- $D_{a1} = 91.8$ mm^2/yr after 1 year of exposure
- $D_{ain} = D_{a10} = 58.0$ mm^2/yr after 10 years of exposure.

The initial chloride content of the concrete was measured to $C_i = 0$. The HETEK model is assumed to be valid for the chloride ingress into the concrete. From the specification and the information of the applied concrete composition it is assumed that the threshold value of chloride in the concrete is $C_{cr} = 1.10$ % mass binder.

In Example 5.2.4, page 295, the following governing parameters were found by estimation:

$$p = 0.20 \text{ non-dimensional},$$

$$\alpha = 0.30 \text{ non-dimensional},$$

$$S_p = 1.47 \text{ \% mass binder}.$$

Prediction of the initiation period is carried out as follows:

Step 1. The parameter ν_{cr} is

$$\nu_{cr} = \frac{1.10 - 0}{1.47 \times (0.5 \times 75)^{2\times 0.30}} = 0.0850 \qquad \text{cf. (5.132).}$$

Step 2. The inverse function inv $\Lambda_{0.30}(0.085)$ is, cf. Table 8.6,

$$\text{inv } \Lambda_{0.30}(0.085) = 1.65.$$

Step 3. The initiation period of time is calculated to, cf. (5.133),

$$t_{\text{cr}} = 1 \times \left\{ \frac{0.5 \times 75}{\sqrt{1 \times 91.8} \times 1.65} \right\}^{\frac{2}{1-0.2}} = 8.7 \text{ yr} \approx 10 \text{ yr}. \qquad \Diamond$$

Example 5.2.9 A concrete for the marine splash zone of a bridge pillar has the following composition:

- Portland cement: 330 kg/mm^3 of concrete
- Fly ash: 40 kg/m^3 of concrete
- Silica fume: 20 kg/m^3 of concrete
- Mixing water: 133 kg/m^3 of concrete

with a sufficient amount of admixtures, e.g. air-entraining and plasticising admixtures.

The initial chloride content of the concrete is assumed to be $C_i = 0$. The HETEK model is assumed to be valid for the chloride ingress into the concrete. The concrete cover of the reinforcement is planned to be $c = 75$ mm.

In Example 5.2.5, page 297, the following governing parameters were found by estimation:

$$p = 0.327 \text{ non-dimensional}$$

$$\alpha = 0.365 \text{ non-dimensional}$$

$$S_p = 1.48 \text{ \% mass binder.}$$

Prediction of the initiation period of corrosion is carried out as follows:

Step 1. The equivalent water/binder ratio with respect to corrosion is

$$\text{eqv}\,\{w/b_{cr}\} = \frac{133}{330 - 4.7 \times 20 - 1.4 \times 40} = 0.739 \qquad \text{cf. (5.25).}$$

Step 2. The threshold value of chloride in the concrete is

$$C_{cr} = 1.25 \times \exp(-1.5 \times 0.739) = 0.41 \text{ \% mass binder cf. (5.126).}$$

Step 3. The parameter ν_{cr} is

$$\nu_{cr} = \frac{0.41 - 0}{1.48 \times (0.5 \times 75)^{2 \times 0.327}} = 0.0259 \qquad \text{cf. (5.132).}$$

Step 4. The inverse function inv $\Lambda_{0.327}(0.0259)$ is, cf. Table 8.6,

$$\text{inv } \lambda_{0.327}(0.0259) = 1.355.$$

Step 5. Finally, we get the initiation period of time, cf. (5.133),

$$t_{\text{cr}} = 1 \times \left\{ \frac{0{,}5 \times 75}{\sqrt{1 \times 30.75} \times 1.355} \right\}^{\frac{2}{1-0.365}} = 158 \text{ yr.} \qquad \Diamond$$

Corrosion domain

Section 5.1.4 describes the corrosion domain in a (t, D)-diagram assuming that the LIGHTCON model is valid. Here, the HETEK model is assumed to be valid, and the influences are studied.

The chloride content of a marine concrete at the level of the reinforcing bar (distance c from the exposed surface) when following the HETEK model is given by the following equation:

$$C(c,t) = C_i + S_p \times \left(\frac{t}{t_1}\right)^p \times \Psi_p\left(\frac{0.5\,c}{\sqrt{t \times D_a}}\right), \qquad (5.134)$$

where $t_1 = 1$ yr and

- C_i is the initial chloride content of the concrete (equally distributed).
- C_{s1} is the surface chloride content of the concrete after 1 yr of exposure.
- S_p is a factor, which here is given by $S_p = C_{s1} - C_i$.
- p is an exponent which depends on the concrete composition and the environment. It is assumed that $C_s = C_i + S_p \times (t/t_1)^p$.
- D_{a1} is the achieved chloride diffusion coefficient after 1 yr of exposure.
- α is an age parameter which depends on the concrete composition and the environment.
- D_a is the achieved chloride diffusion coefficient after t yr of exposure. It is assumed that $D_a = D_{a1} \times (t/t_1)^{-\alpha}$.

When plotting t and D_a in a (t, D)-diagram we may ask what the value of D_a should be if corrosion should occur under the conditions that the values of S_p, p, α and D_{a1} are kept constant. The value is named D_{acr}.

The value of D_{acr} is determined by solving the following equation:

$$C_{cr} = C_i + S_p \times \left(\frac{t}{t_1}\right)^p \times \Psi_p \left(\frac{0.5\, c}{\sqrt{t \times D_{acr}}}\right). \tag{5.135}$$

Thus, the value of $D_a = D_{acr}$ which at the time t leads to corrosion can be determined by

$$D_{acr} = \left\{\frac{0.5\, c}{\text{inv } \Psi_p\left(\nu_{cr}\right)}\right\}^2 \times \frac{1}{t}, \tag{5.136}$$

where

$$\nu_{cr} = \frac{C_{cr} - C_i}{S_p \times t^p}. \tag{5.137}$$

Since inv Ψ_p here depends on t, the border of the corrosion domain, given by Equation (5.136), is not a straight line when plotted as a logarithmic graph, as in the case when the LIGHTCON model is assumed to be valid. The deviation from a straight line is, however, small when $t > 10$ yr, cf. Example 5.2.10.

Hence, when plotting t and D_a in a (t, D_a)-diagram the following 'state of corrosion' will occur:

$D_a < D_{acr}$ implies 'no corrosion',

$D_a = D_{acr}$ implies 'initiation of corrosion',

$D_a > D_{acr}$ implies 'ongoing corrosion'.

Table 5.13: Observations from a marine concrete bridge, cf. Example 5.2.10.

Inspection year	Age years	Diffusion coefficient mm^2/yr	Surface chloride % mass binder
1989	7	22.1	6.04
1991	9	18.9	7.62
1994	12	15.8	8.51

The region where $D_a > D_{acr}$ for any $t > 0$ is called the corrosion domain or the domain of corrosion.

Example 5.2.10 An area of a concrete bridge exposed to a marine splash environment is inspected and tested. The observations are shown in Table 5.13.

It is assumed that the initial chloride content of the concrete is $C_i \approx 0$. The cover of the reinforcement is $c = 50$ mm and the concrete mixture proportions in the area are as stated in Table 5.14, cf. Stoltzner (1995) and Frederiksen *et al.* (1997).

The determination of the corrosion domain and estimation of the initiation period of time is carried out in the following way:

Step 1. The water/binder ratio with respect to corrosion is:

$$\text{eqv}\,\{w/b_{cr}\} = \frac{140}{330 - 4.7 \times 0 - 1.4 \times 40} = 0.51 \qquad \text{cf. (5.25).}$$

Thus, the threshold value of chloride in the concrete is

$$C_{cr} = 1.25 \times \exp(-1.5 \times 0.51) = 0.58 \text{ \% mass binder cf. (5.126).}$$

Step 2. From the observation of the achieved chloride diffusion coefficient versus time, cf. Table 5.13, a regression analysis gives

$$\log D_a = 1.871 - 0.6226 \times \log t,$$

where $R^2 = 1.000$. It follows from this that the achieved diffusion coefficient after 1 yr of exposure is given by $\log D_{a1} = 1.871$, i.e. $D_{a1} = 73.30$ mm^2/yr and that the age parameter is $\alpha = 0.623$.

Step 3. We get by a regression analysis of the observations of the surface chloride content versus time, cf. Table 5.13,

$$\log C_s = 0.2596 + 0.6300 \times \log t,$$

where $R^2 = 0.943$. Hence the surface diffusion coefficient after 1 yr of exposure is given by $\log C_{s1} = 0.2598$, i.e. $C_{s1} = 1.82$ % mass binder, and the parameter $p = 0.630$. Since $C_i = 0$, we get the factor $S_p = 1.82$ mass binder, cf. Equation (5.85).

Table 5.14: Mixture proportions, cf. Example 5.2.10.

Constituents	Contents	Unit
Portland cement	330	kg/m^3 concrete
Fly ash	40	kg/m^3 concrete
Aggregates	1827	kg/m^3 concrete
Water	140	kg/m^3 concrete

Step 4. In order to plot the corrosion domain in a (t, D)-diagram the following parameter is determined.

$$\nu_{cr} = \frac{0.58 - 0}{1.82 \times t^{0.63}} = 0.3187 \times t^{-0.63}, \qquad \text{cf. (5.137).}$$

Thus, the border line of the corrosion domain is given by, cf. Figure 5.40,

$$D_{acr} = \left\{ \frac{0.5 \times 50}{\text{inv } \Psi_{0.63}(\nu_{cr})} \right\}^2 \times \frac{1}{t} \text{ mm}^2/\text{yr} \qquad \text{cf. (5.136).}$$

The border line of the corrosion domain is calculated by means of a spreadsheet.

Step 5. By plotting the corrosion domain it is observed that no corrosion takes place at the moment, cf. Figure 5.40. In order to predict the future chloride ingress the movement of the point (t, D_a) in time should be estimated. This can be done by the straight line through the observation of the measured achieved chloride diffusion coefficient versus time, cf.

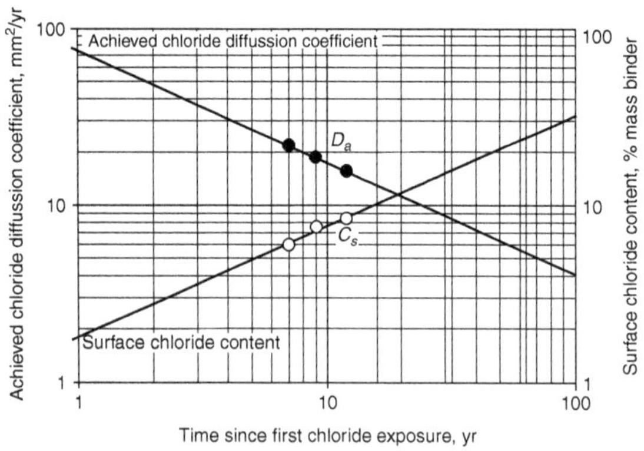

Figure 5.39: Observations from an inspection area of an RC bridge, cf. Example 5.2.10. The best fit of straight lines through the observations are found by a regression analysis.

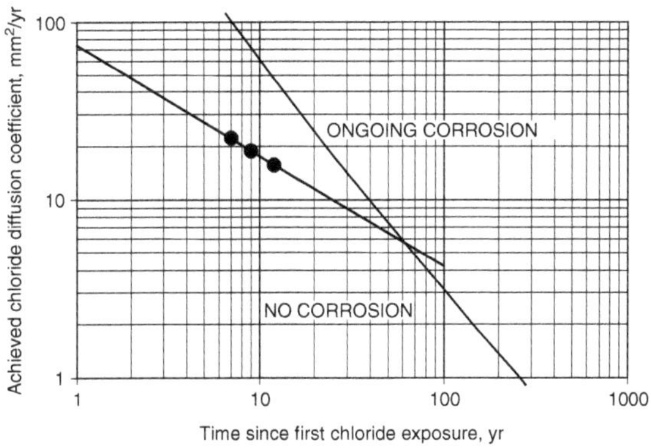

Figure 5.40: Corrosion domain of an inspection area of an RC bridge, cf. Example 5.2.10. The observations of the achieved chloride diffusion coefficient at 7, 9 and 12 yr after the first chloride exposure are shown by •. The best fit of a straight line through the observations is shown up to 100 yr. If the tendency observed is valid in the future together with the HETEK model, the initiation period of time is approximately 60 yr.

the regression analysis, **Step 2.** It is seen that this line has a point of intersection with the border of the corrosion domain at the time of approximately 60 yr. Thus, the best estimate of the initiation period of time is $t_{\mathrm{cr}} = 60$ yr.

Discussion and conclusion. Even if the three observations at 7 yr, 9 yr and 12 yr represent more information than usually is the practice, the predicted initiation period of time is rather uncertain. An inspection plan of testing the concrete at e.g. 1 yr, 4 yr and 12 yr would have been more reliable.

The concrete ought to be tested again at least in 2020 to see if D_a and C_s still follow the predicted relations.

Notice that the estimated initiation period of time according to Equation (5.133) do not agree with the observations made. This may be due to surface treatment of the concrete or the like.

At the end of the initiation period of time, here 60 yr, the reinforcement will start corrosion, but it will take several years before the formation of rust will cause the concrete cover to spall, cf. Section 1.3.1.

A non-linear regression analysis of the observations in Table 5.16 is carried out on the following equation by means of an *Excel* spreadsheet:

$$x = \frac{1.66 + 0.3p}{B} \times t^C \times \left\{ 1 - \left(\frac{A}{t^p} \right)^{\frac{0.5}{1+p}} \right\}.$$

We find the solutions $A = 0.1406$, $B = 0.05353$, $C = 0.3602$, and $p = 0.3607$. It follows that $\alpha = 0.27958$. The relation t versus x is shown on Figure 5.41.

Table 5.15: Conditions for the application of the HETEK approximation.

Anode, No.	Position, x mm	Reaction, t yr	Argument, u
Anode 1	7	0.53	$0.47 < 1.1$
Anode 2	13	1.89	$0.55 < 1.1$
Anode 3	19	4.42	$0.60 < 1.1$
Anode 4	26	8.81	$0.64 < 1.1$
Reinforcing bar	50	41	$0.70 < 1.1$

When the cover of the reinforcing bar is $c = 50$ mm, the initiating period of time is determined by solving the following equation:

$$0.14062 - t_{\mathrm{cr}}^{0.36071} \times \left\{1 - \frac{0.05363 \times 50}{(1.66 + 0.03 \times 0.36071) \times t_{\mathrm{cr}}^{0.36021}}\right\}^{2.72142} = 0.$$

On the other hand, the initiation period of time, $t_{\mathrm{cr}} = 43$ yr may also be found graphically from Figure 5.41. We see that $u \leq 1.1$, cf. Table 5.15. ◊

Corrosion multiprobe (macrocell technique)

Section 5.1.5 describes the mathematical model of a corrosion multiprobe assuming that the LIGHTCON model is valid. Here, the HETEK model is assumed to be valid, and the influences are studied.

The assumptions of the HETEK model lead to the solution of a rather complicated equation. Therefore, we apply instead the HETEK model by approximation, cf. (3.73). Before applying the HETEK model it is, however,

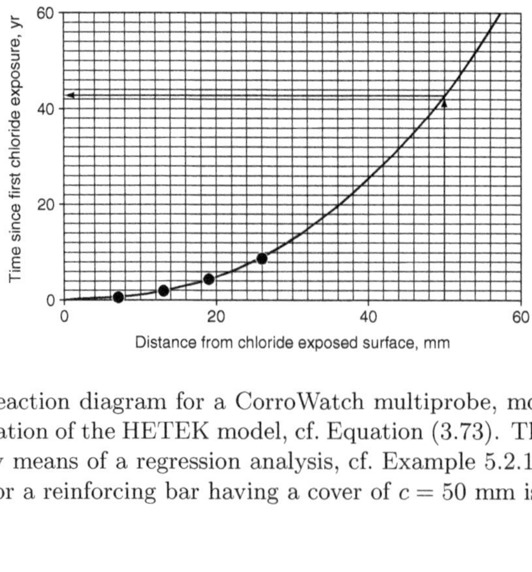

Figure 5.41: Reaction diagram for a CorroWatch multiprobe, modelled according to the approximation of the HETEK model, cf. Equation (3.73). The relation shown is determined by means of a regression analysis, cf. Example 5.2.10. The initiation period of time for a reinforcing bar having a cover of $c = 50$ mm is $t_{\mathrm{cr}} = 43$ yr.

Table 5.16: The border line of the corrosion domain, cf. Example 5.2.10.

$\Psi_{0.60}$	$\Psi_{0.63}$	$\Psi_{0.65}$	u	$t^{0.63}$	$\ln t$	t	D_{acr}
-	-	-	-	-	-	yr	mm^2/yr
0.26012	0.25561	0.25261	0.60	1.2467	0.3500	1.4	1223.37
0.22761	0.22336	0.22053	0.65	1.4267	0.5641	1.8	841.51
0.19846	0.19449	0.19184	0.70	1.6386	0.7838	2.2	582.46
0.17243	0.16876	0.16631	0.75	1.8884	1.0091	2.7	405.06
0.14927	0.14590	0.14365	0.80	2.1843	1.2401	3.5	282.56
0.12876	0.12569	0.12364	0.85	2.5355	1.4768	4.4	197.55
0.11066	0.10788	0.10603	0.90	2.9540	1.7193	5.6	138.27
0.09475	0.09225	0.09059	0.95	3.4544	1.9677	7.2	96.80
0.08082	0.07859	0.07711	1.00	4.0548	2.2221	9.2	67.74
0.06868	0.06671	0.06539	1.05	4.7774	2.4824	12.0	47.36
0.05814	0.05640	0.05524	1.10	5.6504	2.7488	15.6	33.06
0.04904	0.04750	0.04648	1.15	6.7085	3.0212	20.5	23.03
0.04119	0.03986	0.03897	1.20	7.9954	3.2998	27.1	16.01
0.03446	0.03331	0.03254	1.25	9.5677	3.5848	36.0	11.10
0.02872	0.02772	0.02706	1.30	11.4948	3.8760	48.2	7.67
0.02384	0.02299	0.02242	1.35	13.8629	4.1734	64.9	5.28
0.01871	0.01858	0.01850	1.40	17.1482	4.5109	91.0	3.50
0.01623	0.01562	0.01521	1.45	20.4047	4.7869	119.9	2.48
0.01331	0.01279	0.01245	1.50	24.9087	5.1035	164.6	1.69
0.01087	0.01044	0.01015	1.55	30.5309	5.4266	227.4	1.14

recommended first to apply the LIGHTCON model, cf. Section 5.1.5, in order to get a first approximation of the solution.

The approximation formula for the $\Psi_p(u)$ function, cf. Equation (3.73), is

$$\Psi_p(u) \approx \left\{1 - \frac{u}{1.66 + 0.3p}\right\}^{2(1+p)}, \tag{5.138}$$

where the relative error is less that 10 % for all $0 \leq u \leq 1.1$ and $0 \leq p \leq 1$.

The chloride ingress of the threshold value into the concrete is according to the HETEK model

$$\begin{aligned} C_{cr} &= C_i + S_p \times \tau^p \times \Psi_p(u) \\ &\approx C_i + S_p \times \left\{\frac{t}{t_1}\right\}^p \times \left\{1 - \frac{u}{1.66 + 0.3p}\right\}^{2(1+p)}, \end{aligned} \tag{5.139}$$

where t is the time since the first chloride exposure, and $t_1 = 1$ yr, and C_i is the initial chloride content of the concrete, and

$$u = \frac{0.5\,x}{\sqrt{t \times D_a}}. \tag{5.140}$$

Here, D_a denotes the chloride diffusion coefficient of the concrete at time t,

$$D_a = D_1 \times \left\{\frac{t}{t_1}\right\}^{-\alpha} = D_1 \times t^{-\alpha}, \tag{5.141}$$

Table 5.17: Observations from a CorroWatch Multiprobe.

Anode n, nos	Position x, mm	Time of corrosion t, yr
1	7	0.53
2	13	1.89
3	19	4.42
4	26	8.81

where D_1 is the chloride diffusion coefficient of the concrete after 1 yr of exposure, and $t_1 = 1$ yr. We may write Equations (5.139)–(5.141) as

$$\frac{C_{cr} - C_i}{S_p} = t^p \times \left\{1 - \frac{0.5\,x}{(1.66 + 0.3p) \times \sqrt{D_1 t^{1-\alpha}}}\right\}^{2(1+p)}, \tag{5.142}$$

or

$$A = t^p \times \left\{1 - \frac{B \times x}{(1.66 + 0.3p) \times t^C}\right\}^{2(1+p)}, \tag{5.143}$$

where

$$A = \frac{C_{cr} - C_i}{S_p}, \quad B = \frac{0.5}{\sqrt{D_1}}, \quad \text{and} \quad C = 0.5 \times (1 - \alpha). \tag{5.144}$$

When Equation (5.143) is solved with respect to x, we get

$$x = \frac{1.66 + 0.3p}{B} \times t^C \times \left\{1 - \left(\frac{A}{t^p}\right)^{\frac{0.5}{1+p}}\right\}. \tag{5.145}$$

Equation (5.145) is very suitable for a non-linear regression analysis of the observations made, using the spreadsheet *Excel* version 5 or a later version. When the parameters A, B, C and p have been determined, the initiation period of time t_{cr} for a reinforcement bar with a cover c may be determined by solving the following equation with respect to t_{cr},

$$f\left(t_{\text{cr}}\right) = A - t_{\text{cr}}^p \times \left\{1 - \frac{B \times c}{(1.66 + 0.3p) \times \tau^C}\right\}^{2(1+p)}. \tag{5.146}$$

A simple numerical solution method is the iteration method *regula falsi*, though a graphical solution is also possible from the relation t versus x.

Example 5.2.11 A corrosion macrocell, the multiprobe CorroWatch from FORCE with $n = 4$ anodes is cast into concrete exposed to marine splash. The

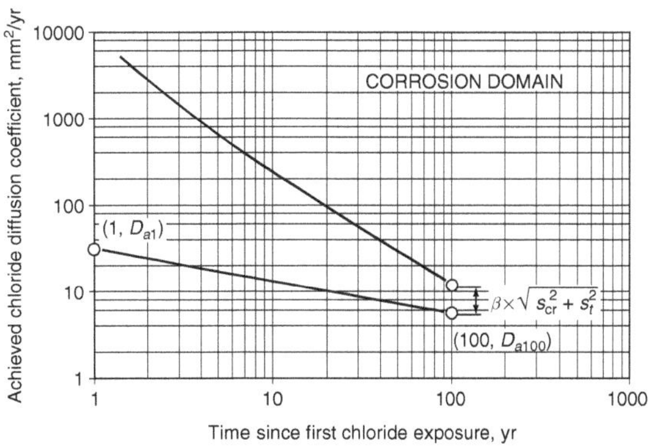

Figure 5.42: Corrosion domain for an RC structure. Due to the uncertainties of the decisive parameters the achieved diffusion coefficient must lie in some distance from the corrosion domain.

positions of the anodes and there times of reaction are shown in Table 5.17. The concrete cover of the reinforcement is $c = 50$ mm, and there is no special surface treatment of the concrete.

These observations have already been studied in Example 5.1.5, page 262, and the initiation period of time was found to be $t_{cr} = 35$ yr, when we assume that LIGHTCON model is valid for the chloride ingress into concrete. $\diamondsuit$

Design of concrete cover

The concrete cover of a reinforcing bar has a structural purpose as well as a purpose of durability. In order to achieve a good protection of the reinforcement, the cover must be cast of concrete with a low chloride diffusivity and a low initial chloride content. Furthermore, the thickness of the cover should be adequate and must not vary in thickness (have a small tolerance). There will be a balance between durability and cost. However, in general the following means are important for the durability of the concrete, i.e. to protect the reinforcement from corrosion:

- High content of binders (cement, fly ash, silica fume, etc.)
- Minimum defects of the cover (cracks and voids), i.e. proper casting, compaction and curing.

The use of a surface coating (e.g. like siloxane) is not a durable solution and has to be repeated. Such a surface treatment will reduce the surface chloride content C_s but may not be stable. However, an effect in maybe five years will delay the chloride ingress into the concrete.

The cover must be designed so that it will protect the reinforcement at least in a period of time longer than a prescribed initiation period of time. There must, however, like in structural design be a certain safety against failure (i.e. corrosion). In a structural failure there will be a risk of loss of human life. However, a failure of durability (initiation of corrosion) will only damage the RC structure and there is not a risk of loss of human life. Therefore, the safety against durability may be less than the safety against structural failure. This is left to the designer or the building owner. The Code of Practice requires a certain thickness of the cover, which depends on the concrete composition and the environment, but it does not correspond to the safety level against structural failure, cf. Karlsson *et al.* (1995).

In a given period of time (e.g. $50 \leq t \leq 100$ years) the point (t, D_a) must not enter the domain of corrosion. Here it must be taken into account that D_a is a stochastic variable. This means that there should be a certain distance between the point (t, D_a) and the border line D_{acr} of the corrosion domain, cf. Figure 5.42. Cornell (1969) has proposed a method, *Cornell's method of reliability index*, which is used in structural design and which is also applicable to durability problems, cf. Section 2.4.6.

The thickness of a cover may be designed according to Cornell's method of reliability in the following way:

- The expectation value of the achieved chloride diffusion coefficient $D_a(t)$ is determined at the specified initiation period of time t.

- The expectation value of the border line $E\{D_{acr}\}$ of the corrosion domain at the specified initiation period of time t is determined as

$$E\{D_{acr}\} = E\{D_a(t)\} + \beta \times \sqrt{s_{cr}^2 + s_t^2}, \tag{5.147}$$

where

s_{cr} is the standard deviation of the border line of the corrosion domain,

s_t is the standard deviation of $D_a(t)$,

β is the reliability index.

- The border line is given by the formula, cf. Equation (5.136),

$$D_{acr} = \left\{ \frac{c}{\sqrt{4t} \times \text{ inv } \Psi_p(\nu_{cr})} \right\}^2 = \left\{ \frac{c}{k_{cr}} \right\}^2, \tag{5.148}$$

where

$$k_{cr} = \sqrt{4t} \times \text{ inv } \Psi_p(\nu_{cr}), \tag{5.149}$$

and, cf. Equation (5.137),

$$\nu_{cr} = \frac{C_{cr} - C_i}{S_p \times t^p}. \tag{5.150}$$

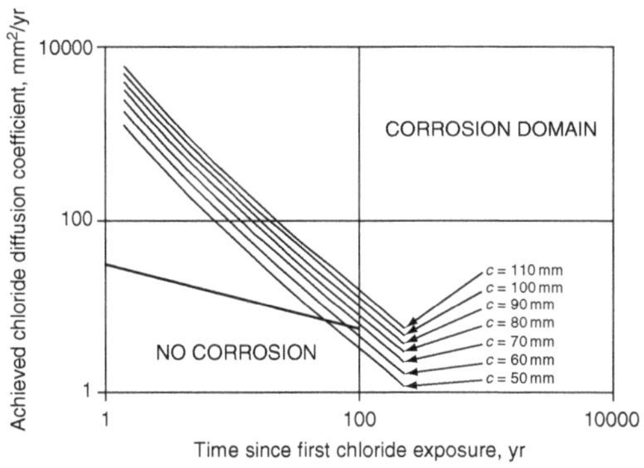

Figure 5.43: Corrosion domain for the reinforced concrete of Example 5.2.12. The border line of the corrosion domain depends on the cover. It is seen that a cover of 70 mm is sufficient, if the parameters are deterministic. It is shown in Example 5.2.12 that a cover of 100 mm is necessary in case of stochastic parameters.

- The necessary cover is

$$\text{nec}\{c\} \geq k_{cr} \times \sqrt{E\{D_{acr}\}}. \tag{5.151}$$

For the time being there is no requirement for the reliability index, and only little information of the standard deviations s_{cr} and s_t. When the reinforcement starts to corrode, there is an early warning before the failure of the structure, and there is no risk of loss of human life. Therefore the reliability index would be smaller than that for structural failures, cf. Table 2.7, page 101.

Example 5.2.12 A concrete for the marine splash zone of a bridge pillar has the following composition:

- Portland cement : 330 kg/m^3 of concrete,
- Fly ash : 40 kg/m^3 of concrete,
- Silica fume : 20 kg/m^3 of concrete,
- Mixing water : 133 kg/m^3 of concrete,

with a sufficient amount of admixtures, e.g. air-entraining and plasticising admixtures.

The initial chloride content of the concrete is assumed to be $C_i = 0$. The HETEK model is assumed to be valid for the chloride ingress into the concrete. The concrete cover of the reinforcement has to be $c = 75$ mm designed.

It was shown in Example 5.2.5, page 297 that the expectation value of the initiation period of time is $E\{t_{\text{cr}}\} = 158$ yr, when the cover is $c = 75$ mm.

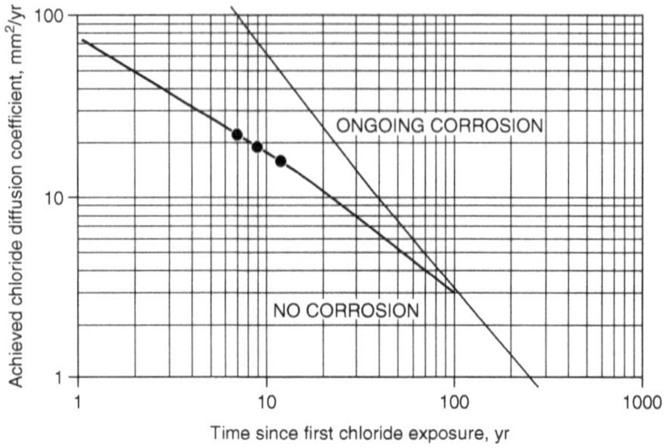

Figure 5.44: Corrosion domain for an RC structure. At the time 15 yr after the first chloride exposure the concrete surface is protected, cf. Example 5.2.13.

The following governing parameters were found there by estimation:

$p = 0.327$ non-dimensional,
$\alpha = 0.365$ non-dimensional,
$S_p = 1.48$ % mass binder,
$D_1 = 30.75$ mm^2/yr,
$C_{cr} = 0.41$ % mass binder.

With a prescribed initiation period of time $t_{\text{cr}} = 100$ yr the determination of the necessary cover of the reinforcing bars according to Cornell's method of reliability index with $\beta = 2.0$ is carried out as follows:

Step 1. The achieved chloride diffusion coefficient at time $t = 100$ yr is

$$D_{a100} = 30.75 \times 100^{-0.365} = 5.73 \text{ mm}^2/\text{yr} \qquad \text{cf. (5.20).}$$

Step 2. The standard deviations are estimated to $s_{cr} = s_t = 4$ mm^2/yr. Thus the necessary expectation value of the border line at time $t = 100$ yr is, cf. Figure 5.43,

$$E\{D_{acr}\} = 5.73 + 2.0 \times \sqrt{4^2 + 4^2} = 17.04 \text{ mm}^2/\text{yr} \qquad \text{cf. (5.147).}$$

Step 3. The value of ν_{cr} is

$$\nu_{cr} = \frac{0.41 - 0}{1.48 \times 100^{0.327}} = 0.0615 \qquad \text{cf. (5.150).}$$

Step 4. The value of inv $\Psi_{0.327}(0.0615) = 1.178$, cf. Table 8.4. Thus, the constant k_{cr} is

$$k_{cr} = \sqrt{4 \times 100} \times 1.178 = 17.04 \qquad \text{cf. (5.149).}$$

Step 5. Finally, the necessary cover is

$$\text{nec}\{c\} \geq 23.56 \times \sqrt{17.04} = 97 \text{ mm} \approx 100 \text{ mm} \qquad \text{cf. (5.151).} \quad \Diamond$$

Example 5.2.13 In Example 5.2.10, page 317, it was found that an area of a concrete bridge exposed to a marine splash environment had a predicted initiation period of time, $t_{\text{cr}} = 60$ yr. The cover of the reinforcement is 50 mm. Now, the question is: What should we do, if the predicted initiation period of time ought to be 100 yr instead?

There are several methods to increase the initiation period of time, from surface protection (e.g. coating) to cathodic protection, cf. Section 1.6, page 53. Here we suggest that the surface of the area in question is protected in such a way that the initiation period of time is increased to 100 yr. It is seen from Figure 5.2.12 that the surface protection should be able to cause the achieved diffusion coefficient at 100 yr to achieve the value

$$D_a(100 \text{ yr}) \leq 3 \text{ mm}^2/\text{yr},$$

instead of approximately 4.2 mm^2/yr, cf. Figure 5.40.

Notice that the surface protection must be renovated according to the recommendation of the producer. A financial study will decide if other methods like a cathodic protection is a more economic method. $\Diamond$

5.2.7 Summary

According to the HETEK model the chloride ingress into concrete by diffusion gives at time t a chloride distribution $C = C(x,t)$, which obeys

$$C = C_i + (C_{sz} - C_i) \times \Psi_p \left(\frac{0.5\, x}{\sqrt{t\, D_a}} \right) \qquad \text{cf. (5.87),}$$

under the following conditions

- The chloride parameters: The achieved chloride diffusion coefficient D_a at time t, and the surface chloride content C_{sa}, both follow power functions in time.
- The concrete is exposed to chloride for the first time at time $t = 0$.
- The initial chloride content of the concrete C_i is constant, i.e. independent of location x, time t, and chloride concentration C.
- The functions Ψ_p are defined in Section 3.3.1 and tabulated i Section 8.3.

The achieved chloride diffusion coefficient depends on time in the following way,

$$D_a = D_1 \times \left\{ \frac{t}{t_1} \right\}^{-\alpha} \qquad \text{cf. (5.6),}$$

where α is the age parameter, $t_1 = 1$ yr, and D_1 is the achieved chloride diffusion coefficient at time $t_1 = 1$ (given below).

The achieved chloride content of the exposed concrete surface is

$$C_{sa} = C_i + S_1 \times \left\{\frac{t}{t_1}\right\}^{(1-\alpha)p}, \qquad \text{cf. (5.81)},$$

where S_1, α and p are parameters in the following.

The age parameter, α

The age parameter may be estimated by the following formula,

$$\alpha = k_\alpha \times (1 - 1.5 \times \text{ eqv}\{w/c_D\}) = \frac{\ln(D_1/D_{in})}{\ln(t_{\text{in}}/t_1)} \qquad \text{cf.} \begin{cases} (5.7) \\ (5.9) \end{cases}$$

where D_{in} is the achieved chloride diffusion coefficient at the inspection time t_{in} and

$$\text{eqv}\{w/c_D\} = \frac{W}{PS + FA + 7 \times SF}, \qquad \text{cf. (5.3)},$$

and (cf. (5.8)),

$$k_\alpha = \begin{cases} 1.0 \text{ for concrete exposed to marine atmosphere (ATM)} \\ 0.1 \text{ for concrete exposed to marine splash zone (SPL)} \\ 0.6 \text{ for concrete submerged in seawater (SUB).} \end{cases}$$

The symbols W, PC, FA and SF are the composition of the concrete by mass of mixing water, Portland cement, fly ash and silica fume, respectively.

Diffusion coefficient after 1 year of exposure, D_1

The achieved chloride diffusion coefficient after 1 year of chloride exposure may be estimated by the following formula

$$D_1 = k_D \times 25{,}000 \times \exp\left(-\sqrt{\frac{10}{\text{eqv}\{w/c_D\}}}\right) \qquad \text{cf.} \begin{cases} (5.2) \\ (5.5) \end{cases}$$

where

$$\text{eqv}\{w/c_D\} = \frac{W}{PC + FA + 7 \times SF} \qquad \text{cf. (5.67).}$$

and, cf. (5.4),

$$k_D = \begin{cases} 0.4 \text{ for concrete exposed to marine atmosphere (ATM)} \\ 0.6 \text{ for concrete exposed to marine splash zone (SPL)} \\ 1.0\text{for concrete submerged in seawater (SUB)} \end{cases}$$

The symbols W, PC, FA and SF are the composition of the concrete by mass of mixing water, Portland cement, fly ash and silica fume, respectively.

Surface chloride content after 1 year of exposure, C_1

The achieved chloride surface content after 1 year of chloride exposure may be estimated by the following formula

$$C_1 = k_b \times \text{ eqv}\{w/c_b\} \qquad \text{cf. (5.78),}$$

where

$$\text{eqv}\{w/k_b\} = \frac{W}{PC + 0.75 \times FA - 1.55 \times SF} \qquad \text{cf. (5.79),}$$

and, cf. (5.80),

$$k_b = \begin{cases} 2.20 \text{ for concrete exposed to marine atmosphere (ATM)} \\ 3.67 \text{ for concrete exposed to marine splash zone (SPL)} \\ 5.13 \text{ for concrete submerged in seawater (SUB)} \end{cases}$$

The symbols W, PC, FA and SF are the composition of the concrete by mass of mixing water, Portland cement, fly ash and silica fume, respectively.

Chloride diffusion coefficient after 1 year of exposure, D_1

The achieved chloride diffusion coefficient after 1 year of exposure may be estimated by the following formula

$$D_1 = \frac{1}{2} k_{cr} D_{pex}.$$

Surface chloride content after 100 years of exposure, C_{100}

The achieved chloride surface content after 100 years of chloride exposure may be estimated by the following formula,

$$C_{100} = k_{100} \times C_1 \qquad \text{cf. (5.82)}$$

where, cf. (5.83),

$$k_{100} = \begin{cases} 7.0 \text{ for concrete exposed to marine atmosphere (ATM)} \\ 4.5 \text{ for concrete exposed to marine splash zone (SPL)} \\ 1.5 \text{ for concrete submerged in seawater (SUB)} \end{cases}$$

The symbols W, PC, FA and SF are the composition of the concrete by mass of mixing water, Portland cement, fly ash and silica fume, respectively.

The exponent p

The exponent p may be estimated by either of the following two formulæ

$$p = \frac{\log_{10}(C_{100} - C_i) - \log_{10}(C_1 - C_i)}{2 \times (1 - \alpha)} = \frac{\ln(C_{in} - C_i) - \ln(C_1 - C_i)}{(1 - \alpha) \times \ln t_{in}},$$

cf. (5.84) and (5.86), where C_{in} is the achieved surface chloride content at the inspection time t_{in}.

The factor S_1

The factor S_1 is estimated by the following formula

$$S_1 = C_1 - C_i \qquad \text{cf. (5.85).}$$

The threshold value of chloride in concrete, C_{cr}

The threshold value of chloride in concrete is estimated by

$$C_{cr} = k_{cr} \times \exp\left(-1.5 \times \ \text{eqv}\left\{w/k_{cr}\right\}\right) \qquad \text{cf. (5.24),}$$

where

$$\text{eqv}\left\{w/k_{cr}\right\} = \frac{W}{PC - 1.4 \times FA - 4.7 \times SF} \qquad \text{cf. (5.25)}$$

and, cf. (5.26),

$$k_{cr} = \begin{cases} 1.25 \text{ for concrete exposed to marine atmosphere (ATM)} \\ 1.25 \text{ for concrete exposed to marine splash zone (SPL)} \\ 3.35 \text{ for concrete submerged in seawater (SUB)} \end{cases}$$

The symbols W, PC, FA and SF are the composition of the concrete by mass of mixing water, Portland cement, fly ash and silica fume, respectively.

NT Build 443

The HETEK model may be supplemented by test results applying the test method NT Build 443. It is possible to predict the chloride parameters of the HETEK model from the test results D_{pex} and C_{sp}. However, it is only possible to get information on the magnitude of the chloride parameters, but all information counts.

Chapter 6

Location-Dependent Chloride Diffusivity

In this chapter we shall briefly consider the case where the diffusion coefficient $D(\mathbf{x})$ also depends on location. This is not an easy matter, and the reader may instead consult Section 7.2 for a simpler model.

In one dimension Fick's second law becomes

$$\frac{\partial C}{\partial t} = \frac{\partial}{\partial x}\left\{D(x)\frac{\partial C}{\partial x}\right\},$$

which will be considered briefly in Section 6.1.

The generalization of this equation to higher dimensions is in the isotropic case given by

$$\frac{\partial C}{\partial t} = \operatorname{div}\left\{D(\mathbf{x})\mathbf{grad}\ C\right\} = \nabla \cdot (D \nabla C),$$

cf. also Crank (1975).

In this connection it should be mentioned that Crank (1975), correctly states in his preface: *"When we come to systems in which the diffusion coefficients is not constant but variable $\cdots$ we find that strictly formal mathematical solutions no longer exist."* Unfortunately this phrase may be misinterpreted so that one would believe that the problem cannot be solved mathematically. It can! There do exist very old and almost forgotten iteration methods which solve this problem mathematically, though it must be admitted that they in general require the use of computer programmes which did not exist when Crank wrote his book.

We have been able to solve the iteration problem in simple cases, but the method still needs some reconsideration before it can be included in a text book, so we shall only consider the problems from a theoretical point of view in this chapter.

As mentioned above the mathematical methods in this chapter are very old. Titchmarsh (1962), refers vaguely to Sturm and Liouville, i.e. approximately about 1850. Section 6.1 on the one-dimensional case is based on ideas from Titchmarsh (1962), and Mejlbro (1995). The results of Mejlbro (1995) were limited to the case where the diffusion coefficient had the structure

$$\frac{1}{D_k(x)} = \frac{1}{k+1}\left\{k + \left(\frac{x-1}{x+1}\right)^2\right\}, \qquad x \geq 0, \quad k > 0,$$

which can also be written in the form

$$D_k(x) = \frac{(k+1)(x+1)^2}{k(x+1)^2 + (x-1)^2}, \qquad x \geq 0, \quad k > 0.$$

The guidelines of this choice were that $D(0) = \lim_{x\to\infty} D(x) = 1$ and that $D(x)$ has a global maximum $D(1) = 1 + \frac{1}{k}$, and last but not least that the necessary calculations of the iteration process could be carried out using only calculus taught at any university, i.e. even without the help of a pocket calculator. The family $D_k(x)$, $k > 0$, was only meant to illustrate that the problem in principle could be solved. However, Østerdal (1995) showed that even this poor family could be used successfully in practice. She measured $D(x)$ at different locations, adjusted the length unit so that the peak of $D(x)$ occurred at $x = 1$, and finally chose the parameter $k > 0$, such that $D_k(x)$ matched the measured values by a least square approach.

6.1 The one-dimensional case

It has been possible in general solve the one-dimensional problem

$$\begin{aligned} &\frac{\partial C}{\partial t} = \frac{\partial}{\partial x}\left\{D(x)\frac{\partial C}{\partial x}\right\}, \quad \text{for } 0 < x < \ell, \\ &C(x,0) = 0, \quad \text{for } 0 < x < \ell, \\ &C(0,t) = c_0(t) \text{ and } C(\ell,t) = c_\ell(t) \text{ for } t > 0, \end{aligned} \tag{6.1}$$

by generalizing the method of eigenfunctions from Section 4.2. This method was investigated by Vindeløv (2003) in his thesis. Although the mathematical theory claims that the iteration process is converging uniformly, it was shown that the method is extremely slow and elaborated. We therefore instead recommend the alternative approximative approach by step functions given in Section 7.2.

The general theory is therefore left for further study, where one should look for a better iteration scheme than already known (and not produced here).

Another alternative is of course to use the method of *Finite Elements*. This method was developed when it was realized that the method of eigenfunctions was too difficult to use in practice, when only paper and pencil were at hand and the software of the computers did not account for a general solution of the more complicated eigenvalue problems. Notice that *if* one uses the method of finite elements, one will inevitably lose some information at an early state of the solution procedure. Therefore, if it is possible to use the original method of eigenfunction development (i.e. generalized Fourier series), it is highly recommended that one should use this instead of the method of finite elements.

6.2 The isotropic case in higher dimensions

In higher dimensions the problem of solving Fick's second law with varying diffusion coefficient is much more difficult than in one dimension, even in the isotropic case. It is worth here briefly to give some hints on the problems which may occur.

Consider e.g. the rectangular three dimensional space $\mathbb{R}^3$, and let $D(x, y, x)$ $[> 0]$ denote the diffusion coefficient. According to Crank (1975), Fick's second law in the isotropic case is written

$$\begin{aligned} \frac{\partial C}{\partial t} &= \operatorname{div}\{D(x,y,z)\ \mathbf{grad}\ C(x,y,z)\} = \nabla \cdot (D \nabla C) \\ &= \frac{\partial}{\partial x}\left\{D\frac{\partial C}{\partial x}\right\} + \frac{\partial}{\partial y}\left\{D\frac{\partial C}{\partial y}\right\} + \frac{\partial}{\partial z}\left\{D\frac{\partial C}{\partial z}\right\}. \end{aligned} \tag{6.2}$$

If the diffusion coefficient D only depends on one of the rectangular coordinates, $D = D(x)$ say, then we can find the eigenfunctions with respect to x by using the methods of Section 6.1, assuming that $\Omega =]\,0, A\,[\times]\,0, B\,[\times]\,0, C\,[$ is a rectangular box. Since D is constant in y and z we just multiply these eigenfunctions in x by $\sin(\pi n y/B)\sin(\pi p z/C)$, $n, p \in \mathbb{N}$, in order to obtain the right orthogonal system of eigenfunctions associated with (6.2) on Ω.

Unfortunately, when D depends on at least two of the rectangular coordinates we have not in general been able to derive an exact solution formula by known methods, so numerical methods should be applied instead. Although the results are negative it is worth to present the arguments in the rectangular case for the following reason. The problem (6.2) is reduced to a simpler one, on which numerical methods like e.g. the method of finite elements are easier to apply.

Now, rectangular pillars may not seem too important. In Section 6.2.2 we shortly consider the more relevant circular pillars represented by the unit circle in two dimensions. The pillar is thus described in semi-polar coordinates and we assume that the diffusion coefficient D is independent of the height z.

Assuming furthermore that $D = D(r)$ only depends on r, the two dimensional Fick's second law is reduced to a Bessel-like differential operator and the problem can be solved by iteration.

If the boundary conditions do not depend on φ, e.g. when the pillar is placed in seawater, this iteration problem is fairly simple and we can to some extend copy the methods of Section 6.1, or approximate by step functions as in Section 7.2. We consider this as the most important case.

When the boundary conditions also depend on φ, the iteration process becomes more messy. It converges, but requires a lot of computational power in practice. We shall therefore not proceed further into this case either.

6.2.1 The rectangular case

Consider the isotropic Fick's second law (6.2) in three dimensions on $\Omega = \,]\,0,A\,[\,\times\,]\,0,B\,[\,\times\,]\,0,C\,[$ and assume that the diffusion coefficient $D(x,y,z) > 0$ depends on at least two of the variables. We have only three methods at hand:

1. Transform (6.2) into a known case by some divine inspiration.
2. Perform a change of variables which does the same.
3. Introduce a multiplying factor $\alpha(x,y,z) > 0$, such that

   $$C(x,y,z,t) = \alpha(x,y,z)u(x,y,z,t),$$

   and derive a simpler equation in u which can be solved.

In practical situations where $D(x,y,z)$ is given by e.g. measurements no one would rely on a divine inspirations, so case (1) is ruled out.

Concerning case (2) a very long investigation which shall not be given here shows that the only reasonable candidates are

$$\xi(x,y,z) = \int_0^x \frac{ds}{\sqrt{D(s,y,z)}}, \qquad (x,y,z) \in \Omega,$$

$$\eta(x,y,z) = \int_0^y \frac{ds}{\sqrt{D(x,s,z)}}, \qquad (x,y,z) \in \Omega,$$

$$\zeta(x,y,z) = \int_0^z \frac{ds}{\sqrt{D(x,y,s)}}, \qquad (x,y,z) \in \Omega.$$

This change of variables will give the right hand side of (6.2) the right structure, but unfortunately $(\xi,\eta,\zeta) \in \omega$, where the parameter space ω no longer is a rectangular box, which it should be if the method of separation of the variables should be applied. Therefore, case (2) must in general be abandoned too.

Finally, let

$$C(x,y,z,t) = \alpha(x,y,z)u(x,y,z,t) \tag{6.3}$$

for some positive function $\alpha(x,y,z)$. When we apply (6.2) on (6.3) it is seen that the result becomes simplest when we choose

$$\alpha(x,y,z) = \frac{1}{\sqrt{D(x,y,z)}}. \tag{6.4}$$

In the light of case (2) this choice should not be surprising.

With $\alpha(x, y, z)$ given by (6.4) we get

$$\frac{\partial C}{\partial d} = \frac{1}{\sqrt{D}} \frac{\partial u}{\partial t},$$

and

$$\frac{\partial C}{\partial x} = \frac{\partial}{\partial x} \left\{ \frac{u}{\sqrt{D}} \right\} = \frac{1}{\sqrt{D}} \frac{\partial u}{\partial x} - \frac{1}{2} \frac{1}{D\sqrt{D}} \frac{\partial D}{\partial x} \cdot u,$$

hence

$$D \frac{\partial C}{\partial x} = \sqrt{D} \frac{\partial u}{\partial x} - \frac{1}{2} \cdot \frac{1}{\sqrt{D}} \cdot \frac{\partial D}{\partial x} \cdot u,$$

and

$$\frac{\partial}{\partial x} \left\{ D \frac{\partial C}{\partial x} \right\} = \sqrt{D} \frac{\partial^2 u}{\partial x^2} - \frac{\partial^2 D}{\partial x^2} \cdot u.$$

Since we have similar results in the other variables we conclude that

$$\bigtriangledown \cdot \{D \bigtriangledown C\} = \sqrt{D} \cdot \bigtriangledown^2 u - \bigtriangledown^2 (\sqrt{D}) \cdot u$$

and Fick's second law

$$\frac{1}{\sqrt{D}} \frac{\partial u}{\partial t} = \sqrt{D} \cdot \bigtriangledown^2 u - \bigtriangledown^2 (\sqrt{D}) \cdot u$$

is reduced to

$$\frac{\partial u}{\partial t} = D \cdot \left\{ \bigtriangledown^2 u - \frac{1}{\sqrt{D}} \bigtriangledown^2 (\sqrt{D}) \cdot u \right\}. \tag{6.5}$$

Reviewing Section 4.2.4 and Section 4.2.5 on the eigenfunction method applied to Fick's second law we see that the method only works when the left hand side is of the form $\partial u / \partial t$. Hence we cannot divide (6.5) by $D(x, y, z)$, when we apply this method. The eigenvalue problem derived from (6.5) must therefore have the form

$$D \cdot \left\{ \bigtriangledown^2 u - \frac{1}{\sqrt{D}} \bigtriangledown^2 (\sqrt{D}) \cdot u \right\} = \lambda u, \tag{6.6}$$

where we want to find all pairs (λ, u), where $u \neq 0$ and $u(x, y, z) = 0$ on the boundary of Ω. Such solutions do exist, but they cannot be found by separation of the variables. The reason is the following: Since $D > 0$, Equation (6.6) is equivalent to

$$\bigtriangledown^2 u = \left\{ \frac{1}{\sqrt{D}} \bigtriangledown^2 (\sqrt{D}) + \frac{\lambda}{D} \right\} \cdot u.$$

According to Morse *et al.* (1953), this problem can only be solved by the method of separating the variables if there exist functions $f(x)$, $g(y)$, $h(z)$, $f_1(x)$, $g_1(y)$, $h_1(z)$, such that

$$\frac{1}{D} = f(x) + g(y) + h(z) \quad \text{and} \quad \frac{1}{\sqrt{D}} \nabla^2 (\sqrt{D}) = f_1(x) + g_1(y) + h_1(z),$$

and these two requirements can only be met, when D depends on just one of the variables x, y, z. Therefore, the rectangular case with varying diffusion coefficient cannot in general be solved by the chosen simple methods, and we suggest numerical methods instead.

Notice that numerical methods are easier to apply on the derived Equation (6.5) than on the original one (6.2).

6.2.2 The unit circle

Consider a circular pillar of radius R, where the diffusion coefficient $D(r)$ only depends on r. If also the boundary conditions do not depend on the height z, it suffices to consider Fick's second law in a two-dimensional disc of radius R. Finally, by changing the unit of length we may assume that $R = 1$. Hence we consider the initial/boundary value problem

$$\begin{aligned} \frac{\partial C}{\partial t} &= \operatorname{div}\{D(r)\,\mathbf{grad}\ C\}, && 0 \le r < 1, \\ C(r,\varphi,0) &= 0, && 0 \le r < 1, \\ C(1,\varphi,t) &= f(\varphi,t), && r = 1. \end{aligned} \tag{6.7}$$

This problem can be reduced to a problem similar to (6.1) in the one-dimensional case, so it can in principle be solved by the method of iteration. We have, however, in the draft of this book already carried out the huge calculations in a very simple example, and we estimate that this method of classical iteration is outside the realm of this book, as long as simpler methods do not exist.

Chapter 7

Special Topics

This chapter is devoted to miscellaneous topics which occurred during the writing of this book and which could not easily be incorporated in earlier chapters.

Corrosion macro cells are studied from a mathematical point of view in Section 7.1. It follows that both the LIGHTCON model and the more general HETEK model contain a singularity, when one tries to find the unknown constants from the measurements. In principle, two equations are missing, and it is not possible to derive them by integrating more cells into the multi cell. However, some compatibility equations help us to narrow the interval of some of the variables, but even within these constraints the other variables vary violently. Fortunately, the critical time t_{cr} is very stable, even under such violent changes of the basic variables, so concerning the critical time, any set of variables satisfying these compatibility conditions can be used.

We finally study the theory of coating in Section 7.2. The idea is very simple. We may assume that $D(x)$ is a step function so we can use the classical solution inside each interval, where the approximation is constant. Finally, the solution is glued together. This method can be used in general, when $D(x)$ is varying with location, cf. Chapter 6. We use here that the error will be smaller in time than the error of the approximation. This is one of the fundamental properties of Fick's second law.

7.1 Prediction of reinforcement corrosion

Chloride induced corrosion of reinforced steel is one of the major causes of deterioration of marine RC structures and de-iced infrastructures. When corrosion of reinforcing steel bars due to chloride attack occurs, the concrete will spall. However, this "warning" is often too late for an optimum repair and maintenance of the structure. The rehabilitation of RC structures due to corroding steel reinforcing bars is quite expensive compared with preventive maintenance. Therefore, a warning system is needed so that preventive maintenance can be chosen at the right time and repair expenses minimized.

There are several such macro cells on the market. Following Klinghoffer *et al.* (1999). we shall apply advanced mathematical modes of chloride ingress, such as the LIGHTCON model and the HETEK model for the interpretation of observations from installed macro cells or multi probes in marine RC structures.

7.1.1 Principle of a Corrosion Macro cell

The development of corrosion macro cells has been based on an extensive research programme on the main factors influencing chloride induced macro cell corrosion of steel in concrete. These investigations have been carried out by using macro cell current measurements between anodically and cathodically acting steel surface areas. It turned up that the corrosion rate of the reinforcement could easily be monitored continuously by these electric current measurements.

A macro cell consists of a piece of black steel (the anode) and a noble metal (the cathode). In chloride-free and non-carbonated concrete both electrodes are protected against corrosion due to the alkanity of the pore solution of the concrete (passive state). Due to this fact both metals have very similar electrochemical potentials, and the electric current that flows between the two electrodes is under these conditions negligible.

If, however, a critical chloride concentration is attained, or if the pH value of the concrete decreases due to carbonation, the steel surface of the anode (black steel, i.e. the same material as the reinforcement), is no longer protected against corrosion. If the cathode material is corrosion resistant in the chloride contaminated or carbonated concrete (if e.g. is made of titanium or stainless steel), and there are sufficient moisture and oxygen at hand, then an oxygen reduction takes place on the surface of the cathode.

The local separation of the anodically and cathodically acting areas leads to an electron flow between the black steel and the noble cathode, which easily can be measured by means of a low resistance ampere meter.

Now, integrate a number of macro cells to a multi probe. The multi probe is typically designed in such a way that it is possible to register the time of corrosion initiation at different depths in the concrete cover. These sensors are then embedded in sensitive areas of the concrete structure. This non-destructive testing enables the owner to take cost saving preventive protection measures, before the readings show a risk of corrosion of the reinforcement, provided that one can use a suitable mathematical model which can predict when such a corrosion probably starts.

7.1.2 Mathematical models of a corrosion multi probe

Since a mathematical model for chloride ingress into a marine RC structure was published by Collepardi *et al.* (1970), (1972), the mathematical model of chloride ingress into marine concrete has been developed in time with observations from field studies and by laboratory experiments. At present it is known that the chloride transport in concrete, free from gross cracks, can be modelled by diffusion, i.e. by Fick's laws of diffusion. The present knowledge is

- The achieved chloride diffusion coefficient is decreasing in time as shown by Takewaka *et al.* (1988). It is depending on the composition of the concrete and its environment as shown by Frederiksen *et al.* (1997).
- The chloride content of the near-to-surface layer of the exposed concrete surface is increasing in time as shown by Uji *et al.* (1990). It was shown by Frederiksen *et al.* (1997), that it also depends on the composition of the concrete and its environment.
- The chloride content of the near-to-surface layer of the exposed concrete surface in marine splash or submerged in seawater may be taken as a constant after approximately two years as shown by Maage *et al.* (1999).

These conditions may be taken into account by Fick's generalized laws of diffusion as shown by Mejlbro (1996).

On the basis of the given conditions above two mathematical models have been formulated, namely the LIGHTCON model by Maage *et al.* (1995), and the HETEK model by Frederiksen *et al.* (1997). The LIGHTCON model is applicable to concrete in marine splash or submerged in seawater. The HETEK model is applicable in marine splash, submerged in seawater or exposed to marine atmosphere. Section 7.1.3 on the LIGHTCON model is following Klinghoffer *et al.* (1999), while the HETEK model in Section 7.1.4 applied to a multi probe is here published for the first time.

7.1.3 The LIGHTCON model for the multi probe

The LIGHTCON model can be used for concrete in marine splash, or submerged in seawater. It is mathematically described by the following equation

$$C(x,t) = C_i + (C_0 - C_i) \cdot \operatorname{erfc}(u), \tag{7.1}$$

where $C(x,t)$ denotes the chloride content of concrete, C_0 is the chloride content of the near-to-surface layer of concrete, and C_i denotes the initial chloride content of concrete, all by mass binder. The argument u of the complementary error function is defined by

$$u = \frac{0.5x}{\sqrt{\tau t_{\mathrm{ex}} D_{aex}}}, \quad \text{where} \quad \tau = \left(\frac{t}{t_{\mathrm{ex}}} - 1\right) \cdot \left(\frac{t}{t_{\mathrm{ex}}}\right)^{-\alpha}, \tag{7.2}$$

and $t_{\text{ex}} D_{aex}$ is a constant depending on the composition and environment of the concrete and the time t_{ex} of its first chloride exposure.

We shall assume that the corrosion macro cell has $n \geq 4$ anodes. These anodes are placed at the distances

$$x_1, x_2, x_3, \ldots, x_n, \qquad \text{where } n \geq 4, \tag{7.3}$$

from the concrete surface exposed by chloride. The measurements of the macro cell are the time instants

$$t_1, t_2, t_3, \ldots, t_n, \qquad \text{where } n \geq 4, \tag{7.4}$$

when these anodes react, respectively, by the chloride ingress. Denote by C_{cr} the threshold value of chloride in concrete (by mass binder). When the chloride content of concrete at depth x_j of the j-th anode is $C \geq C_{cr}$, we assume that this particular anode reacts. This means that we obtain the following equations

$$C_{cr} = C_i + (C_0 - C_i) \cdot \operatorname{erfc}\left(\frac{0.5 x_j}{\sqrt{\tau_j t_{ex} D_{aex}}} \right) \qquad \text{for } j = 1, 2, 3, \ldots, n, \tag{7.5}$$

where

$$\tau_j = \left(\frac{t_j}{t_{ex}} - 1 \right) \cdot \left(\frac{t_j}{t_{ex}} \right)^{-\alpha} \qquad \text{for } j = 1, 2, 3, \ldots, n, \tag{7.6}$$

When (7.5) is solved with respect to τ_j we get

$$\tau_j = \left(\frac{x_j}{\xi_L \cdot \sqrt{t_{ex} D_{aex}}} \right)^2 \qquad \text{for } j = 1, 2, 3, \ldots \tag{7.7}$$

where

$$\xi_L = 2 \cdot \operatorname{inv\ erfc}\left(\frac{C_{cr} - C_i}{C_0 - C_i} \right). \tag{7.8}$$

Hence by (7.6),

$$\left(\frac{t_j}{t_{\text{ex}}} - 1 \right) \cdot \left(\frac{t_j}{t_{\text{ex}}} \right)^{-\alpha} = \left(\frac{x_j}{\xi_L \cdot \sqrt{t_{\text{ex}} D_{aex}}} \right)^2 \qquad \text{for } j = 1, 2, 3, \ldots, n. \tag{7.9}$$

Assuming that the observations of the reaction time instants t_j versus the anode location x_j are given, we shall determine the unknown parameters α and $\xi_L \cdot \sqrt{t_{ex} D_{aex}}$ by means of regression, i.e. by the method of least squares (the *Gauss-Markov* theorem).

Since the anodes can be placed with great accuracy the uncertainties of the independent variables x_j are negligible. Thus, the method of least squares is carried out in such a way that all uncertainties are found in the time of reaction of the anodes. If therefore the observed reaction times of the anodes are

$$t_{01}, t_{02}, t_{03}, \ldots, t_{0j}, \ldots, t_{0n}, \qquad \text{where } j = 1, 2, 3, \ldots, n \text{ and } n \geq 4,$$

while the estimated reaction times of the anodes are

$$t_{e1}, t_{e2}, t_{e3}, \ldots, t_{ej}, \ldots, t_{en}, \qquad \text{where } k = 1, 2, 3, \ldots, n \text{ and } n \geq 4,$$

according to the LIGHTCON model, then the unknown parameters are determined by solving (7.9) simultaneously by minimizing the least-squares estimates

$$L = \sum_{j=1}^{n} (t_{ej} - t_{0j})^2 = \text{ minimum}, \qquad \text{where } n \geq 4. \tag{7.10}$$

By means of the spreadsheet Excel Version 5.0 or later versions a non-linear regression can be carried out on (7.9), i.e. we determine the unknown parameters α and $\xi_L \cdot \sqrt{t_{ex} D_{aex}}$ by the method of least squares corresponding to the following sets of observations,

$$(x_1, t_1), (x_2, t_2), \ldots, (x_n, t_n), \qquad \text{for } j = 1, 2, \ldots, n \text{ and } n \geq 4.$$

It is here convenient to rewrite (7.9) in the form

$$(Y_j - 1) \cdot Y_j^{-\alpha} = \beta \cdot X_j^2, \tag{7.11}$$

$$\text{where } Y_j = \frac{t_j}{t_{\text{ex}}} \text{ and } X_j = \frac{x_j}{c}, \quad j = 1, \ldots, n, \text{ and } \beta = \left(\frac{c}{\xi \sqrt{t_{\text{ex}} D_{aex}}} \right)^2.$$

Here, c denotes the cover of the reinforcing bar. When β is determined we finally get the unknown constant

$$\xi_L \cdot \sqrt{t_{\text{ex}} D_{aex}} = \frac{c}{\sqrt{\beta}}. \tag{7.12}$$

The solution by Excel spreadsheet requires a certain knowledge of the solution, either an approximate solution or certain limits. Information of this kind can be achieved by finding the (approximate) solution determined by means of the observations from two arbitrarily chosen anodes. It is recommended to use the n-th anode at location X_n and any other anode (of number m, say) at location X_m ($\geq 0.5 X_n$). Using only the two equations for $j = n$ and m from (7.11) we get

$$(Y_n - 1) \cdot Y_n^{-\alpha} = \beta \cdot X_n^2 \quad \text{and} \quad (Y_m - 1) \cdot Y_m^{-\alpha} = \beta \cdot X_m^2.$$

These two equations are easily solved with respect to α and β, giving the following approximate solutions of the system (7.11) with the minimum constraint (7.10),

$$\alpha_m = \frac{\ln\left((Y_n - 1)/(Y_m - 1)\right) - 2 \cdot \ln(X_n / X_m)}{\ln(Y_n / Y_m)}, \tag{7.13}$$

and

$$\beta_m = \frac{Y_n - 1}{Y_n^{\alpha} X_n^2} = \frac{Y_m - 1}{Y_m^{\alpha} X_m^2}, \qquad \text{where } \alpha = \alpha_m. \tag{7.14}$$

These values of α_m and β_m perform in most cases a reasonable first approximation of the roots for each X_m, $m = 1, 2, \ldots, m-1$.

Proceed by solving the equations

$$Y_{m,p}^{1-\alpha_m} - Y_{m,p}^{-\alpha_m} - \beta_m X_p^2 = 0, \qquad m = 1, 2, \ldots, m-1, \quad p = 1, 2, \ldots, n, \; x = c, \tag{7.15}$$

and calculate

$$L = \sum_{p=1}^{n} (Y_{m,p} - Y_p)^2 \qquad \text{for } m = 1, 2, 3, \ldots, n-1. \tag{7.16}$$

Finally, choose m such that (7.16) is minimum (the least-square estimates). Then α_m and β_m are good approximations of the parameters α and β. Thus an estimate of the initiation period of time is $t_{cr} = Y_{m,cr} \cdot t_{ex}$, where $Y_{m,cr}$ is the solution of (7.15) and (7.16) for $x = c$. Anyone of these estimates (α_m, β_m) may be used as initial value for a solution by Excel spreadsheet.

It is possible to use a subtle maximum likelihood method instead. As the derivation is fairly long, the description of this alternative method is here omitted.

When the chloride content of the concrete in contact with a reinforcing bar reaches the threshold value of chloride in the concrete, then the reinforcing bar starts to corrode. The time of initiation may be determined from the general Equation (7.11), or (7.9), by putting $x_j = c$, i.e. $X_j = 1$. This means that the initiation time t_{cr} of the reinforcing bar is a solution of the following equation

$$\left(\frac{t_{cr}}{t_{ex}} - 1\right) \cdot \left(\frac{t_{cr}}{t_{\text{ex}}}\right)^{-\alpha} = \beta, \quad \text{i.e.} \quad \left(\frac{t_{cr}}{t_{\text{ex}}}\right)^{1-\alpha} - \left(\frac{t_{cr}}{t_{\text{ex}}}\right)^{-\alpha} - \beta = 0. \tag{7.17}$$

This equation is solved by means of the *Newton-Raphson* method. Applied to this particular problem the *Newton-Raphson* method is reduced to the following iterative sequence,

$$Y_{n+1} = Y_n \cdot \left(1 + \frac{\beta Y_n^{\alpha} - Y_n + 1}{(1-\alpha) Y_n + \alpha}\right). \tag{7.18}$$

Remark 7.1.1 In some cases it is convenient first to make the following estimates of the unknown parameters. If we assume that $t_1 \gg t_{\text{ex}}$, then (7.9) is reduced to the following approximate formula,

$$\left(\frac{t_j}{t_{\text{ex}}}\right)^{1-\alpha} \approx \left(\frac{x_j}{\xi_L \cdot \sqrt{t_{\text{ex}} D_{aex}}}\right)^2,$$

hence

$$\ln\left(\frac{t_j}{t_{\text{ex}}}\right) \approx \frac{2}{1-\alpha}\ln x_j + \frac{2}{1-\alpha}\ln\left(\frac{1}{\xi \cdot \sqrt{t_{\text{ex}} D_{aex}}}\right).$$

Plotting the observations (x_j, t_j) into a diagram with log scales on both axes, these will approximately lie on a straight line. Draw this straight line on the diagram and find the critical time t_{cr} at depth c by extrapolation.

Since the error from the straight line is largest for t_1, where t_j/t_{ex} is smallest, one may expect (x_1, t_1) to lie above the line in the diagram due to the additional term in (7.9).

For a rough estimate of the critical time t_{cr} this method is often sufficient. Concerning the other unknowns α and ξ_L one gets less reliable estimates from the diagram. If these are wanted one should instead use the full method, cf. also Example 7.1.1 below. ◊

The theory presented above is based on various assumptions like e.g.

- The chloride diffusion coefficient of the concrete is independent of the locations x_j of the anodes.
- The chloride content of the exposed concrete surface is constant.
- The threshold value of chloride C_{cr} in concrete in contact with an anode is independent of the location x_j of the anode.

Only laboratory experiments or field tests can answer whether these assumptions are permissible or not. Such experiments were taking place while this book was written, but it is still too early on the basis of these experiments to conclude whether these simple assumptions are reasonable or not. If not, we are in a situation where Fick's second law cannot be applied, so the mathematical model should be changed. For the time being the LIGHTCON model and the HETEK model in the next section, are the only models available which can predict the critical time t_{cr}.

Example 7.1.1 For the purpose of illustration we show by an example how the calculations are carried out in practice. A corrosion macro cell with $n = 4$ anodes is cast into concrete exposed to marine splash. We have chosen the multi probe CorroWatch developed by the FORCE Institute, Denmark, so the positions x_j are known, i.e. can be measured from the existing device. The experiments are not finished yet, so we have taken the liberty of estimating the reaction times t_3 and t_4 which have not yet occurred. Thus we shall *assume* that we have observed the reaction times given in Table 7.1.

The X and Y columns are based on a concrete cover of the reinforcing bar of $c = 50$ mm, so $X_j = x_j/c$, and on the assumption that the concrete was exposed to chloride for the first time at $t_{ex} = 14.5$ days ≈ 0.04 years, so $Y_j = t_j/t_{ex}$. Notice that X and Y have no physical dimension. We assume

Table 7.1: Observations from a CorroWatch Multi probe.

Anode no.	Position x mm	Reaction time t yr.	X	Y
1	7	0.53	0.14	13.25
2	13	1.89	0.26	47.25
3	19	4.42	0.38	110.50
4	26	8.81	0.52	220.25

furthermore that there is no special surface treatment of the concrete (e.g. coat, paint or membrane).

Assuming that the LIGHTCON model may be applied one can prove the following *compatibility condition*

$$Y_{n+1} \geq Y_n \cdot \left(\frac{X_{n+1}}{X_n}\right)^2. \tag{7.19}$$

In the present case (7.19) gives the following rough estimate

$$Y_{cr} \geq Y_y \cdot \left(\frac{X_5}{X_4}\right)^2 = 220.25 \cdot \left(\frac{50}{26}\right)^2 \approx 814, \tag{7.20}$$

where we have added $x_5 = c = 50$ mm and $t_5 = t_{cr}$ yr. This gives a lower bound for the initiation period of time

$$t_{cr} = t_{ex} \cdot Y_{cr} \geq 32.5 \text{ years.} \tag{7.21}$$

We shall only give all the details of the calculations when $m = 2$ and $n = 4$. All the other possibilities are treated similarly. First we calculate α_2,

$$\begin{aligned} \alpha_2 &= \frac{\ln\left((Y_4 - 1)/(Y_2 - 1)\right) - 2\ln(X_4/X_2)}{\ln(Y_4/Y_2)} \\ &= \frac{\ln(219.25/46.25) - 2\ln(0.52/0.26)}{\ln(220.25/47.25)} = 0.11. \end{aligned}$$

Using this value of α we obtain

$$\beta_2 = \frac{Y_4 - 1}{Y_4^\alpha X_4^2} = \frac{219.25}{220.25^{0.11} \cdot 0.52^2} = 448,$$

and as a check (β_2 calculated by the other formula)

$$\beta_2 = \frac{Y_2 - 1}{Y_2^\alpha X_2^2} = \frac{46.25}{47.25^{0.11} \cdot 0.26^2} = 448,$$

i.e. the same value.

Since $X_5 = x_5/c = c/c = 1$ we get for $m = 2$ and $n = 4$ the following equations, cf. (7.15),

$$Y_{2,j}^{0.89} - Y_{2,j}^{-0.11} - 448 \cdot X_j^2 = 0, \qquad \text{where } j = 1, 2, 3, 4, 5, \tag{7.22}$$

from which we get the estimated initiation period of time $t_{cr} = 0.04 \cdot Y_{2,5}$.

In order to obtain the least-squares estimates we set up Table 7.2.

From Table 7.2 we get the least-squares estimate

$$L_{2,4} = \sum_{j=1}^{4} (Y_{2,j} - Y_j)^2 = 0.646^2 + 1.005^2 + 1.070^2 + 0.137^2 = 2.591,$$

and the estimate of the initiation period of time is for $m = 2$ and $n = 4$

$$t_{cr} = 0.04 \cdot 953.841 = 37.2 \text{ years.}$$

Repeating this process for all $m = 1, 2, 3$ and $n = 4$ (details omitted) we derive Table 7.3.

We conclude from these first estimates that $t_{cr} \geq 37.3$ yr.

A better estimate is obtained by a non-linear regression analysis using Excel spreadsheet. To start the Excel programme we choose the first approximations $\alpha_2 = 0.11$ and $\beta_2 = 448$ as above. By means of the Excel spreadsheet the following values are found:

$$\alpha = 0.1006 \qquad \text{and} \qquad \beta = 471.$$

Thus we get the following equations for the estimate of the initiation period of time of the reinforcing bar,

$$Y^{0.8994} - Y^{-0.1006} - 471 = 0 \quad \text{where} \quad t_{cr} = t_{ex} \cdot Y = 0.04 \cdot Y.$$

By the *Newton-Raphson* iteration, cf. (7.18) we obtain $Y = 938.47$, hence $t_{cr} = 0.04 \cdot 938.47 = 37.5$ yr.

Notice that all the estimates of t_{cr} lie above 37 yr, and well above the necessary lower bound 32.5 yr given by (7.21). Notice also that the estimates are rather robust, meaning that the estimate found by the simple method of only considering $m = 2$ and $n = 4$ differ very little from the estimate found by the more subtle non-linear regression analysis on the Excel spreadsheet. We shall in the next section see that this nice result is even more striking when we consider the more general HETEK model. $\diamond$

Table 7.2: Determination of least-squares estimates for $m = 2$ and $n = 4$.

X_j	0.140	0.260	0.380	0.520	1.000
$\beta_2 X_j^2$	8.781	30.285	64.691	121.139	448
$Y_{2,j}$	12.604	48.225	109.430	220.287	953.841
Y_j	13.250	47.250	110.500	220.250	–
$\lvert Y_{2,j} - Y_j \rvert$	0.646	1.005	1.070	0.137	–

Table 7.3: Calculation of least-squares and initiation time.

Anode m Anode n	1 4	2 4	3 4
α_m	0.093	0.11	0.097
β_m	491	448	480
$L_{m,n}$	2.040	2.591	1.22
t_{cr}	37 yr	37.2 yr	37.3 yr

7.1.4 The HETEK model for the multi probe

We shall consider Fick's second law with the diffusion coefficient $D(t) > 0$ varying in time,

$$\frac{\partial C}{\partial t} = D(t)\frac{\partial^2 C}{\partial x^2}, \qquad x > 0, \quad t > t_{\text{ex}}, \tag{7.23}$$

in a half-infinite isotropic media. By introducing the monotone change of variables

$$T = T(t) = \int_{t_{ex}}^{t} F(\tau)\,d\tau = (t - t_{ex}) \cdot \left\{\frac{1}{t - t_{ex}}\int_{t_{ex}}^{t} D(\tau)\,d\tau\right\}, \tag{7.24}$$

Equation (7.23) is transformed into

$$\frac{\partial C}{\partial T} = \frac{\partial^2 C}{\partial x^2}.$$

Thus, by (7.24) we obtain that the diffusion coefficient $D(t)$ disappears from the problem, which makes the problem easier to solve from a mathematical point of view. There is also a gain from a technical point of view. Pointwise given functions like $D(t)$ are in practice difficult to measure. What can be measured is the *average*

$$D_a(t) = \frac{1}{t - t_{ex}}\int_{t_{ex}}^{t} D(\tau)\,d\tau, \tag{7.25}$$

which occurs as a very convenient factor in (7.24). However, although it is easier to measure (7.25) than $D(t)$ directly, it still requires a huge number of observations. Therefore, Takewaka *et al.* (1988), suggested the following model for $D_a(t)$,

$$D_a(t) = D_{aex}\left(\frac{t_{ex}}{t}\right)^{\alpha}, \tag{7.26}$$

with a parameter $\alpha \geq 0$. this assumption was then used in the HETEK model by Frederiksen *et al.* (1997), where another parameter $p \geq 0$ was introduced,

$$\frac{\partial C}{\partial t} = (t - t_{\text{ex}})D_a(t)\frac{\partial^2 C}{\partial x^2}, \quad x > 0,\, t > t_{\text{ex}},$$

$$C(x, t_{\text{ex}}) = C_i, \qquad x \geq 0, \tag{7.27}$$
$$C(0,t) = C_i + S_p \cdot \{(t - t_{ex})D_a(t)\}^p, \quad t > t_{\text{ex}}.$$

By varying both α and p one could describe other problems than the classical ones, where only the complementary error function solution is applied. The problem (7.27) had been solved earlier by Mejlbro (1996), cf. also Section 4.3.1, where the diffusion coefficient for simplicity is put equal to a constant. The solution of (7.27) is given by

$$C_p(x,t) = C_i + S_p \cdot \{(t - t_{ex})D_a(t)\}^p \cdot \Psi_p\left(\frac{x}{\sqrt{4(t - t_{ex})D_a(t)}}\right), \tag{7.28}$$

cf. e.g. Mejlbro (1996).

Notice that if $p = 0$ in (7.27), then (7.28) is reduced to the well-known solution

$$C_0(x,t) = C_i + S_0 \cdot \operatorname{erfc}\left(\frac{x}{\sqrt{t(t - t_{\text{ex}})D_a(t)}}\right),$$

on which the LIGHTCON model is based, cf. Section 7.1.3.

Since the HETEK model contains more parameters than the LIGHTCON model, and even includes it as a special case, we can expect better results by also letting p vary. On the other hand we may also expect more difficult calculations. Furthermore, we discovered that both models contain a singularity when applied to the multi probe. In the case of LIGHTCON this is deeply hidden, because statistical methods are also involved in Section 7.1.3. However, there are three unknowns in the LIGHTCON model, and the linear approximation of Remark 7.1.1 can of course only specify two parameters.

The nature of the singularity of the problem is more easily described in the general HETEK model. Therefore we have postponed this discussion to the present section.

It will be convenient to let $t = 0$ denote the time of first exposure, hence the concrete was cast at time $-t_{ex}$. We assume that $D_a(t)$ is given by (7.26) according to Takewaka *et al.* (1988). Let C_{cr} denote the critical concentration of chloride, measured by the multi probe. This means that the observations

$$(x_1, t_1), \quad (x_2, t_2), \quad (x_3, t_3), \quad (x_4, t_4)$$

satisfy the equations

$$C_p(x_i, t_i) = C_{cr}, \qquad i = 1, 2, 3, 4.$$

When $C_p(x,t)$ is given by (7.28), i.e. when we assume the model (7.27), these equations can be written

$$\left(\frac{t_{\text{ex}}}{t_{\text{ex}}+t_i}\right)^{\alpha p}\left(\frac{t_i}{t_{\text{ex}}}\right)^p \Psi_p\left(\frac{x_i}{2\sqrt{t_i}}\left\{\frac{t_{\text{ex}}+t_i}{t_{\text{ex}}}\right\}^{\frac{\alpha}{2}}\cdot\frac{1}{\sqrt{D_{aex}}}\right)$$

$$=\frac{C_{cr}-C_i}{S_p}, \qquad i=1,2,3,4. \tag{7.29}$$

The structure of (7.29) shows that the problem is more easily described, if we introduce

$$\xi=\frac{1}{\sqrt{D_{aex}}}, \qquad \text{and} \qquad \tau_i=\frac{t_i}{t_{ex}}, \qquad i=1,2,3,4.$$

Then (7.29) is transformed into

$$\left\{\frac{\tau_i}{(1+\tau_i)^\alpha}\right\}^p \Psi_p\left(\frac{x_i\xi}{2\sqrt{t_{\text{ex}}}}\sqrt{\frac{(1+\tau_i)^\alpha}{\tau_i}}\right)=\frac{C_{cr}-C_i}{S_p}, \qquad i=1,2,3,4. \tag{7.30}$$

Here, t_{ex}, x_i and τ_i are known, because they have been measured. Hence, (7.30) is a nonlinear system of four equations in the four unknowns

$$p, \quad \alpha, \quad \xi \quad \text{and} \quad \frac{C_{cr}-C_i}{S_p}. \tag{7.31}$$

Once the values of (7.31) are determined, we can find t_{cr} at depth x_{cr} by putting $x_i = x_{cr}$ and $t_i = t_{cr}$ in (7.29) and then solve the equation with respect to t_{cr}.

Using that $\Psi_p(u)$ is decreasing in u, cf. Theorem 3.3.4, it is not hard to prove from (7.30) that the following important *compatibility conditions* must hold,

$$\left(\frac{x_i}{x_{i+1}}\right)^2\frac{\tau_{i+1}}{\tau_i}<\left\{\frac{1+\tau_{i+1}}{1+\tau_i}\right\}^\alpha<\frac{\tau_{i+1}}{\tau_i}, \qquad i=1,2,3. \tag{7.32}$$

Since x_i and $\tau_i = t_i/t_{ex}$ are known, we get the following bounds for α,

$$\max_i \frac{\ln(t_{i+1}/t_i)-2\ln(x_{i+1}/x_i)}{\ln(t_{\text{ex}}+t_{i+1})-\ln(t_{\text{ex}}+t_i)}<\alpha$$

$$<\min_i \frac{\ln(t_{i+1}/t_i)}{\ln(t_{\text{ex}}+t_{i+1})-\ln(t_{\text{ex}}+t_i)}. \tag{7.33}$$

In this way we obtain limits $\alpha_1 < \alpha < \alpha_2$ of α. In the chosen examples the difference $\alpha_2 - \alpha_1\ (\geq 0)$ is small. If $\alpha_2 - \alpha_1 < 0$, then either (at least) one of the measurements is wrong, or the model (7.27) does not apply. Therefore, (7.32) and (7.33) are called compatibility conditions.

Let us then try to solve (7.30). First eliminate the right hand side by subtracting the equations from each other. Using $i = 1$ for reference we introduce for $i = 2, 3, 4$,

$$\begin{aligned}\Phi_{i-1}(p,\alpha,\xi) &= \left\{\frac{\tau_i}{(1+\tau_i)^\alpha}\right\}^p \Psi_p\left(\frac{x_i\xi}{2\sqrt{t_{\text{ex}}}}\sqrt{\frac{(1+\tau_i)^\alpha}{\tau_i}}\right) \\ &\quad -\left\{\frac{\tau_1}{(1+\tau_1)^\alpha}\right\}^p \Psi_p\left(\frac{x_1\xi}{2\sqrt{t_{ex}}}\right). \qquad (7.34)\end{aligned}$$

If $(p, \alpha, \xi, (C_{cr} - C_i)/S_p)$ is a solution of (7.30), then we must have

$$(\Phi_1(p,\alpha,\xi), \Phi_2(p,\alpha,\xi), \Phi_3(p,\alpha,\xi)) = (0,0,0). \qquad (7.35)$$

Now, introduce

$$F(p,\alpha,\xi) = \Phi_1(p,\alpha,\xi)^2 + \Phi_2(p,\alpha,\xi)^2 + \varphi_3(p,\alpha,\xi)^3 \geq 0. \qquad (7.36)$$

Then (7.35) is obviously *equivalent to*

$$F(p,\alpha,\xi) = 0, \qquad (7.37)$$

because every addend in $F(p, \alpha, \xi)$ is ≥ 0.

Conversely, if (p, α, ξ) is a solution of (7.37), then all $\Phi_i(p, \alpha, \xi) = 0$, hence the left hand side of (7.30) is a constant, namely

$$(C_{cr} - C_i)/S_p.$$

Thus, the problem of solving the system (7.30) of four nonlinear equations is *equivalent to* the solution of only one Equation (7.37) in three variables. By comparing the dimension 3 with the number of constraints we see that we in general may expect a set of solutions of dimension $3 - 1 = 2$. The conclusion is that the solutions of (7.30) in the four-dimensional space in general form a two-dimensional submanifold. This is what we mean by claiming that the problem in principle is singular.

Now, if the compatibility conditions (7.33) give very narrow bounds for α, then we can fix α as well, so the set of solutions is reduced to a curve in the HETEK model. This was implicitly used in the LIGHTCON model, where statistical methods were used to specify α. Since $p = 0$ also is implicitly chosen in the LIGHTCON model, we get the right number of constraints, hence usually a unique solution of the problem. In general, both assumptions are questionable in the LIGHTCON model, were it not for the fact that the wanted critical time t_{cr} varies very little in α. Some explanation of this strange phenomenon is best given by solving the HETEK problem for the multi probe. Notice that the HETEK model contains the LIGHTCON model as a special case.

The solution of the exact Equation (7.37) requires a lot of computer time, and since there are rounding off errors, we may instead use the following faster

Table 7.4: Fictive data from a CorroWatch

Anode no.	x_i mm	t_i yr
1	7	0.244
2	13	0.786
3	19	1.639
4	26	3.035

method. Choose an upper limit $\varepsilon > 0$ for the error and accept (p, α, ξ) as an approximate solution, if

$$F(p, \alpha, \xi) < \frac{1}{4}\varepsilon^2. \tag{7.38}$$

When (7.38) is satisfied, the four numbers

$$\left\{\frac{\tau_i}{(1+\tau_i)^\alpha}\right\}^p \Psi_p\left(\frac{x_i\xi}{2\sqrt{t_{ex}}}\sqrt{\frac{(1+\tau_i)^\alpha}{\tau_i}}\right), \qquad i = 1, 2, 3, 4,$$

will differ from each other with less than ε.

Fix any $\xi \in [40^{-1}, 10^{-1}]$, corresponding to $1600 \geq D_{aex} \geq 100$, so all cases in practice are covered. Then let $\alpha \in [\alpha_1, \alpha_2]$ vary in the compatibility interval found by (7.33). Since $\alpha_2 - \alpha_1$ in general is small, only a small number of different values of α is necessary. For any chosen (α, ξ) minimize $F(p, \alpha, \xi)$ in p. If the requirement (7.38) is met, (p, α, ξ) is an acceptable approximation. Still for fixed ξ, minimize $F(p, \alpha, \xi)$ among the acceptable approximations. This gives one value of α and an upper and a lower bound for p, thus a small p-interval. Calculate $(C_{cr} - C_i)/S_p$ by (7.29), and calculate finally t_{cr} by solving (7.29), when $x_i = x_{cr}$ and $t_i = t_{cr}$. Repeat this procedure for another choice of ξ, etc.

At a first glance this procedure may seem to be very complicated, but once the functions $\Psi_p(z)$ and $F(p, \alpha, \xi)$ have been programmed it works pretty fast. We shall demonstrate this by the following example.

Example 7.1.2 Since multi probes like e.g. *CorroWatch* have not existed long enough to produce real observations, one of us constructed some fictive data given by Table 7.4.

We assume that $t_{\text{ex}} = 0.5$ year. The critical time at $c = x_{cr} = 50$ mm was calculated, $t_{cr} = 11.19$ yr, but not mentioned to the other one of us, who should try to find t_{cr} using the theory above and a pocket calculator (TI-92). Since the other parameters are not uniquely determined – cf. the previous discussion – they were soon forgotten and cannot be reproduced.

Using the method described above, Table 7.5 was set up for $\varepsilon = 10^{-3}$, where only the minimizing α is given. Actually, the table was calculated for

$.025 \leq \xi \leq .100$, but the missing entries of t_{cr} are just a repetition of the upper and lower bounds 11.20 and 11.10 yr, so they do not contribute anything new, and they were left out.

The conclusion of these calculations is that $11.05 < t_{cr} < 11.25$, and more likely, $11.10 < t_{cr} < 11.20$. This is very satisfactory, since the "measurements" were constructed from $t_{cr} = 11.19$.

Notice also that $0.081 \leq \alpha \leq 0.084$, so α does not change much either, while p and $(C_{cr} - C_i)/S_p$ are very sensible of changes of ξ, and more moderate in changes of D_{aex}.

To convince ourselves of the efficacy of this method it was applied on two other examples as well. In both cases t_{cr} was not given in advance, and in both cases t_{cr} was found with a very small error. We shall not give the tables here, because they do not add any further information, and the data is not based on measurements. ◊

Table 7.5: Estimates of t_{cr}, given the data of Table 7.4.

ξ	D_{aex}	α	p	$(C_{cr} - C_i)/S_p$	t_{cr} yr
.025	1600	.084	.012742	.788584	11.25
			.013544	.787939	11.05
.030	1111	.083	.015868	.747048	11.20
			.016718	.746367	11.05
.035	816	.083	.019317	.705905	11.20
			.020218	.705191	11.05
.040	625	.083	.022993	.665404	11.20
			.023951	.664657	11.05
.045	494	.082	.026691	.625897	11.20
			.027711	.625119	11.05
.050	400	.082	.030801	.587065	11.20
			.031891	.586254	11.10
.055	331	.082	.035148	.549218	11.20
			.036316	.548376	11.10
.060	278	.081	.039423	.512763	11.20
			.040675	.511892	11.10
.065	237	.081	.044215	.477204	11.20
			.045563	.476302	11.10
.070	204	.081	.049251	.442903	11.20
			.050707	.441971	11.10
.075	178	.081	.054534	.409933	11.20
			.056109	.408972	11.10

7.2 Coating

Chloride penetration into a coated medium of concrete is also an important subject. So far, the only known solution deals with a semi-infinite medium with a surface layer, which has a different (constant) diffusion coefficient from the (constant) diffusion coefficient of the concrete itself. This was briefly indicated by Crank (1975), who refers back to Carslaw *et al.* (1947), concerning the derivation. We shall here without proofs give the essentials of this model in Section 7.2.1.

The results of Section 6.2.1 show that one in general still does not know any exact solution of chloride penetration into a column of rectangular shape. However, if the column is cylindrical with a circle as its base, we can generalize Section 6.2.2 to coating of this cylindrical column: Just follow the same procedure as in the continuous case. In practice the calculations become very complicated, so a PC is necessary. They are therefore left to the reader.

It should be noticed that there is a good reason why we do not apply the Finite Element Method here. From a historical point of view, the Eigenvalue Method was introduced in the beginning of the previous century, but since one only rarely was able to find the eigenfunctions, the Finite Element Method was invented as an approximation. With the methods developed here we are again able to construct the eigenfunctions explicitly, so there is no need at all to use a less accurate method, which at the same time takes longer time. But when the eigenfunctions cannot be found, one should of course try the Finite Element Method instead.

7.2.1 The semi-infinite composite medium

The problem considered by Crank (1975), is in our terminology described in the following way,

$$\begin{cases} \dfrac{\partial C}{\partial t} = \dfrac{\partial}{\partial x}\left\{D(x)\dfrac{\partial C}{\partial x}\right\}, & -\ell < x < \infty,\, t > 0 \\ C(-\ell, t) = C_0, & t > 0, \\ C(x, 0) = 0, & -\ell < x < \infty, \end{cases} \tag{7.39}$$

where

$$D(x) = \begin{cases} D_1 > 0 & \text{for } -\ell < x < 0, \\ D_2 > 0 & \text{for } 0 < x < \infty. \end{cases} \tag{7.40}$$

For convenience, the coating has been represented by the interval $]-\ell, 0[$ on the negative axis. This is done to simplify the solution formulæ (7.44) and (7.45) in the following.

Since $D(x)$ is given by (7.40) is discontinuous at $x = 0$, Crank (1975), was forced to assume the following conditions at the interface between the two media (i.e. at $x = 0$),

$$\lim_{x \to 0-} C(x,t) = \lim_{x \to 0+} C(x,t), \tag{7.41}$$

$$\lim_{x \to 0-} D_1 \frac{\partial C}{\partial x}(x,t) = \lim_{x \to 0+} D_2 \frac{\partial C}{\partial x}(x,t). \tag{7.42}$$

They express that both the solution $C(x,t)$ and $D(x)\dfrac{\partial C}{\partial x}$ are continuous at $x = 0$.

Putting

$$k = \sqrt{\frac{D_1}{D_2}} \quad \text{and} \quad \alpha = \frac{1-k}{1+k}, \tag{7.43}$$

Crank (1975), indicates the following solution formulæ of (7.39) and (7.40),

$$C(x,t) = C_0 \sum_{n=0}^{\infty} \alpha^n \left\{ \operatorname{erfc} \frac{(2n+1)\ell + x}{2\sqrt{D_1 t}} - \alpha \cdot \operatorname{erfc} \frac{(2n+1)\ell - x}{2\sqrt{D_1 t}} \right\}$$

$$\text{for } -\ell < x < 0, \tag{7.44}$$

and

$$C(x,t) = \frac{2kC_0}{k+1} \sum_{n=0}^{\infty} \alpha^n \operatorname{erfc} \frac{(2n+1)\ell + kx}{2\sqrt{D_1 t}}$$

$$\text{for } 0 < x < \infty. \tag{7.45}$$

Concerning the proofs of (7.44) and (7.45) the reader is referred to Carslaw *et al.* (1947).

Although the formulæ are straightforward, one may be a little disturbed by the interface conditions (7.41) and (7.42). However, they do not harm, if one keeps strictly to (7.44) and (7.45).

7.2.2 The finite composite medium

In many cases Crank's model gives a quite satisfactory first approximation of the chloride profiles, in particular because very few terms are needed from the series (7.45) in order to obtain a reasonable numerical accuracy. On the other hand, it only handles media consisting of two different constituents, one of which is semi-infinite. In the following we shall develop a theory for finite media consisting of n different constituents.

The right way seems first to consider the finite 2-composite medium in order to get the right ideas. Then turn to the finite 3-composite medium, from which it is in principle easy to generalize to n-composite media. The theory is applicable to all finite composite media, but we shall mainly consider coating, the problem of which actually caused this investigation.

Coating is characterized by a thin layer with a very small diffusion coefficient compared with the rest of the medium. Furthermore, the coating usually has a different density from the concrete, and the porosities of the two parts are also different, so it seems difficult to unite the theory of chloride ingress into one model. On the other hand, it is also difficult to give up Fick's second law, on which this book is based.

Since we have a non-Fickean penetration of chloride, when the surface of the concrete is carbonated, we can at least say that this case is not covered by the theory presented here. Though it is also interesting, we are here forced to leave it for further study, in particular because even the Fickean model may cause some difficulties, until they are fully understood.

The problem of the different densities and porosities of two neighbouring layers is in practice solved by replacing them with equivalent layers of the same density and porosity, but with transformed thickness and diffusion coefficients, so we can use Fick's second law. For coating we here meet another practical problem. How does one measure the thickness of the (thin) coating? One way could be to strip the coated material for its paint over an area A, find its weight and divide it by A times the density. We assume here that the paint is evenly distributed. However, the weight of the stripped paint may still be small giving a high uncertainty of the measurement.

We shall not here go further into the practical engineering problems in finding the parameters needed. We assume in the following that we can use (the 1-dimensional) Fick's second law as a model, where the diffusion coefficient $D(x)$ is piecewise constant. We shall also in this initial state of the art assume that the boundary conditions are constant, while the initial condition is zero. This gives the model in the variables (X, T),

$$\begin{cases} \dfrac{\partial \tilde{C}}{\partial T} = \dfrac{\partial}{\partial X}\left\{\tilde{D}(X)\dfrac{\partial \tilde{C}}{\partial X}\right\}, & 0 < X < \ell, \\ \tilde{C}(0,T) = C_0 \quad \text{and} \quad \tilde{C}(\ell,T) = C_\ell, & T > 0, \\ \tilde{C}(X,0) = 0, & 0 < X < a, \end{cases} \tag{7.46}$$

where $\tilde{D}(X) > 0$ is defined by constants $D_j > 0$, i.e.

$$\begin{cases} \tilde{D}(X) = D_j \quad \text{for } X \in]\ell_j, \ell_{j+1}[, \quad j = 1, \cdots, n, \\ 0 = \ell_1 < \ell_2 < \cdots < \ell_{n+1} = \ell. \end{cases} \tag{7.47}$$

The main ideas are given in the case of a 2-composite medium, where (7.47) is reduced to

$$\tilde{D}(X) = \begin{cases} D_1 > 0 & \text{for } 0 < X < \ell_1, \\ \\ D_2 > 0 & \text{for } \ell_1 < X < \ell_2. \end{cases} \tag{7.48}$$

Once the problem has been solved for (7.48) it is not difficult to extend the methods to 3-composite media, from which we get the pattern of how to solve the problem for the general diffusion coefficient given by (7.47). We shall therefore start by assuming (7.47).

The first step is to transform the problem (7.46) into a physically dimensionless standard problem, where $\tilde{D}(X)$ is given by (7.48). If we introduce the dimensionless variables (x, t) and constants a and c by

$$X = \ell \cdot x, \quad T = \frac{\ell^2}{D_1} t, \quad \ell_1 = \ell \cdot a, \quad c = \sqrt{\frac{D_1}{D_0}}, \tag{7.49}$$

where $0 \le a \le 1$ and $c > 0$, we see that (7.46) is transformed into

$$\begin{cases} \dfrac{\partial C}{\partial t} = \dfrac{\partial}{\partial x}\left\{D(x)\,\dfrac{\partial C}{\partial x}\right\}, & 0 < x < 1, \quad t > 0, \\ \\ C(0,t) = C_0 \text{ and } C(1,t) = C_\ell, & t > 0 \\ \\ C(x,0) = 0, & 0 < x < 1, \end{cases} \tag{7.50}$$

where

$$D(x) = \begin{cases} \dfrac{1}{c^2} & \text{for } 0 < x < a, \\ \\ 1 & \text{for } a < x < 1, \end{cases} \qquad 0 \le a \le 1, \tag{7.51}$$

and

$$C(x,t) = \tilde{C}(X,T).$$

Once $C(x,t)$ has been found, we use (7.49) to get the solution of (7.46),

$$\tilde{C}(X,T) = C(x,t) = C\left(\frac{X}{\ell}, \frac{D_1}{\ell^2} T\right). \tag{7.52}$$

It therefore suffices in the following to solve the standard problem (7.50), where $D(x)$ is given by (7.51).

We are usually interested in $\tilde{C}(X,T)$ for $T \ge T_0 > 0$, so (7.52) gives that the solution $C(x,t)$ should be reliable for

$$t \ge t_0 = \frac{D_1}{\ell^2} T_0. \tag{7.53}$$

The definition (7.53) of $t_0 > 0$ will later give a clue to the error estimates of the error of the truncated solutions.

In the further development we have here come to a crossroad. We might use the method of successive integration, and in principle it should work. In practice the discontinuity of $D(x)$ given by (7.51) causes severe numerical problems, to put it mildly, and only the first 3 – 4 eigenfunctions can be constructed within an excepted error, and in most cases we shall need at least 10 and probably 20 eigenfunctions in order to obtain a reasonable approximation. We shall therefore use a different approach, in which we use the special structure of $D(x)$ being piecewise constant. The two methods supplement each other, so it depends on the structure of $D(x)$, whether one should use the method from Chapter 6 or the method developed in the following.

Boundary conditions. Due to the linearity of Fick's second law there are essentially only two types of boundary conditions, namely

$$\begin{cases} C(0,t) = 1 \text{ and } C(1,t) = 1, & \text{solution } C_1(x,t), \\ C(0,t) = 1 \text{ and } C(1,t) = 0 & \text{solution } C_2(x,t). \end{cases} \tag{7.54}$$

In fact,

$$C(x,t) = C_\ell \cdot C_1(x,t) + (C_0 - C_\ell) \cdot C_2(x,t)$$

is a solution of Fick's second law, for which

$$C(0,t) = C_0 \quad \text{and} \quad C(1,t) = C_\ell,$$

so it suffices to find $C_1(x,t)$ and $C_2(x,t)$.

In practice, a truncated solution of $C_2(x,t)$ requires twice as many terms as a truncated solution of $C_1(x,t)$ in order to get the same error estimate, so we have decided in the examples below only to find approximations of $C_1(x,t)$, because the work for the approximations of $C_2(x,t)$ will be twice as big, although they can be performed even on a pocket calculator.

Stationary solutions. When the boundary conditions are constant, and the general diffusion coefficient $D(x)$ – not necessarily piecewise constant – does not depend on t, it is easy to find the stationary solution (the steady state) $C_\infty(x) = \lim_{t\to\infty} C(x,t)$ of (7.50). In fact, $\partial C_\infty/\partial t = 0$, so (7.50) is reduced to the ordinary differential equation

$$\frac{d}{dx}\left\{D(x)\,\frac{dC_\infty}{dx}\right\} = 0, \quad C_\infty(0) = C_0 \text{ and } C_\infty(1) = C_\ell.$$

The general solution is found by two integrations,

$$\tilde{C}_\infty(x) = a + b\int_0^x \frac{d\xi}{D(\xi)}, \quad a, b \text{ arbitrary constants.}$$

By using the boundary conditions we get

$$C_\infty(x) = C_0 + (C_1 - C_\ell) \int_0^x \frac{d\xi}{D(\xi)} \bigg/ \int_0^1 \frac{d\xi}{D(\xi)}. \tag{7.55}$$

Since any contribution from the initial condition dies out by diffusion for $t \to \infty$, this stationary solution does not depend on the initial condition.

For the two types of boundary conditions (7.54) we get in particular

$$C_{1,\infty}(x) = 1, \qquad C_1(0,t) = 1, \quad C_1(1,t) = 1,$$

and

$$C_{2,\infty}(x) = \int_x^1 \frac{d\xi}{D(\xi)} \bigg/ \int_0^1 \frac{d\xi}{D(\xi)}, \qquad C_2(0,t) = 1, \quad C_2(1,t) = 0.$$

We shall in the examples only be concerned with $C_{1,\infty}(x) = 1$. It is, however, worth to mention a seemingly strange phenomenon concerning $C_{2,\infty}(x)$. If $D(x)$ is given by (7.51), one gets

$$C_{2,\infty}(x) = \begin{cases} 1 - \dfrac{c^2 x}{1 + (c^2 - 1)a} & \text{for } 0 \le x \le a, \\[2ex] \dfrac{1 - x}{1 + (c^2 - 1)a}, & \text{for } a \le x \le 1. \end{cases} \tag{7.56}$$

If instead we consider a symmetric 3-composite medium with

$$D(x) = \begin{cases} \dfrac{1}{c^2} & \text{for } 0 < x < a, \\[2ex] 1 & \text{for } a < x < 1 - a, \qquad 0 < a < \dfrac{1}{2}, \\[2ex] \dfrac{1}{c^2} & \text{for } 1 - a < x < 1, \end{cases}$$

we instead get the stationary solution

$$C_{2,\infty}^*(x) = \begin{cases} 1 - \dfrac{c^2 x}{1 + 2(c^2 - 1)a} & \text{for } 0 \le x \le a, \\[2ex] \dfrac{1}{2} - \dfrac{1}{1 + 2(c^2 - 1)a}\left(x - \dfrac{1}{2}\right) & \text{for } a \le x \le 1 - a, \\[2ex] \dfrac{c^2(1 - x)}{1 + 2(c^2 - 1)a} & \text{for } 1 - a \le x \le 1. \end{cases}$$

It is easily seen that $C_{2,\infty}^*(x) > C_{2,\infty}(x)$ for all $x \in]0,1[$, when $c > 1$ and $0 < a < \dfrac{1}{2}$, although the symmetric 3-composite medium has a better coating than the 2-composite medium. This may look strange at a first glance, until we realize that the boundary condition $C(1,t) = 0$ is actually a drain. In the 3-composite medium the coating to the right will bar the chloride ions inside the concrete, so they can only escape at a slower rate. Hence, $C_\infty^* > C_\infty(x)$ is no mystery.

Elimination of Gibbs's phenomenon. Generalized Fourier analysis is usually pestered by Gibbs's phenomenon, where any finite eigenfunction expansion has an overswing of approximately 8.5 % of the true value somewhere in the neighbourhood of the boundary.

It can be proved that when the boundary conditions are constant, then the structure of the solution is

$$C(x,t) = \sum_{n=1}^{\infty} a_n u_n(x) \cdot \left\{1 - \exp\left(-\mu_n^2 t\right)\right\}, \tag{7.57}$$

where the $u_n(x)$ are the eigenfunctions corresponding to the eigenvalues $-\mu_n^2$ (these are always negative).

Taking the limit $t \to \infty$ we get from (7.57)

$$C_\infty(x) = \sum_{n=1}^{\infty} a_n u_n(x) = \sum_{n=1}^{\infty} \frac{1}{\|u_n\|^2} \langle C_\infty, u_n \rangle \cdot u_n(x), \tag{7.58}$$

where $C_\infty(x)$ is the stationary solution given by (7.55). The stationary solution is easy to find, and since the coefficients of (7.58) are unique, we see that (7.57) can be written in form

$$C(x,t) = C_\infty(x) - \sum_{n=1}^{\infty} \frac{1}{\|u_n\|^2} \langle C_\infty, u_n \rangle \, u_n(x) \cdot \exp(-\mu_n^2 t). \tag{7.59}$$

It is not hard to show that when $t > 0$ is fixed, then the series in (7.59) is absolutely and uniformly convergent in x, and Gibbs's phenomenon has disappeared.

A similar technique of extracting the term carrying Gibbs's phenomenon was used in Example 4.3.1, page 216.

7.2.3 The 2-composite finite medium

We assume that

$$D(x) = \begin{cases} \dfrac{1}{c^2} & \text{for } 0 < x < a, \\[2ex] 1 & \text{for } a < x < 1. \end{cases}$$

The solution is built up by eigenfunctions, so we shall solve the eigenvalue problem

$$\begin{cases} \dfrac{d}{dx}\left\{D(x)\,\dfrac{du}{dx}\right\} = -\mu^2 u(x), \qquad 0 < x < 1, \\ \\ u(0) = 0 \text{ and } u(1) = 0. \end{cases} \tag{7.60}$$

Here we have used that all eigenvalues $-\mu^2$ are negative, when $D(x) > 0$. It is convenient, though an abuse of the terminology, to call in the following $\mu > 0$ an eigenvalue, whenever (7.60) has a nontrivial solution $u(x) \neq 0$.

Using the structure of $D(x)$ we derive from (7.60) that any eigenfunction necessarily must satisfy

$$\begin{cases} \dfrac{d^2u}{dx^2} + (c\mu)^2 u = 0 \quad \text{for } 0 < x < a \text{ and } u(0) = 0, \\ \\ \dfrac{d^2u}{dx^2} + \mu^2 u = 0 \qquad \text{for } a < x < 1 \text{ and } u(1) = 0. \end{cases} \tag{7.61}$$

Solving (7.61) separately in the two intervals we see that any solution of (7.60) necessarily must have the structure (a constant times)

$$u(x) = \begin{cases} \sin(c\mu x) & \text{for } 0 \leq x < a, \\ \\ k \cdot \sin(\mu\{1-x\}) & \text{for } a < x \leq 1. \end{cases} \tag{7.62}$$

Furthermore, $u(x)$ must be continuous at $x = a$. So

$$u(a) = \sin(c\mu a) = k \cdot \sin(\mu\{1-a\}). \tag{7.63}$$

Assuming (7.62) and (7.63) we get

$$u'(x) = \begin{cases} c\mu \cdot cos(c\mu x) & \text{for } 0 < x < a, \\ \\ -k\mu \cdot \cos(\mu\{1-x\}) & \text{for } a < x < 1, \end{cases}$$

hence,

$$D(x)u'(x) = \begin{cases} \dfrac{\mu}{c}\cos(c\mu x) & \text{for } 0 < x < a, \\ \\ -k\mu\cos(\mu\{1-x\}) & \text{for } a < x < 1. \end{cases}$$

According to (7.60) the function $D(x)u'(x)$ is differentiable, and in particular continuous at $x = a$. By removing the common factor $\mu > 0$ this gives

$$\frac{1}{c}\cos(c\mu a) = -k \cdot \cos(\mu\{1-a\}). \tag{7.64}$$

When we eliminate k from (7.63) and (7.64) we get after some trigonometric calculations the following equation in $\mu > 0$,

$$(c-1)\sin(\{1-(c+1)a\}\mu) - (c+1)\sin(\{1+(c-1)a\}\mu) = 0. \qquad (7.65)$$

When $\mu > 0$ is a solution of (7.65) the corresponding constant is found from either (7.63) or (7.64), or more nicely by the formula

$$k_\mu = \frac{1}{2c}\{(c-1)\cos(\{1-(c+1)a\}\mu) - (c+1)\cos(\{1+c-1)a\}\mu)\}. \quad (7.66)$$

It is not hard to prove from (7.65) and (7.66) that

$$k_\mu^2 = \sin^2(ac\mu) + \frac{1}{c^2}\cos^2(ac\mu),$$

but we do not find the sign of k_μ by this formula, so we shall not use it.

An extremely tedious check of (7.60), omitted here, shows that if $\mu > 0$ is a solution of (7.65), and $k = k_\mu$ is given by (7.66), then (7.62) is indeed an eigenfunction. Hence we have found all eigenfunctions, if we can solve (7.65).

One may find the solutions of (7.65) by Newton-Raphson's iteration formula, but it is here easier just to table the left hand side of (7.65) on a pocket calculator and then decrease the steps of the table. This primitive method is extremely efficient on the pocket calculator TI-92.

In the search for the eigenvalues $0 < \mu_1 < \mu_2 < \cdots$ in increasing order it may be helpful to know that it can be proved that

$$\frac{\pi(n-1)}{1+(c-1)a} < \mu_n < \frac{\pi(n+1)}{1+(c-1)a}$$

or

$$\frac{1+(c-1)a}{\pi}\,\mu_n \in \,]n-1, n+1[, n \in \mathbb{N}. \qquad (7.67)$$

Calculation of $\|u_\mu\|^2$. This is straightforward,

$$\|u_\mu\|^2 = \frac{a}{2} - \frac{1}{4c\mu}\sin(2ac\mu) + k_\mu^2\left\{\frac{1-a}{2} - \frac{1}{4\mu}\sin(2\{1-a\}\mu)\right\}, \quad (7.68)$$

where μ and k_μ are given by (7.65) and (7.66).

Calculation of $\langle 1, u_\mu\rangle$. This coefficient enters the solution formula (7.59), when $C(0,t) = 1$ and $C(1,t) = 1$, hence $C_\infty(x) = 1$. We find

$$\langle 1, u_\mu\rangle = \frac{1}{c\mu}\{1-\cos(ac\mu)\} + \frac{k_\mu}{\mu}\{1-\cos(\{1-a\}\mu)\}. \qquad (7.69)$$

Calculation of $\langle C_\infty, u_\mu\rangle$, *when* $C(0,t) = 1$ *and* $C(1,t) = 0$. Here, $C_\infty(x)$ is given by (7.56). We find

$$\begin{aligned}
&\langle C_{2,\infty}, u_\mu\rangle \\
&= \frac{1-\cos(ac\mu)}{c\mu} + \frac{1}{1+(c^2-1)a}\cdot\frac{1}{\mu^2}\{ac\cos(ac\mu) - \sin(ac\mu)\} \\
&\quad + \frac{1}{1+(c^2-1)a}\cdot\frac{k_\mu}{\mu^2}\{\sin(\{1-a\}\mu) - (1-a)\mu\cdot\cos(\{1-a\}\mu)\}.
\end{aligned}$$

7.2.4 The 2-composite finite medium, examples

We shall here compare the classical case where $D(x) = 1$ with the 2-composite medium, when $a = 0.01$ and $c = 20$. In the latter case the coating is thin, $1/100$ of the total length, while the factor $c^2 = 400$ between the two layers is realistic for coating. The numbers have been chosen as simple as possible in order not to make the calculations too difficult, and yet they should be close to what is met in real life. As mentioned previously we shall only consider the simplest boundary conditions $C(0,t) = C(1,t) = 1$, where the stationary solution is $C_\infty(x) = 1$, because the calculations for the case $C(0,t) = 1$ and $C(1,t) = 0$ will be approximately twice as big.

No observations have been available, so the results can only be interpreted as what can be predicted, if we assume Fickean diffusion.

Example 7.2.1 Cf. also Example 4.3.1 and Section 3.5.1. When $D(x) = 1$ and $C(0,t) = C(1,t) = 1$ we have previously found the solution

$$C(x,t) = 1 - \frac{4}{\pi}\sum_{n=0}^{\infty}\frac{1}{2n+1}\sin((2n+1)\pi x)\cdot\exp\left(-(2n+1)^2\pi^2 t\right).$$

We shall approximate this solution by the truncated series

$$C(x,t) \approx 1 - \frac{4}{\pi}\sum_{n=0}^{N-1}\frac{1}{2n+1}\sin((2n+1)\pi x)\cdot\exp\left(-(2n+1)^2\pi^2 t\right),$$

consisting of the first N terms of the series.

In order to show the principle we choose in all examples $x_0 = 0.1$ as our reference point. In the present case $\sin((2n+1)\pi/10)$ changes its sign, when n is replaced by $n+5$. When we group the terms with five indices in each group,

$$C(x,t) =$$

$$1 - \frac{4}{\pi}\sum_{n=0}^{\infty}(-1)^n\left\{\sum_{p=0}^{4}\frac{\sin((2p+1)\pi/10)}{10n+2p+1}\exp\left(-(10n+2p+1)^2\pi^2 t\right)\right\},$$

we obviously get an alternating series, where the terms numerically tend decreasing towards zero. Hence, the error of the truncated series is smaller than the absolute value of the first deleted term. For e.g. $N = 5$ the error is numerically smaller than

$$\frac{4}{\pi}\sum_{n=5}^{9}\left|\sin\frac{(2n+1)\pi}{10}\right|\cdot\exp(-(11\pi)^2 t) < \frac{7}{20}\exp(-(11\pi)^2 t),$$

and for $N = 10$ the error is smaller that $\dfrac{7}{20}\exp(-(21\pi)^2 t)$, etc.

Choosing $N = 5$ we see that the error is $< 1\%$ if $t > t_0 = 0.003$, and it is smaller than 5% if $t > t_0 = 0.0017$. These values may seem small, but we must keep in mind that the real time variable has a very large unit.

We have according to (7.53)

$$T > T_0 = \frac{\ell^2}{D_1} t_0, \quad \text{i.e.} \quad t_0 = \frac{D_1 T_0}{\ell^2}.$$

Given D_1, ℓ and the time T_0, for which we want the error of the truncated series to be smaller that ε when $T > T_0$, we first calculate t_0 and then find $N = 5m$, $m \in \mathbb{N}$, such that

$$\frac{7}{20} \exp\left(-(10m+1)^2\pi^2 t_0\right) < \varepsilon,$$

i.e.

$$N = 5m > \frac{1}{2\pi}\sqrt{\frac{1}{t_0}\ln\left(\frac{7}{20\varepsilon}\right)} - \frac{1}{2}.$$

The closer we are to the boundary the more terms are needed to obtain a given accuracy for given t_0. When $x = \frac{1}{2}$ and we are as far as possible from the boundary, we get

$$C\left(\frac{1}{2}, t\right) = 1 - \frac{4}{\pi}\sum_{n=0}^{\infty} \frac{(-1)^n}{2n+1} \exp\left(-(2n+1)^2\pi^2 t\right),$$

which is alternating in itself without grouping. The error of the truncated series

$$C\left(\frac{1}{2}, t\right) \approx 1 - \frac{4}{\pi}\sum_{n=0}^{N} \frac{(-1)^n}{2n+1} \exp\left(-(2n+1)^2\pi^2 t\right)$$

is therefore smaller than

$$\frac{4}{\pi} \cdot \frac{1}{2N+1} \exp\left(-(2N+1)^2\pi^2 t\right) < \varepsilon.$$

Notice that this inequality holds for all $t \geq 0$, when $N > 2/(\pi\varepsilon) - 1/2$ and $x = 1/2$. $\Diamond$

Example 7.2.2 Let us consider the 2-composite medium with $a = 0.01$ and $c = 20$. We aim for the results at the control point $x_0 = 0.1$.

By (7.62) the eigenfunctions have the structure

$$u(x) = \begin{cases} \sin(0.2\mu x) & \text{for } 0 \leq x \leq 0.01, \\ \\ k_\mu \cdot \sin(\mu\{1 - x\}) & \text{for } 0.01 < x \leq 1, \end{cases}$$

According to (7.65) the eigenvalues μ are the positive solutions of the transcendental equation

$$19\sin(0.79\mu) - 21\sin(1.19\mu) = 0,$$

and by (7.66),

$$k_\mu = \frac{1}{40}\{19\cos(0.79\mu) - 21\cos(1.19\mu)\}.$$

From (7.67) follows that

$$\frac{\pi}{1.19}(n-1) < \mu_n < \frac{\pi}{1.19}(n+1), \quad \text{or} \quad \frac{1.19}{\pi}\,\mu_n \in]n-1, n+1[.$$

Finally, when $C(0,t) = C(1,t) = 1$, the stationary solution is $C_\infty(x) = 1$, and the solution is given by (7.59), i.e.

$$C(x,t) = 1 - \sum_{n=1}^{\infty} \frac{1}{\|u_n\|^2}\langle 1, u_n\rangle\, u_n(x)\, \exp\left(-\mu_n^2 t\right).$$

Furthermore, by (7.68),

$$\|u_\mu\|^2 = \frac{1}{200} - \frac{1}{80\mu}\sin(0.4\mu) + k_\mu^2\left\{\frac{99}{200} - \frac{1}{4\mu}\sin(1.98\mu)\right\},$$

and by (7.69),

$$\langle 1, u_\mu\rangle = \frac{1}{20\mu}\{1 - \cos(0.2\mu)\} + \frac{k_\mu}{\mu}\{1 - \cos(0.99\mu)\}.$$

For completeness we table all the constants μ_n, μ_n^2, k_n, $\langle 1, u_n\rangle/\|u_n\|^2$ and $k_n\langle 1, u_n\rangle/\|u_n\|^2$ in Table 7.6. By these we can give a full description of the solution at any point $x \in [0,1]$. If we are only interested in a specific control point, e.g. $x_0 = 0.1$, they are not all necessary.

We notice that k_n and $\langle 1, u_n\rangle/\|u_n\|^2$ both have the sign $(-1)^{n+1}$, while $k_n\langle 1, u_n\rangle/\|u_n\|^2$ is positive for $n = 1, \dots, 27$. Furthermore, the trend of $k_n\langle 1, u_n\rangle/\|u_n\|^2$ is that it is slowly decreasing with regular anomalies. It is in particular small when $n = 6p$.

In Table 7.7 we specify the calculations to the control point $x = 0.1$. In the first column we demonstrate that $1.19\mu_n/\pi$ is always close to n. In the second column we have calculated

$$k_n\langle 1, u_n\rangle\frac{1}{\|u_n\|^2}\sin(\{1-x\}\mu_n) = k_n\langle 1, u_n\rangle\frac{1}{\|u_n\|^2}\sin(0.9\mu_n),$$

using the values of Table 7.6. The third column gives the sum of the first n entries of the second column. It corresponds to the truncated series for $t = 0$, so it should be close to $C_\infty(x) = 1$. By subtracting $C_\infty(x) = 1$ we get an

Table 7.6: Table of the constants μ_n, μ_n^2, k_n, $\langle 1, u_n\rangle/\|u_n\|^2$ and $k_n\langle 1, u_n\rangle/\|u_n\|^2$.

n	μ_n	μ_n^2	k_n	$\langle 1, u_n\rangle/\|u_n\|^2$	$k_n\langle 1, u_n\rangle/\|u_n\|^2$
1	1.7261994	2.9797644	0.341677	3.613774	1.234745
2	4.7954024	22.9958842	-0.819167	-0.475726	0.389699
3	7.9325214	62.9248958	0.999877	0.264766	0.264734
4	11.0688216	122.5188116	-0.800881	-0.210350	0.168465
5	14.1268022	199.5665404	0.314598	0.478023	0.150385
6	15.7395760	247.7342401	-0.050397	-0.004033	0.000203
7	17.5810574	309.0935793	0.368865	0.329298	0.121466
8	20.6597279	426.8243569	-0.836669	-0.108326	0.090633
9	23.7975626	566.3239857	0.998893	0.088345	0.088247
10	26.9329327	725.3828638	-0.781831	-0.088344	0.069070
11	29.9775063	898.6508840	0.287755	0.245001	0.070500
12	31.4799620	990.9880075	-0.051610	-0.008095	0.000418
13	33.4374271	1118.0615311	0.396052	0.161413	0.063928
14	36.5241444	1334.0131242	-0.853366	-0.060183	0.051358
15	39.6625992	1573.1217753	0.996925	0.053114	0.052951
16	42.7969175	1831.5761475	-0.762041	-0.056891	0.043353
17	45.8253466	2099.9623910	0.261306	0.175090	0.045752
18	47.2220209	2229.9192579	-0.053706	-0.012205	0.000655
19	49.2950383	2430.0008010	0.423150	0.102510	0.043377
20	52.3886428	2744.5698944	-0.869240	-0.041259	0.035864
21	55.5276279	3083.3174602	0.993977	0.038055	0.037825
22	58.6607610	3441.0848811	-0.741533	-0.042530	0.031537
23	61.6696006	3803.1396382	0.235445	0.142682	0.033594
24	62.9667146	3964.8071475	-0.056794	-0.016356	0.000929
25	65.1536775	4245.0016918	0.450081	0.072909	0.032815
26	68.2532146	4658.5013032	-0.884272	-0.031177	0.027569
27	71.3926457	5096.9098601	0.990050	0.029719	0.029423

idea of the errors of the truncated series. Finally, in the fourth column we have summed the groups of entries of the same sign from the second column. These numbers may be used for an error estimate. In fact, choose an n from Table 7.7, and find the numerically largest number $c_n = \max_{k\geq n} |\sum \text{groups}|$. Then the error of the truncated approximation

$$C(0.1, t) \approx 1 - \sum_{k=1}^{n-1} k_n \langle 1, u_k\rangle \frac{1}{\|u_k\|^2} \sin(0.9x) \exp\left(-\mu_k^2 t\right)$$

with $n-1$ terms is less than

$$c_n \cdot \exp\left(-\mu_n^2 t\right) < \varepsilon \quad \text{for} \quad t > t_0 = \frac{1}{\mu_n^2} \ln\left(\frac{c_n}{\varepsilon}\right).$$

Table 7.7: Specification of the calculations to the control point $x = 0.1$.

n	$1.19\mu_n/\pi$	$\frac{k_n\langle 1, u_n\rangle/}{\|u_n\|^2}\sin(0.9\mu_n)$	$\sum_{j=1}^{n} k_j \cdots \sin(0.9\mu_n)$	$\sum$ groups
1	0.6538649	1.234562	1.234562	1.234562
2	1.8154446	-0.359461	0.875101	-0.359461
3	3.0047500	0.199949	1.075050	0.199949
4	4.1927453	-0.086204	0.988846	-0.086204
5	5.3510740	0.022139	1.010985	0.022342
6	5.9619744	0.000203	1.011188	
7	6.6595070	-0.013936	0.997252	-0.036868
8	7.8256728	-0.022932	0.974320	
9	9.0142493	0.047869	1.022189	0.047869
10	10.2018923	-0.053807	0.968382	-0.053807
11	11.3551426	0.067828	1.036210	0.067828
12	11.9242559	-0.000024	1.036186	-0.061988
13	12.6657217	-0.061964	0.974222	
14	13.8349355	0.051019	1.025241	0.051019
15	15.0237470	-0.048087	0.977154	-0.048087
16	16.2109915	0.031642	1.008796	0.031642
17	17.3581264	-0.017905	0.990891	-0.018558
18	17.8871710	-0.000653	0.990238	
19	18.6724067	0.016220	1.006458	0.016220
20	19.8442293	-0.000928	1.005530	-0.011767
21	21.0332416	-0.010839	0.994691	
22	22.2200372	0.018128	1.012819	0.018128
23	23.3597518	-0.029073	0.983746	-0.029073
24	23.8510840	0.000112	0.983858	0.028608
25	24.6794810	0.028496	1.012354	
26	25.8535508	-0.027186	0.985168	-0.027186
27	27.0427321	0.029096	1.014264	0.029096

It is easily shown that $t_0 = 0.001$ for $n = 15$ and $\varepsilon = 1\%$, and similarly that $t_0 = 0.0001$ for $n = 22$ and $\varepsilon = 1\%$.

When we want to minimize the sums for other points x we may get different values of t_0 and n. There is, however, due to the uniform convergence, a (very large) $n_0 = n_0(\varepsilon)$ depending only on ε, such that we can choose $t_0 = 0$.

Let us compare the two examples. In Example 7.2.1 without coating we found the approximation

$$C_1(0.1, t) = 1 - \frac{4}{\pi}\sum_{n=0}^{4} \frac{1}{2n+1} \sin\left((2n+1)\frac{\pi}{10}\right) \cdot \exp(-(2n+1)^2\pi^2 t)$$

with an error $< 1\%$ for $t > t_0 = 0.003$. The sum only contains five terms. If all terms are collected we have nine terms, but every second term (for even indices) is zero.

In the present coating example we find $n = 10$ by Table 7.7, if we want an error $< 1\%$ for $t > t_0 = 0.003$. Hence the approximation needs only contain $n - 1 = 9$ terms, in which case

$$\begin{aligned} C_2(0.1,t) = \; & 1 - 1.23452\exp(-2.9797644t) + 0.359461\exp(-22.9958842t) \\ & -0.199949\exp(-62.9248958t) + 0.086204\exp(-122.5188116t) \\ & -0.022139\exp(-199.5665404t) - 0.000203\exp(-247.734201t) \\ & +0.013936\exp(-309.0935793t) + 0.022932\exp(-426.8243569t) \\ & -0.047869\exp(-566.3239857t). \end{aligned}$$

Even though $C_2(0.1,t)$ only represents coating to the left, so the chloride ions are ingressing into the concrete at the same rate from the right hand side, it is seen that the effect of the coating is quite dramatic. ◊

The case $C(0,t) = 1$ and $C(1,t) = 0$ is treated similarly. It can be expected that we shall need more terms to obtain the same degree of accuracy, and it is indeed so. We therefore leave these examples to the reader, the principles being the same as in the Examples 7.2.1 and 7.2.2.

7.2.5 The 3-composite finite medium

We assume that

$$D(x) = \begin{cases} \dfrac{1}{c^2} & \text{for } 0 < x < a, \\[2ex] 1 & \text{for } a < x < b, \qquad 0 < a < b < 1, \\[2ex] \dfrac{1}{k^2} & \text{for } b < x < 1, \end{cases}$$

and we shall solve the eigenvalue problem (7.60).

Using the structure of $D(x)$ we find that any eigenfunction at least must satisfy the analogue of (7.61), i.e.

$$\begin{cases} \dfrac{d^2u}{dx^2} + (c\mu)^2 u = 0, & \text{for } 0 < x < a, \qquad u(0) = 0, \\[2ex] \dfrac{d^2u}{dx^2} + \mu^2 u = 0, & \text{for } a < x < b, \\[2ex] \dfrac{d^2u}{dx^2} + (k\mu)^2 u = 0, & \text{for } b < x < 1, \qquad u(1) = 0. \end{cases}$$

Hence, the eigenfunctions must have the structure

$$u(x) = \begin{cases} \sin(c\mu x), & \text{for } 0 \le x < a, \\[2ex] c_1 \sin(\mu x) + c_2 \cos(\mu x), & \text{for } a < x < b, \\[2ex] c_3 \sin(k\mu\{1 - x\}), & \text{for } b < x \le a, \end{cases} \tag{7.70}$$

in the four unknown constants μ, c_1, c_2, c_3.

Since $u(x)$ must be continuous at $x = a$ and $x = b$, we get the requirements

$$\begin{cases} u(a) = c_1 \sin(a\mu) + c_2 \cos(a\mu) = \sin(ac\mu), \\ \\ u(b) = c_1 \sin(b\mu) + c_2 \cos(b\mu) = c_3 \sin(k\{1-b\}\mu). \end{cases} \tag{7.71}$$

Furthermore,

$$D(x)u'(x) = \begin{cases} \dfrac{\mu}{c} \cos(c\mu x), & \text{for } 0 < x < a, \\ \\ \mu c_1 \cos(\mu x) - \mu c_2 \sin(\mu x), & \text{for } a < x < b, \\ \\ -\dfrac{\mu}{k} c_3 \cos(k\mu\{1-x\}), & \text{for } b < x < 1, \end{cases}$$

must also be continuous at $x = a$ and $x = b$, so by cancelling the common factor $\mu > 0$ we get

$$\begin{cases} c_1 \cos(a\mu) - c_2 \sin(a\mu) = \dfrac{1}{c} \cos(ac\mu), \\ \\ c_1 \cos(b\mu) - c_2 \sin(b\mu) = -\dfrac{1}{k} c_3 \cos(k\{1-b\}\mu). \end{cases} \tag{7.72}$$

From (7.71) and (7.72) we get by some Linear Algebra,

$$\begin{cases} c_1(a,c,\mu) = \sin(a\mu)\sin(ac\mu) + \dfrac{1}{c} \cos(a\mu)\cos(ac\mu), \\ \\ c_2(a,c,\mu) = \cos(a\mu)\sin(ac\mu) - \dfrac{1}{c} \sin(a\mu)\cos(ac\mu), \end{cases} \tag{7.73}$$

and by eliminating c_3,

$$kc_2 \sin(k\{1-b\}\mu)\sin(b\mu) - kc_1 \sin(k\{1-b\}\mu)\cos(b\mu)$$

$$= c_1 \cos(k\{1-b\}\mu)\sin(b\mu) + c_2 \cos(k\{1-b\}\mu)\cos(b\mu).$$

When we insert (7.73) into this equation, we get after a long and tedious trigonometric reduction that the equation for the eigenvalues $\mu > 0$ becomes

$$\begin{aligned} 0 \;=\;& (k-1)(c+1)\sin((a\{c-1\} + b\{k+1\} - k)\mu) \\ & -(k+1)(c-1)\sin((a\{c+1\} + b\{k-1\} - k)\mu) \\ & -(k+1)(c+1)\sin((a\{c-1\} - b\{k-1\} + k)\mu) \\ & +(k-1)(c-1)\sin((a\{c+1\} - b\{k+1\} + k)\mu). \end{aligned} \tag{7.74}$$

Notice that when $k = 1$ we have a 2-composite medium, and that (7.74) is reduced to (a constant times) formula (7.65), which is the equation for the eigenvalues of the 2-composite medium.

Finally, we get c_3 from either (7.71) or (7.72), e.g.

$$c_3(a,b,c,\mu) = c_1(a,b,\mu)\,\frac{\sin(b\mu)}{\sin(k\{1-b\}\mu)} + c_2(a,c,\mu)\,\frac{\cos(b\mu)}{\sin(k\{1-b\}\mu)}, \tag{7.75}$$

when $\sin(k\{1-b\}\mu) \neq 0$.

A fairly long check, which is omitted here, shows that if $\mu > 0$ is a solution of (7.74), and c_1, c_2, c_3 are given by (7.73) and (7.75), then $u(x)$ given by (7.70) is indeed an eigenfunction.

The squared norm of an eigenfunction is found to be

$$\begin{aligned}\|u\|^2 \;=\;& \frac{a}{2} - \frac{\sin(2ac\mu)}{4c\mu} + \frac{1}{2}\,c_3^2(1-b) - \frac{c_3^2}{4k\mu}\sin(2k\{1-b\}\mu) \\ & - \frac{c_1^2 - c_2^2}{4\mu}\{\sin(2b\mu) - \sin(2a\mu)\} - \frac{c_1 c_2}{2\mu}\{\cos(2b\mu) - \cos(2a\mu)\} \\ & + \frac{c_1^2 + c_2^2}{2}\,(b-a).\end{aligned}$$

It is obvious from these expressions that the general case considered here requires a computer. Since most of this book only is based on a pocket calculator, we shall later only give an example of a symmetric 3-composite medium, where the calculations become easier, hence possible on a pocket calculator.

Notice, however, that the method above immediately can be extended to n-composite media. First one guesses the right structure like in (7.70), and then one writes down the continuity equations for $u(x)$ and $D(x)u'(x)$. This procedure will give us $2n-2$ equations in $2n-2$ unknown constants.

Remark 7.2.1 One can use this construction to give an approximative solution of the case, when $D(x)$ varies in $x \in [0,1]$. For any given $\varepsilon > 0$ one constructs a step function $D_1(x)$, such that

$$|D(x) - D_1(x)| < \varepsilon \qquad \text{for all } x \in [0,1].$$

Since $D_1(x)$ corresponds to some n-composite medium, one can solve for $D_1(x)$ by generalizing the method above. Finally, one of the properties of Fick's second law assures that the error of this approximative solution is $< \varepsilon$ for all $x \in [0,1]$ and all $t \geq 0$. $\Diamond$

7.2.6 Symmetric 3-composite media

This case is particularly important, because it represents symmetric coating. We assume that $k = c$ and $b = 1 - a$, so $0 < a < \frac{1}{2}$. Equation (7.74) for the eigenvalues is reduced to

$$(c+1)^2 \sin(\{1+2(c-1)a\}\mu) + (c-1)^2 \sin(\{1-2(c+1)a\}\mu)$$

$$= 2(c^2-1)\sin(\{1-2a\}\mu). \tag{7.76}$$

It can be proved that if $0 < \mu_1 < \mu_2 < \cdots < \mu_n < \cdots$ are the eigenvalues in increasing order, then

$$\frac{\pi(n-1)}{1+2(c-1)a} < \mu_n < \frac{\pi(n+1)}{1+2(c-1)a}, \qquad n \in \mathbb{N},$$

which is (7.67) with a replaced by $a + a = 2a$.

We still have (7.73), i.e.

$$\begin{cases} c_1(a,c,\mu_n) = \sin(a\mu_n)\sin(ac\mu_n) + \dfrac{1}{c}\cos(a\mu_n)\cos(ac\mu_n), \\ c_2(a,c,\mu_n) = \cos(a\mu_n)\sin(ac\mu_n) - \dfrac{1}{c}\sin(a\mu_n)\cos(ac\mu_n). \end{cases}$$

Since the medium is symmetric, the eigenfunctions must also show some symmetry. When $u_n(x)$ is the eigenfunction corresponding to the n-th eigenvalue μ_n, we have explicitly

$$u_n\left(\frac{1}{2}-x\right) = (-1)^{n+1} u_n\left(\frac{1}{2}+x\right), \qquad x \in \left[-\frac{1}{2}, \frac{1}{2}\right].$$

Using this symmetry condition we see that (7.75) is reduced to

$$c_3(a,c,\mu_n) = (-1)^{n+1}, \qquad n \in \mathbb{N}. \tag{7.77}$$

Example 7.2.3 We consider the symmetric 3-composite medium with $a = 0.01$ and $c = 20$, cf. also the Examples 7.2.1 and 7.2.2. We shall again find an approximate solution of the boundary problem $C(0,t) = C(1,t) = 1$, i.e. the stationary solution is $C_\infty(x) = 1$.

The diffusion coefficient is

$$D(x) = \begin{cases} \dfrac{1}{400} & \text{for } 0 < x < 0.01, \\ 1 & \text{for } 0.01 < x < 0.99, \\ \dfrac{1}{400} & \text{for } 0.99 < x < 1. \end{cases}$$

The equation (7.76) for the eigenvalues is

$$441\sin(1.38\mu) + 361\sin(0.58\mu) - 798\sin(0.98\mu) = 0,$$

and we have

$$\frac{\pi}{1.38}(n-1) < \mu_n < \frac{\pi}{1.38}(n+1), \quad \text{i.e.} \quad n-1 < \frac{1.38}{\pi}\mu_n < n+1.$$

If μ_n is an eigenvalue, then by (7.73) and (7.77),

$$\begin{aligned} c_1(\mu_n) &= \sin(0.01\mu_n)\sin(0.2\mu_n) + \frac{1}{20}\cos(0.01\mu_n)\cos(0.2\mu_n) \\ &= \frac{21}{40}\cos(0.19\mu_n) - \frac{19}{40}\cos(0.21\mu_n), \end{aligned}$$

$$\begin{aligned} c_2(\mu_n) &= \cos(0.01\mu_n)\sin(0.2\mu_n) - \frac{1}{20}\sin(0.01\mu_n)\cos(0.2\mu_n) \\ &= \frac{21}{40}\sin(0.19\mu_n) + \frac{19}{40}\sin(0.21\mu_n), \end{aligned}$$

and $c_3(\mu_n) = (-1)^{n+1}$. In particular,

$$c_1\sin(\mu_n x) + c_2\cos(\mu_n x)$$

$$\begin{aligned} &= \frac{21}{40}\sin(\{x+0.19\}\mu_n) - \frac{19}{40}\sin(\{x-0.21\}\mu_n) \\ &= \frac{21}{40}\sin(\{0.69+(x-0.5)\}\mu_n) - \frac{19}{40}\sin(\{0.29+(x-0.5)\}\mu_n), \end{aligned}$$

and the eigenfunctions can be written

$$u_n(x) = \begin{cases} \sin(20\mu_n x), & 0 \le x \le 0.01, \\ \left.\begin{aligned} &\frac{21}{40}\sin(\{0.69+(x-0.5)\}\mu_n) \\ &-\frac{19}{40}\sin(\{0.29+(x-0.5)\}\mu_n), \end{aligned}\right\} & 0.01 \le x \le 0.99, \\ (-1)^{n+1}\sin(20\mu_n\{1-x\}), & 0.99 \le x \le 1, \end{cases}$$

where furthermore $u_n(0.5-x) = (-1)^{n+1}u_n(0.5+x)$.

A fairly long calculation gives

$$\begin{aligned} \|u_n\|^2 &= \frac{1}{80000}\{20449 - 19551\cos(0.4\mu_n)\} - \frac{1}{40\mu_n}\sin(0.4\mu_n) \\ &\quad + \frac{1}{3200\mu_n}\left(798\{\sin(1.96\mu_n) - \sin(0.98\mu_n)\}\right. \\ &\quad -441\{\sin(2.36\mu_n) - \sin(1.38\mu_n)\} \\ &\quad \left.-361\{\sin(1.56\mu_n) - \sin(0.58\mu_n)\}\right). \end{aligned}$$

Finally, the symmetries of the eigenfunctions give $\langle 1, u_{2n} \rangle$ and

$$\begin{aligned}\langle 1, u_{2n+1} \rangle &= \frac{1}{20\mu_{2n+1}} (2\{1 - \cos(0.2\mu_{2n+1})\} \\ &\quad +21\{\cos(0.69\mu_{2n+1}) - \cos(1.18\mu_{2n+1})\} \\ &\quad -19\{\cos(0.29\mu_{2n+1}) - \cos(0.78\mu_{2n+1})\}) .\end{aligned}$$

As before we choose the control point $x = 0.1$. Then

$$u_n(0.1) = \frac{21}{40} \sin(0.29\mu_n) + \frac{19}{40} \sin(0.11\mu_n).$$

According to (7.59) the solution at $x = 0.1$ is given by

$$C(0.1, t) = 1 - \sum_{n=1}^{\infty} \frac{1}{\|u_n\|^2} \langle 1, u_n \rangle u_n(0.1) \exp(-\mu_n^2 t).$$

Putting $a_n = \|u_n\|^{-2} \langle 1, u_n \rangle u_n(0.1)$ we easily get Table 7.8.

By using the same method as in the previous examples we can also get an error estimate. A first clue is given by the last column in Table 7.8, which corresponds to $t = 0$ in the truncated series in the approximations

$$C_n(0.1, t) = 1 - \sum_{k=1}^{n} \frac{1}{\|u_k\|^2} \langle 1, u_k \rangle u_k(0.1) \exp(-\mu_k^2 t),$$

i.e. the values of the last column should be close to 1. $\Diamond$

If we want a reasonable estimate of the error for small t, at rule of thumb is that one should find all eigenvalues < 50. Since for the 2-composite medium $\frac{\pi(n-1)}{1+(c-1)a} < \mu_n$, we get $\mu_n > 50$ for

$$n > 1 + \frac{50}{\pi} \{1 + (c - 1)a\}.$$

In Example 7.2.2 we find $n = 20$.

For the symmetric 3-composite medium we get

$$n > 1 + \frac{50}{\pi} \{1 + 2(c - 1)a\},$$

so $n = 23$ in Example 7.2.3.

Usually fewer terms are sufficient.

Finally it should be mentioned that although the one-sided coating with thickness $2a$ and the two-sided coating with thickness $a + a$ and the same constant c both have roughly the same distribution of eigenvalues,

$$\frac{\pi(n-1)}{1 + 2(c-1)a} < \mu_n < \frac{\pi(n+1)}{1 + 2(c-1)a}, \qquad n \in \mathbb{N},$$

Table 7.8: Control point $x = 0.1$, i.e. $a_n = \langle 1, u_n \rangle / \|u_n\|^2 \cdot u_n(0.1)$.

n	μ_n	$1.38\mu_n/\pi$	μ_n^2	a_n	$\sum_{j=1}^n a_j$.
1	0.6978365	0.3065370	0.4869758	0.983415	0.983415
2	3.3351215	1.4650110	11.1230354	0	
3	6.4410399	2.8293404	41.4869950	0.027256	1.010671
4	9.5804205	4.2083687	91.7844570	0	
5	12.6756583	5.5680065	160.6723133	0.009292	1.019963
6	15.1592123	6.6589515	229.8017176	0	
7	15.7431517	6.9154571	247.8468254	-0.028508	0.991455
8	16.5879074	7.2865310	275.1586719	0	
9	19.3485753	8.4992031	374.3673661	0.001729	0.993184
10	22.4627477	9.8671582	504.5750340	0	
11	25.6015627	11.2459381	655.4400127	0.005809	0.998993
12	28.6839505	12.5999313	822.7690163	0	
13	30.9854639	13.6109117	960.0989731	0.021293	1.020286
14	31.4875561	13.8314645	991.4661892	0	
15	32.5088882	14.2801027	1056.8278120	-0.018107	1.002179
16	35.3638806	15.5342085	1250.6040511	0	
17	38.4846401	16.9050571	1481.0675236	-0.005001	0.997178
18	41.6223046	18.2833316	1732.4162402	0	
19	44.6878627	19.6299321	1997.0050727	-0.004994	0.992184
20	46.7862810	20.5516994	2188.9560898	0	
21	47.2346120	20.7486367	2231.1085708	0.010086	1.002270
22	48.4555330	21.2849478	2347.9386783	0	
23	51.3806480	22.5698561	2639.9709889	0.000670	1.002940

one cannot conclude that the eigenvalues of e.g. the 3-composite medium is always smaller than the corresponding eigenvalues of the 2-composite medium, or vice versa. This is never true.

7.3 Further developments

It is possible to develop some results for concrete containing cracks by means of the eigenfunction method. Although this subject is also very important, we feel that a discussion of mathematically ill-posed problems and approximative solution formulæ with 5-tuple summations should not occur in this book, so we have omitted the theory. Instead we refer to Mejlbro (2001) for further details.

Chapter 8

Tables

8.1 Tables of erfc(u)

u	0	1	2	3	4	5	6	7	8	9
0.00	1	.99887	.99774	.99661	.99549	.99436	.99323	.99210	.99097	.98984
0.01	.98872	.98759	.98646	.98533	.98420	.98308	.98195	.98082	.97969	.97856
0.02	.97744	.97631	.97518	.97405	.97292	.97180	.97067	.96954	.96841	.96729
0.03	.96616	.96503	.96390	.96278	.96165	.96052	.95940	.95827	.95714	.95602
0.04	.95489	.95376	.95264	.95151	.95038	.94926	.94813	.94701	.94588	.94475
0.05	.94363	.94250	.94138	.94025	.93913	.93800	.93688	.93575	.93463	.93350
0.06	.93238	.93125	.93013	.92901	.92788	.92676	.92563	.92451	.92339	.92227
0.07	.92114	.92002	.91890	.91777	.91665	.91553	.91441	.91329	.91216	.91104
0.08	.90992	.90880	.90768	.90656	.90544	.90432	.90320	.90208	.90096	.89984
0.09	.89872	.89760	.89648	.89536	.89424	.89313	.89201	.89089	.88977	.88865
0.10	.88754	.88642	.88530	.88419	.88307	.88195	.88084	.87972	.87861	.87749
0.11	.87638	.87526	.87415	.87303	.87192	.87081	.86969	.86858	.86747	.86635
0.12	.86524	.86413	.86302	.86191	.86079	.85968	.85857	.85746	.85635	.85524
0.13	.85413	.85302	.85191	.85081	.84970	.84859	.84748	.84637	.84527	.84416
0.14	.84305	.84195	.84084	.83973	.93863	.83752	.83642	.83532	.83421	.83311
0.15	.83200	.83090	.82980	.82870	.82759	.82649	.82539	.82429	.82319	.82209
0.16	.82099	.81989	.81879	.81769	.81659	.81549	.81440	.81330	.81220	.81110
0.17	.81001	.80891	.80782	.80672	.80563	.80453	.80344	.80234	.80125	.80016
0.18	.79906	.79797	.79688	.79579	.79470	.79361	.79252	.79143	.79034	.78925
0.19	.78816	.78707	.78598	.78490	.78381	.78272	.78164	.78055	.77947	.77838
0.20	.77730	.77621	.77513	.77405	.77296	.77188	.77080	.76972	.76864	.76756
0.21	.76648	.76540	.76432	.76324	.76216	.76109	.76001	.75893	.75786	.75678
0.22	.75570	.75463	.75355	.75248	.75141	.75033	.74926	.74819	.74712	.74605
0.23	.74498	.74391	.74284	.74177	.74070	.73963	.73856	.73750	.73643	.73537
0.24	.73430	.73323	.73217	.73111	.73004	.72898	.72792	.72686	.72579	.72473
0.25	.72367	.72261	.72155	.72050	.71944	.71838	.71732	.71627	.71521	.71416
0.26	.71310	.71205	.71099	.70994	.70889	.70783	.70678	.70573	.70468	.70363
0.27	.70258	.70153	.70048	.69944	.69839	.69734	.69630	.69525	.69421	.69316
0.28	.69212	.69108	.69003	.68899	.68795	.68691	.68587	.68483	.68379	.68275
0.29	.68172	.68068	.67964	.67861	.67757	.67654	.67550	.67447	.67344	.67240
0.30	.67137	.67034	.66931	.66828	.66725	.66622	.66520	.66417	.66314	.66212

u	0	1	2	3	4	5	6	7	8	9
0.31	.66109	.66007	.65904	.65802	.65700	.65597	.65495	.65393	.65291	.65189
0.32	.65087	.64986	.64884	.64782	.64681	.64579	.64477	.64376	.64275	.64173
0.33	.64072	.63971	.63870	.63769	.63668	.63567	.63466	.63365	.63265	.63164
0.34	.63064	.62963	.62863	.62762	.62662	.62562	.62462	.62362	.62262	.62162
0.35	.62062	.61962	.61862	.61763	.61663	.61564	.61464	.61365	.61265	.61166
0.36	.61067	.60968	.60869	.60770	.60671	.60572	.60474	.60375	.60276	.60178
0.37	.60079	.59981	.59785	.59785	.59686	.59588	.59490	.59392	.59295	.59197
0.38	.59099	.59001	.58904	.58806	.58709	.58612	.58514	.58417	.58320	.58223
0.39	.58126	.58029	.57932	.57836	.57739	.57642	.57546	.57450	.57353	.57257
0.40	.57161	.57065	.56969	.56873	.56777	.56681	.56585	.56490	.56394	.56299
0.41	.56203	.56108	.56012	.55917	.55822	.55727	.55632	.55537	.55443	.55348
0.42	.55253	.55159	.55064	.54970	.54876	.54781	.54687	.54593	.54499	.54405
0.43	.54311	.54218	.54124	.54030	.53937	.53843	.53750	.53657	.53564	.53471
0.44	.53377	.53285	.53192	.53099	.53006	.52914	.52821	.52729	.52636	.52544
0.45	.52452	.52360	.52268	.52176	.52084	.51992	.51900	.51809	.51717	.51626
0.46	.51534	.51443	.51352	.51261	.51170	.51079	.50988	.50897	.50807	.50716
0.47	.50625	.50535	.50445	.50354	.50264	.50174	.50084	.49994	.49904	.49815
0.48	.49725	.49635	.49546	.49457	.49367	.49278	.49189	.49100	.49011	.48922
0.49	.48833	.48744	.48656	.48567	.48479	.48391	.48302	.48214	.48126	.48038
0.50	.47950	.47862	.47774	.47687	.47599	.47512	.47424	.47337	.47250	.47163
0.51	.47076	.46989	.46902	.46815	.46728	.46642	.46555	.46469	.46383	.46296
0.52	.46210	.46124	.46038	.45952	.45866	.45781	.45695	.45610	.45524	.45439
0.53	.45354	.45268	.45183	.45098	.45013	.44929	.44844	.44759	.44675	.44590
0.54	.44506	.44422	.44338	.44254	.44170	.44086	.44002	.43918	.43835	.43751
0.55	.43668	.43584	.43501	.43418	.43335	.43252	.43169	.43086	.43004	.42921
0.56	.42838	.42756	.42674	.42591	.42509	.42427	.42345	.42263	.42182	.42100
0.57	.42018	.41937	.41856	.41774	.41693	.41612	.41531	.41450	.41369	.41288
0.58	.41208	.41127	.41047	.40966	.40886	.40806	.40726	.40646	.40566	.40486
0.59	.40406	.40327	.40247	.40168	.40088	.40009	.39930	.39851	.39772	.39693
0.60	.39614	.39536	.39457	.39379	.39300	.39222	.39144	.39066	.38988	.38910
0.61	.38832	.38754	.38677	.38599	.38522	.38444	.38367	.38290	.38213	.38136
0.62	.38059	.37982	.37905	.37829	.37752	.37676	.37600	.37523	.37447	.37371
0.63	.37295	.37220	.37144	.37068	.36993	.36917	.36842	.36767	.36691	.36616
0.64	.36541	.36467	.36392	.36317	.36243	.36168	.36094	.36019	.35945	.35871
0.65	.35797	.35723	.35649	.35576	.35502	.35428	.35355	.35282	.35209	.35135
0.66	.35062	.34989	.34917	.34844	.34771	.34699	.34626	.34554	.34481	.34409
0.67	.34337	.34265	.34193	.34122	.34050	.33978	.33907	.33835	.33764	.33693
0.68	.33622	.33551	.33480	.33409	.33338	.33268	.33197	.33127	.33056	.32986
0.69	.32916	.32846	.32776	.32706	.32636	.32567	.32497	.32428	.32358	.32289
0.70	.32220	.32151	.32082	.32013	.31944	.31875	.31807	.31738	.31670	.31602
0.71	.31533	.31465	.31397	.31329	.31262	.31194	.31126	.31059	.30991	.30924
0.72	.30857	.30790	.30722	.30656	.30589	.30522	.30455	.30388	.30322	.30256
0.73	.30190	.30123	.30057	.29991	.29925	.29860	.29794	.29728	.29663	.29597
0.74	.29532	.29467	.29402	.29337	.29272	.29207	.29142	.29078	.29013	.28949
0.75	.28884	.28820	.28756	.28692	.28628	.28564	.28500	.28437	.28373	.28310
0.76	.28246	.28183	.28120	.28057	.27994	.27931	.27868	.27805	.27743	.27680
0.77	.27618	.27556	.27493	.27431	.27369	.27307	.27245	.27184	.27122	.27060
0.78	.26999	.26938	.26876	.26815	.26754	.26693	.26632	.26571	.26511	.26450
0.79	.26390	.26329	.26269	.26209	.26149	.26089	.26029	.25969	.25909	.25849
0.80	.25790	.25730	.25671	.25612	.25553	.25494	.25435	.25376	.25317	.25258

u	0	1	2	3	4	5	6	7	8	9
0.81	.25200	.25141	.25083	.25024	.24966	.24908	.24850	.24792	.24734	.24677
0.82	.24619	.24561	.24504	.24447	.24389	.24332	.24275	.24218	.24161	.24104
0.83	.24048	.23991	.23934	.23878	.23822	.23766	.23709	.23653	.23597	.23541
0.84	.23486	.23430	.23374	.23319	.23264	.23208	.23153	.23098	.23043	.22988
0.85	.22933	.22878	.22824	.22769	.22715	.22660	.22606	.22552	.22498	.22444
0.86	.22390	.22336	.22282	.22229	.22175	.22122	.22068	.22015	.21962	.21909
0.87	.21856	.21803	.21750	.21698	.21645	.21592	.21540	.21488	.21435	.21383
0.88	.21331	.21279	.21227	.21176	.21124	.21072	.21021	.20969	.20918	.20867
0.89	.20816	.20765	.20714	.20663	.20612	.20561	.20511	.20460	.20410	.20359
0.90	.20309	.20259	.20209	.20159	.20109	.20059	.20010	.19960	.19910	.19861
0.91	.19812	.19762	.19713	.19664	.19615	.19566	.19518	.19469	.19420	.19372
0.92	.19323	.19275	.19227	.19178	.19130	.19082	.19034	.18987	.18939	.18891
0.93	.18844	.18796	.18749	.18701	.18654	.18607	.18560	.18513	.18466	.18420
0.94	.18373	.18326	.18280	.18233	.18187	.18141	.18095	.18049	.18003	.17957
0.95	.17911	.17865	.17820	.17774	.17729	.17683	.17638	.17593	.17548	.17503
0.96	.17458	.17423	.17368	.17323	.17279	.17234	.17190	.17145	.17101	.17057
0.97	.17013	.16969	.16925	.16881	.16838	.16794	.16750	.16707	.16663	.16620
0.98	.16577	.16534	.16491	.16448	.16405	.16362	.16319	.16277	.16234	.16192
0.99	.16192	.16107	.16065	.16023	.15980	.15939	.15897	.15855	.15813	.15771
1.00	.15730	.15688	.15647	.15606	.15565	.15523	.15482	.15441	.15400	.15360
1.01	.15319	.15278	.15238	.15197	.15157	.15117	.15076	.15036	.14996	.14956
1.02	.14916	.14876	.14837	.14797	.14757	.14718	.14678	.14639	.14600	.14561
1.03	.14522	.14483	.14444	.14405	.14366	.14327	.14289	.14250	.14212	.14173
1.04	.14135	.14097	.14059	.14021	.13983	.13945	.13907	.13869	.13831	.13794
1.05	.13756	.13719	.13682	.13644	.13607	.13570	.13533	.13496	.13459	.13422
1.06	.13386	.13349	.13312	.13276	.13240	.13203	.13167	.13131	.13095	.13059
1.07	.13023	.12987	.12951	.12915	.12880	.12844	.12809	.12773	.12738	.12703
1.08	.12667	.12632	.12597	.12562	.12527	.12493	.12458	.12423	.12389	.12354
1.09	.12320	.12285	.12251	.12217	.12183	.12149	.12115	.12081	.12047	.12013
1.10	.11979	.11946	.11912	.11879	.11845	.11812	.11779	.11746	.11713	.11680
1.11	.11647	.11614	.11581	.11548	.11516	.11483	.11451	.11418	.11386	.11353
1.12	.11321	.11289	.11257	.11225	.11193	.11161	.11129	.11098	.11066	.11034
1.13	.11003	.10971	.10940	.10909	.10878	.10846	.10815	.10784	.10753	.10722
1.14	.10692	.10661	.10630	.10600	.10569	.10539	.10508	.10478	.10448	.10418
1.15	.10388	.10358	.10328	.10298	.10268	.10238	.10208	.10179	.10149	.10120
1.16	.10090	.10061	.10032	.10003	.09973	.09944	.09915	.09886	.09858	.09829
1.17	.09800	.09771	.09743	.09714	.09686	.09657	.09629	.09601	.09572	.09544
1.18	.09516	.09488	.09460	.09432	.09405	.09377	.09349	.09322	.09294	.09267
1.19	.09239	.09212	.09185	.09157	.09130	.09103	.09076	.09049	.09022	.08995
1.20	.08969	.08942	.08915	.08889	.08862	.08836	.08809	.08783	.08757	.08731
1.21	.08704	.08678	.08652	.08626	.08601	.08575	.08549	.08523	.08498	.08472
1.22	.08447	.08421	.08396	.08370	.08345	.08320	.08295	.08270	.08245	.08220
1.23	.08195	.08170	.08145	.08121	.08096	.08071	.08047	.08022	.07998	.07974
1.24	.07949	.07925	.07901	.07877	.07853	.07829	.07805	.07781	.07757	.07734
1.25	.07710	.07686	.07663	.07639	.07616	.07592	.07569	.07546	.07523	.07500
1.26	.07476	.07453	.07430	.07407	.07385	.07362	.07339	.07316	.07294	.07271
1.27	.07249	.07226	.07204	.07181	.07159	.07137	.07115	.07093	.07071	.07049
1.28	.07027	.07005	.06983	.06961	.06939	.06918	.06896	.06874	.06853	.06832
1.29	.06810	.06789	.06768	.06746	.06725	.06704	.06683	.06662	.06641	.06620
1.30	.06599	.06578	.06558	.06537	.06516	.06496	.06475	.06455	.06434	.06414

u	0	1	2	3	4	5	6	7	8	9
1.31	.06394	.06373	.06353	.06333	.06313	.06293	.06273	.06253	.06233	.06213
1.32	.06193	.06174	.06154	.06134	.06115	.06095	.06076	.06056	.06037	.06018
1.33	.05998	.05979	.05960	.05941	.05922	.05903	.05884	.05865	.05846	.05827
1.34	.05809	.05790	.05771	.05753	.05734	.05716	.05697	.05679	.05660	.05642
1.35	.05624	.05606	.05587	.05569	.05551	.05533	.05515	.05497	.05479	.05462
1.36	.05444	.05426	.05408	.05391	.05373	.05356	.05338	.05321	.05303	.05286
1.37	.05269	.05252	.05234	.05217	.05200	.05183	.05166	.05149	.05132	.05115
1.38	.05098	.05082	.05065	.05048	.05032	.05015	.04998	.04982	.04965	.04949
1.39	.04933	.04916	.04900	.04884	.04868	.04852	.04835	.04819	.04803	.04787
1.40	.04771	.04756	.04740	.04724	.04708	.04693	.04677	.04661	.04646	.04630
1.41	.04615	.04599	.04584	.04569	.04553	.04538	.04523	.04508	.04493	.04477
1.42	.04462	.04447	.04432	.04418	.04403	.04388	.04373	.04358	.04344	.04329
1.43	.04314	.04300	.04285	.04271	.04256	.04242	.04227	.04213	.04199	.04185
1.44	.04170	.04156	.04142	.04128	.04114	.04100	.04086	.04072	.04058	.04044
1.45	.04030	.04017	.04003	.03989	.03976	.03962	.03949	.03935	.03922	.03908
1.46	.03895	.03881	.03868	.03855	.03841	.03828	.03815	.03802	.03789	.03776
1.47	.03763	.03750	.03737	.03724	.03711	.03698	.03685	.03673	.03660	.03647
1.48	.03635	.03622	.03609	.03597	.03584	.03572	.03560	.03547	.03535	.03522
1.49	.03510	.03498	.03486	.03474	.03461	.03449	.03437	.03425	.03413	.03401
1.50	.03389	.03378	.03366	.03354	.03342	.03330	.03319	.03307	.03295	.03284
1.51	.03272	.03261	.03249	.03238	.03226	.03215	.03204	.03192	.03181	.03170
1.52	.03159	.03147	.03136	.03125	.03114	.03103	.03092	.03081	.03070	.03059
1.53	.03048	.03038	.03027	.03016	.03005	.02994	.02984	.02973	.02963	.02952
1.54	.02941	.02931	.02920	.02910	.02900	.02889	.02879	.02869	.02858	.02848
1.55	.02838	.02828	.02817	.02807	.02797	.02787	.02777	.02767	.02757	.02747
1.56	.02737	.02727	.02717	.02708	.02698	.02688	.02678	.02669	.02659	.02649
1.57	.02640	.02630	.02621	.02611	.02602	.02592	.02583	.02573	.02564	.02555
1.58	.02545	.02536	.02527	.02518	.02508	.02499	.02490	.02481	.02472	.02463
1.59	.02454	.02445	.02436	.02427	.02418	.02409	.02400	.02391	.02383	.02374
1.60	.02365	.02356	.02348	.02339	.02330	.02322	.02313	.02305	.02296	.02288
1.61	.02279	.02271	.02262	.02254	.02246	.02237	.02229	.02221	.02213	.02204
1.62	.02196	.02188	.02180	.02172	.02164	.02156	.02148	.02140	.02132	.02124
1.63	.02116	.02108	.02100	.02092	.02084	.02076	.02069	.02061	.02053	.02045
1.64	.02038	.02030	.02023	.02015	.02007	.02000	.01992	.01985	.01977	.01970
1.65	.01962	.01955	.01948	.01940	.01933	.01926	.01918	.01911	.01904	.01897
1.66	.01890	.01882	.01875	.01868	.01861	.01854	.01847	.01840	.01833	.01826
1.67	.01819	.01812	.01805	.01798	.01791	.01785	.01778	.01771	.01764	.01757
1.68	.01751	.01744	.01737	.01731	.01724	.01717	.01711	.01705	.01698	.01691
1.69	.01685	.01678	.01672	.01665	.01659	.01653	.01646	.01640	.01634	.01627
1.70	.01621	.01615	.01608	.01602	.01596	.01590	.01584	.01578	.01571	.01565
1.71	.01559	.01553	.01547	.01541	.01535	.01529	.01523	.01517	.01511	.01506
1.72	.01500	.01494	.01488	.01482	.01476	.01471	.01465	.01459	.01454	.01448
1.73	.01442	.01437	.01431	.01425	.01420	.01414	.01409	.01403	.01398	.01392
1.74	.01387	.01381	.01376	.01370	.01365	.01359	.01354	.01349	.01343	.01338
1.75	.01333	.01327	.01322	.01317	.01312	.01307	.01301	.01296	.01291	.01286
1.76	.01281	.01276	.01271	.01266	.01261	.01256	.01251	.01246	.01241	.01236
1.77	.01231	.01226	.01221	.01216	.01211	.01207	.01202	.01197	.01192	.01187
1.78	.01183	.01178	.01173	.01168	.01164	.01159	.01154	.01150	.01145	.01141
1.79	.01136	.01131	.01127	.01122	.01118	.01113	.01109	.01104	.01100	.01095
1.80	.01091	.01087	.01082	.01078	.01073	.01069	.01065	.01060	.01056	.01052

8.2 Tables of inv erfc(u)

u	0	1	2	3	4	5	6	7	8	9
.01	1.82139	1.79796	1.77635	1.75629	1.73755	1.71995	1.70336	1.68766	1.67274	1.65854
.02	1.64498	1.63199	1.61953	1.60756	1.59603	1.58491	1.57417	1.56378	1.55372	1.54396
.03	1.53449	1.52528	1.51633	1.50761	1.49912	1.49083	1.48275	1.47486	1.46714	1.45960
.04	1.45222	1.44499	1.43792	1.43098	1.42418	1.41750	1.41096	1.40453	1.39821	1.39201
.05	1.38530	1.37931	1.37400	1.36814	1.36248	1.35685	1.35131	1.34584	1.34046	1.33515
.06	1.32992	1.32476	1.31967	1.31465	1.30969	1.30479	1.29996	1.29519	1.29047	1.28582
.07	1.28121	1.27667	1.27217	1.26772	1.26333	1.25898	1.25468	1.25042	1.24621	1.24205
.08	1.23792	1.23384	1.22980	1.22580	1.22183	1.21791	1.21402	1.21017	1.20635	1.20257
.09	1.19883	1.19511	1.19143	1.18778	1.18417	1.18058	1.17702	1.17350	1.17000	1.16653
.10	1.16309	1.15967	1.15628	1.15292	1.14959	1.14628	1.14299	1.13973	1.13650	1.13328
.11	1.13009	1.12693	1.12378	1.12066	1.11756	1.11448	1.11142	1.10838	1.10537	1.10237
.12	1.09939	1.09643	1.09349	1.09057	1.08767	1.08479	1.08192	1.07907	1.07624	1.07343
.13	1.07063	1.06785	1.06509	1.06234	1.05961	1.05689	1.05419	1.05151	1.04884	1.04618
.14	1.04354	1.04092	1.03830	1.03571	1.03312	1.03055	1.02800	1.02545	1.02292	1.02041
.15	1.01790	1.01541	1.01293	1.01047	1.00801	1.00557	1.00314	1.00072	0.99831	0.99592
.16	0.99354	0.99116	0.98880	0.98645	0.98411	0.98178	0.97946	0.97716	0.97486	0.97257
.17	0.97029	0.96803	0.96577	0.96352	0.96129	0.95906	0.95684	0.95463	0.95243	0.95024
.18	0.94806	0.94588	0.94372	0.94157	0.93942	0.93728	0.93515	0.93303	0.93092	0.92882
.19	0.92672	0.92463	0.92255	0.92048	0.91842	0.91636	0.91431	0.91227	0.91024	0.90821
.20	0.90619	0.90418	0.90218	0.90018	0.89819	0.89621	0.89424	0.89227	0.89031	0.88835
.21	0.88640	0.88446	0.88253	0.88060	0.87868	0.87677	0.87486	0.87295	0.87106	0.86917
.22	0.86729	0.86541	0.86354	0.86167	0.85981	0.85796	0.85611	0.85427	9.85244	0.85061
.23	0.84878	0.84696	0.84515	0.84334	0.84154	0.83974	0.83795	0.83617	0.83439	0.83261
.24	0.83084	0.82908	0.82732	0.82556	0.82381	0.82207	0.82033	0.81859	0.81686	0.81514
.25	0.81342	0.81170	0.80999	0.80829	0.80659	0.80489	0.80320	0.80151	0.79983	0.79815
.26	0.79648	0.79481	0.79315	0.79148	0.78983	0.78818	0.78653	0.78489	0.78325	0.78161
.27	0.77998	0.77836	0.77673	0.77512	0.77350	0.77189	0.77029	0.76868	0.76709	0.76549
.28	0.76390	0.76231	0.76073	0.75915	0.75758	0.75601	0.75444	0.75287	0.75131	0.74976
.29	0.74820	0.74666	0.74511	0.74357	0.74203	0.74049	0.73896	0.73743	0.73591	0.73439
.30	0.73287	0.73135	0.72984	0.72834	0.72683	0.72533	0.72383	0.72234	0.72084	0.71936
.31	0.71787	0.71639	0.71491	0.71343	0.71196	0.71049	0.70902	0.70756	0.70610	0.70464
.32	0.70319	0.70174	0.70029	0.69884	0.69740	0.69596	0.69452	0.69309	0.69166	0.69023
.33	0.68880	0.68738	0.68596	0.68454	0.68313	0.68172	0.68031	0.67890	0.67750	0.67610
.34	0.67470	0.67330	0.67191	0.67052	0.66913	0.66774	0.66636	0.66498	0.66360	0.66223
.35	0.66085	0.65948	0.65812	0.65675	0.65539	0.65403	0.65267	0.65131	0.64996	0.64861
.36	0.64726	0.64591	0.64457	0.64323	0.64189	0.64055	0.63922	0.63789	0.63656	0.63523
.37	0.63390	0.63258	0.63126	0.62994	0.62862	0.62731	0.62599	0.62468	0.62338	0.62207
.38	0.62077	0.61946	0.61816	0.61687	0.61557	0.61428	0.61299	0.61170	0.61041	0.60912
.39	0.60784	0.60656	0.60528	0.60400	0.60273	0.60145	0.60018	0.59891	0.59765	0.59638
.40	0.59512	0.59385	0.59259	0.59134	0.59008	0.58883	0.58757	0.58632	0.58507	0.58383
.41	0.58258	0.58134	0.58010	0.57886	0.57763	0.57638	0.57515	0.57391	0.57268	0.57145
.42	0.57023	0.56900	0.56778	0.56655	0.56533	0.56411	0.56290	0.56168	0.56047	0.55925
.43	0.55804	0.55683	0.55563	0.55442	0.55322	0.55201	0.55081	0.54961	0.54841	0.54722
.44	0.54602	0.54483	0.54364	0.54245	0.54126	0.54007	0.53889	0.53770	0.53652	0.53534
.45	0.53416	0.53298	0.53180	0.53063	0.52946	0.52828	0.52711	0.52594	0.52478	0,52361
.46	0.52244	0.52128	0.52012	0.51896	0.51780	0.51664	0.51548	0.51433	0.51317	0.51202
.47	0.51087	0.50972	0.50857	0.50742	0.50628	0.50513	0.50399	0.50285	0.50171	0.50057
.48	0.49943	0,49829	0.49716	0.49603	0.49489	0.49376	0.49263	0.49150	0.49037	0.48925
.49	0.48812	0.48700	0.48588	0.48475	0.48363	0.48251	0.48140	0.48028	0.47916	0.47805

u	0	1	2	3	4	5	6	7	8	9
.50	.47694	.47582	.47471	.47360	.47250	.47139	.47028	.46918	.46807	.46697
.51	.46587	.46477	.46367	.46257	.46147	.46038	.45928	.45819	.45710	.45600
.52	.45491	.45382	.45274	.45165	.45056	.44948	.44839	.44731	.44623	.44515
.53	.44407	.44299	.44191	.44083	.43976	.43868	.43761	.43654	.43546	.43439
.54	.43332	.43226	.43119	.43012	.42905	.42799	.42693	.42586	.42480	.42374
.55	.42268	.42162	.42056	.41951	.41845	.41739	.41634	.41529	.41423	.41318
.56	.41213	.41108	.41003	.40898	.40794	.40689	.40585	.40480	.40376	.40271
.57	.40167	.40063	.39959	.39855	.39751	.39648	.39544	.39440	.39337	.39234
.58	.39130	.39027	.38924	.38821	.38718	.38615	.38512	.38409	.38307	.38204
.59	.38101	.37999	.37897	.37794	.37692	.37590	.37488	.37386	.37284	.37182
.60	.37081	.36979	.36877	.36776	.36675	.36573	.36472	.36371	.36270	.36169
.61	.36068	.35967	.35866	.35765	.35664	.35564	.35463	.35363	.36262	.35162
.62	.35062	.34962	.34862	.34762	.34662	.34562	.34462	.34362	.34262	.34163
.63	.34063	.33964	.33864	.33765	.33666	.33566	.33467	.33368	.33269	.33170
.64	.33071	.32972	.32874	.32775	.32676	.32578	.32479	.32381	.32282	.32184
.65	.32086	.31988	.31889	.31791	.31693	.31595	.31498	.31400	.31302	.31204
.66	.31107	.31009	.30911	.30814	.30717	.30619	.30522	.30425	.30327	.30230
.67	.30133	.30036	.29939	.29842	.29745	.29649	.29552	.29455	.29359	.29262
.68	.29166	.29069	.28973	.28876	.28780	.28684	.28588	.28491	.28395	.28299
.69	.28203	.28107	.28011	.27916	.27820	.27724	.27628	.27533	.27437	.27342
.70	.27246	.27151	.27055	.26960	.26865	.26770	.26674	.26579	.26484	.26389
.71	.26294	.26199	.26104	.26010	.25915	.25820	.25725	.25631	.25536	.25441
.72	.25347	.25252	.25158	.25064	.24969	.24875	.24781	.24686	.24592	.24498
.73	.24404	.24310	.24216	.24122	.24028	.23934	.23840	.23747	.23653	.23559
.74	.23466	.23372	.23278	.23185	.23091	.22998	.22904	.22811	.22718	.22624
.75	.22531	.22438	.22345	.22252	.22159	.22065	.21972	.21879	.21787	.21694
.76	.21601	.21508	.21415	.21322	.21230	.21137	.21044	.20952	.20859	.20767
.77	.20674	.20582	.20489	.20397	.20304	.20212	.20120	.20027	.19935	.19843
.78	.19751	.19659	.19567	.19475	.19383	.19291	.19199	.19107	.19015	.18923
.79	.18831	.18739	.18647	.18556	.18464	.18372	.18281	.18189	.18097	.18006
.80	.17914	.17823	.17731	.17640	.17549	.17457	.17366	.17274	.17183	.17092
.81	.17001	.16909	.16818	.16727	.16636	.16545	.16454	.16363	.16272	.16181
.82	.16090	.15999	.15908	.15817	.15726	.15635	.15545	.15454	.15363	.15272
.83	.15182	.15091	.15000	.14910	.14819	.14729	.14638	.14547	.14457	.14366
.84	.14276	.14186	.14095	.14005	.13914	.13824	.13734	.13643	.13553	.13463
.85	.13373	.13282	.13192	.13102	.13012	.12922	.12832	.12742	.12652	.12562
.86	.12472	.12382	.12292	.12202	.12112	.12022	.11932	.11842	.11752	.11662
.87	.11572	.11483	.11393	.11303	.11213	.11124	.11034	.10944	.10854	.10765
.88	.10675	.10586	.10496	.10406	.10317	.10227	.10138	.10048	.09959	.09869
.89	.09780	.09690	.09601	.09511	.09422	.09332	.09243	.09154	.09064	.08975
.90	.08886	.08796	.08707	.08618	.08528	.08439	.08350	.08261	.08171	.08082
.91	.07993	.07904	.07815	.07726	.07636	.07547	.07458	.07369	.07280	.07191
.92	.07102	.07013	.06924	.06835	.06746	.06657	.06568	.06479	.06390	.06301
.93	.06212	.06123	.06034	.05945	.05856	.05767	.05678	.05589	.05500	.05411
.94	.05322	.05234	.05145	.05056	.04967	.04878	.04789	.04700	.04612	.04523
.95	.04434	.04345	.04256	.04168	.04079	.03990	.03901	.03813	.03724	.03635
.96	.03546	.03458	.03369	.03280	.03192	.03103	.03014	.02925	.02837	.02748
.97	.02659	.02571	.02482	.02393	.02305	.02216	.02127	.02039	.01950	.01861
.98	.01772	.01684	.01595	.01507	.01418	.01329	.01241	.01152	.01064	.00975
.99	.00886	.00798	.00709	.00620	.00532	.00443	.00354	.00266	.00177	.00089

8.3 Tables of $\Psi_p(u)$

u	$\Psi_{-.50}$	$\Psi_{-.45}$	$\Psi_{-.40}$	$\Psi_{-.35}$	$\Psi_{-.30}$	$\Psi_{-.25}$	$\Psi_{-.20}$	$\Psi_{-.15}$
0.01	.99990	.99825	.99679	.99548	.99429	.99319	.99218	.99123
0.02	.99960	.99632	.99342	.99082	.98845	.98628	.98428	.98241
0.03	.99910	.99421	.98989	.98602	.98250	.97928	.97630	.97352
0.04	.99840	.99193	.98621	.98109	.97644	.97218	.96825	.96459
0.05	.99750	.98947	.98237	.97602	.97026	.96499	.96013	.95561
0.06	.99641	.98683	.97838	.97082	.96398	.95772	.95194	.94657
0.07	.99511	.98402	.97424	.96550	.95758	.95035	.94368	.93749
0.08	.99362	.98103	.96995	.96004	.95108	.94290	.93536	.92837
0.09	.99193	.97787	.96550	.95446	.94448	.93537	.92698	.91921
0.10	.99005	.97455	.96092	.94876	.93778	.92776	.91854	.91000
0.11	.98797	.97105	.95619	.94293	.93097	.92007	.91005	.90076
0.12	.98570	.96739	.95131	.93699	.92407	.91231	.90150	.89149
0.13	.98324	.96356	.94630	.93093	.91708	.90447	.89289	.88219
0.14	.98059	.95957	.94114	.92475	.90999	.89656	.88424	.87286
0.15	.97775	.95542	.93586	.91847	.90282	.88859	.87555	.86350
0.16	.97472	.95111	.93043	.91207	.89555	.88055	.86680	.85412
0.17	.97151	.94664	.92488	.90556	.88821	.87245	.85802	.84471
0.18	.96812	.94202	.91920	.89896	.88078	.86429	.84920	.83529
0.19	.96454	.93724	.91339	.89225	.87327	.85607	.84034	.82585
0.20	.96079	.93231	.90746	.88544	.86569	.84780	.83145	.81640
0.21	.95686	.92724	.90140	.87853	.85803	.83948	.82253	.80694
0.22	.95275	.92202	.89523	.87153	.85031	.83110	.81358	.79747
0.23	.94847	.91666	.88894	.86444	.84251	.82269	.80460	.78799
0.24	.94403	.91116	.88254	.85726	.83465	.81422	.79560	.77850
0.25	.93941	.90552	.87603	.85000	.82673	.80572	.78659	.76902
0.26	.93463	.89974	.86941	.84265	.81875	.79718	.77755	.75954
0.27	.92969	.89384	.86269	.83522	.81071	.78861	.76850	.75006
0.28	.92459	.88780	.85586	.82772	.80262	.78000	.75943	.74059
0.29	.91934	.88164	.84894	.82014	.79448	.77136	.75036	.73113
0.30	.91393	.87536	.84192	.81250	.78629	.76269	.74127	.72168
0.31	.90837	.86896	.83481	.80478	.77805	.75401	.73219	.71224
0.32	.90267	.86244	.82761	.79700	.76977	.74529	.72310	.70282
0.33	.89682	.85581	.82032	.78916	.76145	.73657	.71401	.69341
0.34	.89083	.84907	.81295	.78126	.75310	.72782	.70492	.68403
0.35	.88471	.84222	.80551	.77331	.74471	.71906	.69584	.67467
0.36	.87845	.83527	.79798	.76530	.73630	.71030	.68677	.66533
0.37	.87206	.82822	.79038	.75724	.72785	.70151	.67771	.65602
0.38	.86554	.82107	.78271	.74914	.71938	.69273	.66866	.64674
0.39	.85890	.81383	.77498	.74099	.71089	.68395	.65962	.63749
0.40	.85214	.80650	.76718	.73281	.70238	.67516	.65061	.62828
0.41	.84527	.79908	.75932	.72458	.69385	.66638	.64161	.61910
0.42	.83828	.79158	.75141	.71633	.68531	.65761	.63264	.60996
0.43	.83119	.78400	.74344	.70804	.67676	.64884	.62369	.60086
0.44	.82399	.77635	.73542	.69972	.66820	.64008	.61476	.59180
0.45	.81669	.76862	.72735	.69138	.65964	.63133	.60587	.58278
0.46	.80929	.76083	.71924	.68302	.65107	.62261	.59701	.57381
0.47	.80180	.75297	.71109	.67464	.64251	.61389	.58818	.56489
0.48	.79422	.74505	.70291	.66624	.63395	.60520	.57602	.55602
0.49	.78655	.73707	.69469	.65783	.62539	.59653	.57063	.54720

u	$\Psi_{-.50}$	$\Psi_{-.45}$	$\Psi_{-.40}$	$\Psi_{-.35}$	$\Psi_{-.30}$	$\Psi_{-.25}$	$\Psi_{-.20}$	$\Psi_{-.15}$
0.50	.77880	.72904	.68644	.64942	.61685	.58789	.56192	.53844
0.51	.77097	.72095	.67816	.64099	.60831	.57928	.55324	.52972
0.52	.76307	.71282	.66986	.63256	.59979	.57069	.54461	.52107
0.53	.75510	.70465	.66153	.62413	.59129	.56214	.53603	.51247
0.54	.74707	.69644	.65319	.61571	.58280	.55362	.52750	.50394
0.55	.73897	.68819	.64484	.60728	.57434	.54513	.51901	.49546
0.56	.73081	.67990	.63647	.59887	.56590	.53669	.51058	.48705
0.57	.72260	.67159	.62810	.59046	.55748	.52829	.50220	.47870
0.58	.71434	.66325	.61972	.58207	.54910	.51992	.49387	.47042
0.59	.70603	.65489	.61134	.57369	.54075	.51161	.48560	.46221
0.60	.69768	.64651	.60296	.56534	.53243	.50334	.47739	.45407
0.61	.68929	.63811	.59458	.55700	.42414	.49512	.46924	.44599
0.62	.68086	.62970	.58621	.54869	.51590	.48695	.46115	.43799
0.63	.67240	.62129	.57786	.54040	.50769	.47883	.45313	.43006
0.64	.66392	.61286	.56951	.53214	.49953	.47076	.44516	.42220
0.65	.65541	.60444	.56118	.52391	.49141	.46276	.43727	.41442
0.66	.64688	.59601	.55287	.51572	.48334	.45480	.42944	.40671
0.67	.63833	.58759	.54458	.50756	.47531	.44691	.42168	.39908
0.68	.62977	.57918	.53631	.49944	.46734	.43908	.41399	.39153
0.69	.62120	.57077	.52807	.49136	.45941	.43131	.40637	.38406
0.70	.61263	.56238	.51986	.48332	.45154	.42361	.39882	.37667
0.71	.60405	.55401	.51168	.47533	.44373	.41597	.39135	.36935
0.72	.59547	.54565	.50353	.46738	.43597	.40839	.38395	.36212
0.73	.58690	.53732	.49542	.45949	.42828	.40089	.37663	.35497
0.74	.57834	.52901	.48735	.45164	.42064	.39345	.36938	.34790
0.75	.56978	.52073	.47932	.44384	.41307	.38608	.36221	.34092
0.76	.56124	.51248	.47133	.43610	.40556	.37879	.35512	.33402
0.77	.55272	.50426	.46339	.42842	.39811	.37156	.34810	.32720
0.78	.54422	.49608	.45550	.42079	.39073	.36441	.34117	.32047
0.79	.53574	.48793	.44766	.41323	.38342	.35734	.33431	.31382
0.80	.52729	.47983	.43987	.40572	.37618	.35034	.32754	.30726
0.81	.51887	.47177	.43213	.39828	.36900	.34342	.32085	.30079
0.82	.51048	.46375	.42445	.39090	.36190	.33657	.31424	.29440
0.83	.50213	.45578	.41683	.38359	.35487	.32980	.30771	.28809
0.84	.49381	.44786	.40926	.37634	.34792	.32311	.30127	.28188
0.85	.48554	.44000	.40176	.36916	.34104	.31650	.29491	.27575
0.86	.47730	.43219	.39432	.36206	.33423	.30997	.28863	.26970
0.87	.46912	.42443	.38694	.35502	.32750	.30352	.28244	.26375
0.88	.46098	.41673	.37963	.34806	.32085	.29716	.27633	.25788
0.89	.45289	.40909	.37239	.34117	.31428	.29087	.27031	.25210
0.90	.44486	.40152	.36521	.33435	.30778	.28466	.26437	.24640
0.91	.43688	.39400	.35811	.32761	.30136	.27854	.25851	.24080
0.92	.42896	.38656	.35108	.32094	.29503	.27250	.25275	.23528
0.93	.42109	.37917	.34412	.31435	.28877	.26655	.24706	.22984
0.94	.41329	.37186	.33723	.30784	.28260	.26067	.24146	.22449
0.95	.40555	.36462	.33042	.30141	.27650	.25488	.23595	.21923
0.96	.39788	.35745	.32368	.29506	.27049	.24918	.23052	.21406
0.97	.39028	.35035	.31702	.28879	.26456	.24356	.22518	.20897
0.98	.38274	.34333	.31044	.28259	.25871	.23802	.21992	.20396
0.99	.37527	.33638	.30394	.27648	.25295	.23257	.21475	.19904
1.00	.36788	.32950	.29751	.27045	.24727	.22720	.20966	.19421

u	$\Psi_{-.50}$	$\Psi_{-.45}$	$\Psi_{-.40}$	$\Psi_{-.35}$	$\Psi_{-.30}$	$\Psi_{-.25}$	$\Psi_{-.20}$	$\Psi_{-.15}$
1.01	.36056	.32271	.29117	.26450	.24167	.22191	.20465	.18946
1.02	.35331	.31599	.28491	.25864	.23615	.21671	.19973	.18479
1.03	.34614	.30935	.27873	.25285	.23072	.21159	.19489	.18021
1.04	.33905	.30279	.27263	.24715	.22537	.20655	.19014	.17571
1.05	.33204	.29632	.26661	.24154	.22011	.20160	.18547	.17129
1.06	.32511	.28992	.26068	.23600	.21493	.19673	.18088	.16695
1.07	.31826	.28361	.25482	.23055	.20983	.19195	.17637	.16270
1.08	.31149	.27738	.24906	.22519	.20481	.18724	.17194	.15852
1.09	.30480	.27123	.24337	.21990	.19988	.18262	.16760	.15443
1.10	.29820	.26517	.23777	.21470	.19503	.17808	.16333	.15041
1.11	.29168	.25920	.23226	.20959	.19026	.17362	.15915	.14647
1.12	.28525	.25330	.22683	.20455	.18558	.16924	.15505	.14261
1.13	.27890	.24750	.22148	.19960	.18097	.16494	.15102	.13883
1.14	.27264	.24178	.21622	.19473	.17645	.16072	.14707	.13512
1.15	.26647	.23614	.21104	.18995	.17201	.15658	.14320	.13149
1.16	.26038	.23059	.20595	.18525	.16765	.15252	.13940	.12793
1.17	.25439	.22513	.20094	.18063	.16337	.14854	.13569	.12445
1.18	.24848	.21976	.19601	.17609	.15917	.14464	.13204	.12104
1.19	.24266	.21447	.19117	.17163	.15504	.14081	.12847	.11770
1.20	.23693	.20926	.18641	.16726	.15100	.13705	.12498	.11444
1.21	.23129	.20414	.18174	.16296	.14704	.13338	.12156	.11124
1.22	.22573	.19911	.17715	.15875	.14315	.12978	.11821	.10812
1.23	.22027	.19417	.17264	.15461	.13934	.12625	.11493	.10506
1.24	.21490	.18931	.16821	.15056	.13560	.12279	.11172	.10207
1.25	.20961	.18453	.16387	.14658	.13194	.11941	.10858	.09915
1.26	.20442	.17984	.15960	.14268	.12836	.11610	.10551	.09630
1.27	.19931	.17524	.15542	.13886	.12485	.11286	.10251	.09351
1.28	.19429	.17072	.15132	.13511	.12141	.10969	.09958	.09078
1.29	.18936	.16628	.14730	.13145	.11804	.10659	.09671	.08812
1.30	.18452	.16193	.14336	.12785	.11475	.10356	.09391	0.8552
1.31	.17977	.15766	.13949	.12433	.11153	.10059	.09117	.08299
1.32	.17510	.15347	.13570	.12089	.10837	.09769	.08850	.08051
1.33	.17052	.14937	.13199	.11751	.10529	.09486	.08589	.07809
1.34	.16603	.14534	.12836	.11421	.10228	.09210	.08334	.07573
1.35	.16162	.14140	.12481	.11098	.09933	.08939	.08085	.07343
1.36	.15730	.13754	.12132	.10783	.09645	.08675	.07842	.07119
1.37	.15306	.13375	.11792	.10474	.09363	.08418	.07605	.06900
1.38	.14891	.13005	.11458	.10172	.09088	.08166	.07374	.06687
1.39	.14484	.12642	.11132	.09877	.08820	.07921	.07148	.06479
1.40	.14086	.12287	.10813	.09588	.08558	.07681	.06928	.06277
1.41	.13696	.11939	.10501	.09307	.08302	.07447	.06714	.06080
1.42	.13313	.11599	.10196	.09031	.08052	.07219	.06505	.05887
1.43	.12939	.11267	.09898	.08763	.07808	.06997	.06301	.05700
1.44	.12573	.10941	.09607	.08500	.07570	.06780	.06103	.05518
1.45	.12215	.10624	.09323	.08244	.07338	.06569	.05910	.05341
1.46	.11865	.10313	.09045	.07994	.07112	.06363	.05722	.05168
1.47	.11522	.10010	.08774	.07750	.06891	.06163	.05539	.05000
1.48	.11187	.09713	.08510	.07513	.06676	.05967	.05360	.04837
1.49	.10860	.09423	.08251	.07281	.06467	.05777	.05187	.04678
1.50	.10540	.09141	.07999	.07055	.06263	.05592	.05018	.04524

u	$\Psi_{-.50}$	$\Psi_{-.45}$	$\Psi_{-.40}$	$\Psi_{-.35}$	$\Psi_{-.30}$	$\Psi_{-.25}$	$\Psi_{-.20}$	$\Psi_{-.15}$
1.51	.10227	.08865	.07754	.06834	.06064	.05411	.04854	.04373
1.52	.09922	.08595	.07514	.06619	.05870	.05236	.04694	.04227
1.53	.09624	.08332	.07280	.06410	.05682	.05065	.04539	.04086
1.54	.09333	.08076	.07052	.06206	.05498	.04899	.04388	.03948
1.55	.09049	.07826	.06830	.06008	.05319	.04738	.04241	.03814
1.56	.08772	.07582	.06614	.05814	.05146	.04680	.04098	.03684
1.57	.08502	.07345	.06403	.05626	.04977	.04428	.03960	.03558
1.58	.08238	.07113	.06198	.05443	.04812	.04279	.03825	.03435
1.59	.07981	.06887	.05998	.05265	.04652	.04135	.03695	.03316
1.60	.07730	.06667	.05804	.05092	.04497	.03995	.03568	.03201
1.61	.07486	.06453	.05614	.04923	.04346	.03859	.03444	.03089
1.62	.07248	.06245	.05430	.04759	.04199	.03727	.03325	.02980
1.63	.07017	.06042	.05251	.04600	.04056	.03599	.03209	.02875
1.64	.06791	.05845	.05077	.04445	.03918	.03474	.03096	.02773
1.65	.06571	.05652	.04908	.04294	.03783	.03353	.02987	.02674
1.66	.06357	.05466	.04743	.04148	.03653	.03236	.02881	.02578
1.67	.06149	.05284	.04583	.04006	.03526	.03122	.02779	.02485
1.68	.05946	.05107	.04427	.03868	.03403	.03012	.02679	.02395
1.69	.05749	.04935	.04276	.03735	.03284	.02905	.02583	.02308
1.70	.05558	.04768	.04129	.03605	.03168	.02801	.02490	.02223
1.71	.05371	.04606	.03987	.03479	.03056	.02700	.02399	.02141
1.72	.05190	.04449	.03849	.03356	.02947	.02603	.02312	.02062
1.73	.05014	.04296	.03714	.03238	.02841	.02509	.02227	.01986
1.74	.04843	.04147	.03584	.03123	.02739	.02417	.02145	.01912
1.75	.04677	.04003	.03458	.03011	.02640	.02329	.02065	.01840
1.76	.04516	.03863	.03335	.02903	.02544	.02243	.01988	.01771
1.77	.04359	.03727	.03216	.02798	.02451	.02160	.01914	.01704
1.78	.04207	.03595	.03101	.02697	.02361	.02080	.01842	.01639
1.79	.04060	.03467	.02990	.02598	.02274	.02002	.01772	.01576
1.80	.03916	.03343	.02881	.02503	.02190	.01927	.01705	.01516
1.81	.03778	.03223	.02776	.02411	.02108	.01854	.01640	.01457
1.82	.03643	.03107	.02675	.02322	.02029	.01784	.01577	.01401
1.83	.03512	.02994	.02577	.02235	.01953	.01716	.01516	.01346
1.84	.03386	.02885	.02481	.02152	.01879	.01650	.01458	.01294
1.85	.03263	.02779	.02389	.02071	.01807	.01587	.01401	.01243
1.86	.03144	.02677	.02300	.01993	.01738	.01526	.01346	.01194
1.87	.03029	.02577	.02214	.01916	.01671	.01466	.01293	.01146
1.88	.02918	.02481	.02130	.01844	.01607	.01409	.01242	.01101
1.89	.02810	.02388	.02050	.01773	.01545	.01354	.01193	.01057
1.90	.02705	.02298	.01971	.01705	.01485	.01301	.01146	.01014
1.91	.02604	.02211	.01896	.01639	.01426	.01249	.01100	.00973
1.92	.02506	.02127	.01823	.01575	.01370	.01200	.01056	.00934
1.93	.02412	.02046	.01753	.01514	.01316	.01152	.01013	.00896
1.94	.02320	.01968	.01685	.01454	.01264	.01106	.00972	.00859
1.95	.02231	.01892	.01619	.01397	.01214	.01061	.00933	.00824
1.96	.02146	.01818	.01555	.01341	.01165	.01018	.00895	.00790
1.97	.02063	.01747	.01494	.01288	.01118	.00977	.00858	.00757
1.98	.01983	.01679	.01435	.01236	.01073	.00937	.00822	.00726
1.99	.01906	.01613	.01378	.01187	.01029	.00898	.00788	.00695
2.00	.01832	.01549	.01323	.01139	.00987	.00861	.00756	.00666

u	$\Psi_{-.10}$	$\Psi_{-.05}$	erfc	$\Psi_{.05}$	$\Psi_{.10}$	$\Psi_{.15}$	$\Psi_{.20}$	$\Psi_{.25}$
0.01	.99035	.98951	.98872	.98796	.98724	.98655	.98589	.98526
0.02	.98065	.97900	.97744	.97595	.97453	.97317	.97187	.97062
0.03	.97092	.96848	.96616	.96396	.96186	.95985	.95793	.95608
0.04	.96116	.95794	.95489	.95199	.94923	.94660	.94407	.94164
0.05	.95137	.94739	.94363	.94006	.93666	.93341	.93030	.92731
0.06	.94155	.93683	.93238	.92815	.92413	.92029	.91662	.91309
0.07	.93171	.92627	.92114	.91628	.91166	.90725	.90303	.89898
0.08	.92184	.91571	.90992	.90444	.89924	.89427	.88953	.88497
0.09	.91195	.90514	.89872	.89264	.88687	.88137	.87612	.87108
0.10	.90204	.89457	.88754	.88088	.87457	.86855	.86280	.85730
0.11	.89211	.88401	.87638	.86916	.86232	.85580	.84958	.84363
0.12	.88217	.87345	.86524	.85749	.85013	.84314	.83643	.83008
0.13	.87222	.86290	.85413	.84586	.83801	.83055	.82344	.81664
0.14	.86227	.85236	.84305	.83427	.82595	.81805	.81052	.80332
0.15	.85230	.84183	.83200	.82274	.81396	.80569	.79769	.79011
0.16	.84233	.83132	.82099	.81125	.80204	.79330	.78497	.77703
0.17	.83236	.82082	.81001	.79982	.79019	.78105	.77236	.76406
0.18	.82239	.81035	.79906	.78844	.77841	.76889	.75984	.75122
0.19	.81242	.79989	.78816	.77712	.76670	.75682	.74743	.73849
0.20	.80245	.78946	.77730	.76586	.75507	.74484	.73513	.72589
0.21	.79250	.77906	.76648	.75466	.74351	.73296	.72294	.71341
0.22	.78255	.76868	.75570	.74352	.73203	.72117	.71086	.70105
0.23	.77262	.75833	.74498	.73244	.72063	.70947	.69888	.68882
0.24	.76270	.74802	.73430	.72143	.70932	.69787	.68702	.67671
0.25	.75280	.73774	.72367	.71049	.69808	.68637	.67527	.66473
0.26	.74292	.72749	.71310	.69962	.68693	.67496	.66363	.65287
0.27	.73306	.71729	.70258	.68881	.67587	.66366	.65210	.64114
0.28	.72325	.70712	.69212	.67808	.66489	.65245	.64069	.62954
0.29	.71342	.69700	.68172	.66742	.65400	.64135	.62939	.61806
0.30	.70363	.68692	.67137	.65684	.64319	.63035	.61821	.60672
0.31	.69388	.67689	.66109	.64633	.63248	.61945	.60715	.59550
0.32	.68416	.66691	.65087	.63590	.62186	.60866	.59620	.58441
0.33	.67448	.65698	.64072	.62555	.61134	.59797	.58537	.57345
0.34	.66483	.64710	.63064	.61528	.60090	.58739	.57466	.56262
0.35	.65523	.63728	.62062	.60509	.59056	.57692	.56406	.55191
0.36	.64566	.62751	.61067	.59499	.58032	.56655	.55358	.54134
0.37	.63613	.61779	.60079	.58497	.57017	.55629	.54323	.53090
0.38	.62665	.60814	.59099	.57503	.56012	.54614	.53299	.52059
0.39	.61722	.59855	.58126	.56518	.55017	.53610	.52287	.51040
0.40	.60784	.58902	.57161	.55542	.54032	.52617	.51288	.50035
0.41	.59850	.57955	.56203	.54575	.53057	.51635	.50300	.49043
0.42	.58922	.57015	.55253	.53617	.52092	.50664	.49325	.48063
0.43	.58000	.56082	.54311	.52668	.51137	.49705	.48361	.47097
0.44	.57082	.55156	.53377	.51728	.50192	.48756	.47410	.46144
0.45	.56171	.54237	.52452	.50797	.49257	.47819	.46471	.45203
0.46	.55266	.53325	.51534	.49876	.48333	.46893	.45543	.44276
0.47	.54366	.52420	.50625	.48964	.47419	.45978	.44628	.43361
0.48	.53473	.51522	.49725	.48062	.46516	.45075	.43726	.42459
0.49	.52587	.50632	.48833	.47169	.45623	.44182	.42835	.41571
0.50	.51707	.49750	.47950	.46286	.44741	.43302	.41956	.40694

u	$\Psi_{-.10}$	$\Psi_{-.05}$	erfc	$\Psi_{.05}$	$\Psi_{.10}$	$\Psi_{.15}$	$\Psi_{.20}$	$\Psi_{.25}$
0.51	.50833	.48876	.47076	.45412	.43869	.42432	.41089	.39831
0.52	.49967	.48009	.46210	.44549	.43008	.41574	.40235	.38981
0.53	.49107	.47151	.45354	.43695	.42157	.40727	.39392	.38143
0.54	.48254	.46300	.44506	.42851	.41317	.39892	.38562	.37317
0.55	.47409	.45458	.43668	.42017	.40488	.39068	.37743	.36505
0.56	.46571	.44624	.42838	.41193	.39670	.38255	.36937	.35705
0.57	.45741	.43799	.42018	.40379	.38862	.37454	.36142	.34917
0.58	.44918	.42982	.41208	.39575	.38065	.36664	.35360	.34141
0.59	.44103	.42173	.40406	.38781	.37278	.35885	.34589	.33378
0.60	.43296	.41374	.39614	.37997	.36503	.35118	.33830	.32628
0.61	.42496	.40583	.38832	.37223	.35738	.34362	.33082	.31889
0.62	.41705	.39800	.38059	.36459	.34983	.33617	.32347	.31163
0.63	.40922	.39027	.37295	.35706	.34240	.32883	.31623	.30448
0.64	.40146	.38262	.36541	.34962	.33507	.32160	.30910	.29746
0.65	.39380	.37507	.35797	.34229	.32784	.31449	.30209	.29056
0.66	.38621	.36760	.35062	.33506	.32073	.30748	.29520	.28377
0.67	.37871	.36023	.34337	.32793	.31372	.30059	.28842	.27710
0.68	.37129	.35294	.33622	.32090	.30681	.29380	.28175	.27055
0.69	.36396	.34575	.32916	.31397	.30001	.28713	.27519	.26411
0.70	.35672	.33865	.32220	.30715	. 29332	.28056	.26875	.25779
0.71	.34956	.33164	.31533	.30042	.28673	.27410	.26242	.25158
0.72	.34249	.32473	.30857	.29380	.28024	.26775	.25620	.24548
0.73	.33550	.31790	.30190	.28727	.27386	.26150	.25008	.23949
0.74	.32861	.31117	.29532	.28085	.26758	.25536	.24408	.23362
0.75	.32180	.30453	.28884	.27453	.26141	.24933	.23818	.22785
0.76	.31508	.29798	.28246	.26830	.25533	.24340	.23239	.22219
0.77	.30845	.29153	.27618	.26218	.24936	.23758	.22671	.21664
0.78	.30191	.28517	.26999	.25615	.24349	.23186	.22113	.21120
0.79	.29546	.27890	.26390	.25023	.23772	.22624	.21565	.20586
0.80	.28910	.27273	.25790	.24440	.23205	.22072	.21028	.20063
0.81	.28282	.26665	.25200	.23867	.22648	.21530	.20501	.19549
0.82	.27664	.26066	.24619	.23303	.22101	.20998	.19984	.19047
0.83	.27055	.25476	.24048	.22749	.21563	.20477	.19477	.18554
0.84	.26454	.24895	.23486	.22205	.21036	.19965	.18980	.18071
0.85	.25863	.24324	.22933	.21670	.20518	.19462	.18492	.17598
0.86	.25280	.23762	.22390	.21145	.20009	.18970	.18015	.17134
0.87	.24707	.23209	.21856	.20629	.19510	.18486	.17546	.16680
0.88	.24142	.22645	.21331	.20122	.19020	.18013	.17088	.16236
0.89	.23586	.22130	.20816	.19624	.18540	.17548	.16639	.15801
0.90	.23039	.21604	.20309	.19136	.18069	.17093	.16199	.15376
0.91	.22501	.21087	.19812	.18657	.17607	.16647	.15768	.14959
0.92	.21972	.20579	.19323	.18187	.17154	.16210	.15346	.14551
0.93	.21452	.20079	.18844	.17726	.16709	.15782	.14933	.14153
0.94	.20940	.19589	.18373	.17273	.16274	.15363	.14529	.13763
0.95	.20437	.19107	.17911	.16830	.15848	.14953	.14133	.13381
0.96	.19942	.18634	.17458	.16395	.15430	.14551	.13747	.13009
0.97	.19457	.18170	.17013	.15968	.15021	.14157	.13368	.12644
0.98	.18980	.17714	.16577	.15550	.14620	.13772	.12998	.12288
0.99	.18511	.17266	.16149	.15141	.14227	.13396	.12636	.11940
1.00	.18051	.16828	.15730	.14740	.13843	.13027	.12282	.11600

u	$\Psi_{-.10}$	$\Psi_{-.05}$	erfc	$\Psi_{.05}$	$\Psi_{.10}$	$\Psi_{.15}$	$\Psi_{.20}$	$\Psi_{.25}$
1.01	.17599	.16397	.15319	.14347	.13467	.12667	.11936	.11268
1.02	.17155	.15975	.14916	.13962	.13099	.12314	.11598	.10943
1.03	.16720	.15561	.14522	.13586	.12739	.11970	.11268	.10626
1.04	.16293	.15155	.14135	.13217	.12387	.11633	.10946	.10317
1.05	.15874	.14757	.13756	.12856	.12042	.11303	.10630	.10015
1.06	.15463	.14367	.13386	.12503	.11705	.10982	.10322	.09721
1.07	.15061	.13985	.13023	.12157	.11376	.10667	.10022	.09433
1.08	.14666	.13611	.12667	.11820	.11054	.10360	.09729	.09153
1.09	.14279	.13244	.12320	.11489	.10740	.10060	.09443	.08879
1.10	.13900	.12885	.11979	.11166	.10432	.09768	.09163	.08612
1.11	.13528	.12534	.11647	.10850	.10132	.09482	.08891	.08352
1.12	.13164	.12190	.11321	.10541	.09839	.09203	.08625	.08099
1.13	.12808	.11854	.11003	.10240	.09552	.08930	.08366	.07851
1.14	.12459	.11525	.10692	.09945	.09273	.08665	.08113	.07611
1.15	.12117	.11203	.10388	.09657	.09000	.08406	.07867	.07376
1.16	.11783	.10888	.10090	.09376	.08733	.08153	.07626	.07147
1.17	.11456	.10580	.09800	.09102	.08473	.07906	.07392	.06925
1.18	.11136	.10279	.09516	.08834	.08220	.07666	.07164	.06708
1.19	.10823	.09985	.09239	.08572	.07973	.07493	.06942	.06497
1.20	.10517	.09698	.08969	.08317	.07731	.07203	.06725	.06291
1.21	.10218	.09417	.08704	.08068	.07496	.06981	.06514	.06091
1.22	.09926	.09143	.08447	.07825	.07267	.06764	.06309	.05896
1.23	.09640	.08875	.08195	.07588	.07043	.06553	.06109	.05707
1.24	.09361	.08613	.07949	.07357	.06826	.06347	.05915	.05523
1.25	.09088	.08358	.07710	.07132	.06614	.06147	.05726	.05344
1.26	.08822	.08109	.07476	.06912	.06407	.05952	.05542	.05169
1.27	.08562	.07866	.07249	.06698	.06206	.05763	.05363	.05000
1.28	.08308	.07629	.07027	.06490	.06010	.05578	.05188	.04836
1.29	.08060	.07397	.06810	.06287	.05819	.05398	.05019	.04676
1.30	.07818	.07172	.06599	.06089	.05633	.05224	.04854	.04520
1.31	.07582	.06952	.06394	.05897	.05453	.05054	.04694	.04369
1.32	.07352	.06738	.06193	.05709	.05277	.04880	.04539	.04222
1.33	.07128	.06529	.05998	.05527	.05106	.04728	.04388	.04080
1.34	.06909	.06325	.05809	.05349	.04939	.04572	.04241	.03942
1.35	.06606	.06127	.05624	.05177	.04778	.04420	.04098	.03808
1.36	.06488	.05934	.05444	.05009	.04621	.04273	.03960	.03677
1.37	.06286	.05746	.05269	.04845	.04468	.04129	.03825	.03551
1.38	.06088	.05563	.05098	.04686	.04319	.03990	.03695	.03424
1.39	.05896	.05384	.04933	.04532	.04175	.03855	.03568	.03309
1.40	.05709	.05211	.04771	.04382	.04035	.03724	.03445	.03194
1.41	.05527	.05042	.04615	.04236	.03899	.03597	.03326	.03082
1.42	.05349	.04878	.04462	.04094	.03766	.03473	.03210	.02973
1.43	.05177	.04718	.04314	.03956	.03638	.03353	.03098	.02868
1.44	.05009	.04563	.04170	.03823	.03513	.03237	.02989	.02766
1.45	.04846	.04412	.04030	.03693	.03392	.03124	.02884	.02668
1.46	.04687	.04265	.03895	.03567	.03275	.03015	.02782	.02572
1.47	.04532	.04123	.03763	.03444	.03161	.02909	.02683	.02479
1.48	.04382	.03984	.03635	.03325	.03051	.02806	.02587	.02390
1.49	.04236	.03850	.03510	.03210	.02944	.02706	.02494	.02303
1.50	.04094	.03719	.03389	.03098	.02840	.02610	.02404	.02219

u	$\Psi_{-.10}$	$\Psi_{-.05}$	erfc	$\Psi_{.05}$	$\Psi_{.10}$	$\Psi_{.15}$	$\Psi_{.20}$	$\Psi_{.25}$
1.51	.03956	.03592	.03272	.02990	.02739	.02516	.02317	.02137
1.52	.03823	.03469	.03159	.02885	.02642	.02426	.02232	.02059
1.53	.03692	.03349	.03048	.02783	.02547	.02338	.02150	.01982
1.54	.03566	.03233	.02941	.02684	.02456	.02253	.02071	.01909
1.55	.03444	.03121	.02838	.02588	.02367	.02170	.01995	.01837
1.56	.03325	.03012	.02737	.02495	.02281	.02091	.01921	.01769
1.57	.03209	.02906	.02640	.02405	.02198	.02014	.01849	.01702
1.58	.03097	.02803	.02545	.02318	.02118	.01939	.01780	.01638
1.59	.02989	.02703	.02454	.02234	.02040	.01867	.01713	.01573
1.60	.02883	.02607	.02365	.02152	.01964	.01797	.01648	.01515
1.61	.02781	.02513	.02279	.02073	.01891	.01730	.01586	.01457
1.62	.02682	.02423	.02196	.01997	.01821	.01665	.01526	.01401
1.63	.02586	.02335	.02116	.01923	.01753	.01601	.01467	.01347
1.64	.02493	.02238	.02038	.01851	.01687	.01541	.01411	.01294
1.65	.02403	.02168	.01962	.01782	.01623	.01482	.01356	.01244
1.66	.02316	.02088	.01890	.01715	.01561	.01425	.01304	.01195
1.67	.02231	.02011	.01819	.01650	.01502	.01370	.01253	.01148
1.68	.02149	.01937	.01751	.01588	.01444	.01317	.01204	.01103
1.69	.02070	.01864	.01685	.01527	.01388	.01266	.01156	.01059
1.70	.01994	.01795	.01621	.01469	.01335	.01216	.01111	.01017
1.71	.01919	.01727	.01559	.01412	.01283	.01168	.01067	.00976
1.72	.01848	.01662	.01499	.01358	.01233	.01122	.01024	.00937
1.73	.01778	.01599	.01442	.01305	.01184	.01078	.00983	.00899
1.74	.01711	.01538	.01387	.01254	.01138	.01035	.00944	.00863
1.75	.01646	.01479	.01333	.01205	.01093	.00994	.00906	.00828
1.76	.01584	.01422	.01281	.01158	.01049	.00954	.00869	.00794
1.77	.01523	.01367	.01231	.01112	.01008	.00916	.00834	.00761
1.78	.01464	.01314	.01183	.01068	.00967	.00878	.00800	.00730
1.79	.01408	.01262	.01136	.01025	.00928	.00843	.00767	.00700
1.80	.01353	.01213	.01091	.00984	.00891	.00808	.00735	.00671
1.81	.01301	.01165	.01048	.00945	.00855	.00775	.00705	.00643
1.82	.01250	.01119	.01006	.00907	.00820	.00743	.00676	.00616
1.83	.01200	.01075	.00965	.00870	.00786	.00713	.00648	.00590
1.84	.01153	.01032	.00926	.00835	.00754	.00683	.00621	.00565
1.85	.01107	.00990	.00889	.00800	.00723	.00655	.00595	.00541
1.86	.01063	.00950	.00853	.00768	.00693	.00627	.00569	.00518
1.87	.01021	.00912	.00818	.00736	.00664	.00601	.00545	.00496
1.88	.00979	.00875	.00784	.00705	.00636	.00576	.00522	.00475
1.89	.00940	.00839	.00752	.00676	.00610	.00551	.00500	.00454
1.90	.00902	.00805	.00721	.00648	.00584	.00528	.00478	.00435
1.91	.00865	.00772	.00691	.00621	.00559	.00505	.00458	.00416
1.92	.00830	.00740	.00662	.00595	.00536	.00484	.00438	.00398
1.93	.00795	.00709	.00634	.00569	.00513	.00463	.00419	.00380
1.94	.00763	.00680	.00608	.00545	.00491	.00443	.00401	.00364
1.95	.00731	.00651	.00582	.00522	.00470	.00424	.00383	.00348
1.96	.00701	.00624	.00557	.00500	.00449	.00405	.00366	.00332
1.97	.00671	.00597	.00534	.00478	.00430	.00388	.00350	.00317
1.98	.00643	.00572	.00511	.00458	.00411	.00371	.00335	.00303
1.99	.00616	.00548	.00489	.00438	.00393	.00354	.00320	.00290
2.00	.00590	.00524	.00468	.00419	.00376	.00339	.00306	.00277

u	$\Psi_{.30}$	$\Psi_{.35}$	$\Psi_{.40}$	$\Psi_{.45}$	$\Psi_{.50}$	$\Psi_{.55}$	$\Psi_{.60}$	$\Psi_{.65}$
0.01	.98464	.98405	.98347	.98292	.98238	.98185	.98134	.98084
0.02	.96941	.96824	.96711	.96601	.96495	.96392	.96291	.96193
0.03	.95429	.95257	.95091	.94929	.94773	.94620	.94472	.94328
0.04	.93930	.93705	.93486	.93275	.93070	.92871	.92678	.92489
0.05	.92444	.92166	.91898	.91639	.91388	.91144	.90906	.90675
0.06	.90970	.90643	.90327	.90021	.89725	.89438	.89159	.88887
0.07	.89509	.89133	.88771	.88421	.88082	.87754	.87434	.87124
0.08	.88060	.87639	.87232	.86840	.86460	.86091	.85734	.85386
0.09	.86624	.86159	.85710	.85276	.84857	.84450	.84056	.83673
0.10	.85202	.84694	.84204	.83731	.83274	.82831	.82402	.81985
0.11	.83792	.83243	.82714	.82204	.81711	.81233	.80770	.80321
0.12	.82396	.81807	.81241	.80695	.80167	.79657	.79162	.78682
0.13	.81012	.80387	.79784	.79204	.78643	.78101	.77577	.77068
0.14	.79642	.78981	.78344	.77731	.77139	.76567	.76014	.75477
0.15	.78286	.77590	.76921	.76276	.75655	.75054	.74474	.73911
0.16	.76943	.76214	.75514	.74840	.74190	.73562	.72956	.72369
0.17	.75613	.74853	.74123	.73421	.72744	.72091	.71460	.70850
0.18	.74297	.73507	.72749	.72020	.71318	.70641	.69987	.69355
0.19	.72995	.72177	.71392	.70608	.69912	.69212	.68536	.67883
0.20	.71706	.70861	.70051	.69273	.68524	.67803	.67107	.66434
0.21	.70431	.69561	.68727	.67926	.67156	.66415	.65699	.65008
0.22	.69170	.68276	.67419	.66597	.65807	.65047	.64314	.63606
0.23	.67922	.67005	.66128	.65286	.64477	.63699	.62949	.62225
0.24	.66689	.65751	.64853	.63993	.63166	.62372	.61606	.60868
0.25	.65469	.64511	.63595	.62717	.61874	.61064	.60284	.59532
0.26	.64263	.63286	.62353	.61459	.60601	.59777	.58984	.58219
0.27	.63071	.62077	.61127	.60218	.59346	.58509	.57704	.56928
0.28	.61894	.60883	.59918	.58995	.58110	.57261	.56444	.55658
0.29	.60730	.59704	.58726	.57790	.56893	.56033	.55206	.54410
0.30	.59780	.58541	.57549	.56601	.55694	.54823	.53987	.53183
0.31	.58444	.57392	.56389	.55430	.54513	.53634	.52789	.51978
0.32	.57322	.56258	.55244	.54276	.53350	.52463	.51611	.50793
0.33	.56214	.55140	.54116	.53140	.52206	.51311	.50453	.49629
0.34	.55121	.54037	.53004	.52020	.51079	.50178	.49314	.48485
0.35	.54041	.52948	.51908	.50917	.49970	.49064	.48195	.47362
0.36	.52975	.51875	.50828	.49831	.48879	.47968	.47096	.46259
0.37	.51923	.50816	.49764	.48762	.47805	.46891	.46015	.45176
0.38	.50885	.49773	.48716	.47709	.46749	.45832	.44954	.44112
0.39	.49861	.48744	.47683	.46673	.45710	.44791	.43911	.43068
0.40	.48851	.47730	.46666	.45653	.44688	.43767	.42887	.42043
0.41	.47855	.46731	.45664	.44650	.43684	.42762	.41881	.41038
0.42	.46873	.45746	.44678	.43663	.42696	.41774	.40893	.40051
0.43	.45904	.44776	.43707	.42692	.41725	.40804	.39924	.39083
0.44	.44950	.43821	.42752	.41736	.40771	.39851	.38972	.38133
0.45	.44009	.42880	.41811	.40797	.39833	.38914	.38038	.37201
0.46	.43082	.41954	.40886	.39874	.38911	.37995	.37122	.36287
0.47	.42168	.41042	.39976	.38966	.38006	.37093	.36222	.35391
0.48	.41268	.40144	.39081	.38073	.37117	.36207	.35340	.34513
0.49	.40381	.39260	.38200	.37196	.36243	.35337	.34475	.33652
0.50	.39508	.38390	.37334	.36334	.35385	.34484	.33626	.32808

u	$\Psi_{.30}$	$\Psi_{.35}$	$\Psi_{.40}$	$\Psi_{.45}$	$\Psi_{.50}$	$\Psi_{.55}$	$\Psi_{.60}$	$\Psi_{.65}$
0.51	.38649	.37535	.36483	.35487	.34543	.33647	.32794	.31981
0.52	.37802	.36693	.35646	.34655	.33717	.32825	.31978	.31171
0.53	.36969	.35865	.34823	.33838	.32905	.32020	.31178	.30377
0.54	.36150	.35051	.34015	.33036	.32109	.31230	.30394	.29599
0.55	.35343	.34250	.33221	.32248	.31327	.30455	.29656	.28837
0.56	.34549	.33463	.32440	.31474	.30561	.29695	.28873	.28091
0.57	.33768	.32690	.31674	.30715	.29809	.28950	.28135	.27361
0.58	.33001	.31929	.30921	.29970	.29071	.28220	.27413	.26646
0.59	.32246	.31182	.30182	.29239	.28348	.27505	.26705	.25946
0.60	.31503	.30448	.29456	.28521	.27639	.26804	.26012	.25261
0.61	.30773	.29727	.28744	.27818	.26944	.26117	.25334	.24591
0.62	.30056	.29019	.28045	.27128	.26262	.25444	.24670	.23935
0.63	.29352	.28324	.27359	.26451	.25594	.24785	.24020	.23294
0.64	.28659	.27641	.26686	.25787	.24940	.24140	.23383	.22666
0.65	.27979	.26971	.26026	.25137	.24299	.23508	.22761	.22052
0.66	.27311	.26313	.25378	.24499	.23671	.22890	.22152	.21452
0.67	.26655	.25668	.24743	.23874	.23056	.22285	.21556	.20866
0.68	.26010	.25034	.24120	.23262	.22454	.21692	.20973	.20292
0.69	.25378	.24413	.23510	.22662	.21864	.21113	.20403	.19732
0.70	.24757	.23804	.22911	.22074	.21287	.20545	.19846	.19184
0.71	.24148	.23206	.22325	.21498	.20722	.19991	.19301	.18649
0.72	.23551	.22620	.21750	.20935	.20169	.19448	.18769	.18127
0.73	.22964	.22046	.21187	.20383	.19628	.18916	.18248	.17616
0.74	.22389	.21483	.20636	.19843	.19099	.18399	.17740	.17117
0.75	.21825	.20931	.20096	.19314	.18581	.17892	.17243	.16631
0.76	.21272	.20390	.19567	.18797	.18075	.17396	.16757	.16155
0.77	.20730	.19861	.19050	.18291	.17580	.16912	.16283	.15691
0.78	.20199	.19342	.18543	.17796	.17096	.16439	.15821	.15238
0.79	.19678	.18834	.18047	.17311	.16623	.15976	.15369	.14797
0.80	.19168	.18336	.17561	.16838	.16160	.15525	.14927	.14365
0.81	.18668	.17849	.17087	.16375	.15708	.15084	.14497	.13945
0.82	.18179	.17373	.16622	.15922	.15267	.14653	.14077	.13534
0.83	.17699	.16906	.16168	.15479	.14835	.14232	.13666	.13134
0.84	.17230	.16449	.15724	.15047	.14414	.13822	.13266	.12744
0.85	.16770	.16003	.15289	.14624	.14003	.13421	.12876	.12364
0.86	.16320	.15566	.14865	.14211	.13601	.13030	.12496	.11993
0.87	.15880	.15139	.14450	.13808	.13209	.12649	.12124	.11632
0.88	.15449	.14721	.14044	.13414	.12826	.12277	.11763	.11280
0.89	.15028	.14312	.13648	.13029	.12453	.11914	.11410	.10937
0.90	.14616	.13913	.13261	.12654	.12088	.11560	.11066	.10603
0.91	.14213	.13523	.12883	.12287	.11733	.11215	.10731	.10277
0.92	.13819	.13141	.12513	.11930	.11386	.10879	.10404	.09960
0.93	.13433	.12769	.12153	.11581	.11048	.10551	.10086	.09651
0.94	.13057	.12405	.11801	.11240	.10718	.10231	.09777	.09351
0.95	.12689	.12049	.11457	.10908	.10396	.09920	.09475	.09059
0.96	.12329	.11702	.11122	.10584	.10083	.09617	.09182	.08774
0.97	.11978	.11363	.10795	.10268	.09778	.09321	.08895	.08497
0.98	.11635	.11033	.10476	.09960	.09480	.09033	.08617	.08228
0.99	.11300	.10710	.10165	.09659	.09190	.08753	.08346	.07966
1.00	.10973	.10395	.09861	.09366	.08907	.08480	.08082	.07711

u	$\Psi_{.30}$	$\Psi_{.35}$	$\Psi_{.40}$	$\Psi_{.45}$	$\Psi_{.50}$	$\Psi_{.55}$	$\Psi_{.60}$	$\Psi_{.65}$
1.01	.10653	.10088	.09565	.09081	.08632	.08215	.07826	.07463
1.02	.10342	.09788	.09276	.08803	.08364	.07956	.07576	.07222
1.03	.10037	.09495	.08995	.08532	.08103	.07705	.07334	.06988
1.04	.09741	.09210	.08721	.08269	.07849	.07460	.07098	.06760
1.05	.09451	.08932	.08454	.08012	.07602	.07222	.06868	.06539
1.06	.09169	.08661	.08194	.07762	.07362	.06990	.06645	.06324
1.07	.08893	.08397	.07940	.07518	.07128	.06765	.06428	.06115
1.08	.08625	.08140	.07694	.07281	.06900	.06546	.06218	.05912
1.09	.08363	.07889	.07453	.07051	.06679	.06334	.06013	.05715
1.10	.08108	.07645	.07219	.06827	.06463	.06127	.05814	.05524
1.11	.07859	.07407	.06992	.06608	.06254	.05926	.05621	.05338
1.12	.07617	.07176	.06770	.06396	.06050	.05730	.05434	.05158
1.13	.07381	.06950	.06554	.06190	.05853	.05541	.05252	.04983
1.14	.07152	.06731	.06345	.05989	.05660	.05356	.05075	.04813
1.15	.06928	.06517	.06141	.05794	.05474	.05177	.04903	.04648
1.16	.06710	.06310	.05942	.05604	.05292	.05004	.04737	.04489
1.17	.06498	.06107	.05749	.05420	.05116	.04835	.04575	.04334
1.18	.06292	.05911	.05562	.05241	.04945	.04671	.04418	.04184
1.19	.06091	.05720	.05379	.05067	.04778	.04512	.04266	.04038
1.20	.05895	.05534	.05202	.04898	.04617	.04358	.04119	.03897
1.21	.05705	.05353	.05030	.04733	.04460	.04209	.03976	.03760
1.22	.05520	.05177	.04863	.04574	.04308	.04063	.03837	.03627
1.23	.05341	.05006	.04700	.04419	.04161	.03923	.03703	.03499
1.24	.05166	.04840	.04543	.04269	.04018	.03786	.03572	.03374
1.25	.04996	.04679	.04389	.04124	.03879	.03654	.03446	.03254
1.26	.04831	.04523	.04241	.03982	.03745	.03526	.03324	.03137
1.27	.04671	.04371	.04096	.03845	.03614	.03402	.03205	.03024
1.28	.04515	.04223	.03956	.03712	.03488	.03281	.03091	.02915
1.29	.04364	.04080	.03820	.03583	.03365	.03164	.02980	.02809
1.30	.04217	.03941	.03689	.03458	.03246	.03051	.02872	.02706
1.31	.04074	.03806	.03561	.03337	.03131	.02942	.02768	.02607
1.32	.03936	.03675	.03437	.03219	.03019	.02836	.02667	.02511
1.33	.03801	.03548	.03317	.03105	.02911	.02733	.02570	.02419
1.34	.03671	.03425	.03200	.02995	.02807	.02634	.02475	.02329
1.35	.03544	.03305	.03087	.02888	.02705	.02538	.02384	.02242
1.36	.03422	.03189	.02978	.02784	.02607	.02445	.02296	.02158
1.37	.03302	.03077	.02872	.02684	.02512	.02355	.02210	.02077
1.38	.03187	.02968	.02769	.02587	.02421	.02268	.02128	.01999
1.39	.03075	.02863	.02669	.02493	.02332	.02184	.02048	.01923
1.40	.02967	.02760	.02573	.02402	.02246	.02103	.01971	.01850
1.41	.02861	.02661	.02480	.02314	.02163	.02024	.01897	.01780
1.42	.02760	.02566	.02389	.02229	.02082	.01948	.01825	.01711
1.43	.02661	.02473	.02302	.02146	.02004	.01874	.01755	.01645
1.44	.02565	.02383	.02217	.02067	.01929	.01803	.01688	.01582
1.45	.02473	.02296	.02136	.01990	.01856	.01735	.01623	.01521
1.46	.02383	.02212	.02056	.01915	.01786	.01668	.01560	.01461
1.47	.02296	.02130	.01980	.01843	.01718	.01604	.01500	.01404
1.48	.02212	.02052	.01906	.01773	.01653	.01543	.01442	.01349
1.49	.02131	.01975	.01834	.01706	.01590	.01483	.01385	.01296
1.50	.02052	.01902	.01765	.01641	.01528	.01425	.01331	.01245

u	$\Psi_{.30}$	$\Psi_{.35}$	$\Psi_{.40}$	$\Psi_{.45}$	$\Psi_{.50}$	$\Psi_{.55}$	$\Psi_{.60}$	$\Psi_{.65}$
1.51	.01976	.01830	.01698	.01578	.01469	.01370	.01279	.01195
1.52	.01902	.01761	.01634	.01518	.01412	.01316	.01228	.01147
1.53	.01831	.01695	.01571	.01459	.01357	.01264	.01179	.01102
1.54	.01762	.01631	.01511	.01403	.01304	.01214	.01132	.01057
1.55	.01696	.01568	.01453	.01348	.01253	.01166	.01087	.01015
1.56	.01632	.01508	.01397	.01296	.01204	.01120	.01043	.00973
1.57	.01570	.01450	.01342	.01245	.01156	.01075	.01001	.00934
1.58	.01510	.01394	.01290	.01196	.01110	.01032	.00961	.00896
1.59	.01452	.01340	.01240	.01148	.01066	.00990	.00922	.00859
1.60	.01396	.01288	.01191	.01103	.01023	.00950	.00884	.00824
1.61	.01342	.01238	.01144	.01059	.00982	.00912	.00848	.00790
1.62	.01289	.01189	.01098	.01017	.00942	.00875	.00813	.00757
1.63	.01239	.01142	.01055	.00976	.00904	.00839	.00780	.00725
1.64	.01190	.01097	.01013	.00936	.00867	.00804	.00747	.00695
1.65	.01144	.01053	.00972	.00898	.00832	.00771	.00716	.00666
1.66	.01098	.01011	.00933	.00862	.00798	.00739	.00686	.00638
1.67	.01055	.00971	.00895	.00827	.00765	.00709	.00658	.00611
1.68	.01013	.00932	.00859	.00793	.00733	.00679	.00630	.00585
1.69	.00972	.00894	.00824	.00760	.00703	.00651	.00603	.00560
1.70	.00933	.00858	.00790	.00729	.00673	.00623	.00578	.00536
1.71	.00895	.00823	.00757	.00698	.00645	.00597	.00553	.00513
1.72	.00859	.00789	.00726	.00669	.00618	.00572	.00530	.00491
1.73	.00824	.00757	.00696	.00641	.00592	.00547	.00507	.00470
1.74	.00790	.00725	.00667	.00614	.00567	.00524	.00485	.00450
1.75	.00758	.00695	.00639	.00589	.00543	.00502	.00464	.00430
1.76	.00727	.00666	.00612	.00564	.00520	.00480	.00444	.00411
1.77	.00697	.00639	.00587	.00540	.00497	.00459	.00425	.00393
1.78	.00668	.00612	.00562	.00517	.00476	.00439	.00406	.00376
1.79	.00640	.00586	.00538	.00495	.00456	.00420	.00388	.00359
1.80	.00613	.00561	.00515	.00473	.00436	.00402	.00371	.00343
1.81	.00587	.00537	.00493	.00453	.00417	.00384	.00355	.00328
1.82	.00562	.00515	.00472	.00433	.00399	.00367	.00339	.00313
1.83	.00538	.00493	.00451	.00414	.00381	.00351	.00324	.00299
1.84	.00516	.00471	.00432	.00396	.00364	.00336	.00310	.00286
1.85	.00494	.00451	.00413	.00379	.00348	.00321	.00296	.00273
1.86	.00472	.00432	.00395	.00362	.00333	.00306	.00282	.00261
1.87	.00452	.00413	.00378	.00346	.00318	.00293	.00270	.00249
1.88	.00432	.00395	.00361	.00331	.00304	.00279	.00257	.00237
1.89	.00414	.00378	.00345	.00316	.00290	.00267	.00246	.00227
1.90	.00396	.00361	.00330	.00302	.00277	.00255	.00234	.00216
1.91	.00378	.00345	.00315	.00289	.00265	.00243	.00224	.00206
1.92	.00362	.00330	.00301	.00276	.00253	.00232	.00213	.00197
1.93	.00346	.00315	.00288	.00263	.00241	.00221	.00204	.00187
1.94	.00330	.00301	.00275	.00251	.00230	.00211	.00194	.00179
1.95	.00316	.00288	.00262	.00240	.00220	.00202	.00185	.00170
1.96	.00302	.00275	.00251	.00229	.00210	.00192	.00177	.00162
1.97	.00288	.00262	.00239	.00218	.00200	.00183	.00168	.00155
1.98	.00275	.00250	.00228	.00208	.00191	.00175	.00160	.00147
1.99	.00263	.00239	.00218	.00199	.00182	.00167	.00153	.00140
2.00	.00251	.00228	.00207	.00190	.00173	.00159	.00146	.00134

u	$\Psi_{.70}$	$\Psi_{.75}$	$\Psi_{.80}$	$\Psi_{.85}$	$\Psi_{.90}$	$\Psi_{.95}$	$\Psi_{1.00}$
0.01	.98035	.97987	.97940	.97895	.97850	.97806	.97763
0.02	.96097	.96004	.95913	.95823	.95736	.95650	.95566
0.03	.94188	.94050	.93916	.93785	.93657	.93531	.93408
0.04	.92305	.92126	.91951	.91780	.91613	.91449	.91288
0.05	.90450	.90231	.90017	.89808	.89603	.89403	.89207
0.06	.88623	.88365	.88113	.87868	.87628	.87393	.87163
0.07	.86822	.86528	.86240	.85960	.85686	.85419	.85157
0.08	.85048	.84719	.84398	.84085	.83779	.83480	.83187
0.09	.83301	.82938	.82585	.82240	.81904	.81576	.81254
0.10	.81580	.81186	.80802	.80427	.80062	.79706	.79357
0.11	.79885	.79461	.79048	.78645	.78253	.77870	.77496
0.12	.78217	.77764	.77323	.76894	.76476	.76068	.75669
0.13	.76574	.76094	.75628	.75173	.74730	.74299	.73877
0.14	.74957	.74452	.73960	.73482	.73016	.72562	.72119
0.15	.73366	.72836	.72321	.71821	.71333	.70858	.70395
0.16	.71799	.71247	.70710	.70189	.69681	.69187	.68705
0.17	.70258	.69684	.69127	.68586	.68059	.67546	.67047
0.18	.68742	.68148	.67572	.67012	.66467	.65937	.65421
0.19	.67250	.66637	.66043	.65466	.64905	.64359	.63827
0.20	.65783	.65153	.64541	.63948	.63371	.62811	.62265
0.21	.64340	.63693	.63066	.62458	.61867	.61293	.60734
0.22	.62921	.62259	.61618	.60995	.60391	.59805	.59234
0.23	.61526	.60850	.60195	.59560	.58944	.58346	.57764
0.24	.60155	.59465	.58798	.58151	.57524	.56915	.56324
0.25	.58807	.58105	.57426	.56769	.56132	.55513	.54913
0.26	.57482	.56769	.56080	.55413	.54767	.54140	.53531
0.27	.56180	.55457	.54759	.54083	.53428	.52793	.52178
0.28	.54900	.54169	.53462	.52778	.52116	.51474	.50852
0.29	.53643	.52904	.52189	.51498	.50830	.50182	.49555
0.30	.52407	.51662	.50941	.50244	.49570	.48917	.48284
0.31	.51196	.50443	.49716	.49014	.48335	.47677	.47041
0.32	.50005	.49246	.48514	.47808	.47125	.46464	.45824
0.33	.48836	.48072	.47336	.46626	.45939	.45275	.44633
0.34	.47688	.46921	.46181	.45467	.44778	.44112	.43467
0.35	.46561	.45790	.45048	.44332	.43641	.42973	.42327
0.36	.45455	.44682	.43937	.43220	.42527	.41859	.41212
0.37	.44370	.43595	.42849	.42130	.41439	.40768	.40121
0.38	.43305	.42529	.41782	.41063	.40369	.39700	.39054
0.39	.42260	.41483	.40736	.40017	.39325	.38656	.38011
0.40	.41235	.40458	.39712	.38994	.38302	.37635	.36991
0.41	.40229	.39454	.38708	.37991	.37301	.36636	.35993
0.42	.39244	.38469	.37725	.37010	.36322	.35658	.35019
0.43	.38277	.37504	.36763	.36050	.35364	.34703	.34066
0.44	.37329	.36559	.35820	.35110	.34427	.33769	.33135
0.45	.36400	.35633	.34897	.34190	.33510	.32856	.32226
0.46	.35489	.34725	.33993	.33289	.32614	.31963	.31337
0.47	.34597	.33837	.33108	.32409	.31737	.31091	.30469
0.48	.33723	.32967	.32242	.31547	.30880	.30239	.29622
0.49	.32866	.32115	.31395	.30705	.30043	.29406	.28794
0.50	.32027	.31281	.30566	.29881	.29224	.28593	.27986

u	$\Psi_{.70}$	$\Psi_{.75}$	$\Psi_{.80}$	$\Psi_{.85}$	$\Psi_{.90}$	$\Psi_{.95}$	$\Psi_{1.00}$
0.51	.31205	.30464	.29755	.29075	.28424	.27798	.27197
0.52	.30401	.29665	.28961	.28288	.27642	.27022	.26427
0.53	.29613	.28883	.28185	.27518	.26878	.26264	.25675
0.54	.28841	.28118	.27427	.26765	.26132	.25524	.24941
0.55	.28086	.27369	.26685	.26030	.25403	.24802	.24226
0.56	.27347	.26637	.25959	.25311	.24691	.24097	.23527
0.57	.26624	.25921	.25250	.24610	.23996	.23409	.22846
0.58	.25916	.25221	.24558	.23924	.23318	.22738	.22182
0.59	.25224	.24536	.23880	.23254	.22656	.22083	.21534
0.60	.24547	.23867	.23219	.22600	.22009	.21444	.20902
0.61	.23885	.23213	.22573	.21962	.21378	.20820	.20286
0.62	.23237	.22574	.21941	.21338	.20763	.20212	.19686
0.63	.22604	.21949	.21325	.20730	.20162	.19620	.19101
0.64	.21985	.21339	.20723	.20136	.19576	.19042	.18531
0.65	.21381	.20742	.20135	.19557	.19005	.18478	.17975
0.66	.20790	.20160	.19561	.18991	.18448	.17929	.17434
0.67	.20212	.19591	.19001	.18440	.17905	.17394	.16907
0.68	.19648	.19036	.18455	.17902	.17375	.16873	.16393
0.69	.19097	.18494	.17921	.17377	.16859	.16365	.15893
0.70	.18558	.17965	.17401	.16865	.16356	.15870	.15406
0.71	.18033	.17448	.16894	.16367	.15865	.15388	.14932
0.72	.17519	.16944	.16399	.15880	.15387	.14918	.14471
0.73	.17018	.16453	.15916	.15406	.14922	.14461	.14022
0.74	.16529	.15973	.15445	.14945	.14469	.14016	.13585
0.75	.16052	.15505	.14987	.14495	.14027	.13583	.13160
0.76	.15587	.15049	.14539	.14056	.13597	.13161	.12746
0.77	.15132	.14604	.14103	.13629	.13179	.12751	.12344
0.78	.14689	.14170	.13679	.13213	.12771	.12352	.11952
0.79	.14257	.13747	.13265	.12808	.12375	.11963	.11572
0.80	.13835	.13335	.12862	.12414	.11989	.11585	.11202
0.81	.13424	.12933	.12469	.12030	.11613	.11218	.10843
0.82	.13024	.12542	.12087	.11656	.11248	.10861	.10493
0.83	.12633	.12161	.11714	.11292	.10892	.10513	.10153
0.84	.12253	.11789	.11352	.10938	.10547	.10176	.09823
0.85	.11882	.11428	.10999	.10594	.10211	.09847	.09503
0.86	.11521	.11076	.10656	.10259	.09884	.09528	.09191
0.87	.11169	.10733	.10322	.09933	.09566	.09218	.08889
0.88	.10826	.10399	.09996	.09616	.09257	.08917	.08595
0.89	.10492	.10074	.09680	.09308	.08957	.08625	.08310
0.90	.10167	.09758	.09372	.09009	.08665	.08340	.08033
0.91	.09851	.09450	.09073	.08718	.08382	.08065	.07764
0.92	.09543	.09151	.08782	.08435	.08107	.07797	.07503
0.93	.09243	.08860	.08500	.08160	.07839	.07537	.07250
0.94	.08952	.08577	.08225	.07893	.07580	.07284	.07005
0.95	.08668	.08302	.07958	.07633	.07328	.07039	.06766
0.96	.08392	.08034	.07698	.07381	.07083	.06801	.06535
0.97	.08124	.07774	.07446	.07137	.06845	.06571	.06311
0.98	.07863	.07522	.07201	.06899	.06615	.06347	.06094
0.99	.07610	.07276	.06963	.06668	.06391	.06130	.05883
1.00	.07363	.07037	.06732	.06444	.06174	.05919	.05679

u	$\Psi_{.70}$	$\Psi_{.75}$	$\Psi_{.80}$	$\Psi_{.85}$	$\Psi_{.90}$	$\Psi_{.95}$	$\Psi_{1.00}$
1.01	.07123	.06805	.06507	.06227	.05963	.05715	.05481
1.02	.06890	.06580	.06289	.06016	.05759	.05517	.05289
1.03	.06664	.06361	.06078	.05811	.05561	.05326	.05104
1.04	.06444	.06149	.05873	.05613	.05369	.05140	.04924
1.05	.06231	.05943	.05673	.05420	.05183	.04960	.04749
1.06	.06023	.05743	.05480	.05234	.05003	.04785	.04580
1.07	.05822	.05549	.05293	.05053	.04828	.04616	.04417
1.08	.05627	.05360	.05111	.04877	.04658	.04452	.04259
1.09	.05437	.05177	.04935	.04707	.04494	.04294	.04105
1.10	.05253	.05000	.04764	.04542	.04335	.04140	.03957
1.11	.05074	.04828	.04598	.04383	.04181	.03992	.03814
1.12	.04901	.04661	.04437	.04228	.04032	.03848	.03675
1.13	.04733	.04500	.04282	.04078	.03887	.03708	.03540
1.14	.04570	.04343	.04131	.03933	.03748	.03574	.03411
1.15	.04412	.04191	.03985	.03792	.03612	.03443	.03285
1.16	.04258	.04044	.03843	.03656	.03481	.03317	.03164
1.17	.04110	.03901	.03706	.03525	.03355	.03195	.03046
1.18	.03966	.03763	.03574	.03397	.03232	.03077	.02933
1.19	.03826	.03629	.03445	.03274	.03113	.02963	.02823
1.20	.03691	.03499	.03321	.03154	.02999	.02853	.02717
1.21	.03560	.03374	.03200	.03039	.02888	.02747	.02614
1.22	.03433	.03252	.03084	.02927	.02781	.02644	.02515
1.23	.03310	.03134	.02971	.02819	.02677	.02544	.02420
1.24	.03191	.03021	.02862	.02714	.02577	.02448	.02328
1.25	.03076	.02910	.02757	.02613	.02480	.02355	.02239
1.26	.02964	.02804	.02655	.02516	.02386	.02265	.02153
1.27	.02856	.02701	.02556	.02421	.02296	.02179	.02070
1.28	.02752	.02601	.02461	.02330	.02209	.02095	.01989
1.29	.02651	.02505	.02369	.02242	.02124	.02015	.01912
1.30	.02553	.02411	.02280	.02157	.02043	.01937	.01837
1.31	.02459	.02321	.02193	.02075	.01964	.01862	.01766
1.32	.02367	.02234	.02110	.01995	.01889	.01789	.01696
1.33	.02279	.02150	.02030	.01919	.01815	.01719	.01629
1.34	.02194	.02069	.01953	.01845	.01745	.01652	.01565
1.35	.02111	.01990	.01878	.01773	.01677	.01587	.01503
1.36	.02031	.01914	.01805	.01705	.01611	.01524	.01443
1.37	.01954	.01841	.01736	.01638	.01548	.01463	.01385
1.38	.01880	.01770	.01668	.01574	.01486	.01405	.01329
1.39	.01808	.01702	.01603	.01512	.01427	.01349	.01276
1.40	.01739	.01636	.01541	.01452	.01371	.01295	.01224
1.41	.01672	.01572	.01480	.01395	.01316	.01242	.01174
1.42	.01607	.01511	.01422	.01339	.01263	.01192	.01126
1.43	.01545	.01452	.01366	.01286	.01212	.01144	.01080
1.44	.01484	.01394	.01311	.01234	.01163	.01097	.01036
1.45	.01426	.01339	.01259	.01185	.01116	.01052	.00993
1.46	.01370	.01286	.01209	.01137	.01071	.01009	.00952
1.47	.01316	.01235	.01160	.01091	.01027	.00968	.00912
1.48	.01264	.01186	.01113	.01047	.00985	.00928	.00874
1.49	.01214	.01138	.01068	.01004	.00944	.00889	.00838
1.50	.01165	.01092	.01025	.00963	.00905	.00852	.00803

u	$\Psi_{.70}$	$\Psi_{.75}$	$\Psi_{.80}$	$\Psi_{.85}$	$\Psi_{.90}$	$\Psi_{.95}$	$\Psi_{1.00}$
1.51	.01119	.01048	.00983	.00923	.00868	.00816	.00769
1.52	.01074	.01005	.00943	.00885	.00832	.00782	.00736
1.53	.01030	.00965	.00904	.00848	.00797	.00749	.00705
1.54	.00988	.00925	.00867	.00813	.00763	.00716	.00675
1.55	.00948	.00887	.00831	.00779	.00731	.00687	.00646
1.56	.00909	.00851	.00796	.00746	.00700	.00658	.00618
1.57	.00872	.00815	.00763	.00715	.00671	.00630	.00592
1.58	.00836	.00781	.00731	.00685	.00642	.00603	.00566
1.59	.00802	.00749	.00700	.00656	.00615	.00577	.00542
1.60	.00768	.00718	.00671	.00628	.00588	.00552	.00518
1.61	.00736	.00687	.00643	.00601	.00563	.00528	.00496
1.62	.00706	.00658	.00615	.00575	.00539	.00505	.00474
1.63	.00676	.00631	.00589	.00551	.00515	.00483	.00453
1.64	.00647	.00604	.00564	.00527	.00493	.00462	.00433
1.65	.00620	.00578	.00540	.00504	.00472	.00442	.00414
1.66	.00594	.00553	.00516	.00482	.00451	.00422	.00395
1.67	.00568	.00530	.00494	.00461	.00431	.00403	.00378
1.68	.00544	.00507	.00472	.00441	.00412	.00385	.00361
1.69	.00521	.00485	.00452	.00422	.00394	.00368	.00345
1.70	.00498	.00464	.00432	.00403	.00376	.00352	.00329
1.71	.00477	.00444	.00413	.00385	.00360	.00336	.00314
1.72	.00456	.00424	.00395	.00368	.00344	.00321	.00300
1.73	.00436	.00406	.00377	.00352	.00328	.00306	.00286
1.74	.00417	.00388	.00361	.00336	.00313	.00293	.00273
1.75	.00399	.00371	.00345	.00321	.00299	.00279	.00261
1.76	.00381	.00354	.00329	.00306	.00286	.00266	.00249
1.77	.00364	.00338	.00314	.00293	.00273	.00254	.00237
1.78	.00348	.00323	.00300	.00279	.00260	.00243	.00226
1.79	.00333	.00309	.00287	.00267	.00248	.00231	.00216
1.80	.00318	.00295	.00274	.00255	.00237	.00221	.00206
1.81	.00304	.00282	.00261	.00243	.00226	.00210	.00196
1.82	.00290	.00269	.00249	.00232	.00215	.00201	.00187
1.83	.00277	.00257	.00238	.00221	.00205	.00191	.00178
1.84	.00264	.00245	.00227	.00211	.00196	.00182	.00170
1.85	.00252	.00234	.00217	.00201	.00187	.00174	.00162
1.86	.00241	.00223	.00207	.00192	.00178	.00165	.00154
1.87	.00230	.00213	.00197	.00183	.00170	.00158	.00147
1.88	.00219	.00203	.00188	.00174	.00162	.00150	.00140
1.89	.00209	.00193	.00179	.00166	.00154	.00143	.00133
1.90	.00199	.00184	.00171	.00158	.00147	.00136	.00127
1.91	.00190	.00176	.00163	.00151	.00140	.00130	.00120
1.92	.00181	.00167	.00155	.00143	.00133	.00123	.00115
1.93	.00173	.00160	.00148	.00137	.00127	.00117	.00109
1.94	.00165	.00152	.00141	.00130	.00120	.00112	.00104
1.95	.00157	.00145	.00134	.00124	.00115	.00106	.00099
1.96	.00150	.00138	.00127	.00118	.00109	.00101	.00094
1.97	.00142	.00131	.00121	.00112	.00104	.00096	.00089
1.98	.00136	.00125	.00115	.00107	.00099	.00091	.00085
1.99	.00129	.00119	.00110	.00101	.00094	.00087	.00081
2.00	.00123	.00113	.00105	.00097	.00089	.00083	.00077

8.4 Tables of inv $\Psi_p(u)$

u	$\Psi^{-1}_{-.50}$	$\Psi^{-1}_{-.45}$	$\Psi^{-1}_{-.40}$	$\Psi^{-1}_{-.35}$	$\Psi^{-1}_{-.30}$	$\Psi^{-1}_{-.25}$	$\Psi^{-1}_{-.20}$	$\Psi^{-1}_{-.15}$
0.01	2.6283	2.1056	2.0673	2.0312	1.9969	1.9643	1.9332	1.9034
0.02	1.9779	1.9358	1.8963	1.8590	1.8237	1.7903	1.7584	1.7281
0.03	1.8726	1.8295	1.7891	1.7510	1.7151	1.6811	1.6488	1.6181
0.04	1.7941	1.7502	1.7091	1.6704	1.6340	1.5996	1.5671	1.5362
0.05	1.7308	1.6862	1.6445	1.6054	1.5686	1.5339	1.5011	1.4700
0.06	1.6773	1.6322	1.5899	1.5504	1.5133	1.4783	1.4453	1.4141
0.07	1.6307	1.5850	1.5423	1.5024	1.4650	1.4299	1.3967	1.3654
0.08	1.5893	1.5430	1.5000	1.4598	1.4221	1.3867	1.3535	1.3221
0.09	1.5518	1.5051	1.4617	1.4212	1.3833	1.3477	1.3143	1.2829
0.10	1.5174	1.4703	1.4266	1.3858	1.3477	1.3120	1.2785	1.2471
0.11	1.4857	1.4382	1.3941	1.3531	1.3148	1.2790	1.2455	1.2140
0.12	1.4561	1.4082	1.3639	1.3226	1.2842	1.2482	1.2146	1.1831
0.13	1.4284	1.3801	1.3355	1.2940	1.2554	1.2194	1.1857	1.1542
0.14	1.4022	1.3536	1.3087	1.2670	1.2282	1.1921	1.1584	1.1269
0.15	1.3774	1.3284	1.2833	1.2414	1.2025	1.1663	1.1326	1.1010
0.16	1.3537	1.3045	1.2591	1.2170	1.1780	1.1417	1.1080	1.0764
0.17	1.3311	1.2816	1.2359	1.1937	1.1546	1.1183	1.0845	1.0530
0.18	1.3095	1.2597	1.2138	1.1714	1.1321	1.0957	1.0619	1.0305
0.19	1.2887	1.2386	1.1924	1.1499	1.1106	1.0741	1.0403	1.0089
0.20	1.2686	1.2182	1.1719	1.1292	1.0898	1.0533	1.0194	0.9880
0.21	1.2493	1.1986	1.1520	1.1092	1.0697	1.0331	0.9993	0.9680
0.22	1.2305	1.1795	1.1328	1.0898	1.0502	1.0137	0.9798	0.9485
0.23	1.2123	1.1611	1.1141	1.0710	1.0313	0.9948	0.9610	0.9297
0.24	1.1946	1.1431	1.0960	1.0528	1.0130	0.9764	0.9426	0.9114
0.25	1.1774	1.1257	1.0784	1.0350	0.9952	0.9586	0.9248	0.8937
0.26	1.1606	1.1086	1.0611	1.0177	0.9778	0.9412	0.9074	0.8764
0.27	1.1443	1.0920	1.0443	1.0008	0.9608	0.9242	0.8905	0.8595
0.28	1.1283	1.0758	1.0279	0.9842	0.9442	0.9076	0.8740	0.8430
0.29	1.1126	1.0599	1.0119	0.9681	0.9280	0.8914	0.8578	0.8270
0.30	1.0973	1.0443	0.9961	0.9522	0.9121	0.8755	0.8420	0.8112
0.31	1.0822	1.0290	0.9807	0.9367	0.8966	0.8600	0.8265	0.7958
0.32	1.0674	1.0140	0.9655	0.9214	0.8813	0.8447	0.8113	0.7807
0.33	1.0529	0.9993	0.9506	0.9064	0.8663	0.8297	0.7964	0.7659
0.34	1.0387	0.9848	0.9360	0.8917	0.8515	0.8150	0.7817	0.7513
0.35	1.0246	0.9705	0.9215	0.8772	0.8370	0.8005	0.7673	0.7370
0.36	1.0108	0.9564	0.9073	0.8629	0.8227	0.7862	0.7531	0.7230
0.37	0.9971	0.9426	0.8933	0.8488	0.8086	0.7722	0.7391	0.7091
0.38	0.9837	0.9289	0.8795	0.8349	0.7947	0.7583	0.7254	0.6955
0.39	0.9704	0.9154	0.8658	0.8212	0.7810	0.7447	0.7118	0.6820
0.40	0.9572	0.9020	0.8524	0.8077	0.7675	0.7312	0.6984	0.6688
0.41	0.9442	0.8888	0.8390	0.7943	0.7541	0.7179	0.6852	0.6557
0.42	0.9314	0.8757	0.8258	0.7810	0.7408	0.7047	0.6722	0.6428
0.43	0.9187	0.8628	0.8128	0.7679	0.7278	0.6917	0.6593	0.6301
0.44	0.9061	0.8500	0.7998	0.7550	0.7148	0.6788	0.6465	0.6175
0.45	0.8936	0.8373	0.7870	0.7421	0.7020	0.6661	0.6339	0.6050
0.46	0.8812	0.8247	0.7743	0.7293	0.6893	0.6535	0.6214	0.5927
0.47	0.8689	0.8122	0.7617	0.7167	0.6767	0.6410	0.6091	0.5805
0.48	0.8567	0.7998	0.7492	0.7042	0.6641	0.6286	0.5968	0.5684
0.49	0.8446	0.7875	0.7367	0.6917	0.6517	0.6163	0.5847	0.5565
0.50	0.8325	0.7752	0.7243	0.6793	0.6394	0.6041	0.5726	0.5446

u	$\Psi^{-1}_{-.50}$	$\Psi^{-1}_{-.45}$	$\Psi^{-1}_{-.40}$	$\Psi^{-1}_{-.35}$	$\Psi^{-1}_{-.30}$	$\Psi^{-1}_{-.25}$	$\Psi^{-1}_{-.20}$	$\Psi^{-1}_{-.15}$
0.51	0.8206	0.7630	0.7121	0.6670	0.6272	0.5919	0.5607	0.5329
0.52	0.8087	0.7509	0.6998	0.6548	0.6150	0.5799	0.5488	0.5212
0.53	0.7968	0.7388	0.6877	0.6426	0.6029	0.5680	0.5371	0.5097
0.54	0.7850	0.7268	0.6755	0.6305	0.5909	0.5561	0.5254	0.4982
0.55	0.7732	0.7148	0.6635	0.6184	0.5789	0.5443	0.5138	0.4868
0.56	0.7615	0.7028	0.6514	0.6064	0.5670	0.5325	0.5022	0.4755
0.57	0.7497	0.6909	0.6394	0.5944	0.5551	0.5208	0.4907	0.4643
0.58	0.7381	0.6790	0.6274	0.5825	0.5433	0.5092	0.4793	0.4531
0.59	0.7264	0.6671	0.6155	0.5705	0.5315	0.4976	0.4679	0.4420
0.60	0.7147	0.6553	0.6035	0.5587	0.5198	0.4860	0.4566	0.4309
0.61	0.7031	0.6434	0.5916	0.5468	0.5080	0.4745	0.4454	0.4200
0.62	0.6914	0.6315	0.5797	0.5349	0.4963	0.4630	0.4341	0.4090
0.63	0.6797	0.6196	0.5677	0.5230	0.4846	0.4515	0.4229	0.3981
0.64	0.6680	0.6078	0.5558	0.5112	0.4729	0.4401	0.4118	0.3873
0.65	0.6563	0.5958	0.5438	0.4993	0.4613	0.4287	0.4007	0.3765
0.66	0.6446	0.5839	0.5318	0.4874	0.4496	0.4173	0.3896	0.3657
0.67	0.6328	0.5719	0.5198	0.4755	0.4379	0.4059	0.3785	0.3550
0.68	0.6210	0.5599	0.5078	0.4636	0.4262	0.3945	0.3675	0.3443
0.69	0.6091	0.5478	0.4957	0.4517	0.4145	0.3831	0.3564	0.3336
0.70	0.5972	0.5357	0.4835	0.4397	0.4028	0.3717	0.3454	0.3230
0.71	0.5852	0.5235	0.4713	0.4276	0.3910	0.3603	0.3344	0.3124
0.72	0.5732	0.5112	0.4591	0.4156	0.3793	0.3489	0.3234	0.3018
0.73	0.5610	0.4988	0.4467	0.4034	0.3675	0.3375	0.3124	0.2912
0.74	0.5487	0.4863	0.4343	0.3912	0.3556	0.3261	0.3014	0.2806
0.75	0.5364	0.4738	0.4218	0.3789	0.3437	0.3146	0.2904	0.2701
0.76	0.5239	0.4611	0.4091	0.3666	0.3317	0.3031	0.2794	0.2595
0.77	0.5112	0.4482	0.3964	0.3541	0.3197	0.2916	0.2683	0.2490
0.78	0.4985	0.4352	0.3835	0.3416	0.3076	0.2800	0.2573	0.2384
0.79	0.4855	0.4221	0.3705	0.3289	0.2955	0.2684	0.2462	0.2279
0.80	0.4724	0.4088	0.3573	0.3162	0.2832	0.2567	0.2351	0.2173
0.81	0.4590	0.3952	0.3440	0.3032	0.2709	0.2450	0.2240	0.2068
0.82	0.4455	0.3815	0.3304	0.2902	0.2584	0.2332	0.2128	0.1962
0.83	0.4317	0.3675	0.3167	0.2770	0.2459	0.2213	0.2016	0.1856
0.84	0.4176	0.3532	0.3027	0.2636	0.2332	0.2094	0.1904	0.1750
0.85	0.4031	0.3386	0.2885	0.2500	0.2204	0.1973	0.1791	0.1644
0.86	0.3884	0.3237	0.2740	0.2362	0.2074	0.1852	0.1678	0.1537
0.87	0.3732	0.3084	0.2591	0.2222	0.1943	0.1730	0.1564	0.1431
0.88	0.3575	0.2926	0.2439	0.2079	0.1810	0.1607	0.1449	0.1323
0.89	0.3414	0.2764	0.2283	0.1933	0.1676	0.1482	0.1334	0.1216
0.90	0.3246	0.2596	0.2123	0.1784	0.1539	0.1357	0.1217	0.1108
0.91	0.3071	0.2421	0.1957	0.1632	0.1400	0.1230	0.1101	0.1000
0.92	0.2888	0.2238	0.1786	0.1476	0.1258	0.1101	0.0983	0.0891
0.93	0.2694	0.2046	0.1608	0.1315	0.1114	0.0971	0.0864	0.0782
0.94	0.2487	0.1843	0.1422	0.1150	0.0967	0.0839	0.0744	0.0672
0.95	0.2265	0.1625	0.1226	0.0978	0.0817	0.0705	0.0624	0.0562
0.96	0.2020	0.1389	0.1020	0.0801	0.0662	0.0569	0.0502	0.0451
0.97	0.1745	0.1129	0.0799	0.0616	0.0504	0.0431	0.0378	0.0340
0.98	0.1421	0.0833	0.0560	0.0422	0.0342	0.0290	0.0254	0.0227
0.99	0.1003	0.0479	0.0297	0.0217	0.0174	0.0146	0.0128	0.0114
1.00	0.0000	0.0000	0.0000	0.0000	0.0000	0.0000	0.0000	0.0000

u	$\Psi^{-1}_{-.10}$	$\Psi^{-1}_{-.05}$	inv erfc	$\Psi^{-1}_{.05}$	$\Psi^{-1}_{.10}$	$\Psi^{-1}_{.15}$	$\Psi^{-1}_{.20}$	$\Psi^{-1}_{.25}$
0.01	1.8750	1.8476	1.8214	1.7962	1.7719	1.7485	1.7259	1.7041
0.02	1.6992	1.6715	1.6450	1.6196	1.5952	1.5718	1.5493	1.5277
0.03	1.5889	1.5611	1.5345	1.5091	1.4847	1.4614	1.4390	1.4175
0.04	1.5068	1.4789	1.4522	1.4268	1.4025	1.3793	1.3571	1.3358
0.05	1.4405	1.4125	1.3859	1.3605	1.3363	1.3132	1.2911	1.2700
0.06	1.3846	1.3565	1.3299	1.3046	1.2805	1.2575	1.2356	1.2146
0.07	1.3358	1.3078	1.2812	1.2560	1.2320	1.2091	1.1874	1.1666
0.08	1.2925	1.2645	1.2379	1.2128	1.1889	1.1662	1.1446	1.1240
0.09	1.2533	1.2253	1.1988	1.1738	1.1500	1.1274	1.1060	1.0856
0.10	1.2174	1.1895	1.1631	1.1381	1.1145	1.0920	1.0708	1.0505
0.11	1.1843	1.1564	1.1301	1.1052	1.0817	1.0594	1.0383	1.0182
0.12	1.1535	1.1256	1.0994	1.0746	1.0512	1.0291	1.0081	0.9883
0.13	1.1246	1.0968	1.0706	1.0460	1.0227	1.0007	0.9799	0.9602
0.14	1.0973	1.0696	1.0435	1.0190	0.9959	0.9741	0.9534	0.9339
0.15	1.0715	1.0439	1.0179	0.9935	0.9705	0.9488	0.9284	0.9090
0.16	1.0470	1.0194	0.9935	0.9692	0.9464	0.9249	0.9046	0.8854
0.17	1.0235	0.9960	0.9703	0.9461	0.9234	0.9021	0.8819	0.8629
0.18	1.0011	0.9737	0.9481	0.9240	0.9015	0.8803	0.8603	0.8415
0.19	0.9796	0.9522	0.9267	0.9028	0.8804	0.8594	0.8396	0.8209
0.20	0.9588	0.9316	0.9062	0.8824	0.8602	0.8393	0.8197	0.8012
0.21	0.9388	0.9117	0.8864	0.8628	0.8407	0.8200	0.8005	0.7822
0.22	0.9195	0.8924	0.8673	0.8438	0.8219	0.8013	0.7820	0.7639
0.23	0.9007	0.8738	0.8488	0.8254	0.8037	0.7833	0.7642	0.7462
0.24	0.8825	0.8557	0.8308	0.8077	0.7860	0.7658	0.7469	0.7291
0.25	0.8649	0.8382	0.8134	0.7904	0.7689	0.7489	0.7301	0.7126
0.26	0.8477	0.8211	0.7965	0.7736	0.7523	0.7324	0.7139	0.6965
0.27	0.8309	0.8045	0.7800	0.7573	0.7361	0.7164	0.6980	0.6808
0.28	0.8145	0.7882	0.7639	0.7413	0.7204	0.7009	0.6827	0.6656
0.29	0.7986	0.7724	0.7482	0.7258	0.7050	0.6857	0.6676	0.6508
0.30	0.7828	0.7569	0.7329	0.7106	0.6900	0.6709	0.6530	0.6364
0.31	0.7677	0.7418	0.7179	0.6958	0.6754	0.6564	0.6387	0.6223
0.32	0.7527	0.7269	0.7032	0.6813	0.6610	0.6422	0.6248	0.6085
0.33	0.7380	0.7124	0.6888	0.6671	0,6470	0.6284	0.6111	0.5950
0.34	0.7235	0.6981	0.6747	0.6531	0.6333	0.6148	0.5977	0.5818
0.35	0.7094	0.6841	0.6609	0.6395	0.6198	0.6015	0.5846	0.5689
0.36	0.6955	0.6703	0.6473	0.6261	0.6066	0.5885	0.5718	0.5563
0.37	0.6818	0.6568	0.6339	0.6129	0.5936	0.5757	0.5592	0.5439
0.38	0.6683	0.6435	0.6208	0.6000	0.5808	0.5632	0.5469	0.5317
0.39	0.6550	0.6303	0.6078	0.5872	0.5683	0.5508	0.5347	0.5198
0.40	0.6419	0.6174	0.5951	0.5747	0.5560	0.5387	0.5228	0.5080
0.41	0.6290	0.6047	0.5826	0.5624	0.5438	0.5268	0.5110	0.4965
0.42	0.6163	0.5922	0.5702	0.5502	0.5319	0.5150	0.4995	0.4851
0.43	0.6037	0.5798	0.5580	0.5382	0.5201	0.5035	0.4881	0.4740
0.44	0.5913	0.5676	0.5460	0.5264	0.5085	0.4921	0.4769	0.4630
0.45	0.5790	0.5555	0.5342	0.5148	0.4971	0.4808	0.4659	0.4522
0.46	0.5669	0.5436	0.5224	0.5033	0.4858	0.4698	0.4551	0.4415
0.47	0.5549	0.5318	0.5109	0.4919	0.4746	0.4588	0.4443	0.4310
0.48	0.5430	0.5201	0.4994	0.4807	0.4636	0.4481	0.4338	0.4207
0.49	0.5312	0.5086	0.4881	0.4696	0.4528	0.4374	0.4234	0.4104
0.50	0.5196	0.4972	0.4769	0.4586	0.4420	0.4269	0.4131	0.4004

u	$\Psi^{-1}_{-.10}$	$\Psi^{-1}_{-.05}$	inv erfc	$\Psi^{-1}_{.05}$	$\Psi^{-1}_{.10}$	$\Psi^{-1}_{.15}$	$\Psi^{-1}_{.20}$	$\Psi^{-1}_{.25}$
0.51	0.5081	0.4859	0.4659	0.4478	0.4314	0.4165	0.4029	0.3904
0.52	0.4967	0.4747	0.4549	0.4371	0.4210	0.4063	0.3929	0.3806
0.53	0.4853	0.4636	0.4441	0.4265	0.4106	0.3961	0.3829	0.3709
0.54	0.4741	0.4526	0.4333	0.4160	0.4003	0.3861	0.3731	0.3613
0.55	0.4629	0.4417	0.4227	0.4056	0.3902	0.3762	0.3634	0.3518
0.56	0.4519	0.4309	0.4121	0.3953	0.3801	0.3665	0.3539	0.3424
0.57	0.4409	0.4202	0.4017	0.3851	0.3702	0.3567	0.3444	0.3332
0.58	0.4300	0.4095	0.3913	0.3750	0.3603	0.3470	0.3350	0.3240
0.59	0.4192	0.3990	0.3810	0.3650	0.3505	0.3375	0.3257	0.3149
0.60	0.4084	0.3885	0.3708	0.3550	0.3409	0.3281	0.3165	0.3060
0.61	0.3977	0.3781	0.3607	0.3452	0.3313	0.3188	0.3074	0.2971
0.62	0.3870	0.3677	0.3506	0.3354	0.3218	0.3095	0.2984	0.2883
0.63	0.3765	0.3574	0.3406	0.3257	0.3123	0.3003	0.2895	0.2796
0.64	0.3659	0.3472	0.3307	0.3161	0.3030	0.2912	0.2806	0.2710
0.65	0.3555	0.3371	0.3209	0.3065	0.2937	0.2822	0.2718	0.2624
0.66	0.3450	0.3270	0.3111	0.2970	0.2845	0.2733	0.2631	0.2540
0.67	0.3346	0.3169	0.3013	0.2876	0.2753	0.2644	0.2545	0.2456
0.68	0.3243	0.3069	0.2917	0.2782	0.2663	0.2556	0.2460	0.2373
0.69	0.3140	0.2969	0.2820	0.2689	0.2572	0.2468	0.2375	0.2290
0.70	0.3037	0.2870	0.2725	0.2596	0.2483	0.2382	0.2291	0.2209
0.71	0.2935	0.2772	0.2629	0.2504	0.2394	0.2295	0.2207	0.2127
0.72	0.2833	0.2673	0.2535	0.2413	0.2306	0.2210	0.2124	0.2047
0.73	0.2731	0.2575	0.2440	0.2322	0.2218	0.2125	0.2042	0.1967
0.74	0.2630	0.2478	0.2347	0.2232	0.2131	0.2041	0.1960	0.1888
0.75	0.2528	0.2381	0.2253	0.2142	0.2044	0.1957	0.1879	0.1810
0.76	0.2427	0.2284	0.2160	0.2052	0.1958	0.1874	0.1799	0.1732
0.77	0.2326	0.2187	0.2067	0.1963	0.1872	0.1791	0.1719	0.1654
0.78	0.2226	0.2091	0.1975	0.1875	0.1786	0.1709	0.1639	0.1577
0.79	0.2125	0.1995	0.1883	0.1786	0.1702	0.1627	0.1560	0.1501
0.80	0.2025	0.1899	0.1791	0.1698	0.1617	0.1546	0.1482	0.1425
0.81	0.1924	0.1803	0.1700	0.1611	0.1533	0.1465	0.1404	0.1350
0.82	0.1824	0.1708	0.1609	0.1524	0.1450	0.1384	0.1327	0.1275
0.83	0.1724	0.1613	0.1518	0.1437	0.1366	0.1304	0.1250	0.1201
0.84	0.1623	0.1517	0.1428	0.1350	0.1284	0.1225	0.1173	0.1127
0.85	0.1525	0.1422	0.1337	0.1264	0.1201	0.1146	0.1097	0.1053
0.86	0.1423	0.1328	0.1247	0.1178	0.1119	0.1067	0.1021	0.0980
0.87	0.1322	0.1233	0.1157	0.1093	0.1037	0.0989	0.0946	0.0908
0.88	0.1222	0.1138	0.1068	0.1008	0.0956	0.0911	0.0871	0.0836
0.89	0.1121	0.1043	0.0978	0.0922	0.0875	0.0833	0.0796	0.0764
0.90	0.1021	0.0949	0.0889	0.0838	0.0794	0.0756	0.0722	0.0693
0.91	0.0920	0.0854	0.0799	0.0753	0.0713	0.0679	0.0649	0.0622
0.92	0.0819	0.0759	0.0710	0.0669	0.0633	0.0602	0.0575	0.0551
0.93	0.0717	0.0665	0.0621	0.0584	0.0553	0.0526	0.0502	0.0481
0.94	0.0616	0.0570	0.0532	0.0500	0.0473	0.0450	0.0429	0.0411
0.95	0.0514	0.0475	0.0443	0.0417	0.0394	0.0374	0.0357	0.0342
0.96	0.0412	0.0380	0.0355	0.0333	0.0315	0.0299	0.0285	0.0273
0.97	0.0309	0.0286	0.0266	0.0250	0.0236	0.0224	0.0213	0.0204
0.98	0.0207	0.0190	0.0177	0.0166	0.0157	0.0149	0.0142	0.0136
0.99	0.0104	0.0095	0.0089	0.0083	0.0078	0.0074	0.0071	0.0068
1.00	0.0000	0.0000	0.0000	0.0000	0.0000	0.0000	0.0000	0.0000

u	$\Psi^{-1}_{.30}$	$\Psi^{-1}_{.35}$	$\Psi^{-1}_{.40}$	$\Psi^{-1}_{.45}$	$\Psi^{-1}_{.50}$	$\Psi^{-1}_{.55}$	$\Psi^{-1}_{.60}$	$\Psi^{-1}_{.65}$
0.01	1.6831	1.6627	1.6431	1.6240	1.6056	1.5877	1.5703	1.5535
0.02	1.5068	1.4867	1.4673	1.4486	1.4306	1.4131	1.3962	1.3799
0.03	1.3969	1.3770	1.3579	1.3395	1.3218	1.3047	1.2881	1.2722
0.04	1.3153	1.2957	1.2768	1.2587	1.2413	1.2245	1.2083	1.1927
0.05	1.2498	1.2304	1.2118	1.1939	1.1767	1.1602	1.1443	1.1290
0.06	1.1946	1.1754	1.1571	1.1394	1.1225	1.1063	1.0907	1.0756
0.07	1.1467	1.1278	1.1096	1.0922	1.0756	1.0596	1.0442	1.0295
0.08	1.1043	1.0856	1.0676	1.0505	1.0340	1.0183	1.0032	0.9887
0.09	1.0661	1.0475	1.0298	1.0129	0.9967	0.9812	0.9663	0.9520
0.10	1.0312	1.0129	0.9954	0.9787	0.9627	0.9474	0.9328	0.9187
0.11	0.9992	0.9810	0.9637	0.9472	0.9314	0.9164	0.9019	0.8881
0.12	0.9694	0.9514	0.9343	0.9180	0.9025	0.8876	0.8734	0.8598
0.13	0.9415	0.9238	0.9069	0.8908	0.8754	0.8608	0.8468	0.8334
0.14	0.9154	0.8978	0.8811	0.8652	0.8501	0.8356	0.8218	0.8087
0.15	0.8907	0.8733	0.8568	0.8411	0.8262	0.8119	0.7983	0.7854
0.16	0.8673	0.8501	0.8338	0.8183	0.8035	0.7895	0.7761	0.7633
0.17	0.8450	0.8280	0.8118	0.7965	0.7820	0.7682	0.7550	0.7424
0.18	0.8237	0.8069	0.7910	0.7758	0.7615	0.7478	0.7349	0.7225
0.19	0.8033	0.7867	0.7710	0.7560	0.7419	0.7284	0.7156	0.7034
0.20	0.7838	0.7674	0.7518	0.7371	0.7231	0.7098	0.6972	0.6852
0.21	0.7650	0.7487	0.7334	0.7188	0.7051	0.6920	0.6795	0.6677
0.22	0.7469	0.7308	0.7156	0.7013	0.6877	0.6748	0.6625	0.6509
0.23	0.7294	0.7135	0.6985	0.6843	0.6709	0.6582	0.6461	0.6347
0.24	0.7125	0.6968	0.6820	0.6680	0.6547	0.6422	0.6303	0.6190
0.25	0.6961	0.6806	0.6659	0.6521	0.6391	0.6267	0.6150	0.6039
0.26	0.6802	0.6648	0.6504	0.6368	0.6239	0.6117	0.6002	0.5892
0.27	0.6647	0.6496	0.6353	0.6219	0.6092	0.5972	0.5858	0.5750
0.28	0.6497	0.6347	0.6206	0.6074	0.5949	0.5831	0.5719	0.5612
0.29	0.6351	0.6203	0.6064	0.5933	0.5810	0.5693	0.5583	0.5478
0.30	0.6208	0.6062	0.5925	0.5796	0.5674	0.5560	0.5451	0.5348
0.31	0.6069	0.5925	0.5789	0.5662	0.5542	0.5429	0.5323	0.5221
0.32	0.5933	0.5791	0.5657	0.5532	0.5414	0.5302	0.5197	0.5098
0.33	0.5800	0.5660	0.5528	0.5405	0.5288	0.5179	0.5075	0.4977
0.34	0.5670	0.5532	0.5402	0.5280	0.5166	0.5058	0.4956	0.4859
0.35	0.5543	0.5406	0.5278	0.5158	0.5046	0.4939	0.4839	0.4744
0.36	0.5418	0.5284	0.5157	0.5039	0.4928	0.4824	0.4725	0.4632
0.37	0.5296	0.5163	0.5039	0.4923	0.4813	0.4710	0.4613	0.4522
0.38	0.5177	0.5045	0.4923	0.4808	0.4701	0.4599	0.4504	0.4414
0.39	0.5059	0.4930	0.4809	0.4606	0.4590	0.4491	0.4397	0.4309
0.40	0.4943	0.4816	0.4697	0.4586	0.4482	0.4384	0.4292	0.4205
0.41	0.4830	0.4705	0.4588	0.4478	0.4376	0.4280	0.4189	0.4104
0.42	0.4719	0.4595	0.4480	0.4372	0.4272	0.4177	0.4088	0.4004
0.43	0.4609	0.4487	0.4374	0.4268	0.4169	0.4076	0.3989	0.3907
0.44	0.4501	0.4381	0.4270	0.4166	0.4068	0.3977	0.3891	0.3811
0.45	0.4395	0.4277	0.4167	0.4065	0.3969	0.3880	0.3796	0.3716
0.46	0.4290	0.4174	0.4066	0.3966	0.3872	0.3784	0.3701	0.3624
0.47	0.4187	0.4073	0.3967	0.3868	0.3776	0.3690	0.3609	0.3533
0.48	0.4085	0.3973	0.3869	0.3772	0.3682	0.3597	0.3518	0.3443
0.49	0.3985	0.3875	0.3773	0.3678	0.3589	0.3506	0.3428	0.3355
0.50	0.3886	0.3778	0.3678	0.3584	0.3497	0.3416	0.3340	0.3268

u	$\Psi^{-1}_{.30}$	$\Psi^{-1}_{.35}$	$\Psi^{-1}_{.40}$	$\Psi^{-1}_{.45}$	$\Psi^{-1}_{.50}$	$\Psi^{-1}_{.55}$	$\Psi^{-1}_{.60}$	$\Psi^{-1}_{.65}$
0.51	0.3789	0.3683	0.3584	0.3492	0.3407	0.3327	0.3253	0.3182
0.52	0.3693	0.3588	0.3492	0.3402	0.3318	0.3240	0.3167	0.3098
0.53	0.3598	0.3495	0.3400	0.3312	0.3230	0.3154	0.3082	0.3015
0.54	0.3504	0.3403	0.3310	0.3224	0.3144	0.3069	0.2999	0.2933
0.55	0.3411	0.3313	0.3222	0.3137	0.3059	0.2985	0.2917	0.2853
0.56	0.3319	0.3223	0.3134	0.3051	0.2974	0.2903	0.2836	0.2773
0.57	0.3229	0.3134	0.3047	0.2966	0.2891	0.2821	0.2756	0.2694
0.58	0.3139	0.3047	0.2962	0.2882	0.2809	0.2741	0.2677	0.2617
0.59	0.3051	0.2960	0.2877	0.2800	0.2728	0.2661	0.2599	0.2540
0.60	0.2963	0.2875	0.2793	0.2718	0.2648	0.2583	0.2522	0.2465
0.61	0.2877	0.2790	0.2710	0.2637	0.2569	0.2505	0.2446	0.2390
0.62	0.2791	0.2706	0.2629	0.2557	0.2490	0.2428	0.2371	0.2316
0.63	0.2706	0.2624	0.2548	0.2478	0.2413	0.2352	0.2296	0.2244
0.64	0.2622	0.2542	0.2468	0.2399	0.2336	0.2278	0.2223	0.2172
0.65	0.2539	0.2460	0.2388	0.2322	0.2261	0.2203	0.2150	0.2101
0.66	0.2456	0.2380	0.2310	0.2245	0.2186	0.2130	0.2079	0.2030
0.67	0.2375	0.2300	0.2232	0.2170	0.2112	0.2058	0.2008	0.1961
0.68	0.2294	0.2222	0.2155	0.2094	0.2038	0.1986	0.1937	0.1892
0.69	0.2214	0.2143	0.2079	0.2020	0.1966	0.1915	0.1868	0.1824
0.70	0.2134	0.2066	0.2004	0.1947	0.1894	0.1845	0.1799	0.1757
0.71	0.2055	0.1989	0.1929	0.1874	0.1823	0.1775	0.1731	0.1690
0.72	0.1977	0.1913	0.1855	0.1801	0.1752	0.1706	0.1664	0.1624
0.73	0.1900	0.1838	0.1782	0.1730	0.1682	0.1638	0.1597	0.1559
0.74	0.1823	0.1763	0.1709	0.1659	0.1613	0.1571	0.1531	0.1494
0.75	0.1746	0.1689	0.1637	0.1589	0.1545	0.1504	0.1466	0.1430
0.76	0.1671	0.1616	0.1565	0.1519	0.1477	0.1437	0.1401	0.1367
0.77	0.1596	0.1543	0.1494	0.1450	0.1409	0.1371	0.1337	0.1304
0.78	0.1521	0.1470	0.1424	0.1382	0.1343	0.1307	0.1273	0.1242
0.79	0.1447	0.1399	0.1354	0.1314	0.1276	0.1242	0.1210	0.1181
0.80	0.1374	0.1327	0.1285	0.1246	0.1211	0.1178	0.1148	0.1120
0.81	0.1301	0.1257	0.1216	0.1180	0.1146	0.1115	0.1086	0.1059
0.82	0.1229	0.1187	0.1148	0.1113	0.1081	0.1052	0.1024	0.0999
0.83	0.1157	0.1117	0.1081	0.1048	0.1017	0.0990	0.0964	0.0940
0.84	0.1085	0.1048	0.1014	0.0983	0.0954	0.0928	0.0903	0.0881
0.85	0.1014	0.0979	0.0947	0.0918	0.0891	0.0866	0.0844	0.0822
0.86	0.0944	0.0911	0.0881	0.0853	0.0829	0.0806	0.0784	0.0765
0.87	0.0874	0.0843	0.0815	0.0790	0.0767	0.0745	0.0725	0.0707
0.88	0.0804	0.0776	0.0750	0.0727	0.0705	0.0685	0.0667	0.0650
0.89	0.0735	0.0709	0.0685	0.0664	0.0644	0.0626	0.0609	0.0594
0.90	0.0666	0.0642	0.0621	0.0601	0.0583	0.0567	0.0552	0.0538
0.91	0.0598	0.0576	0.0557	0.0539	0.0523	0.0508	0.0495	0.0482
0.92	0.0530	0.0511	0.0494	0.0478	0.0463	0.0450	0.0438	0.0427
0.93	0.0462	0.0446	0.0431	0.0417	0.0404	0.0393	0.0382	0.0372
0.94	0.0395	0.0381	0.0368	0.0356	0.0345	0.0335	0.0326	0.0318
0.95	0.0329	0.0316	0.0306	0.0296	0.0287	0.0278	0.0271	0.0264
0.96	0.0262	0.0252	0.0244	0.0236	0.0229	0.0222	0.0216	0.0210
0.97	0.0196	0.0189	0.0182	0.0176	0.0171	0.0166	0.0161	0.0157
0.98	0.0130	0.0126	0.0121	0.0117	0.0114	0.0110	0.0107	0.0104
0.99	0.0065	0.0063	0.0060	0.0058	0.0057	0.0055	0.0053	0.0052
1.00	0.0000	0.0000	0.0000	0.0000	0.0000	0.0000	0.0000	0.0000

u	$\Psi^{-1}_{.70}$	$\Psi^{-1}_{.75}$	$\Psi^{-1}_{.80}$	$\Psi^{-1}_{.85}$	$\Psi^{-1}_{.90}$	$\Psi^{-1}_{.95}$	$\Psi^{-1}_{1.00}$
0.01	1.5372	1.5213	1.5059	1.4909	1.4763	1.4622	1.4483
0.02	1.3640	1.3487	1.3338	1.3194	1.3054	1.2919	1.2787
0.03	1.2568	1.2418	1.2274	1.2134	1.1999	1.1868	1.1740
0.04	1.1776	1.1630	1.1489	1.1353	1.1222	1.1094	1.0971
0.05	1.1142	1.1000	1.0863	1.0730	1.0601	1.0477	1.0357
0.06	1.0611	1.0472	1.0338	1.0208	1.0082	0.9961	0.9844
0.07	1.0153	1.0016	0.9884	0.9757	0.9635	0.9516	0.9402
0.08	0.9747	0.9613	0.9484	0.9360	0.9240	0.9124	0.9012
0.09	0.9383	0.9252	0.9125	0.9003	0.8885	0.8772	0.8663
0.10	0.9053	0.8923	0.8799	0.8679	0.8564	0.8453	0.8343
0.11	0.8749	0.8622	0.8500	0.8382	0.8269	0.8161	0.8056
0.12	0.8468	0.8343	0.8223	0.8108	0.7997	0.7890	0.7788
0.13	0.8206	0.8083	0.7965	0.7852	0.7744	0.7639	0.7538
0.14	0.7961	0.7840	0.7724	0.7613	0.7506	0.7404	0.7305
0.15	0.7730	0.7611	0.7497	0.7388	0.7283	0.7182	0.7086
0.16	0.7511	0.7394	0.7282	0.7175	0.7072	0.6973	0.6878
0.17	0.7304	0.7189	0.7079	0.6973	0.6872	0.6775	0.6682
0.18	0.7106	0.6993	0.6885	0.6782	0.6682	0.6587	0.6495
0.19	0.6918	0.6807	0.6700	0.6598	0.6501	0.6407	0.6317
0.20	0.6737	0.6628	0.6523	0.6423	0.6327	0.6236	0.6147
0.21	0.6564	0.6456	0.6354	0.6255	0.6161	0.6071	0.5984
0.22	0.6398	0.6292	0.6191	0.6094	0.6001	0.5913	0.5828
0.23	0.6237	0.6133	0.6034	0.5939	0.5848	0.5761	0.5677
0.24	0.6082	0.5980	0.5882	0.5789	0.5699	0.5614	0.5532
0.25	0.5933	0.5832	0.5736	0.5644	0.5556	0.5472	0.5392
0.26	0.5788	0.5689	0.5594	0.5504	0.5418	0.5335	0.5256
0.27	0.5648	0.5550	0.5457	0.5369	0.5284	0.5203	0.5125
0.28	0.5512	0.5416	0.5324	0.5237	0.5154	0.5074	0.4998
0.29	0.5379	0.5285	0.5195	0.5109	0.5028	0.4950	0.4875
0.30	0.5251	0.5158	0.5070	0.4985	0.4905	0.4828	0.4755
0.31	0.5125	0.5034	0.4947	0.4865	0.4786	0.4711	0.4639
0.32	0.5003	0.4914	0.4828	0.4747	0.4670	0.4596	0.4525
0.33	0.4884	0.4796	0.4712	0.4633	0.4557	0.4484	0.4415
0.34	0.4768	0.4681	0.4599	0.4521	0.4446	0.4375	0.4307
0.35	0.4655	0.4570	0.4489	0.4412	0.4339	0.4269	0.4202
0.36	0.4544	0.4460	0.4381	0.4305	0.4233	0.4165	0.4099
0.37	0.4435	0.4353	0.4275	0.4201	0.4131	0.4063	0.3999
0.38	0.4329	0.4248	0.4172	0.4099	0.4030	0.3964	0.3901
0.39	0.4225	0.4140	0.4071	0.3000	0.3932	0.3867	0.3805
0.40	0.4123	0.4045	0.3972	0.3902	0.3835	0.3772	0.3711
0.41	0.4023	0.3947	0.3875	0.3806	0.3741	0.3679	0.3619
0.42	0.3925	0.3850	0.3779	0.3712	0.3648	0.3587	0.3529
0.43	0.3829	0.3756	0.3686	0.3620	0.3557	0.3498	0.3441
0.44	0.3734	0.3662	0.3594	0.3530	0.3468	0.3410	0.3354
0.45	0.3642	0.3571	0.3504	0.3441	0.3381	0.3323	0.3269
0.46	0.3550	0.3481	0.3416	0.3354	0.3295	0.3239	0.3185
0.47	0.3461	0.3393	0.3329	0.3268	0.3210	0.3156	0.3103
0.48	0.3373	0.3306	0.3243	0.3184	0.3127	0.3074	0.3023
0.49	0.3286	0.3221	0.3159	0.3101	0.3046	0.2993	0.2943
0.50	0.3200	0.3137	0.3077	0.3020	0.2966	0.2914	0.2865

u	$\Psi^{-1}_{.70}$	$\Psi^{-1}_{.75}$	$\Psi^{-1}_{.80}$	$\Psi^{-1}_{.85}$	$\Psi^{-1}_{.90}$	$\Psi^{-1}_{.95}$	$\Psi^{-1}_{1.00}$
0.51	0.3116	0.3054	0.2995	0.2939	0.2887	0.2836	0.2789
0.52	0.3034	0.2973	0.2915	0.2861	0.2809	0.2760	0.2713
0.53	0.2952	0.2892	0.2836	0.2783	0.2732	0.2685	0.2639
0.54	0.2871	0.2813	0.2758	0.2706	0.2657	0.2610	0.2566
0.55	0.2792	0.2735	0.2682	0.2631	0.2583	0.2537	0.2494
0.56	0.2714	0.2658	0.2606	0.2556	0.2510	0.2465	0.2423
0.57	0.2637	0.2583	0.2531	0.2483	0.2437	0.2394	0.2353
0.58	0.2561	0.2508	0.2458	0.2411	0.2366	0.2324	0.2284
0.59	0.2486	0.2434	0.2385	0.2340	0.2296	0.2255	0.2216
0.60	0.2411	0.2361	0.2314	0.2269	0.2227	0.2187	0.2149
0.61	0.2338	0.2289	0.2243	0.2200	0.2159	0.2120	0.2083
0.62	0.2266	0.2218	0.2173	0.2131	0.2091	0.2053	0.2017
0.63	0.2194	0.2148	0.2105	0.2063	0.2025	0.1988	0.1953
0.64	0.2124	0.2079	0.2036	0.1997	0.1959	0.1923	0.1889
0.65	0.2054	0.2010	0.1969	0.1930	0.1894	0.1859	0.1826
0.66	0.1985	0.1943	0.1903	0.1865	0.1830	0.1796	0.1764
0.67	0.1917	0.1876	0.1837	0.1801	0.1766	0.1734	0.1703
0.68	0.1850	0.1810	0.1772	0.1737	0.1704	0.1672	0.1642
0.69	0.1783	0.1744	0.1708	0.1674	0.1642	0.1611	0.1582
0.70	0.1717	0.1680	0.1645	0.1612	0.1581	0.1551	0.1523
0.71	0.1652	0.1616	0.1582	0.1550	0.1520	0.1492	0.1465
0.72	0.1587	0.1552	0.1520	0.1489	0.1460	0.1433	0.1407
0.73	0.1523	0.1490	0.1458	0.1429	0.1401	0.1375	0.1350
0.74	0.1460	0.1428	0.1398	0.1369	0.1342	0.1318	0.1293
0.75	0.1397	0.1366	0.1337	0.1310	0.1284	0.1260	0.1237
0.76	0.1335	0.1306	0.1278	0.1252	0.1228	0.1204	0.1182
0.77	0.1274	0.1246	0.1219	0.1194	0.1170	0.1148	0.1127
0.78	0.1213	0.1186	0.1161	0.1137	0.1114	0.1093	0.1073
0.79	0.1153	0.1127	0.1103	0.1080	0.1058	0.1038	0.1019
0.80	0.1093	0.1069	0.1046	0.1024	0.1003	0.0984	0.0966
0.81	0.1034	0.1011	0.0989	0.0968	0.0949	0.0931	0.0913
0.82	0.0975	0.0953	0.0933	0.0913	0.0895	0.0878	0.0861
0.83	0.0917	0.0896	0.0877	0.0859	0.0841	0.0825	0.0810
0.84	0.0860	0.0840	0.0822	0.0805	0.0788	0.0773	0.0759
0.85	0.0803	0.0784	0.0767	0.0751	0.0736	0.0721	0.0708
0.86	0.0746	0.0729	0.0713	0.0698	0.0684	0.0670	0.0658
0.87	0.0690	0.0674	0.0659	0.0645	0.0632	0.0620	0.0608
0.88	0.0634	0.0620	0.0606	0.0593	0.0581	0.0570	0.0559
0.89	0.0579	0.0566	0.0553	0.0541	0.0530	0.0520	0.0510
0.90	0.0525	0.0512	0.0501	0.0490	0.0480	0.0471	0.0462
0.91	0.0470	0.0459	0.0449	0.0439	0.0430	0.0422	0.0414
0.92	0.0416	0.0407	0.0397	0.0389	0.0381	0.0373	0.0366
0.93	0.0363	0.0354	0.0346	0.0339	0.0332	0.0325	0.0319
0.94	0.0310	0.0303	0.0296	0.0289	0.0283	0.0278	0.0272
0.95	0.0257	0.0251	0.0246	0.0240	0.0235	0.0230	0.0226
0.96	0.0205	0.0200	0.0196	0.0191	0.0187	0.0184	0.0180
0.97	0.0153	0.0150	0.0146	0.0143	0.0140	0.0137	0.0135
0.98	0.0102	0.0099	0.0097	0.0095	0.0093	0.0091	0.0089
0.99	0.0051	0.0049	0.0048	0.0047	0.0046	0.0045	0.0044
1.00	0.0000	0.0000	0.0000	0.0000	0.0000	0.0000	0.0000

8.5 Tables of $\Lambda_p(u)$

u	$\Lambda_{0.00}$	$\Lambda_{0.05}$	$\Lambda_{0.10}$	$\Lambda_{0.15}$	$\Lambda_{0.20}$	$\Lambda_{0.25}$	$\Lambda_{0.30}$
0.01	0.98872	1.56582	2.47984	3.92755	6.22057	9.85257	15.60554
0.02	0.97744	1.44319	2.13103	3.14688	4.64724	6.86329	10.13650
0.03	0.96616	1.36882	1.93949	2.74832	3.89479	5.51991	7.82364
0.04	0.95489	1.31349	1.80701	2.48626	3.42122	4.70821	6.47992
0.05	0.94363	1.26840	1.70525	2.29288	3.08344	4.14707	5.57822
0.06	0.93238	1.22972	1.62220	2.14033	2.82441	3.72768	4.92047
0.07	0.92114	1.19542	1.55172	2.01463	2.61614	3.39782	4.41373
0.08	0.90992	1.16432	1.49024	1.90785	2.44299	3.12886	4.00797
0.09	0.89872	1.13568	1.43553	1.81504	2.29544	2.90360	3.67362
0.10	0.88754	1.10897	1.38609	1.73299	2.16726	2.71102	3.39194
0.11	0.87638	1.08383	1.34088	1.65942	2.05423	2.54365	3.15041
0.12	0.86524	1.06001	1.29912	1.59274	1.95332	2.39623	2.94032
0.13	0.85413	1.03729	1.26026	1.53174	1.86232	2.26495	2.75540
0.14	0.84305	1.01553	1.22385	1.47551	1.77955	2.14696	2.59100
0.15	0.83200	0.99461	1.18956	1.42334	1.70372	2.04007	2.44359
0.16	0.82099	0.97441	1.15710	1.37467	1.63383	1.94257	2.31044
0.17	0.81001	0.95488	1.12626	1.32906	1.56905	1.85312	2.18941
0.18	0.79906	0.93593	1.09686	1.28612	1.50874	1.77063	2.07878
0.19	0.78816	0.91752	1.06874	1.24557	1.45235	1.69421	1.97715
0.20	0.77730	0.89959	1.04179	1.20714	1.39944	1.62313	1.88337
0.21	0.76648	0.88212	1.01588	1.17061	1.34963	1.55678	1.79652
0.22	0.75570	0.86507	0.99094	1.13582	1.30261	1.49464	1.71578
0.23	0.74498	0.84840	0.96687	1.10259	1.25810	1.43628	1.64050
0.24	0.73430	0.83210	0.94362	1.07081	1.21586	1.38133	1.57009
0.25	0.72367	0.81614	0.92113	1.04034	1.17571	1.32945	1.50408
0.26	0.71310	0.80050	0.89933	1.01108	1.13746	1.28039	1.44204
0.27	0.70258	0.78517	0.87819	0.98296	1.10096	1.23388	1.38361
0.28	0.69212	0.77013	0.85766	0.95588	1.06607	1.28972	1.32847
0.29	0.68172	0.75537	0.83771	0.92977	1.03268	1.14772	1.27633
0.30	0.67137	0.74088	0.81831	0.90457	1.00067	1.10771	1.22695
0.31	0.66109	0.72664	0.79942	0.88024	0.96995	1.06955	1.18011
0.32	0.65087	0.71265	0.78102	0.85670	0.94044	1.03310	1.13562
0.33	0.64072	0.69889	0.76309	0.83392	0.91206	0.99824	1.09330
0.34	0.63064	0.68537	0.74560	0.81186	0.88474	0.96488	1.05299
0.35	0.62062	0.67207	0.72854	0.79048	0.85842	0.93291	1.01457
0.36	0.61067	0.65899	0.71188	0.76975	0.83303	0.90224	0.97789
0.37	0.60079	0.64612	0.69561	0.74962	0.80854	0.87279	0.94284
0.38	0.59099	0.63345	0.67971	0.73008	0.78489	0.84450	0.90933
0.39	0.58126	0.62099	0.66418	0.71109	0.76203	0.81730	0.87725
0.40	0.57161	0.60872	0.64899	0.69264	0.73993	0.79112	0.84653
0.41	0.56203	0.59665	0.63414	0.67470	0.71855	0.76592	0.81707
0.42	0.55253	0.58476	0.61961	0.65725	0.69785	0.74163	0.78881
0.43	0.54311	0.57306	0.60539	0.64026	0.67781	0.71822	0.76168
0.44	0.53377	0.56154	0.59148	0.62372	0.65840	0.69564	0.73562
0.45	0.52452	0.55020	0.57787	0.60762	0.63958	0.67385	0.71058
0.46	0.51534	0.53903	0.56454	0.59194	0.62133	0.65281	0.68649
0.47	0.50625	0.52804	0.55149	0.57666	0.60363	0.63249	0.66332
0.48	0.49725	0.51722	0.53871	0.56177	0.58646	0.61285	0.64101
0.49	0.48833	0.50657	0.52620	0.54726	0.56979	0.59387	0.61953
0.50	0.47950	0.49608	0.51394	0.53311	0.55361	0.57551	0.59883

u	$\Lambda_{0.00}$	$\Lambda_{0.05}$	$\Lambda_{0.10}$	$\Lambda_{0.15}$	$\Lambda_{0.20}$	$\Lambda_{0.25}$	$\Lambda_{0.30}$
0.51	0.47076	0.48576	0.50193	0.51931	0.53790	0.55775	0.57889
0.52	0.46210	0.47559	0.49017	0.50585	0.52264	0.54056	0.55965
0.53	0.45354	0.46559	0.47865	0.49272	0.50781	0.52393	0.54110
0.54	0.44506	0.45574	0.46737	0.47992	0.49340	0.50783	0.52320
0.55	0.43668	0.44605	0.45631	0.46742	0.47940	0.49223	0.50592
0.56	0.42838	0.43652	0.44547	0.45523	0.46579	0.47712	0.48924
0.57	0.42018	0.42713	0.43486	0.44334	0.45255	0.46248	0.47314
0.58	0.41208	0.41790	0.42446	0.43173	0.43968	0.44830	0.45758
0.59	0.40406	0.40882	0.41427	0.42040	0.42716	0.43455	0.44255
0.60	0.39614	0.39988	0.40429	0.40934	0.41499	0.42122	0.42802
0.61	0.38832	0.39109	0.39451	0.39854	0.40315	0.40830	0.41398
0.62	0.38059	0.38244	0.38493	0.38800	0.39163	0.39577	0.40041
0.63	0.37295	0.37394	0.37554	0.37772	0.38042	0.38362	0.38728
0.64	0.36541	0.36558	0.36635	0.36767	0.36951	0.37183	0.37459
0.65	0.35797	0.35736	0.35734	0.35787	0.35890	0.36039	0.36231
0.66	0.35062	0.34927	0.34852	0.34830	0.34857	0.34930	0.35043
0.67	0.34337	0.34133	0.33988	0.33896	0.33852	0.33853	0.33894
0.68	0.33622	0.33352	0.33141	0.32984	0.32874	0.32809	0.32783
0.69	0.32916	0.32584	0.32312	0.32094	0.31923	0.31795	0.31707
0.70	0.32220	0.31830	0.31500	0.31224	0.30996	0.30811	0.30665
0.71	0.31533	0.31089	0.30705	0.30376	0.30095	0.29856	0.29657
0.72	0.30857	0.30361	0.29927	0.29548	0.29217	0.28930	0.28682
0.73	0.30190	0.29646	0.29165	0.28740	0.28363	0.28030	0.27737
0.74	0.29532	0.28944	0.28419	0.27951	0.27532	0.27157	0.26823
0.75	0.28884	0.28254	0.27689	0.27180	0.26723	0.26310	0.25937
0.76	0.28246	0.27577	0.26974	0.26429	0.25935	0.25487	0.25080
0.77	0.27618	0.26912	0.26274	0.25696	0.25169	0.24689	0.24250
0.78	0.26999	0.26260	0.25590	0.24980	0.24423	0.23914	0.23446
0.79	0.26390	0.25620	0.24920	0.24281	0.23697	0.23161	0.22668
0.80	0.25790	0.24991	0.24264	0.23600	0.22991	0.22431	0.21914
0.81	0.25200	0.24375	0.23623	0.22935	0.22303	0.21722	0.21184
0.82	0.24619	0.23770	0.22996	0.22286	0.21635	0.21033	0.20477
0.83	0.24048	0.23177	0.22382	0.21654	0.20984	0.20365	0.19793
0.84	0.23486	0.22595	0.21782	0.21037	0.20350	0.19717	0.19130
0.85	0.22933	0.22025	0.21195	0.20435	0.19734	0.19087	0.18488
0.86	0.22390	0.21466	0.20622	0.19848	0.19135	0.18476	0.17866
0.87	0.21856	0.20918	0.20061	0.19275	0.18552	0.17883	0.17264
0.88	0.21331	0.20381	0.19513	0.18717	0.17984	0.17308	0.16681
0.89	0.20816	0.19855	0.18977	0.18173	0.17433	0.16749	0.16116
0.90	0.20309	0.19339	0.18453	0.17642	0.16896	0.16207	0.15570
0.91	0.19812	0.18834	0.17942	0.17125	0.16374	0.15681	0.15040
0.92	0.19323	0.18339	0.17442	0.16621	0.15866	0.15171	0.14528
0.93	0.18844	0.17855	0.16954	0.16130	0.15373	0.14676	0.14031
0.94	0.18373	0.17380	0.16477	0.15651	0.14893	0.14195	0.13551
0.95	0.17911	0.16916	0.16011	0.15184	0.14426	0.13729	0.13085
0.96	0.17458	0.16462	0.15556	0.14730	0.13973	0.13277	0.12635
0.97	0.17013	0.16017	0.15112	0.14287	0.13532	0.12838	0.12199
0.98	0.16577	0.15582	0.14679	0.13856	0.13103	0.12413	0.11777
0.99	0.16149	0.15156	0.14256	0.13436	0.12687	0.12000	0.11368
1.00	0.15730	0.14740	0.13843	0.13027	0.12282	0.11600	0.10973

u	$\Lambda_{0.00}$	$\Lambda_{0.05}$	$\Lambda_{0.10}$	$\Lambda_{0.15}$	$\Lambda_{0.20}$	$\Lambda_{0.25}$	$\Lambda_{0.30}$
1.01	0.15319	0.14333	0.13440	0.12629	0.11889	0.11212	0.10590
1.02	0.14916	0.13935	0.13047	0.12241	0.11507	0.10835	0.10219
1.03	0.14522	0.13546	0.12664	0.11864	0.11136	0.10471	0.09861
1.04	0.14135	0.13165	0.12290	0.11497	0.10775	0.10117	0.09514
1.05	0.13756	0.12793	0.11925	0.11139	0.10425	0.09774	0.09178
1.06	0.13386	0.12430	0.11570	0.10791	0.10085	0.09441	0.08854
1.07	0.13023	0.12075	0.11223	0.10453	0.09755	0.09119	0.08540
1.08	0.12667	0.11729	0.10885	0.10124	0.09434	0.08807	0.08236
1.09	0.12320	0.11390	0.10556	0.09804	0.09123	0.08505	0.07942
1.10	0.11979	0.11060	0.10235	0.09492	0.08821	0.08212	0.07657
1.11	0.11647	0.10737	0.09923	0.09189	0.08527	0.07928	0.07382
1.12	0.11321	0.10423	0.09618	0.08895	0.08243	0.07652	0.07117
1.13	0.11003	0.10115	0.09322	0.08609	0.07967	0.07386	0.06859
1.14	0.10692	0.09816	0.09033	0.08331	0.07699	0.07128	0.06611
1.15	0.10388	0.09523	0.08752	0.08060	0.07439	0.06878	0.06371
1.16	0.10090	0.09238	0.08478	0.07798	0.07187	0.06636	0.06138
1.17	0.09800	0.08960	0.08211	0.07542	0.06942	0.06402	0.05914
1.18	0.09516	0.08689	0.07952	0.07295	0.06705	0.06175	0.05697
1.19	0.09239	0.08424	0.07700	0.07054	0.06475	0.05955	0.05487
1.20	0.08963	0.08166	0.07454	0.06820	0.06252	0.05743	0.05284
1.21	0.08704	0.07915	0.07216	0.06593	0.06036	0.05537	0.05089
1.22	0.08447	0.07671	0.06984	0.06372	0.05827	0.05338	0.04900
1.23	0.08195	0.07432	0.06758	0.06158	0.05624	0.05146	0.04717
1.24	0.07949	0.07200	0.06538	0.05951	0.05427	0.04960	0.04541
1.25	0.07710	0.06974	0.06325	0.05749	0.05237	0.04780	0.04370
1.26	0.07476	0.06754	0.06118	0.05554	0.05052	0.04605	0.04206
1.27	0.07249	0.06540	0.05916	0.05364	0.04874	0.04437	0.04047
1.28	0.07027	0.06332	0.05720	0.05180	0.04701	0.04274	0.03894
1.29	0.06810	0.06129	0.05530	0.05001	0.04533	0.04117	0.03746
1.30	0.06599	0.05932	0.05345	0.04828	0.04371	0.03964	0.03603
1.31	0.06394	0.05740	0.05166	0.04661	0.04214	0.03817	0.03465
1.32	0.06193	0.05553	0.04992	0.04498	0.04062	0.03675	0.03332
1.33	0.05998	0.05372	0.04823	0.04340	0.03915	0.03538	0.03203
1.34	0.05809	0.05195	0.04659	0.04188	0.03772	0.03405	0.03080
1.35	0.05624	0.05024	0.04499	0.04040	0.03635	0.03277	0.02960
1.36	0.05444	0.04857	0.04345	0.03896	0.03501	0.03153	0.02845
1.37	0.05269	0.04695	0.04195	0.03757	0.03373	0.03034	0.02734
1.38	0.05098	0.04538	0.04050	0.03623	0.03248	0.02918	0.02627
1.39	0.04933	0.04385	0.03909	0.03493	0.03128	0.02807	0.02524
1.40	0.04771	0.04237	0.03772	0.03367	0.03011	0.02699	0.02424
1.41	0.04615	0.04093	0.03640	0.03245	0.02899	0.02595	0.02328
1.42	0.04462	0.03953	0.03511	0.03126	0.02790	0.02495	0.02236
1.43	0.04314	0.03817	0.03387	0.03012	0.02685	0.02399	0.02147
1.44	0.04170	0.03686	0.03266	0.02902	0.02584	0.02305	0.02061
1.45	0.04030	0.03558	0.03149	0.02795	0.02486	0.02215	0.01978
1.46	0.03895	0.03434	0.03036	0.02691	0.02391	0.02129	0.01899
1.47	0.03763	0.03314	0.02927	0.02591	0.02300	0.02045	0.01822
1.48	0.03635	0.03198	0.02821	0.02495	0.02211	0.01964	0.01748
1.49	0.03510	0.03085	0.02718	0.02401	0.02126	0.01887	0.01677
1.50	0.03389	0.02975	0.02619	0.02311	0.02044	0.01812	0.01609

u	$\Lambda_{0.00}$	$\Lambda_{0.05}$	$\Lambda_{0.10}$	$\Lambda_{0.15}$	$\Lambda_{0.20}$	$\Lambda_{0.25}$	$\Lambda_{0.30}$
1.51	0.03272	0.02869	0.02523	0.02224	0.01964	0.01739	0.01543
1.52	0.03159	0.02766	0.02430	0.02139	0.01888	0.01670	0.01480
1.53	0.03048	0.02667	0.02340	0.02058	0.01814	0.01603	0.01419
1.54	0.02941	0.02571	0.02253	0.01979	0.01743	0.01538	0.01360
1.55	0.02838	0.02477	0.02168	0.01903	0.01674	0.01476	0.01304
1.56	0.02737	0.02387	0.02087	0.01830	0.01608	0.01416	0.01250
1.57	0.02640	0.02299	0.02008	0.01759	0.01544	0.01358	0.01197
1.58	0.02545	0.02215	0.01932	0.01690	0.01482	0.01303	0.01147
1.59	0.02454	0.02133	0.01859	0.01625	0.01423	0.01249	0.01099
1.60	0.02365	0.02054	0.01788	0.01561	0.01366	0.01198	0.01053
1.61	0.02279	0.01977	0.01719	0.01499	0.01311	0.01148	0.01008
1.62	0.02196	0.01903	0.01653	0.01440	0.01258	0.01101	0.00965
1.63	0.02116	0.01831	0.01589	0.01383	0.01207	0.01055	0.00924
1.64	0.02038	0.01762	0.01528	0.01328	0.01157	0.01011	0.00885
1.65	0.01962	0.01695	0.01468	0.01275	0.01110	0.00968	0.00847
1.66	0.01890	0.01630	0.01411	0.01224	0.01064	0.00928	0.00810
1.67	0.01819	0.01568	0.01355	0.01175	0.01020	0.00889	0.00775
1.68	0.01751	0.01507	0.01302	0.01127	0.00978	0.00851	0.00742
1.69	0.01685	0.01449	0.01250	0.01081	0.00937	0.00815	0.00710
1.70	0.01621	0.01393	0.01200	0.01037	0.00898	0.00800	0.00679
1.71	0.01559	0.01339	0.01152	0.00995	0.00861	0.00747	0.00649
1.72	0.01500	0.01286	0.01106	0.00954	0.00825	0.00714	0.00620
1.73	0.01442	0.01235	0.01062	0.00914	0.00790	0.00684	0.00593
1.74	0.01387	0.01187	0.01019	0.00877	0.00756	0.00654	0.00567
1.75	0.01333	0.01140	0.00977	0.00840	0.00724	0.00626	0.00542
1.76	0.01281	0.01094	0.00937	0.00805	0.00693	0.00598	0.00518
1.77	0.01231	0.01050	0.00899	0.00771	0.00664	0.00572	0.00494
1.78	0.01183	0.01008	0.00862	0.00739	0.00635	0.00547	0.00472
1.79	0.01136	0.00967	0.00826	0.00708	0.00608	0.00523	0.00451
1.80	0.01091	0.00928	0.00792	0.00678	0.00581	0.00500	0.00431
1.81	0.01048	0.00890	0.00759	0.00649	0.00556	0.00478	0.00411
1.82	0.01006	0.00854	0.00727	0.00621	0.00532	0.00456	0.00393
1.83	0.00965	0.00819	0.00697	0.00595	0.00509	0.00436	0.00375
1.84	0.00926	0.00785	0.00667	0.00569	0.00486	0.00417	0.00358
1.85	0.00889	0.00753	0.00639	0.00544	0.00465	0.00398	0.00341
1.86	0.00853	0.00721	0.00612	0.00521	0.00444	0.00380	0.00326
1.87	0.00818	0.00691	0.00586	0.00498	0.00425	0.00363	0.00310
1.88	0.00784	0.00663	0.00561	0.00476	0.00405	0.00346	0.00296
1.89	0.00752	0.00634	0.00537	0.00455	0.00387	0.00330	0.00282
1.90	0.00721	0.00608	0.00514	0.00435	0.00370	0.00315	0.00269
1.91	0.00691	0.00582	0.00491	0.00416	0.00353	0.00301	0.00257
1.92	0.00662	0.00557	0.00470	0.00398	0.00337	0.00287	0.00245
1.93	0.00634	0.00533	0.00450	0.00380	0.00322	0.00274	0.00233
1.94	0.00608	0.00510	0.00430	0.00363	0.00307	0.00261	0.00222
1.95	0.00582	0.00488	0.00411	0.00347	0.00293	0.00249	0.00212
1.96	0.00557	0.00467	0.00393	0.00331	0.00280	0.00237	0.00201
1.97	0.00534	0.00447	0.00375	0.00316	0.00267	0.00226	0.00192
1.98	0.00511	0.00427	0.00359	0.00302	0.00255	0.00215	0.00183
1.99	0.00489	0.00409	0.00343	0.00288	0.00249	0.00205	0.00174
2.00	0.00468	0.00391	0.00327	0.00275	0.00232	0.00196	0.00166

u	$\Lambda_{0.35}$	$\Lambda_{0.40}$	$\Lambda_{0.45}$	$\Lambda_{0.50}$	$\Lambda_{0.55}$	$\Lambda_{0.60}$	$\Lambda_{0.65}$
0.01	24.71820	39.15282	62.01787	98.23755	155.61254	246.50039	390.47766
0.02	14.97138	22.11320	32.66303	48.24754	71.26999	105.28100	155.52651
0.03	11.08951	15.71957	22.28388	31.59087	44.78697	63.49792	90.02930
0.04	8.91906	12.27723	16.90097	23.26754	32.03425	44.10647	60.73125
0.05	7.50402	10.09560	13.58337	18.27753	24.59568	33.10014	44.54801
0.06	6.49571	8.57621	11.32423	14.95418	19.74943	26.08445	34.45419
0.07	5.73419	7.45066	9.68209	12.58320	16.35526	21.26008	27.63825
0.08	5.13492	6.57970	8.43214	10.80746	13.85352	17.76001	22.77038
0.09	4.64868	5.88352	7.44749	9.42854	11.93809	15.11744	19.14568
0.10	4.24473	5.31290	6.65099	8.32738	10.42781	13.05980	16.35815
0.11	3.90277	4.83577	5.99292	7.42823	9.20877	11.41777	14.15863
0.12	3.60882	4.43027	5.43979	6.68059	8.20583	10.08092	12.38633
0.13	3.35292	4.08097	4.96818	6.04949	7.36752	8.97427	10.93319
0.14	3.12774	3.77663	4.56120	5.50995	6.65737	8.04524	9.72413
0.15	2.92779	3.50889	4.20637	5.04365	6.04889	7.25593	8.70544
0.16	2.74884	3.27137	3.89425	4.63687	5.52236	6.57834	7.83780
0.17	2.58759	3.05911	3.61756	4.27908	5.06279	5.99138	7.09178
0.18	2.44140	2.86821	3.37062	3.96214	4.65864	5.47888	6.44497
0.19	2.30818	2.69554	3.14890	3.67957	4.30082	5.02823	5.88003
0.20	2.18618	2.53858	2.94875	3.42622	3.98214	4.62947	5.38334
0.21	2.07401	2.39525	2.76720	3.19792	3.69678	4.27463	4.94407
0.22	1.97047	2.26384	2.60181	2.99124	3.44002	3.95727	4.55351
0.23	1.87457	2.14290	2.45056	2.80337	3.20799	3.67212	4.20458
0.24	1.78547	2.03124	2.31175	2.63194	2.99747	3.41484	3.89147
0.25	1.70246	1.92783	2.18393	2.47497	2.80578	3.18183	3.60937
0.26	1.62491	1.83180	2.06590	2.33081	2.63064	2.97005	3.35429
0.27	1.55231	1.74239	1.95660	2.19802	2.47015	2.77694	3.12285
0.28	1.48418	1.65895	1.85513	2.07537	2.32266	2.60035	2.91221
0.29	1.42013	1.58092	1.76072	1.96182	2.18677	2.43841	2.71995
0.30	1.35979	1.50779	1.67270	1.85646	2.06126	2.28953	2.54400
0.31	1.30286	1.43914	1.59045	1.75848	1.94509	2.15234	2.38257
0.32	1.24906	1.37458	1.51348	1.66719	1.83733	2.02565	2.23412
0.33	1.19814	1.31377	1.44130	1.58199	1.73718	1.90840	2.09732
0.34	1.14988	1.25640	1.37353	1.50232	1.64395	1.79970	1.97100
0.35	1.10409	1.20222	1.30979	1.42771	1.55699	1.69873	1.85414
0.36	1.06058	1.15097	1.24976	1.35774	1.47577	1.60479	1.74583
0.37	1.01921	1.10244	1.19316	1.29203	1.39980	1.51726	1.64529
0.38	0.97981	1.05643	1.13971	1.23024	1.32863	1.43557	1.55182
0.39	0.94227	1.01277	1.08920	1.17205	1.26187	1.35924	1.46478
0.40	0.90646	0.97129	1.04140	1.11721	1.19918	1.28781	1.38363
0.41	0.87228	0.93185	0.99613	1.06546	1.14024	1.22089	1.30787
0.42	0.83962	0.89432	0.95320	1.01657	1.08476	1.15812	1.23705
0.43	0.80839	0.85857	0.91247	0.97035	1.03249	1.09918	1.17078
0.44	0.77851	0.82450	0.87379	0.92661	0.98319	1.04379	1.10869
0.45	0.74991	0.79200	0.83703	0.88517	0.93665	0.99167	1.05046
0.46	0.72250	0.76097	0.80205	0.84590	0.89268	0.94258	0.99580
0.47	0.69623	0.73134	0.76876	0.80864	0.85110	0.89631	0.94443
0.48	0.67104	0.70302	0.73706	0.77326	0.81176	0.85267	0.89613
0.49	0.64686	0.67594	0.70684	0.73966	0.77449	0.81146	0.85066
0.50	0.62365	0.65002	0.67802	0.70771	0.73918	0.77252	0.80783

u	$\Lambda_{0.35}$	$\Lambda_{0.40}$	$\Lambda_{0.45}$	$\Lambda_{0.50}$	$\Lambda_{0.55}$	$\Lambda_{0.60}$	$\Lambda_{0.65}$
0.51	0.60136	0.62522	0.65052	0.67732	0.70570	0.73571	0.76745
0.52	0.57994	0.60146	0.62426	0.64840	0.67392	0.70088	0.72936
0.53	0.55934	0.57869	0.59918	0.62085	0.64375	0.66791	0.69340
0.54	0.53954	0.55687	0.57522	0.59461	0.61508	0.63667	0.65942
0.55	0.52049	0.53594	0.55230	0.56959	0.58783	0.60706	0.62731
0.56	0.50215	0.51586	0.53038	0.54573	0.56192	0.57898	0.59694
0.57	0.48450	0.49659	0.50941	0.52296	0.53727	0.55234	0.56819
0.58	0.46751	0.47809	0.48933	0.50123	0.51379	0.52704	0.54097
0.59	0.45114	0.46033	0.47010	0.48047	0.49144	0.50301	0.51519
0.60	0.43537	0.44326	0.45169	0.46065	0.47014	0.48017	0.49074
0.61	0.42017	0.42686	0.43404	0.44170	0.44984	0.45846	0.46757
0.62	0.40552	0.41109	0.41712	0.42358	0.43049	0.43782	0.44558
0.63	0.39139	0.39594	0.40090	0.40626	0.41202	0.41817	0.42471
0.64	0.37778	0.38136	0.38534	0.38969	0.39440	0.39947	0.40490
0.65	0.36463	0.36734	0.37041	0.37383	0.37759	0.38167	0.38607
0.66	0.35196	0.35385	0.35609	0.35865	0.36153	0.36471	0.36819
0.67	0.33973	0.34087	0.34234	0.34412	0.34620	0.34856	0.35119
0.68	0.32793	0.32838	0.32914	0.33020	0.33155	0.33316	0.33502
0.69	0.31654	0.31635	0.31647	0.31687	0.31755	0.31847	0.31965
0.70	0.30555	0.30477	0.30429	0.30410	0.30416	0.30447	0.30502
0.71	0.29493	0.29362	0.29260	0.29186	0.29137	0.29112	0.29109
0.72	0.28469	0.28288	0.28137	0.28012	0.27913	0.27838	0.27783
0.73	0.27479	0.27253	0.27057	0.26888	0.26743	0.26621	0.26521
0.74	0.26523	0.26257	0.26019	0.25809	0.25623	0.25460	0.25319
0.75	0.25601	0.25297	0.25022	0.24775	0.24552	0.24352	0.24173
0.76	0.24709	0.24371	0.24063	0.23783	0.23527	0.23293	0.23081
0.77	0.23848	0.23480	0.23141	0.22831	0.22545	0.22282	0.22040
0.78	0.23016	0.22620	0.22255	0.21918	0.21605	0.21316	0.21048
0.79	0.22213	0.21792	0.21403	0.21041	0.20705	0.20393	0.20102
0.80	0.21436	0.20994	0.20583	0.20200	0.19844	0.19511	0.19200
0.81	0.20686	0.20224	0.19794	0.19393	0.19018	0.18668	0.18339
0.82	0.19962	0.19482	0.19035	0.18618	0.18228	0.17862	0.17518
0.83	0.19261	0.18767	0.18305	0.17874	0.17470	0.17091	0.16734
0.84	0.18585	0.18077	0.17603	0.17160	0.16744	0.16354	0.15986
0.85	0.17931	0.17412	0.16927	0.16474	0.16049	0.15649	0.15273
0.86	0.17299	0.16771	0.16277	0.15815	0.15382	0.14975	0.14591
0.87	0.16689	0.16152	0.15652	0.15183	0.14743	0.14330	0.13940
0.88	0.16099	0.15556	0.15050	0.14575	0.14131	0.13713	0.13319
0.89	0.15529	0.14981	0.14470	0.13992	0.13544	0.13122	0.12726
0.90	0.14978	0.14427	0.13913	0.13432	0.12981	0.12557	0.12159
0.91	0.14446	0.13892	0.13376	0.12893	0.12441	0.12017	0.11618
0.92	0.13931	0.13376	0.12859	0.12376	0.11924	0.11499	0.11100
0.93	0.13434	0.12879	0.12362	0.11879	0.11428	0.11004	0.10606
0.94	0.12954	0.12400	0.11884	0.11402	0.10952	0.10530	0.10134
0.95	0.12490	0.11937	0.11423	0.10944	0.10496	0.10076	0.09683
0.96	0.12042	0.11491	0.10980	0.10503	0.10058	0.09642	0.09252
0.97	0.11608	0.11061	0.10553	0.10080	0.09639	0.09226	0.08840
0.98	0.11190	0.10646	0.10142	0.09673	0.09236	0.08828	0.08447
0.99	0.10785	0.10247	0.09747	0.09283	0.08850	0.08447	0.08070
1.00	0.10395	0.09861	0.09366	0.08907	0.08480	0.08082	0.07711

u	$\Lambda_{0.35}$	$\Lambda_{0.40}$	$\Lambda_{0.45}$	$\Lambda_{0.50}$	$\Lambda_{0.55}$	$\Lambda_{0.60}$	$\Lambda_{0.65}$
1.01	0.10018	0.09489	0.09000	0.08547	0.08125	0.07733	0.07367
1.02	0.09653	0.09131	0.08648	0.08200	0.07785	0.07398	0.07038
1.03	0.09301	0.08785	0.08308	0.07867	0.07458	0.07078	0.06724
1.04	0.08961	0.08452	0.07982	0.07548	0.07145	0.06771	0.06424
1.05	0.08632	0.08130	0.07668	0.07240	0.06845	0.06478	0.06137
1.06	0.08315	0.07821	0.07365	0.06945	0.06556	0.06196	0.05862
1.07	0.08009	0.07522	0.07074	0.06661	0.06280	0.05927	0.05600
1.08	0.07713	0.07234	0.06794	0.06389	0.06015	0.05669	0.05349
1.09	0.07428	0.06957	0.06525	0.06127	0.05761	0.95422	0.05109
1.10	0.07152	0.06689	0.06265	0.05876	0.05517	0.05186	0.04880
1.11	0.06886	0.06432	0.06016	0.05634	0.05283	0.04960	0.04661
1.12	0.06629	0.06183	0.05776	0.05402	0.05059	0.04743	0.04451
1.13	0.06381	0.05944	0.05545	0.05179	0.04844	0.04535	0.04251
1.14	0.06141	0.05713	0.05323	0.04965	0.04637	0.04336	0.04059
1.15	0.05910	0.05491	0.05109	0.04760	0.04440	0.04146	0.03876
1.16	0.05687	0.05277	0.04903	0.04562	0.04250	0.03964	0.03701
1.17	0.05472	0.05071	0.04705	0.04372	0.04068	0.03789	0.03534
1.18	0.05264	0.04872	0.04515	0.04190	0.03894	0.03622	0.03374
1.19	0.05064	0.04680	0.04332	0.04015	0.03726	0.03462	0.03221
1.20	0.04871	0.04496	0.04156	0.03848	0.03566	0.03309	0.03074
1.21	0.04684	0.04319	0.03987	0.03686	0.03412	0.03163	0.02935
1.22	0.04504	0.04148	0.03825	0.03532	0.03265	0.03022	0.02801
1.23	0.04331	0.03983	0.03668	0.03383	0.03124	0.02888	0.02673
1.24	0.04164	0.03824	0.03518	0.03240	0.02988	0.02760	0.02551
1.25	0.04003	0.03672	0.03373	0.03103	0.02859	0.02637	0.02435
1.26	0.03847	0.03525	0.03234	0.02972	0.02734	0.02519	0.02323
1.27	0.03697	0.03384	0.03101	0.02846	0.02615	0.02406	0.02216
1.28	0.03553	0.03247	0.02972	0.02725	0.02501	0.02298	0.02115
1.29	0.03414	0.03116	0.02849	0.02608	0.02391	0.02195	0.02017
1.30	0.03280	0.02990	0.02731	0.02497	0.02286	0.02096	0.01924
1.31	0.03150	0.02869	0.02617	0.02390	0.02186	0.02002	0.01835
1.32	0.03026	0.02752	0.02507	0.02287	0.02090	0.01911	0.01751
1.33	0.02906	0.02640	0.02402	0.02189	0.01997	0.01825	0.01669
1.34	0.02790	0.02532	0.02301	0.02095	0.01909	0.01742	0.01592
1.35	0.02679	0.02428	0.02204	0.02004	0.01824	0.01663	0.01518
1.36	0.02572	0.02328	0.02111	0.01917	0.01743	0.01587	0.01447
1.37	0.02468	0.02232	0.02022	0.01834	0.01666	0.01515	0.01380
1.38	0.02369	0.02140	0.01936	0.01754	0.01591	0.01446	0.01315
1.39	0.02273	0.02051	0.01854	0.01677	0.01520	0.01380	0.01253
1.40	0.02181	0.01966	0.01774	0.01604	0.01452	0.01316	0.01195
1.41	0.02092	0.01884	0.01698	0.01534	0.01387	0.01256	0.01138
1.42	0.02007	0.01805	0.01626	0.01466	0.01324	0.01198	0.01085
1.43	0.01925	0.01729	0.01556	0.01402	0.01265	0.01143	0.01034
1.44	0.01846	0.01656	0.01488	0.01340	0.01207	0.01090	0.00985
1.45	0.01770	0.01586	0.01424	0.01280	0.01153	0.01039	0.00938
1.46	0.01697	0.01519	0.01362	0.01223	0.01100	0.00991	0.00893
1.47	0.01627	0.01455	0.01303	0.01169	0.01050	0.00945	0.00851
1.48	0.01559	0.01393	0.01246	0.01117	0.01002	0.00901	0.00810
1.49	0.01494	0.01333	0.01192	0.01067	0.00956	0.00858	0.00772
1.50	0.01432	0.01276	0.01139	0.01019	0.00912	0.00818	0.00735

u	$\Lambda_{0.35}$	$\Lambda_{0.40}$	$\Lambda_{0.45}$	$\Lambda_{0.50}$	$\Lambda_{0.55}$	$\Lambda_{0.60}$	$\Lambda_{0.65}$
1.51	0.01372	0.01221	0.01089	0.00973	0.00870	0.00780	0.00699
1.52	0.01314	0.01169	0.01041	0.00929	0.00830	0.00743	0.00666
1.53	0.01258	0.01118	0.00995	0.00887	0.00792	0.00708	0.00634
1.54	0.01205	0.01070	0.00951	0.00847	0.00755	0.00674	0.00603
1.55	0.01154	0.01023	0.00909	0.00808	0.00720	0.00642	0.00574
1.56	0.01105	0.00979	0.00868	0.00772	0.00687	0.00612	0.00546
1.57	0.01058	0.00936	0.00829	0.00736	0.00655	0.00583	0.00520
1.58	0.01012	0.00895	0.00792	0.00703	0.00624	0.00555	0.00494
1.59	0.00969	0.00855	0.00757	0.00670	0.00595	0.00528	0.00470
1.60	0.00927	0.00818	0.00722	0.00639	0.00567	0.00503	0.00447
1.61	0.00887	0.00781	0.00690	0.00610	0.00540	0.00479	0.00425
1.62	0.00848	0.00747	0.00659	0.00582	0.00514	0.00456	0.00404
1.63	0.00811	0.00713	0.00629	0.00555	0.00490	0.00434	0.00384
1.64	0.00776	0.00682	0.00600	0.00529	0.00467	0.00413	0.00365
1.65	0.00742	0.00651	0.00572	0.00504	0.00445	0.00393	0.00347
1.66	0.00709	0.00622	0.00546	0.00481	0.00423	0.00374	0.00330
1.67	0.00678	0.00594	0.00521	0.00458	0.00403	0.00355	0.00314
1.68	0.00648	0.00567	0.00497	0.00436	0.00384	0.00338	0.00298
1.69	0.00619	0.00541	0.00474	0.00416	0.00365	0.00321	0.00283
1.70	0.00592	0.00517	0.00452	0.00396	0.00348	0.00306	0.00269
1.71	0.00565	0.00493	0.00431	0.00377	0.00331	0.00291	0.00256
1.72	0.00540	0.00471	0.00411	0.00359	0.00315	0.00276	0.00243
1.73	0.00515	0.00449	0.00392	0.00342	0.00300	0.00263	0.00230
1.74	0.00492	0.00428	0.00373	0.00326	0.00285	0.00250	0.00219
1.75	0.00470	0.00408	0.00356	0.00310	0.00271	0.00237	0.00208
1.76	0.00449	0.00390	0.00339	0.00295	0.00258	0.00225	0.00197
1.77	0.00428	0.00371	0.00323	0.00281	0.00245	0.00214	0.00187
1.78	0.00409	0.00354	0.00308	0.00267	0.00233	0.00203	0.00178
1.79	0.00390	0.00338	0.00293	0.00254	0.00221	0.00193	0.00169
1.80	0.00372	0.00322	0.00279	0.00242	0.00211	0.00183	0.00160
1.81	0.00355	0.00307	0.00266	0.00230	0.00200	0.00174	0.00152
1.82	0.00338	0.00292	0.00253	0.00219	0.00190	0.00165	0.00144
1.83	0.00323	0.00278	0.00241	0.00208	0.00181	0.00157	0.00136
1.84	0.00308	0.00265	0.00229	0.00198	0.00172	0.00149	0.00129
1.85	0.00293	0.00253	0.00218	0.00188	0.00163	0.00141	0.00123
1.86	0.00280	0.00240	0.00207	0.00179	0.00155	0.00134	0.00116
1.87	0.00266	0.00229	0.00197	0.00170	0.00147	0.00127	0.00110
1.88	0.00254	0.00218	0.00188	0.00162	0.00140	0.00121	0.00104
1.89	0.00242	0.00207	0.00178	0.00154	0.00132	0.00114	0.00099
1.90	0.00230	0.00197	0.00170	0.00146	0.00126	0.00109	0.00094
1.91	0.00219	0.00188	0.00161	0.00139	0.00119	0.00103	0.00089
1.92	0.00209	0.00179	0.00153	0.00132	0.00113	0.00098	0.00084
1.93	0.00199	0.00170	0.00146	0.00125	0.00107	0.00092	0.00080
1.94	0.00189	0.00162	0.00138	0.00119	0.00102	0.00088	0.00076
1.95	0.00180	0.00154	0.00132	0.00113	0.00097	0.00083	0.00072
1.96	0.00171	0.00146	0.00125	0.00107	0.00092	0.00079	0.00068
1.97	0.00163	0.00139	0.00119	0.00101	0.00087	0.00075	0.00064
1.98	0.00155	0.00132	0.00113	0.00096	0.00082	0.00071	0.00061
1.99	0.00148	0.00126	0.00107	0.00091	0.00078	0.00067	0.00057
2.00	0.00140	0.00119	0.00102	0.00087	0.00074	0.00063	0.00054

u	$\Lambda_{0.70}$	$\Lambda_{0.75}$	$\Lambda_{0.80}$	$\Lambda_{0.85}$	$\Lambda_{0.90}$	$\Lambda_{.95}$	$\Lambda_{1.00}$
0.01	618.55734	979.87031	1552.25047	2459.00437	3895.47888	6171.15199	9776.31664
0.02	229.75724	339.42490	501.44961	740.83156	1094.51006	1617.06581	2389.14704
0.03	127.65086	180.99984	256.65295	363.93782	516.08382	731.85453	1037.86327
0.04	83.62626	115.15758	158.58443	218.39650	300.77860	414.25090	570.55096
0.05	59.95874	80.70508	108.63554	146.23936	196.86876	265.03814	356.82723
0.06	45.51275	60.12472	79.43286	104.94770	138.66593	183.22698	242.12002
0.07	35.93289	46.72050	60.75113	79.00072	102.73912	133.61869	173.78960
0.08	29.19699	37.44074	48.01609	61.58333	78.98990	101.32353	129.98038
0.09	24.24985	30.71784	38.91461	49.30294	62.46964	79.15881	100.31405
0.10	20.49193	25.67316	32.16775	40.30922	50.51589	63.31250	79.35727
0.11	17.55969	21.78035	27.01857	33.52018	41.59049	51.60875	64.04600
0.12	15.22109	18.70709	22.99431	28.26736	34.75344	42.73222	52.54798
0.13	13.32174	16.23441	19.78656	24.11898	29.40350	35.84991	43.71429
0.14	11.75529	14.21289	17.18675	20.78566	25.14140	30.41360	36.79560
0.15	10.44635	12.53745	15.04944	18.06736	21.69343	26.05062	31.28680
0.16	9.34012	11.13234	13.27063	15.82207	18.86681	22.50058	26.83771
0.17	8.39593	9.94173	11.77419	13.94668	16.52258	19.57709	23.19951
0.18	7.58299	8.92368	10.50334	12.36477	14.55847	17.14401	20.19168
0.19	6.87764	8.04615	9.41501	11.01876	12.89790	15.09997	17.68074
0.20	6.26142	7.28429	8.47598	9.86451	11.48259	13.36836	15.56636
0.21	5.71973	6.61858	7.66031	8.86777	10.26752	11.89034	13.77199
0.22	5.24090	6.03349	6.94749	8.00162	9.21753	10.62018	12.23846
0.23	4.81550	5.51655	6.32113	7.24464	8.30478	9.52193	10.91949
0.24	4.43582	5.05762	5.76798	6.57960	7.50705	8.56698	9.77844
0.25	4.09553	4.64841	5.27725	5.99257	6.80639	7.73236	8.78607
0.26	3.78936	4.28205	4.84006	5.47214	6.18819	6.99948	7.91878
0.27	3.51293	3.95286	4.44908	5.00888	5.64047	6.35314	7.15741
0.28	3.26252	3.65604	4.09817	4.59499	5.15332	5.78085	6.48625
0.29	3.03500	3.38757	3.78220	4.22394	4.71850	5.27224	5.89234
0.30	2.82770	3.14404	3.49680	3.89024	4.32910	4.81868	5.36491
0.31	2.63834	2.92251	3.23830	3.58924	3.97931	4.41291	4.89497
0.32	2.46493	2.72051	3.00353	3.31698	3.66419	4.04882	4.47496
0.33	2.30579	2.53586	2.78979	3.07009	3.37954	3.72121	4.09849
0.34	2.15943	2.36671	2.59475	2.84567	3.12179	3.42567	3.76014
0.35	2.02455	2.21143	2.41639	2.64120	2.88783	3.15839	3.45527
0.36	1.90002	2.06861	2.25295	2.45453	2.67499	2.91613	3.17991
0.37	1.78486	1.93701	2.10289	2.28376	2.48099	2.69607	2.93066
0.38	1.67818	1.81554	1.96487	2.12724	2.30379	2.49577	2.70456
0.39	1.57920	1.70324	1.83772	1.98353	2.14164	2.31310	2.49905
0.40	1.48723	1.59926	1.72039	1.85137	1.99302	2.14622	2.31191
0.41	1.40167	1.50284	1.61195	1.72963	1.85657	1.99349	2.14119
0.42	1.32196	1.41331	1.51159	1.61733	1.73109	1.85349	1.98519
0.43	1.24761	1.33008	1.41859	1.51358	1.61553	1.72496	1.84241
0.44	1.17819	1.25260	1.33229	1.41761	1.50897	1.60679	1.71154
0.45	1.11328	1.18040	1.25211	1.32872	1.41056	1.49799	1.59140
0.46	1.05254	1.11304	1.17754	1.24629	1.31958	1.39770	1.48097
0.47	0.99564	1.05013	1.10810	1.16976	1.23535	1.30512	1.37933
0.48	0.94229	0.99132	1.04337	1.09863	1.15730	1.21957	1.28567
0.49	0.89223	0.93629	0.98298	1.03246	1.08488	1.14042	1.19925
0.50	0.84520	0.88475	0.92658	0.97083	1.01763	1.06712	1.11944

u	$\Lambda_{0.70}$	$\Lambda_{0.75}$	$\Lambda_{0.80}$	$\Lambda_{0.85}$	$\Lambda_{0.90}$	$\Lambda_{0.95}$	$\Lambda_{1.00}$
0.51	0.80100	0.83643	0.87386	0.91339	0.95511	0.99915	1.04563
0.52	0.75941	0.79111	0.82454	0.85979	0.89694	0.93608	0.97732
0.53	0.72026	0.74856	0.77837	0.80974	0.84275	0.87749	0.91402
0.54	0.68338	0.70858	0.73509	0.76296	0.79225	0.82301	0.85533
0.55	0.64861	0.67100	0.69452	0.71921	0.74513	0.77233	0.80085
0.56	0.61581	0.63563	0.65644	0.67826	0.70114	0.72512	0.75023
0.57	0.58485	0.60234	0.62068	0.63991	0.66004	0.68112	0.70317
0.58	0.55561	0.57098	0.58708	0.60395	0.62161	0.64008	0.65939
0.59	0.52798	0.54142	0.55549	0.57023	0.58566	0.60177	0.61861
0.60	0.50186	0.51354	0.52577	0.53859	0.55199	0.56599	0.58062
0.61	0.47715	0.48723	0.49780	0.50887	0.52045	0.53255	0.54518
0.62	0.45377	0.46239	0.47145	0.48095	0.49089	0.50127	0.51212
0.63	0.43163	0.43894	0.44662	0.45469	0.46315	0.47200	0.48125
0.64	0.41066	0.41677	0.42322	0.43000	0.43713	0.44460	0.45241
0.65	0.39079	0.39581	0.40114	0.40676	0.41269	0.41892	0.42544
0.66	0.37195	0.37599	0.38030	0.38488	0.38973	0.39485	0.40022
0.67	0.35408	0.35723	0.36063	0.36427	0.36815	0.37227	0.37662
0.68	0.33713	0.33948	0.34205	0.34485	0.34786	0.35109	0.35452
0.69	0.32105	0.32267	0.32450	0.32654	0.32877	0.33120	0.33382
0.70	0.30577	0.30674	0.30791	0.30927	0.31081	0.31252	0.31441
0.71	0.29127	0.29165	0.29222	0.29297	0.29389	0.29497	0.29621
0.72	0.27750	0.27735	0.27738	0.27759	0.27795	0.27847	0.27914
0.73	0.26440	0.26379	0.26334	0.26306	0.26293	0.26296	0.26312
0.74	0.25196	0.25092	0.25005	0.24934	0.24878	0.24836	0.24808
0.75	0.24013	0.23872	0.23747	0.23637	0.23543	0.23462	0.23395
0.76	0.22888	0.22713	0.22555	0.22412	0.22284	0.22169	0.22067
0.77	0.21818	0.21614	0.21426	0.21253	0.21096	0.20951	0.20819
0.78	0.20800	0.20570	0.20356	0.20158	0.19974	0.19804	0.19646
0.79	0.19831	0.19578	0.19342	0.19121	0.18915	0.18722	0.18542
0.80	0.18909	0.18636	0.18380	0.18140	0.17915	0.17703	0.17503
0.81	0.18031	0.17741	0.17469	0.17212	0.16970	0.16741	0.16526
0.82	0.17195	0.16891	0.16604	0.16333	0.16077	0.15835	0.15605
0.83	0.16399	0.16082	0.15783	0.15500	0.15233	0.14979	0.14739
0.84	0.15640	0.15313	0.15004	0.14712	0.14435	0.14172	0.13922
0.85	0.14918	0.14582	0.14265	0.13965	0.13680	0.13410	0.13153
0.86	0.14229	0.13887	0.13564	0.13257	0.12966	0.12690	0.12427
0.87	0.13573	0.13226	0.12898	0.12587	0.12291	0.12011	0.11744
0.88	0.12948	0.12597	0.12265	0.11951	0.11652	0.11369	0.11099
0.89	0.12352	0.11998	0.11664	0.11348	0.11047	0.10762	0.10491
0.90	0.11783	0.11429	0.11093	0.10776	0.10475	0.10189	0.09917
0.91	0.11241	0.10887	0.10551	0.10234	0.09933	0.09647	0.09376
0.92	0.10725	0.10370	0.10036	0.09719	0.09419	0.09135	0.08865
0.93	0.10232	0.09879	0.09546	0.09231	0.08933	0.08651	0.08383
0.94	0.09762	0.09411	0.09081	0.08768	0.08473	0.08193	0.07927
0.95	0.09314	0.08966	0.08638	0.08329	0.08036	0.07760	0.07497
0.96	0.08886	0.08542	0.08218	0.07912	0.07623	0.07350	0.07091
0.97	0.08478	0.08138	0.07818	0.07516	0.07231	0.06962	0.06708
0.98	0.08089	0.07753	0.07437	0.07140	0.06860	0.06595	0.06345
0.99	0.07717	0.07386	0.07076	0.06783	0.06508	0.06248	0.06003
1.00	0.07363	0.07037	0.06732	0.06444	0.06174	0.05919	0.05679

u	$\Lambda_{0.70}$	$\Lambda_{0.75}$	$\Lambda_{0.80}$	$\Lambda_{0.85}$	$\Lambda_{0.90}$	$\Lambda_{0.95}$	$\Lambda_{1.00}$
1.01	0.07025	0.06705	0.06404	0.06122	0.05858	0.05608	0.05373
1.02	0.06702	0.06388	0.06093	0.05817	0.05558	0.05314	0.05084
1.03	0.06394	0.06086	0.05797	0.05527	0.05273	0.05035	0.04811
1.04	0.06100	0.05798	0.05515	0.05251	0.05003	0.04771	0.04552
1.05	0.05819	0.05524	0.05247	0.04989	0.04747	0.04520	0.04308
1.06	0.05552	0.05262	0.04992	0.04740	0.04504	0.04284	0.04077
1.07	0.05296	0.05013	0.04750	0.04504	0.04274	0.04059	0.03858
1.08	0.05052	0.04776	0.04519	0.04279	0.04056	0.03847	0.03651
1.09	0.04819	0.04550	0.04299	0.04066	0.03848	0.03645	0.03455
1.10	0.04597	0.04334	0.04090	0.03863	0.03652	0.03454	0.03270
1.11	0.04384	0.04128	0.03891	0.03670	0.03465	0.03274	0.03095
1.12	0.04182	0.03933	0.03702	0.03487	0.03288	0.03102	0.02930
1.13	0.03988	0.03746	0.03521	0.03313	0.03120	0.02940	0.02773
1.14	0.03804	0.03568	0.03350	0.03148	0.02960	0.02786	0.02624
1.15	0.03628	0.03398	0.03186	0.02990	0.02809	0.02640	0.02484
1.16	0.03459	0.03237	0.03031	0.02841	0.02665	0.02502	0.02351
1.17	0.03299	0.03082	0.02883	0.02699	0.02529	0.02371	0.02225
1.18	0.03145	0.02935	0.02742	0.02564	0.02399	0.02247	0.02106
1.19	0.02999	0.02795	0.02608	0.02436	0.02276	0.02129	0.01993
1.20	0.02859	0.02662	0.02480	0.02314	0.02160	0.02018	0.01887
1.21	0.02726	0.02535	0.02359	0.02198	0.02049	0.01912	0.01786
1.22	0.02599	0.02413	0.02243	0.02087	0.01944	0.01812	0.01690
1.23	0.02477	0.02298	0.02133	0.01983	0.01844	0.01717	0.01600
1.24	0.02361	0.02188	0.02029	0.01883	0.01749	0.01627	0.01514
1.25	0.02250	0.02082	0.01929	0.01788	0.01660	0.01541	0.01433
1.26	0.02145	0.01982	0.01834	0.01698	0.01574	0.01460	0.01356
1.27	0.02044	0.01887	0.01744	0.01613	0.01493	0.01384	0.01283
1.28	0.01948	0.01796	0.01658	0.01532	0.01416	0.01311	0.01214
1.29	0.01856	0.01709	0.01576	0.01454	0.01343	0.01242	0.01149
1.30	0.01768	0.01627	0.01498	0.01381	0.01274	0.01176	0.01087
1.31	0.01685	0.01548	0.01424	0.01311	0.01208	0.01114	0.01029
1.32	0.01605	0.01473	0.01353	0.01245	0.01146	0.01056	0.00973
1.33	0.01529	0.01402	0.01286	0.01182	0.01087	0.01000	0.00921
1.34	0.01456	0.01334	0.01222	0.01122	0.01030	0.00947	0.00871
1.35	0.01387	0.01269	0.01162	0.01065	0.00977	0.00897	0.00824
1.36	0.01321	0.01207	0.01104	0.01011	0.00926	0.00850	0.00780
1.37	0.01258	0.01148	0.01049	0.00959	0.00878	0.00805	0.00738
1.38	0.01198	0.01092	0.00996	0.00910	0.00832	0.00762	0.00698
1.39	0.01140	0.01038	0.00947	0.00864	0.00789	0.00721	0.00660
1.40	0.01086	0.00987	0.00899	0.00820	0.00748	0.00683	0.00624
1.41	0.01033	0.00939	0.00854	0.00778	0.00709	0.00647	0.00591
1.42	0.00984	0.00893	0.00811	0.00738	0.00672	0.00612	0.00559
1.43	0.00936	0.00849	0.00770	0.00700	0.00637	0.00580	0.00528
1.44	0.00891	0.00807	0.00732	0.00664	0.00603	0.00549	0.00500
1.45	0.00848	0.00767	0.00695	0.00630	0.00572	0.00519	0.00472
1.46	0.00807	0.00729	0.00660	0.00598	0.00542	0.00492	0.00447
1.47	0.00767	0.00693	0.00626	0.00567	0.00513	0.00465	0.00422
1.48	0.00730	0.00658	0.00595	0.00537	0.00486	0.00440	0.00399
1.49	0.00694	0.00626	0.00564	0.00510	0.00461	0.00417	0.00377
1.50	0.00661	0.00595	0.00536	0.00483	0.00436	0.00394	0.00357

u	$\Lambda_{0.70}$	$\Lambda_{0.75}$	$\Lambda_{0.80}$	$\Lambda_{0.85}$	$\Lambda_{0.90}$	$\Lambda_{0.95}$	$\Lambda_{1.00}$
1.51	0.00628	0.00565	0.00508	0.00458	0.00413	0.00373	0.00337
1.52	0.00597	0.00537	0.00482	0.00434	0.00391	0.00353	0.00319
1.53	0.00568	0.00510	0.00458	0.00412	0.00371	0.00334	0.00301
1.54	0.00540	0.00484	0.00434	0.00390	0.00351	0.00316	0.00285
1.55	0.00513	0.00460	0.00412	0.00370	0.00332	0.00299	0.00269
1.56	0.00488	0.00437	0.00391	0.00350	0.00315	0.00283	0.00254
1.57	0.00464	0.00414	0.00371	0.00332	0.00298	0.00267	0.00240
1.58	0.00441	0.00393	0.00352	0.00315	0.00282	0.00253	0.00227
1.59	0.00419	0.00374	0.00334	0.00298	0.00267	0.00239	0.00214
1.60	0.00398	0.00355	0.00316	0.00282	0.00253	0.00226	0.00202
1.61	0.00378	0.00337	0.00300	0.00268	0.00239	0.00214	0.00191
1.62	0.00359	0.00319	0.00284	0.00253	0.00226	0.00202	0.00181
1.63	0.00341	0.00303	0.00270	0.00240	0.00214	0.00191	0.00170
1.64	0.00324	0.00287	0.00255	0.00227	0.00202	0.00180	0.00161
1.65	0.00308	0.00273	0.00242	0.00215	0.00191	0.00171	0.00152
1.66	0.00292	0.00259	0.00229	0.00204	0.00181	0.00161	0.00144
1.67	0.00277	0.00245	0.00217	0.00193	0.00171	0.00152	0.00135
1.68	0.00263	0.00233	0.00206	0.00183	0.00162	0.00144	0.00128
1.69	0.00250	0.00221	0.00195	0.00173	0.00153	0.00136	0.00121
1.70	0.00237	0.00209	0.00185	0.00164	0.00145	0.00128	0.00114
1.71	0.00225	0.00198	0.00175	0.00155	0.00137	0.00121	0.00107
1.72	0.00213	0.00188	0.00166	0.00146	0.00129	0.00115	0.00101
1.73	0.00203	0.00178	0.00157	0.00139	0.00122	0.00108	0.00096
1.74	0.00192	0.00169	0.00149	0.00131	0.00116	0.00102	0.00090
1.75	0.00182	0.00160	0.00141	0.00124	0.00109	0.00096	0.00085
1.76	0.00173	0.00152	0.00133	0.00117	0.00103	0.00091	0.00080
1.77	0.00164	0.00144	0.00126	0.00111	0.00098	0.00086	0.00076
1.78	0.00155	0.00136	0.00119	0.00105	0.00092	0.00081	0.00071
1.79	0.00147	0.00129	0.00113	0.00099	0.00087	0.00077	0.00067
1.80	0.00140	0.00122	0.00107	0.00094	0.00082	0.00072	0.00064
1.81	0.00132	0.00116	0.00101	0.00089	0.00078	0.00068	0.00060
1.82	0.00125	0.00109	0.00096	0.00084	0.00073	0.00064	0.00056
1.83	0.00119	0.00104	0.00090	0.00079	0.00069	0.00061	0.00053
1.84	0.00113	0.00098	0.00086	0.00075	0.00065	0.00057	0.00050
1.85	0.00107	0.00093	0.00081	0.00071	0.00062	0.00054	0.00047
1.86	0.00101	0.00088	0.00077	0.00067	0.00058	0.00051	0.00045
1.87	0.00096	0.00083	0.00072	0.00063	0.00055	0.00048	0.00042
1.88	0.00091	0.00079	0.00068	0.00060	0.00052	0.00045	0.00040
1.89	0.00086	0.00074	0.00065	0.00056	0.00049	0.00043	0.00037
1.90	0.00081	0.00070	0.00061	0.00053	0.00046	0.00040	0.00035
1.91	0.00077	0.00067	0.00058	0.00050	0.00044	0.00038	0.00033
1.92	0.00073	0.00063	0.00055	0.00047	0.00041	0.00036	0.00031
1.93	0.00069	0.00060	0.00052	0.00045	0.00039	0.00034	0.00029
1.94	0.00065	0.00056	0.00049	0.00042	0.00037	0.00032	0.00028
1.95	0.00062	0.00053	0.00046	0.00040	0.00034	0.00030	0.00026
1.96	0.00058	0.00050	0.00043	0.00038	0.00032	0.00028	0.00024
1.97	0.00055	0.00048	0.00041	0.00035	0.00031	0.00027	0.00023
1.98	0.00052	0.00045	0.00039	0.00033	0.00029	0.00025	0.00022
1.99	0.00049	0.00042	0.00037	0.00032	0.00027	0.00024	0.00020
2.00	0.00047	0.00040	0.00034	0.00030	0.00026	0.00022	0.00019

8.6 Tables of inv $\Lambda_p(u)$

$\Lambda_p^{-1}(u)$ for $0.00 \le p \le 0.30$ and $0.001 \le u \le 0.009$.

u	$\Lambda_{0.00}^{-1}$	$\Lambda_{0.05}^{-1}$	$\Lambda_{0.10}^{-1}$	$\Lambda_{0.15}^{-1}$	$\Lambda_{0.20}^{-1}$	$\Lambda_{0.25}^{-1}$	$\Lambda_{0.30}^{-1}$
0.001	2.3268	2.2859	2.2463	2.2080	2.1709	2.1350	2.1003
0.002	2.1851	2.1444	2.1061	2.0672	2.0306	1.9954	1.9615
0.003	2.0985	2.0579	2.0189	1.9814	1.9453	1.9106	1.8773
0.004	2.0352	1.9948	1.9560	1.9188	1.8830	1.8488	1.8160
0.005	1.9849	1.9447	1.9061	1.8692	1.8338	1.8000	1.7676
0.006	1.9430	1.9029	1.8646	1.8279	1.7929	1.7594	1.7274
0.007	1.9070	1.8671	1.8289	1.7923	1.7578	1.7246	1.6930
0.008	1.8753	1.8356	1.7977	1.7615	1.7270	1.6942	1.6629
0.009	1.8470	1.8074	1.7697	1.7338	1.6996	1.6670	1.6361

$\Lambda_p^{-1}(u)$ for $0.35 \le p \le 0.65$ and $0.001 \le u \le 0.009$.

u	$\Lambda_{0.35}^{-1}$	$\Lambda_{0.40}^{-1}$	$\Lambda_{0.45}^{-1}$	$\Lambda_{0.50}^{-1}$	$\Lambda_{0.55}^{-1}$	$\Lambda_{0.60}^{-1}$	$\Lambda_{0.65}^{-1}$
0.001	2.0668	2.0344	2.0031	1.9728	1.9436	1.9154	1.8882
0.002	1.9289	1.8974	1.8672	1.8381	1.8101	1.7832	1.7573
0.003	1.8453	1.8145	1.7851	1.7568	1.7297	1.7037	1.6787
0.004	1.7845	1.7544	1.7256	1.6980	1.6716	1.6463	1.6221
0.005	1.7366	1.7070	1.6787	1.6517	1.6259	1.6012	1.5777
0.006	1.6969	1.6678	1.6400	1.6134	1.5882	1.5640	1.5410
0.007	1.6629	1.6342	1.6068	1.5808	1.5560	1.5323	1.5098
0.008	1.6331	1.6048	1.5779	1.5522	1.5279	1.5047	1.4826
0.009	1.6067	1.5787	1.5521	1.5269	1.5029	1.4801	1.4585

$\Lambda_p^{-1}(u)$ for $0.70 \le p \le 1.00$ and $0.001 \le u \le 0.009$.

u	$\Lambda_{0.70}^{-1}$	$\Lambda_{0.75}^{-1}$	$\Lambda_{0.80}^{-1}$	$\Lambda_{0.85}^{-1}$	$\Lambda_{0.90}^{-1}$	$\Lambda_{0.95}^{-1}$	$\Lambda_{1.00}^{-1}$
0.001	1.8619	1.8365	1.8120	1.7884	1.7656	1.7436	1.7224
0.002	1.7324	1.7085	1.6855	1.6634	1.6421	1.6217	1.6021
0.003	1.6548	1.6319	1.6099	1.5889	1.5687	1.5493	1.5307
0.004	1.5990	1.5768	1.5557	1.5354	1.5160	1.4974	1.4796
0.005	1.5552	1.5337	1.5132	1.4936	1.4748	1.4569	1.4398
0.006	1.5191	1.4982	1.4783	1.4592	1.4411	1.4237	1.4072
0.007	1.4884	1.4680	1.4486	1.4300	1.4124	1.3955	1.3795
0.008	1.4617	1.4417	1.4227	1.4046	1.3874	1.3710	1.3554
0.009	1.4380	1.4184	1.3998	1.3822	1.3654	1.3494	1.3342

u	$\Lambda^{-1}_{0.00}$	$\Lambda^{-1}_{0.05}$	$\Lambda^{-1}_{0.10}$	$\Lambda^{-1}_{0.15}$	$\Lambda^{-1}_{0.20}$	$\Lambda^{-1}_{0.25}$	$\Lambda^{-1}_{0.30}$
0.01	1.8214	1.7820	1.7444	1.7087	1.6748	1.6425	1.6118
0.02	1.6450	1.6070	1.5711	1.5373	1.5055	1.4755	1.4474
0.03	1.5345	1.4977	1.4633	1.4311	1.4010	1.3729	1.3466
0.04	1.4522	1.4166	1.3835	1.3527	1.3242	1.2976	1.2730
0.05	1.3859	1.3514	1.3195	1.2901	1.2629	1.2378	1.2146
0.06	1.3299	1.2965	1.2658	1.2376	1.2117	1.1879	1.1661
0.07	1.2812	1.2489	1.2193	1.1923	1.1676	1.1451	1.1245
0.08	1.2379	1.2066	1.1781	1.1523	1.1288	1.1074	1.0880
0.09	1.1988	1.1685	1.1412	1.1164	1.0940	1.0738	1.0555
0.10	1.1631	1.1338	1.1075	1.0838	1.0625	1.0434	1.0261
0.11	1.1301	1.1018	1.0766	1.0540	1.0337	1.0156	0.9993
0.12	1.0994	1.0721	1.0479	1.0264	1.0071	0.9900	0.9747
0.13	1.0706	1.0444	1.0212	1.0007	0.9825	0.9663	0.9519
0.14	1.0435	1.0183	0.9962	0.9766	0.9594	0.9442	0.9306
0.15	1.0179	0.9937	0.9726	0.9540	0.9377	0.9234	0.9108
0.16	0.9935	0.9704	0.9502	0.9327	0.9173	0.9039	0.8921
0.17	0.9703	0.9482	0.9290	0.9125	0.8980	0.8855	0.8745
0.18	0.9481	0.9270	0.9089	0.8932	0.8797	0.8680	0.8578
0.19	0.9267	0.9067	0.8896	0.8749	0.8623	0.8514	0.8420
0.20	0.9062	0.8872	0.8711	0.8574	0.8457	0.8356	0.8269
0.21	0.8864	0.8685	0.8534	0.8406	0.8297	0.8205	0.8126
0.22	0.8673	0.8504	0.8364	0.8245	0.8145	0.8060	0.7988
0.23	0.8488	0.8330	0.8199	0.8090	0.7999	0.7922	0.7857
0.24	0.8308	0.8162	0.8041	0.7941	0.7858	0.7789	0.7731
0.25	0.8134	0.7999	0.7888	0.7797	0.7722	0.7661	0.7609
0.26	0.7965	0.7840	0.7740	0.7658	0.7592	0.7537	0.7493
0.27	0.7800	0.7687	0.7596	0.7524	0.7465	0.7418	0.7380
0.28	0.7639	0.7537	0.7457	0.7394	0.7343	0.7303	0.7272
0.29	0.7482	0.7392	0.7322	0.7268	0.7225	0.7192	0.7167
0.30	0.7329	0.7250	0.7191	0.7145	0.7111	0.7085	0.7066
0.31	0.7179	0.7112	0.7063	0.7026	0.7000	0.6981	0.6967
0.32	0.7032	0.6977	0.6938	0.6911	0.6892	0.6880	0.6872
0.33	0.6888	0.6845	0.6817	0.6798	0.6787	0.6781	0.6780
0.34	0.6747	0.6717	0.6699	0.6689	0.6685	0.6686	0.6691
0.35	0.6609	0.6591	0.6583	0.6582	0.6586	0.6594	0.6604
0.36	0.6473	0.6468	0.6470	0.6478	0.6490	0.6503	0.6519
0.37	0.6339	0.6347	0.6360	0.6377	0.6395	0.6416	0.6437
0.38	0.6208	0.6229	0.6252	0.6278	0.6304	0.6330	0.6357
0.39	0.6078	0.6113	0.6147	0.6181	0.6214	0.6247	0.6279
0.40	0.5951	0.5999	0.6044	0.6086	0.6127	0.6166	0.6203
0.41	0.5826	0.5887	0.5943	0.5994	0.6042	0.6087	0.6129
0.42	0.5702	0.5777	0.5844	0.5904	0.5959	0.6009	0.6057
0.43	0.5580	0.5669	0.5746	0.5815	0.5877	0.5934	0.5986
0.44	0.5460	0.5563	0.5651	0.5729	0.5797	0.5860	0.5917
0.45	0.5342	0.5459	0.5558	0.5644	0.5720	0.5788	0.5850
0.46	0.5224	0.5357	0.5466	0.5561	0.5643	0.5717	0.5784
0.47	0.5109	0.5256	0.5376	0.5479	0.5568	0.5648	0.5720
0.48	0.4994	0.5156	0.5288	0.5399	0.5496	0.5581	0.5657
0.49	0.4881	0.5059	0.5201	0.5321	0.5424	0.5515	0.5595
0.50	0.4769	0.4962	0.5116	0.5244	0.5354	0.5450	0.5535

u	$\Lambda^{-1}_{0.00}$	$\Lambda^{-1}_{0.05}$	$\Lambda^{-1}_{0.10}$	$\Lambda^{-1}_{0.15}$	$\Lambda^{-1}_{0.20}$	$\Lambda^{-1}_{0.25}$	$\Lambda^{-1}_{0.30}$
0.51	0.4659	0.4868	0.5033	0.5169	0.5285	0.5386	0.5476
0.52	0.4549	0.4774	0.4950	0.5095	0.5218	0.5324	0.5418
0.53	0.4441	0.4682	0.4869	0.5022	0.5151	0.5263	0.5362
0.54	0.4333	0.4591	0.4790	0.4951	0.5086	0.5203	0.5306
0.55	0.4227	0.4502	0.4712	0.4881	0.5023	0.5145	0.5252
0.56	0.4121	0.4413	0.4635	0.4812	0.4960	0.5087	0.5198
0.57	0.4017	0.4326	0.4559	0.4744	0.4899	0.5031	0.5146
0.58	0.3913	0.4240	0.4484	0.4678	0.4838	0.4975	0.5094
0.59	0.3810	0.4156	0.4411	0.4613	0.4779	0.4921	0.5044
0.60	0.3708	0.4072	0.4339	0.4548	0.4721	0.4867	0.4994
0.61	0.3607	0.3989	0.4267	0.4485	0.4664	0.4815	0.4946
0.62	0.3506	0.3908	0.4197	0.4423	0.4607	0.4763	0.4898
0.63	0.3406	0.3828	0.4128	0.4362	0.4552	0.4712	0.4851
0.64	0.3307	0.3748	0.4060	0.4302	0.4498	0.4663	0.4805
0.65	0.3209	0.3670	0.3993	0.4242	0.4444	0.4614	0.4759
0.66	0.3111	0.3592	0.3927	0.4184	0.4392	0.4565	0.4715
0.67	0.3013	0.3516	0.3862	0.4127	0.4340	0.4518	0.4671
0.68	0.2917	0.3440	0.3798	0.4070	0.4289	0.4471	0.4628
0.69	0.2820	0.3366	0.3735	0.4015	0.4239	0.4426	0.4585
0.70	0.2725	0.3292	0.3673	0.3960	0.4189	0.4380	0.4543
0.71	0.2629	0.3219	0.3611	0.3906	0.4141	0.4336	0.4502
0.72	0.2535	0.3147	0.3551	0.3853	0.4093	0.4292	0.4462
0.73	0.2440	0.3076	0.3491	0.3800	0.4046	0.4249	0.4422
0.74	0.2347	0.3006	0.3433	0.3749	0.4000	0.4207	0.4383
0.75	0.2253	0.2937	0.3375	0.3698	0.3954	0.4165	0.4344
0.76	0.2160	0.2868	0.3317	0.3648	0.3909	0.4124	0.4306
0.77	0.2067	0.2801	0.3261	0.3599	0.3865	0.4084	0.4269
0.78	0.1975	0.2734	0.3206	0.3550	0.3821	0.4044	0.4232
0.79	0.1883	0.2668	0.3151	0.3502	0.3778	0.4004	0.4196
0.80	0.1791	0.2603	0.3097	0.3455	0.3736	0.3966	0.4160
0.81	0.1700	0.2539	0.3044	0.3409	0.3694	0.3928	0.4125
0.82	0.1609	0.2476	0.2991	0.3363	0.3653	0.3890	0.4090
0.83	0.1518	0.2413	0.2939	0.3318	0.3612	0.3853	0.4056
0.84	0.1428	0.2351	0.2888	0.3273	0.3572	0.3816	0.4022
0.85	0.1337	0.2290	0.2838	0.3229	0.3533	0.3780	0.3988
0.86	0.1247	0.2230	0.2788	0.3186	0.3494	0.3745	0.3956
0.87	0.1157	0.2171	0.2740	0.3143	0.3456	0.3710	0.3923
0.88	0.1068	0.2112	0.2691	0.3101	0.3418	0.3675	0.3891
0.89	0.0978	0.2055	0.2644	0.3059	0.3380	0.3641	0.3860
0.90	0.0889	0.1998	0.2597	0.3019	0.3344	0.3607	0.3829
0.91	0.0799	0.1942	0.2551	0.2978	0.3307	0.3574	0.3798
0.92	0.0710	0.1886	0.2505	0.2938	0.3272	0.3542	0.3768
0.93	0.0621	0.1832	0.2460	0.2899	0.3236	0.3509	0.3738
0.94	0.0532	0.1778	0.2416	0.2860	0.3202	0.3477	0.3708
0.95	0.0443	0.1725	0.2372	0.2822	0.3167	0.3446	0.3679
0.96	0.0355	0.1673	0.2329	0.2785	0.3133	0.3415	0.3650
0.97	0.0266	0.1622	0.2287	0.2747	0.3100	0.3384	0.3622
0.98	0.0177	0.1572	0.2245	0.2711	0.3067	0.3354	0.3594
0.99	0.0089	0.1523	0.2204	0.2675	0.3034	0.3324	0.3566
1.00	0.0000	0.1474	0.2163	0.2639	0.3002	0.3295	0.3539

u	$\Lambda_{0.05}^{-1}$	$\Lambda_{0.10}^{-1}$	$\Lambda_{0.15}^{-1}$	$\Lambda_{0.20}^{-1}$	$\Lambda_{0.25}^{-1}$	$\Lambda_{0.30}^{-1}$
1.1	0.10350	0.17891	0.23080	0.27027	0.30198	0.32838
1.2	0.06860	0.14689	0.20191	0.24389	0.27763	0.30569
1.3	0.04281	0.11978	0.17670	0.22057	0.25594	0.28539
1.4	0.02537	0.09709	0.15471	0.19989	0.23654	0.26714
1.5	0.01454	0.07833	0.13555	0.18151	0.21911	0.25064
1.6	0.00823	0.06299	0.11887	0.16513	0.20341	0.23567
1.7	0.00469	0.05058	0.10436	0.15051	0.18922	0.22204
1.8	0.00271	0.04062	0.09175	0.13744	0.17635	0.20958
1.9	0.00160	0.03267	0.08080	0.12574	0.16466	0.19817
2.0	0.00097	0.02635	0.07128	0.11524	0.15400	0.18768
2.1	0.00060	0.02134	0.06302	0.10581	0.14428	0.17801
2.2	0.00037	0.01735	0.05583	0.09732	0.13538	0.16909
2.3	0.00024	0.01418	0.04958	0.08967	0.12722	0.16083
2.4	0.00016	0.01165	0.04414	0.08276	0.11973	0.15317
2.5	0.00010	0.00963	0.03938	0.07652	0.11284	0.14606
2.6	0.00008	0.00800	0.03523	0.07086	0.10648	0.13942
2.7	0.00005	0.00668	0.03159	0.06573	0.10062	0.13324
2.8	0.00003	0.00561	0.02840	0.06107	0.09520	0.12747
2.9	0.00002	0.00473	0.02560	0.05683	0.09017	0.12208
3.0	0.00002	0.00401	0.02312	0.05297	0.08552	0.11702
3.1	0.00001	0.00342	0.02094	0.04910	0.08119	0.11228
3.2	0.00001	0.00292	0.01900	0.04621	0.07717	0.10783
3.3	0.00001	0.00251	0.01728	0.04326	0.07342	0.10364
3.4	0.00000	0.00217	0.01576	0.04055	0.06993	0.09969
3.5	0.00000	0.00188	0.01439	0.03806	0.06666	0.09597
3.6	0.00000	0.00164	0.01318	0.03577	0.06361	0.09246
3.7	0.00000	0.00143	0.01209	0.03366	0.06076	0.08914
3.8	0.00000	0.00125	0.01111	0.03171	0.05808	0.08601
3.9	0.00000	0.00110	0.01023	0.02991	0.05557	0.08303
4.0	0.00000	0.00097	0.00943	0.02825	0.05321	0.08022
4.1	0.00000	0.00086	0.00872	0.02670	0.05100	0.07755
4.2	0.00000	0.00076	0.00807	0.02527	0.04891	0.07501
4.3	0.00000	0.00067	0.00748	0.02394	0.04695	0.07260
4.4	0.00000	0.00060	0.00694	0.02271	0.04509	0.07031
4.5	0.00000	0.00054	0.00646	0.02156	0.04334	0.06812
4.6	0.00000	0.00048	0.00601	0.02048	0.04169	0.06604
4.7	0.00000	0.00043	0.00561	0.01948	0.04012	0.06406
4.8	0.00000	0.00039	0.00524	0.01854	0.03864	0.06217
4.9	0.00000	0.00035	0.00490	0.01767	0.03724	0.06036
5.0	0.00000	0.00032	0.00458	0.01685	0.03591	0.05863
5.1	0.00000	0.00029	0.00430	0.01608	0.03465	0.06598
5.2	0.00000	0.00026	0.00403	0.01536	0.03345	0.05540
5.3	0.00000	0.00024	0.00379	0.01468	0.03231	0.05389
5.4	0.00000	0.00022	0.00356	0.01404	0.03123	0.05244
5.5	0.00000	0.00020	0.00335	0.01344	0.03020	0.05105
5.6	0.00000	0.00018	0.00316	0.01287	0.02922	0.04971
5.7	0.00000	0.00017	0.00298	0.01234	0.02828	0.04843
5.8	0.00000	0.00015	0.00282	0.01183	0.02739	0.04720
5.9	0.00000	0.00014	0.00266	0.01136	0.02654	0.04602
6.0	0.00000	0.00013	0.00252	0.01091	0.02572	0.04489

u	$\Lambda_{0.35}^{-1}$	$\Lambda_{0.40}^{-1}$	$\Lambda_{0.45}^{-1}$	$\Lambda_{0.50}^{-1}$	$\Lambda_{0.55}^{-1}$	$\Lambda_{0.60}^{-1}$	$\Lambda_{0.65}^{-1}$
0.01	1.5828	1.5552	1.5289	1.5041	1.4805	1.4581	1.4368
0.02	1.4209	1.3959	1.3725	1.3505	1.3297	1.3102	1.2918
0.03	1.3221	1.2992	1.2778	1.2578	1.2391	1.2216	1.2053
0.04	1.2502	1.2289	1.2092	1.1909	1.1739	1.1580	1.1432
0.05	1.1933	1.1735	1.1552	1.1383	1.1227	1.1082	1.0947
0.06	1.1461	1.1276	1.1106	1.0950	1.0806	1.0673	1.0549
0.07	1.1057	1.0884	1.0726	1.0581	1.0448	1.0325	1.0212
0.08	1.0703	1.0542	1.0394	1.0260	1.0136	1.0023	0.9919
0.09	1.0388	1.0237	1.0100	0.9975	0.9861	0.9756	0.9661
0.10	1.0105	0.9964	0.9836	0.9719	0.9614	0.9517	0.9429
0.11	0.9847	0.9715	0.9595	0.9487	0.9390	0.9301	0.9220
0.12	0.9609	0.9486	0.9375	0.9275	0.9185	0.9103	0.9029
0.13	0.9390	0.9275	0.9172	0.9080	0.8997	0.8921	0.8853
0.14	0.9186	0.9080	0.8984	0.8899	0.8822	0.8753	0.8691
0.15	0.8996	0.8897	0.8808	0.8730	0.8659	0.8596	0.8539
0.16	0.8817	0.8725	0.8644	0.8572	0.8507	0.8450	0.8398
0.17	0.8649	0.8564	0.8489	0.8423	0.8364	0.8312	0.8266
0.18	0.8489	0.8411	0.8343	0.8283	0.8230	0.8183	0.8141
0.19	0.8338	0.8267	0.8205	0.8150	0.8102	0.8060	0.8023
0.20	0.8195	0.8130	0.8073	0.8024	0.7982	0.7944	0.7911
0.21	0.8058	0.7999	0.7949	0.7905	0.7867	0.7834	0.7805
0.22	0.7927	0.7875	0.7830	0.7791	0.7757	0.7729	0.7704
0.23	0.7802	0.7755	0.7716	0.7682	0.7653	0.7629	0.7608
0.24	0.7682	0.7641	0.7607	0.7578	0.7553	0.7533	0.7516
0.25	0.7567	0.7532	0.7502	0.7478	0.7458	0.7441	0.7427
0.26	0.7456	0.7426	0.7402	0.7382	0.7366	0.7353	0.7343
0.27	0.7350	0.7325	0.7305	0.7290	0.7278	0.7268	0.7261
0.28	0.7247	0.7227	0.7212	0.7201	0.7193	0.7187	0.7183
0.29	0.7148	0.7133	0.7123	0.7116	0.7111	0.7109	0.7108
0.30	0.7052	0.7042	0.7036	0.7033	0.7032	0.7033	0.7035
0.31	0.6959	0.6954	0.6953	0.6953	0.6956	0.6960	0.6965
0.32	0.6869	0.6869	0.6872	0.6876	0.6882	0.6889	0.6898
0.33	0.6782	0.6787	0.6793	0.6801	0.6811	0.6821	0.6832
0.34	0.6696	0.6707	0.6717	0.6729	0.6742	0.6755	0.6769
0.35	0.6616	0.6629	0.6644	0.6657	0.6675	0.6691	0.6707
0.36	0.6536	0.6554	0.6572	0.6591	0.6610	0.6629	0.6648
0.37	0.6459	0.6481	0.6503	0.6525	0.6547	0.6568	0.6590
0.38	0.6383	0.6410	0.6435	0.6461	0.6485	0.6510	0.6533
0.39	0.6310	0.6340	0.6370	0.6398	0.6426	0.6453	0.6479
0.40	0.6239	0.6273	0.6306	0.6337	0.6368	0.6397	0.6426
0.41	0.6169	0.6207	0.6243	0.6278	0.6311	0.6343	0.6374
0.42	0.6101	0.6143	0.6183	0.6220	0.6256	0.6290	0.6323
0.43	0.6035	0.6081	0.6123	0.6164	0.6203	0.6239	0.6274
0.44	0.5970	0.6020	0.6066	0.6109	0.6150	0.6189	0.6226
0.45	0.5907	0.5960	0.6009	0.6056	0.6099	0.6140	0.6179
0.46	0.5845	0.5902	0.5954	0.6003	0.6049	0.6093	0.6134
0.47	0.5785	0.5845	0.5901	0.5952	0.6001	0.6046	0.6089
0.48	0.5726	0.5790	0.5848	0.5902	0.5953	0.6001	0.6046
0.49	0.5668	0.5735	0.5797	0.5854	0.5907	0.5956	0.6003
0.50	0.5612	0.5682	0.5746	0.5806	0.5861	0.5913	0.5962

u	$\Lambda^{-1}_{0.35}$	$\Lambda^{-1}_{0.40}$	$\Lambda^{-1}_{0.45}$	$\Lambda^{-1}_{0.50}$	$\Lambda^{-1}_{0.55}$	$\Lambda^{-1}_{0.60}$	$\Lambda^{-1}_{0.65}$
0.51	0.5557	0.5630	0.5697	0.5759	0.5817	0.5870	0.5921
0.52	0.5503	0.5579	0.5649	0.5713	0.5773	0.5829	0.5881
0.53	0.5450	0.5529	0.5602	0.5669	0.5730	0.5788	0.5842
0.54	0.5398	0.5480	0.5556	0.5625	0.5689	0.5748	0.5804
0.55	0.5347	0.5432	0.5510	0.5582	0.5648	0.5709	0.5766
0.56	0.5297	0.5385	0.5466	0.5540	0.5608	0.5671	0.5730
0.57	0.5248	0.5339	0.5422	0.5498	0.5568	0.5633	0.5694
0.58	0.5200	0.5294	0.5380	0.5458	0.5530	0.5596	0.5658
0.59	0.5153	0.5250	0.5338	0.5418	0.5492	0.5560	0.5624
0.60	0.5106	0.5206	0.5297	0.5379	0.5455	0.5525	0.5590
0.61	0.5061	0.5164	0.5256	0.5341	0.5418	0.5490	0.5556
0.62	0.5016	0.5122	0.5217	0.5303	0.5382	0.5456	0.5524
0.63	0.4972	0.5080	0.5178	0.5266	0.5347	0.5422	0.5491
0.64	0.4929	0.5040	0.5140	0.5230	0.5313	0.5389	0.5460
0.65	0.4887	0.5000	0.5102	0.5194	0.5279	0.5357	0.5429
0.66	0.4845	0.4961	0.5065	0.5159	0.5245	0.5325	0.5398
0.67	0.4804	0.4923	0.5029	0.5125	0.5213	0.5293	0.5368
0.68	0.4764	0.4885	0.4993	0.5091	0.5180	0.5263	0.5339
0.69	0.4724	0.4848	0.4958	0.5058	0.5149	0.5232	0.5310
0.70	0.4685	0.4811	0.4923	0.5025	0.5118	0.5203	0.5281
0.71	0.4647	0.4775	0.4889	0.4993	0.5087	0.5173	0.5253
0.72	0.4609	0.4740	0.4856	0,4961	0,5057	0.5144	0.5225
0.73	0.4572	0.4705	0.4823	0.4930	0.5027	0.5116	0.5198
0.74	0.4536	0.4670	0.4791	0.4899	0.4998	0.5088	0.5171
0.75	0.4500	0.4637	0.4759	0.4869	0.4969	0.5061	0.5145
0.76	0.4464	0.4603	0.4727	0.4839	0.4940	0.5033	0.5119
0.77	0.4429	0.4570	0.4696	0.4809	0.4912	0.5007	0.5094
0.78	0.4395	0.4538	0.4666	0.4781	0.4885	0.4980	0.5068
0.79	0.4361	0.4506	0.4636	0.4752	0.4858	0.4954	0.5043
0.80	0.4328	0.4475	0.4606	0.4724	0.4831	0.4929	0.5019
0.81	0.4295	0.4444	0.4577	0.4696	0.4805	0.4904	0.4995
0.82	0.4262	0.4414	0.4548	0.4669	0.4779	0.4879	0.4971
0.83	0.4230	0.4384	0.4520	0.4642	0.4753	0.4854	0.4947
0.84	0.4199	0.4354	0.4492	0.4615	0.4728	0.4830	0.4924
0.85	0.4168	0.4325	0.4464	0.4589	0.4703	0.4806	0.4901
0.86	0.4137	0.4296	0.4437	0.4563	0.4678	0.4783	0.4879
0.87	0.4107	0.4268	0.4410	0.4538	0.4654	0.4760	0.4857
0.88	0.4078	0.4239	0.4384	0.4513	0.4630	0.4737	0.4835
0.89	0.4048	0.4212	0.4357	0.4488	0.4606	0.4714	0.4813
0.90	0.4019	0.4185	0.4332	0.4464	0.4583	0.4692	0.4792
0.91	0.3990	0.4158	0.4306	0.4439	0.4560	0.4670	0.4771
0.92	0.3962	0.4131	0.4281	0.4416	0.4537	0.4648	0.4750
0.93	0.3934	0.4105	0.4256	0.4392	0.4515	0.4627	0.4729
0.94	0.3906	0.4079	0.4232	0.4369	0.4493	0.4605	0.4709
0.95	0.3879	0.4053	0.4208	0.4346	0.4471	0.4584	0.4689
0.96	0.3852	0.4028	0.4184	0.4323	0.4449	0.4564	0.4669
0.97	0.3826	0.4003	0.4160	0.4301	0.4428	0.4543	0.4649
0.98	0.3800	0.3979	0.4137	0.4279	0.4407	0.4523	0.4630
0.99	0.3774	0.3954	0.4114	0.4257	0.4386	0.4503	0.4611
1.00	0.3748	0.3930	0.4091	0.4235	0.4365	0.4484	0.4592

u	$\Lambda_{0.35}^{-1}$	$\Lambda_{0.40}^{-1}$	$\Lambda_{0.45}^{-1}$	$\Lambda_{0.50}^{-1}$	$\Lambda_{0.55}^{-1}$	$\Lambda_{0.60}^{-1}$	$\Lambda_{0.65}^{-1}$
1.0	0.37481	0.39302	0.40912	0.42352	0.43652	0.44836	0.45921
1.1	0.35092	0.37052	0.38781	0.40326	0.41719	0.42986	0.44145
1.2	0.32962	0.35042	0.36876	0.38512	0.39987	0.41326	0.42551
1.3	0.31052	0.33235	0.35159	0.36875	0.38421	0.39825	0.41108
1.4	0.29327	0.31599	0.33602	0.35389	0.36997	0.38458	0.39793
1.5	0.27762	0.30110	0.32182	0.34030	0.35695	0.37206	0.38587
1.6	0.26336	0.28749	0.30880	0.32783	0.34497	0.36053	0.37476
1.7	0.25031	0.27500	0.29682	0.31633	0.33390	0.34987	0.36447
1.8	0.23832	0.26348	0.28575	0.30567	0.32364	0.33997	0.35490
1.9	0.22727	0.25282	0.27549	0.29578	0.31409	0.33075	0.34598
2.0	0.21707	0.24294	0.26594	0.28655	0.30518	0.32212	0.33763
2.1	0.20761	0.23375	0.25703	0.27793	0.29683	0.31404	0.32980
2.2	0.19882	0.22518	0.24870	0.26984	0.28899	0.30643	0.32242
2.3	0.19064	0.21716	0.24089	0.26225	0.28161	0.29927	0.31546
2.4	0.18301	0.20965	0.23355	0.25510	0.27465	0.29250	0.30888
2.5	0.17588	0.20261	0.22663	0.24835	0.26807	0.28610	0.30264
2.6	0.16919	0.19598	0.22011	0.24197	0.26184	0.28002	0.29672
2.7	0.16292	0.18973	0.21395	0.23592	0.25593	0.27425	0.29109
2.8	0.15703	0.18383	0.20812	0.23019	0.25032	0.26876	0.28573
2.9	0.15148	0.17826	0.20259	0.22474	0.24497	0.26352	0.28061
3.0	0.14626	0.17299	0.19734	0.21956	0.23987	0.25853	0.27572
3.1	0.14132	0.16799	0.19235	0.21462	0.23501	0.25375	0.27104
3.2	0.13666	0.16324	0.18760	0.20990	0.23036	0.24918	0.26656
3.3	0.13224	0.15873	0.18307	0.20540	0.22592	0.24481	0.26226
3.4	0.12806	0.15444	0.17875	0.20110	0.22166	0.24061	0.25813
3.5	0.12409	0.15035	0.17462	0.19698	0.21757	0.23658	0.25417
3.6	0.12032	0.14646	0.17068	0.19303	0.21365	0.23270	0.25035
3.7	0.11674	0.14274	0.16690	0.18924	0.20988	0.22897	0.24667
3.8	0.11333	0.13918	0.16328	0.18560	0.20626	0.22539	0.24313
3.9	0.11009	0.13579	0.15980	0.18210	0.20277	0.22193	0.23971
4.0	0.10699	0.13253	0.15647	0.17874	0.19941	0.21859	0.23641
4.1	0.10404	0.12942	0.15327	0.17550	0.19617	0.21537	0.23322
4.2	0.10122	0.12643	0.15019	0.17239	0.19304	0.21225	0.23014
4.3	0.09853	0.12356	0.14723	0.16938	0.19002	0.20925	0.22715
4.4	0.09595	0.12081	0.14438	0.16648	0.18711	0.20634	0.22426
4.5	0.09348	0.11817	0.14163	0.16368	0.18429	0.20352	0.22146
4.6	0.09112	0.11562	0.13898	0.16097	0.18156	0.20079	0.21875
4.7	0.08885	0.11318	0.13643	0.15836	0.17892	0.19814	0.21611
4.8	0.08667	0.11082	0.13396	0.15583	0.17636	0.19558	0.21356
4.9	0.08459	0.10855	0.13158	0.15338	0.17388	0.19309	0.21107
5.0	0.08258	0.10636	0.12928	0.15101	0.17147	0.19067	0.20866
5.1	0.08066	0.10425	0.12705	0.14871	0.16914	0.18832	0.20631
5.2	0.07880	0.10222	0.12489	0.14649	0.16687	0.18604	0.20403
5.3	0.07702	0.10025	0.12281	0.14433	0.16467	0.18382	0.20181
5.4	0.07530	0.09835	0.12079	0.14223	0.16254	0.18166	0.19964
5.5	0.07365	0.09651	0.11883	0.14020	0.16046	0.17956	0.19754
5.6	0.07205	0.09473	0.11693	0.13822	0.15844	0.17752	0.19548
5.7	0.07051	0.09301	0.11509	0.13630	0.15647	0.17552	0.19348
5.8	0.06903	0.09135	0.11330	0.13444	0.15455	0.17358	0.19153
5.9	0.06760	0.08974	0.11157	0.13262	0.15269	0.17169	0.18962
6.0	0.06621	0.08818	0.10988	0.13086	0.15087	0.16984	0.18776

u	$\Lambda^{-1}_{0.35}$	$\Lambda^{-1}_{0.40}$	$\Lambda^{-1}_{0.45}$	$\Lambda^{-1}_{0.50}$	$\Lambda^{-1}_{0.55}$	$\Lambda^{-1}_{0.60}$	$\Lambda^{-1}_{0.65}$
1	0.37481	0.39302	0.40912	0.42352	0.43652	0.44836	0.45921
2	0.21707	0.24294	0.26594	0.28655	0.30518	0.32212	0.33763
3	0.14626	0.17299	0.19734	0.21956	0.23987	0.25853	0.27572
4	0.10699	0.13253	0.15647	0.17874	0.19941	0.21859	0.23641
5	0.08258	0.10636	0.12928	0.15101	0.17147	0.19067	0.20866
6	0.06621	0.08818	0.10988	0.13086	0.15087	0.16984	0.18776
7	0.05462	0.07488	0.09537	0.11551	0.13498	0.15361	0.17134
8	0.04606	0.06478	0.08412	0.10342	0.12230	0.14053	0.15801
9	0.03953	0.05687	0.07514	0.09364	0.11193	0.12975	0.14694
10	0.03442	0.05054	0.06784	0.08557	0.10328	0.12067	0.13756
11	0.03033	0.04536	0.06175	0.07878	0.09594	0.11292	0.12950
12	0.02699	0.04106	0.05663	0.07300	0.08963	0.10620	0.12248
13	0.02423	0.03744	0.05226	0.06801	0.08415	0.10033	0.11630
14	0.02191	0.03436	0.04849	0.06366	0.07933	0.09514	0.11081
15	0.01995	0.03170	0.04521	0.05983	0.07506	0.09051	0.10591
16	0.01826	0.02938	0.04232	0.05645	0.07126	0.08636	0.10148
17	0.01680	0.02736	0.03976	0.05342	0.06784	0.08261	0.09747
18	0.01553	0.02557	0.03748	0.05071	0.06475	0.07921	0.09381
19	0.01441	0.02398	0.03544	0.04825	0.06194	0.07611	0.09046
20	0.01343	0.02255	0.03359	0.04603	0.05938	0.07326	0.08738
21	0.01255	0.02128	0.03192	0.04400	0.05703	0.07065	0.08454
22	0.01176	0.02012	0.03041	0.04214	0.05487	0.06823	0.08190
23	0.01106	0.01908	0.02902	0.04043	0.05288	0.06599	0.07945
24	0.01042	0.01813	0.02775	0.03886	0.05104	0.06391	0.07717
25	0.00984	0.01726	0.02658	0.03740	0.04932	0.06196	0.07503
26	0.00932	0.01646	0.02550	0.03605	0.04772	0.06015	0.07303
27	0.00884	0.01572	0.02450	0.03480	0.04623	0.05845	0.07114
28	0.00840	0.01505	0.02357	0.03363	0.04484	0.05685	0.06937
29	0.00800	0.01442	0.02271	0.03253	0.04353	0.05535	0.06770
30	0.00762	0.01384	0.02190	0.03151	0.04229	0.05393	0.06612
35	0.00614	0.01147	0.01857	0.02721	0.03709	0.04789	0.05934
40	0.00508	0.00974	0.01609	0.02395	0.03307	0.04317	0.05399
45	0.00431	0.00843	0.01417	0.02139	0.02988	0.03938	0.04964
50	0.00371	0.00741	0.01264	0.01932	0.02727	0.03625	0.04603
55	0.00324	0.00659	0.01140	0.01762	0.02510	0.03362	0.04297
60	0.00286	0.00592	0.01037	0.01619	0.02326	0.03138	0.04035
65	0.00256	0.00536	0.00950	0.01498	0.02168	0.02945	0.03807
70	0.00230	0.00489	0.00876	0.01394	0.02032	0.02776	0.03607
75	0.00209	0.00449	0.00813	0.01303	0.01912	0.02627	0.03430
80	0.00191	0.00414	0.00757	0.01223	0.01806	0.02495	0.03272
85	0.00175	0.00384	0.00708	0.01153	0.01712	0.02376	0.03130
90	0.00161	0.00358	0.00665	0.01090	0.01628	0.02270	0.03001
95	0.00149	0.00335	0.00627	0.01033	0.01552	0.02173	0.02884
100	0.00139	0.00314	0.00593	0.00983	0.01483	0.02085	0.02777
105	0.00129	0.00296	0.00562	0.00937	0.01420	0.02004	0.02678
110	0.00121	0.00279	0.00534	0.00895	0.01362	0.01930	0.02588
115	0.00114	0.00264	0.00508	0.00856	0.01310	0.01862	0.02504
120	0.00107	0.00250	0.00485	0.00821	0.01261	0.01799	0.02426
125	0.00101	0.00238	0.00464	0.00789	0.01216	0.01741	0.02354

u	$\Lambda^{-1}_{0.70}$	$\Lambda^{-1}_{0.75}$	$\Lambda^{-1}_{0.80}$	$\Lambda^{-1}_{0.85}$	$\Lambda^{-1}_{0.90}$	$\Lambda^{-1}_{0.95}$	$\Lambda^{-1}_{1.00}$
0.01	1.4167	1.3975	1.3793	1.3620	1.3456	1.3300	1.3151
0.02	1.2745	1.2582	1.2428	1.2283	1.2146	1.2016	1.1894
0.03	1.1899	1.1755	1.1620	1.1494	1.1375	1.1262	1.1157
0.04	1.1294	1.1165	1.1045	1.0932	1.0826	1.0727	1.0634
0.05	1.0822	1.0705	1.0597	1.0496	1.0401	1.0313	1.0230
0.06	1.0435	1.0329	1.0231	1.0139	1.0054	0.9975	0.9901
0.07	1.0108	1.0011	0.9922	0.9839	0.9762	0.9690	0.9623
0.08	0.9823	0.9735	0.9654	0.9578	0.9509	0.9444	0.9384
0.09	0.9573	0.9492	0.9418	0.9349	0.9286	0.9227	0.9173
0.10	0.9349	0.9275	0.9208	0.9145	0.9087	0.9034	0.8985
0.11	0.9146	0.9079	0.9017	0.8960	0.8908	0.8860	0.8816
0.12	0.8961	0.8900	0.8843	0.8792	0.8745	0.8702	0.8662
0.13	0.8791	0.8735	0.8684	0.8638	0.8595	0.8556	0.8521
0.14	0.8634	0.8583	0.8537	0.8495	0.8457	0.8422	0.8390
0.15	0.8488	0.8442	0.8401	0.8363	0.8329	0.8298	0.8269
0.16	0.8352	0.8310	0.8273	0.8239	0.8209	0.8181	0.8156
0.17	0.8224	0.8187	0.8154	0.8124	0.8097	0.8073	0.8051
0.18	0.8104	0.8071	0.8041	0.8015	0.7991	0.7970	0.7951
0.19	0.7990	0.7961	0.7935	0.7912	0.7892	0.7874	0.7858
0.20	0.7882	0.7857	0.7834	0.7815	0.7798	0.7782	0.7769
0.21	0.7780	0.7758	0.7739	0.7723	0.7708	0.7696	0.7685
0.22	0.7683	0.7664	0.7648	0.7635	0.7624	0.7614	0.7605
0.23	0.7590	0.7575	0.7562	0.7551	0.7542	0.7535	0.7529
0.24	0.7501	0.7489	0.7479	0.7471	0.7465	0.7460	0.7456
0.25	0.7416	0.7407	0.7400	0.7395	0.7391	0.7388	0.7387
0.26	0.7335	0.7329	0.7325	0.7322	0.7320	0.7320	0.7320
0.27	0.7257	0.7254	0.7252	0.7252	0.7252	0.7254	0.7256
0.28	0.7181	0.7181	0.7182	0.7184	0.7187	0.7190	0.7195
0.29	0.7109	0.7111	0.7115	0.7119	0.7124	0.7129	0.7136
0.30	0.7039	0.7044	0.7050	0.7056	0.7063	0.7071	0.7079
0.31	0.6972	0.6979	0.6987	0.6996	0.7005	0.7014	0.7024
0.32	0.6907	0.6916	0.6927	0.6937	0.6948	0.6959	0.6971
0.33	0.6844	0.6856	0.6868	0.6881	0.6893	0.6906	0.6919
0.34	0.6783	0.6797	0.6811	0.6826	0.6840	0.6855	0.6869
0.35	0.6724	0.6740	0.6757	0.6773	0.6789	0.6805	0.6821
0.36	0.6666	0.6685	0.6703	0.6722	0.6739	0.6757	0.6775
0.37	0.6611	0.6631	0.6652	0.6672	0.6691	0.6710	0.6729
0.38	0.6557	0.6579	0.6601	0.6623	0.6644	0.6665	0.6685
0.39	0.6504	0.6529	0.6553	0.6576	0.6599	0.6621	0.6643
0.40	0.6453	0.6480	0.0505	0.6530	0.6555	0.6578	0.6601
0.41	0.6403	0.6432	0.6459	0.6486	0.6511	0.6536	0.6560
0.42	0.6355	0.6385	0.6414	0.6442	0.6469	0.6496	0.6521
0.43	0.6308	0.6340	0.6370	0.6400	0.6429	0.6456	0.6483
0.44	0.6262	0.6295	0.6328	0.6359	0.6389	0.6417	0.6445
0.45	0.6217	0.6252	0.6286	0.6319	0.6350	0.6380	0.6409
0.46	0.6173	0.6210	0.6245	0.6279	0.6312	0.6343	0.6373
0.47	0.6120	0.6169	0.6206	0.6241	0.6275	0.6307	0.6338
0.48	0.6088	0.6129	0.6167	0.6204	0.6238	0.6272	0.6304
0.49	0.6047	0.6089	0.6129	0.6167	0.6203	0.6238	0.6271
0.50	0.6007	0.6051	0.6092	0.6131	0.6168	0.6204	0.6238

u	$\Lambda_{0.70}^{-1}$	$\Lambda_{0.75}^{-1}$	$\Lambda_{0.80}^{-1}$	$\Lambda_{0.85}^{-1}$	$\Lambda_{0.90}^{-1}$	$\Lambda_{0.95}^{-1}$	$\Lambda_{1.00}^{-1}$
0.51	0.5968	0.6013	0.6056	0.6096	0.6135	0.6171	0.6207
0.52	0.5930	0.5976	0.6020	0.6062	0.6101	0.6139	0.6176
0.53	0.5893	0.5940	0.5985	0.6028	0.6069	0.6108	0.6145
0.54	0.5856	0.5905	0.5951	0.5995	0.6037	0.6077	0.6115
0.55	0.5820	0.5870	0.5918	0.5963	0.6006	0.6047	0.6086
0.56	0.5785	0.5836	0.5885	0.5931	0.5976	0.6017	0.6057
0.57	0.5750	0.5803	0.5853	0.5901	0.5946	0.5988	0.6029
0.58	0.5716	0.5771	0.5822	0.5870	0.5916	0.5960	0.6002
0.59	0.5683	0.5739	0.5791	0.5841	0.5888	0.5932	0.5975
0.60	0.5650	0.5707	0.5761	0.5811	0.5859	0.5905	0.5948
0.61	0.5618	0.5676	0.5731	0.5783	0.5832	0.5878	0.5922
0.62	0.5587	0.5646	0.5702	0.5755	0.5804	0.5852	0.5896
0.63	0.5556	0.5617	0.5673	0.5727	0.5778	0.5826	0.5871
0.64	0.5526	0.5587	0.5645	0.5700	0.5751	0.5800	0.5847
0.65	0.5496	0.5559	0.5618	0.5673	0.5725	0.5775	0.5822
0.66	0.5467	0.5530	0.5590	0.5647	0.5700	0.5751	0.5799
0.67	0.5438	0.5503	0.5564	0.5621	0.5675	0.5726	0.5775
0.68	0.5409	0.5475	0.5537	0.5596	0.5651	0.5703	0.5752
0.69	0.5382	0.5449	0.5512	0.5571	0.5626	0.5679	0.5729
0.70	0.5354	0.5422	0.5486	0.5546	0.5603	0.5656	0.5707
0.71	0.5327	0.5396	0.5461	0.5522	0.5579	0.5634	0.5685
0.72	0.5301	0.5371	0.5436	0.5498	0.5556	0.5611	0.5663
0.73	0.5275	0.5346	0.5412	0.5475	0.5534	0.5589	0.5642
0.74	0.5249	0.5321	0.5388	0.5452	0.5511	0.5568	0.5621
0.75	0.5223	0.5297	0.5365	0.5429	0.5489	0.5546	0.5600
0.76	0.5199	0.5272	0.5342	0.5407	0.5468	0.5525	0.5580
0.77	0.5174	0.5249	0.5319	0.5385	0.5446	0.5505	0.5560
0.78	0.5150	0.5226	0.5296	0.5363	0.5425	0.5484	0.5540
0.79	0.5126	0.5203	0.5274	0.5341	0.5405	0.5464	0.5521
0.80	0.5102	0.5180	0.5252	0.5320	0.5384	0.5445	0.5502
0.81	0.5079	0.5158	0.5231	0.5299	0.5364	0.5425	0.5483
0.82	0.5056	0.5136	0.5210	0.5279	0.5344	0.5406	0.5464
0.83	0.5034	0.5114	0.5189	0.5259	0.5325	0.5387	0.5446
0.84	0.5011	0.5092	0.5168	0.5239	0.5305	0.5368	0.5427
0.85	0.4990	0.5071	0.5148	0.5219	0.5286	0.5350	0.5409
0.86	0.4968	0.5050	0.5127	0.5200	0.5267	0.5331	0.5392
0.87	0.4946	0.5030	0.5108	0.5180	0.5249	0.5313	0.5374
0.88	0.4925	0.5010	0.5088	0.5161	0.5231	0.5296	0.5357
0.89	0.4905	0.4990	0.5069	0.5143	0.5212	0.5278	0.5340
0.90	0.4884	0.4970	0.5050	0.5124	0.5195	0.5261	0.5323
0.91	0.4864	0.4950	0.5031	0.5106	0.5177	0.5244	0.5307
0.92	0.4844	0.4931	0.5012	0.5088	0.5159	0.5227	0.5290
0.93	0.4824	0.4912	0.4994	0.5070	0.5142	0.5210	0.5274
0.94	0.4804	0.4893	0.4976	0.5053	0.5125	0.5194	0.5258
0.95	0.4785	0.4874	0.4958	0.5035	0.5109	0.5177	0.5242
0.96	0.4766	0.4856	0.4940	0.5018	0.5092	0.5161	0.5227
0.97	0.4747	0.4838	0.4922	0.5001	0.5076	0.5145	0.5211
0.98	0.4729	0.4820	0.4905	0.4985	0.5059	0.5130	0.5196
0.99	0.4710	0.4802	0.4888	0.4968	0.5043	0.5114	0.5181
1.00	0.4692	0.4785	0.4871	0.4952	0.5027	0.5099	0.5166

u	$\Lambda^{-1}_{0.70}$	$\Lambda^{-1}_{0.75}$	$\Lambda^{-1}_{0.80}$	$\Lambda^{-1}_{0.85}$	$\Lambda^{-1}_{0.90}$	$\Lambda^{-1}_{0.95}$	$\Lambda^{-1}_{1.00}$
1.0	0.4692	0.4785	0.4871	0.4952	0.5027	0.5099	0.5166
1.1	0.4521	0.4620	0.4712	0.4798	0.4879	0.4954	0.5026
1.2	0.4368	0.4472	0.4569	0.4660	0.4744	0.4824	0.4899
1.3	0.4229	0.4338	0.4439	0.4534	0.4623	0.4706	0.4784
1.4	0.4102	0.4216	0.4321	0.4419	0.4511	0.4598	0.4679
1.5	0.3986	0.4103	0.4212	0.4314	0.4409	0.4498	0.4582
1.6	0.3878	0.3999	0.4112	0.4216	0.4314	0.4406	0.4493
1.7	0.3779	0.3903	0.4018	0.4126	0.4226	0.4320	0.4409
1.8	0.3686	0.3813	0.3931	0.4041	0.4144	0.4241	0.4331
1.9	0.3600	0.3730	0.3850	0.3962	0.4067	0.4166	0.4259
2.0	0.3519	0.3651	0.3774	0.3888	0.3995	0.4096	0.4190
2.1	0.3443	0.3577	0.3702	0.3818	0.3927	0.4029	0.4125
2.2	0.3371	0.3508	0.3634	0.3752	0.3863	0.3967	0.4065
2.3	0.3304	0.3442	0.3570	0.3690	0.3802	0.3908	0.4007
2.4	0.3240	0.3380	0.3510	0.3631	0.3745	0.3851	0.3952
2.5	0.3179	0.3320	0.3452	0.3575	0.3690	0.3798	0.3900
2.6	0.3121	0.3264	0.3397	0.3521	0.3638	0.3747	0.3850
2.7	0.3066	0.3211	0.3345	0.3470	0.3588	0.3698	0.3802
2.8	0.3014	0.3160	0.3295	0.3422	0.3540	0.3652	0.3757
2.9	0.2964	0.3111	0.3247	0.3375	0.3495	0.3607	0.3713
3.0	0.2916	0.3064	0.3202	0.3330	0.3451	0.3564	0.3671
3.1	0.2870	0.3019	0.3158	0.3287	0.3409	0.3523	0.3631
3.2	0.2827	0.2976	0.3116	0.3246	0.3369	0.3484	0.3592
3.3	0.2784	0.2935	0.3075	0.3207	0.3330	0.3446	0.3555
3.4	0.2744	0.2895	0.3036	0.3168	0.3292	0.3409	0.3519
3.5	0.2705	0.2857	0.2999	0.3132	0.3256	0.3374	0.3485
3.6	0.2667	0.2820	0.2963	0.3096	0.3222	0.3340	0.3451
3.7	0.2631	0.2785	0.2928	0.3062	0.3188	0.3307	0.3419
3.8	0.2596	0.2750	0.2894	0.3029	0.3156	0.3275	0.3388
3.9	0.2563	0.2717	0.2862	0.2997	0.3124	0.3244	0.3357
4.0	0.2530	0.2685	0.2830	0.2966	0.3094	0.3214	0.3328
4.1	0.2499	0.2654	0.2799	0.2936	0.3064	0.3185	0.3300
4.2	0.2468	0.2624	0.2770	0.2907	0.3036	0.3157	0.3272
4.3	0.2439	0.2595	0.2741	0.2879	0.3008	0.3130	0.3245
4.4	0.2410	0.2567	0.2713	0.2851	0.2981	0.3103	0.3219
4.5	0.2382	0.2539	0.2686	0.2825	0.2955	0.3078	0.3194
4.6	0.2355	0.2513	0.2660	0.2799	0.2929	0.3053	0.3169
4.7	0.2329	0.2487	0.2635	0.2774	0.2905	0.3028	0.3145
4.8	0.2304	0.2462	0.2610	0.2749	0.2880	0.3004	0.3122
4.9	0.2279	0.2437	0.2586	0.2725	0.2857	0.2981	0.3099
5.0	0.2255	0.2413	0.2562	0.2702	0.2834	0.2959	0.3077
5.1	0.2232	0.2390	0.2539	0.2679	0.2812	0.2937	0.3055
5.2	0.2209	0.2368	0.2517	0.2657	0.2790	0.2915	0.3034
5.3	0.2187	0.2346	0.2495	0.2636	0.2769	0.2894	0.3013
5.4	0.2165	0.2324	0.2474	0.2615	0.2748	0.2874	0.2993
5.5	0.2144	0.2303	0.2453	0.2594	0.2728	0.2854	0.2973
5.6	0.2124	0.2283	0.2433	0.2574	0.2708	0.2834	0.2954
5.7	0.2104	0.2263	0.2413	0.2555	0.2689	0.2815	0.2935
5.8	0.2084	0.2244	0.2394	0.2536	0.2670	0.2796	0.2917
5.9	0.2065	0.2225	0.2375	0.2517	0.2651	0.2778	0.2899
6.0	0.2047	0.2206	0.2357	0.2499	0.2633	0.2760	0.2881

u	$\Lambda_{0.70}^{-1}$	$\Lambda_{0.75}^{-1}$	$\Lambda_{0.80}^{-1}$	$\Lambda_{0.85}^{-1}$	$\Lambda_{0.90}^{-1}$	$\Lambda_{0.95}^{-1}$	$\Lambda_{1.00}^{-1}$
1	0.4692	0.4785	0.4871	0.4952	0.5027	0.5099	0.5166
2	0.3519	0.3651	0.3774	0.3888	0.3995	0.4096	0.4190
3	0.2916	0.3064	0.3202	0.3330	0.3451	0.3564	0.3671
4	0.2530	0.2686	0.2830	0.2966	0.3094	0.3214	0.3328
5	0.2255	0.2413	0.2562	0.2702	0.2834	0.2959	0.3077
6	0.2047	0.2206	0.2357	0.2499	0.2633	0.2760	0.2881
7	0.1882	0.2041	0.2192	0.2335	0.2471	0.2600	0.2722
8	0.1747	0.1906	0.2057	0.2200	0.2337	0.2466	0.2590
9	0.1634	0.1792	0.1942	0.2086	0.2223	0.2353	0.2477
10	0.1538	0.1695	0.1844	0.1987	0.2124	0.2255	0.2379
11	0.1456	0.1610	0.1759	0.1902	0.2038	0.2169	0.2293
12	0.1383	0.1536	0.1684	0.1826	0.1962	0.2092	0.2217
13	0.1319	0.1471	0.1617	0.1758	0.1893	0.2024	0.2148
14	0.1262	0.1412	0.1557	0.1697	0.1832	0.1962	0.2086
15	0.1211	0.1359	0.1503	0.1642	0.1776	0.1905	0.2030
16	0.1164	0.1311	0.1453	0.1591	0.1725	0.1854	0.1978
17	0.1122	0.1267	0.1408	0.1545	0.1678	0.1807	0.1930
18	0.1083	0.1227	0.1366	0.1503	0.1635	0.1763	0.1886
19	0.1048	0.1189	0.1328	0.1463	0.1595	0.1722	0.1845
20	0.1015	0.1155	0.1293	0.1427	0.1558	0.1684	0.1807
21	0.0985	0.1123	0.1260	0.1393	0.1523	0.1649	0.1771
22	0.0957	0.1094	0.1229	0.1361	0.1490	0.1616	0.1738
23	0.0930	0.1066	0.1200	0.1331	0.1460	0.1585	0.1706
24	0.0906	0.1040	0.1173	0.1303	0.1431	0.1555	0.1676
25	0.0883	0.1016	0.1147	0.1277	0.1404	0.1528	0.1648
26	0.0861	0.0993	0.1123	0.1252	0.1378	0.1501	0.1621
27	0.0841	0.0971	0.1100	0.1228	0.1354	0.1476	0.1596
28	0.0822	0.0950	0.1079	0.1206	0.1331	0.1453	0.1572
29	0.0803	0.0931	0.1058	0.1185	0.1309	0.1430	0.1549
30	0.0786	0.0913	0.1039	0.1164	0.1288	0.1409	0.1527
31	0.0770	0.0895	0.1021	0.1145	0.1268	0.1388	0.1506
32	0.0754	0.0878	0.1003	0.1126	0.1249	0.1368	0.1486
33	0.0740	0.0863	0.0986	0.1109	0.1230	0.1350	0.1466
34	0.0725	0.0847	0.0970	0.1092	0.1213	0.1332	0.1448
35	0.0712	0.0833	0.0955	0.1076	0.1196	0.1314	0.1431
36	0.0699	0.0819	0.0940	0.1060	0.1180	0.1298	0.1413
37	0.0687	0.0806	0.0926	0.1045	0.1164	0.1282	0.1397
38	0.0675	0.0793	0.0912	0.1031	0.1149	0.1266	0.1381
39	0.0664	0.0782	0.0899	0.1017	0.1135	0.1251	0.1366
40	0.0653	0.0769	0.0886	0.1004	0.1121	0.1237	0.1251
41	0.0642	0.0758	0.0874	0.0991	0.1108	0.1223	0.1336
42	0.0632	0.0747	0.0862	0.0979	0.1095	0.1210	0.1323
43	0.0623	0.0736	0.0851	0.0967	0.1082	0.1197	0.1309
44	0.0613	0.0726	0.0840	0.0955	0.1070	0.1184	0.1296
45	0.0604	0.0716	0.0830	0.0944	0.1058	0.1172	0.1284
46	0.0596	0.0707	0.0820	0.0933	0.1047	0.1160	0.1272
47	0.0587	0.0697	0.0810	0.0923	0.1036	0.1149	0.1260
48	0.0579	0.0689	0.0800	0.0913	0.1025	0.1138	0.1248
49	0.0572	0.0680	0.0791	0.0903	0.1015	0.1127	0.1237
50	0.0564	0.0672	0.0782	0.0893	0.1005	0.1116	0.1226

u	$\Lambda_{0.70}^{-1}$	$\Lambda_{0.75}^{-1}$	$\Lambda_{0.80}^{-1}$	$\Lambda_{0.85}^{-1}$	$\Lambda_{0.90}^{-1}$	$\Lambda_{0.95}^{-1}$	$\Lambda_{1.00}^{-1}$
10	0.15384	0.16947	0.18443	0.19874	0.21241	0.22547	0.23794
20	0.10152	0.11552	0.12926	0.14269	0.15576	0.16843	0.18070
30	0.07862	0.09127	0.10390	0.11642	0.12876	0.14086	0.15268
40	0.06529	0.07689	0.08863	0.10040	0.11211	0.12368	0.13507
50	0.05640	0.06717	0.07819	0.08934	0.10051	0.11163	0.12263
60	0.04998	0.06008	0.07050	0.08112	0.09183	0.10255	0.11321
70	0.04509	0.05462	0.06454	0.07471	0.08502	0.09539	0.10575
80	0.04121	0.05027	0.05975	0.06953	0.07949	0.08955	0.09964
90	0.03806	0.04671	0.05581	0.06523	0.07488	0.08467	0.09451
100	0.03544	0.04372	0.05248	0.06160	0.07097	0.08051	0.09013
110	0.03321	0.04117	0.04964	0.05848	0.06760	0.07690	0.08632
120	0.03130	0.03897	0.04716	0.05575	0.06464	0.07374	0.08297
130	0.02963	0.03705	0.04499	0.05335	0.06203	0.07094	0.07999
140	0.02816	0.03534	0.04307	0.05122	0.05970	0.06843	0.07733
150	0.02686	0.03382	0.04134	0.04930	0.05761	0.06617	0.07492
160	0.02569	0.03246	0.03979	0.04757	0.05571	0.06412	0.07272
170	0.02464	0.03123	0.03838	0.04600	0.05398	0.06225	0.07072
180	0.02368	0.03011	0.03710	0.04456	0.05240	0.06053	0.06887
190	0.02281	0.02908	0.03592	0.04324	0.05094	0.05894	0.06717
200	0.02202	0.02814	0.03484	0.04202	0.04959	0.05747	0.06559
210	0.02129	0.02727	0.03384	0.04089	0.04834	0.05611	0.06412
220	0.02061	0.02647	0.03291	0.03984	0.04717	0.05483	0.06274
230	0.01999	0.02572	0.03204	0.03886	0.04608	0.05364	0.06146
240	0.01940	0.02503	0.03123	0.03794	0.04506	0.05252	0.06025
250	0.01886	0.02438	0.03048	0.03708	0.04410	0.05147	0.05911
260	0.01835	0.02377	0.02977	0.03627	0.04320	0.05048	0.05803
270	0.01788	0.02319	0.02910	0.03551	0.04235	0.04954	0.05702
280	0.01743	0.02266	0.02847	0.03479	0.04154	0.04865	0.05605
290	0.01701	0.02215	0.02787	0.03411	0.04078	0.04781	0.05513
300	0.01661	0.02167	0.02731	0.03346	0.04005	0.04701	0.05426
310	0.01624	0.02121	0.02678	0.03285	0.03936	0.04625	0.05343
320	0.01588	0.02078	0.02627	0.03227	0.03871	0.04552	0.05264
330	0.01554	0.02037	0.02578	0.03171	0.03808	0.04483	0.05188
340	0.01522	0.01998	0.02532	0.03118	0.03748	0.04416	0.05115
350	0.01491	0.01961	0.02488	0.03067	0.03691	0.04353	0.05046
360	0.01462	0.01925	0.02446	0.03019	0.03636	0.04292	0.04979
370	0.01435	0.01891	0.02406	0.02972	0.03583	0.04233	0.04915
380	0.01408	0.01859	0.02367	0.02927	0.03533	0.04177	0.04853
390	0.01383	0.01827	0.02330	0.02881	0.03484	0.04123	0.04794
400	0.01358	0.01708	0.02295	0.02843	0.03438	0.04071	0.04737
410	0.01335	0.01769	0.02260	0.02804	0.03393	0.04021	0.04682
420	0.01313	0.01741	0.02228	0.02766	0.03349	0.03972	0.04628
430	0.01291	0.01715	0.02196	0.02729	0.03308	0.03926	0.04577
440	0.01271	0.01689	0.02166	0.02694	0.03267	0.03881	0.04527
450	0.01251	0.01665	0.02136	0.02659	0.03228	0.03837	0.04479
460	0.01232	0.01641	0.02108	0.02626	0.03191	0.03795	0.04433
470	0.01213	0.01618	0.02080	0.02594	0.03154	0.03754	0.04388
480	0.01195	0.01596	0.02054	0.02563	0.03119	0.03714	0.04344
490	0.01178	0.01575	0.02028	0.02533	0.03085	0.03676	0.04301
500	0.01161	0.01554	0.02004	0.02504	0.03051	0.03638	0.04260

8.7 Table of $H(x,t)$

t	$H(0.0,t)$ $H(1.0,t)$	$H(0.05,t)$ $H(0.95,t)$	$H(0.1,t)$ $H(0.9,t)$	$H(0.2,t)$ $H(0.8,t)$	$H(0.3,t)$ $H(0.7,t)$	$H(0.4,t)$ $H(0.6,t)$	$H(0.5,t)$ $H(0.5,t)$
0	0	1	1	1	1	1	1
0.0013	0	0.67320	0.95014	0.99991	1.00000	1.00000	1.00000
0.0014	0	0.65530	0.94122	0.99984	1.00000	1.00000	1.00000
0.0015	0	0.63869	0.93211	0.99974	1.00000	1.00000	1.00000
0.0016	0	0.62324	0.92290	0.99959	1.00000	1.00000	1.00000
0.0017	0	0.60883	0.91365	0.99940	1.00000	1.00000	1.00000
0.0018	0	0.59534	0.90442	0.99914	1.00000	1.00000	1.00000
0.0019	0	0.58270	0.89524	0.99882	1.00000	1.00000	1.00000
0.002	0	0.57080	0.88615	0.99843	1.00000	1.00000	1.00000
0.003	0	0.48139	0.80329	0.99018	0.99989	1.00000	1.00000
0.004	0	0.42385	0.73645	0.97465	0.99920	0.99999	1.00000
0.005	0	0.38292	0.68269	0.95450	0.99730	0.99994	1.00000
0.006	0	0.35192	0.63869	0.93211	0.99383	0.99974	0.99999
0.007	0	0.32740	0.60198	0.90903	0.98877	0.99927	0.99995
0.008	0	0.30737	0.57080	0.88615	0.98229	0.99843	0.99985
0.009	0	0.29061	0.54394	0.86396	0.97465	0.99712	0.99961
0.01	0	0.27633	0.52050	0.84270	0.96610	0.99530	0.99919
0.02	0	0.19741	0.38292	0.68263	0.86592	0.95180	0.97516
0.03	0	0.16166	0.31668	0.58469	0.77506	0.88322	0.91755
0.04	0	0.13974	0.27496	0.51584	0.69783	0.80881	0.84580
0.05	0	0.12387	0.24425	0.46165	0.63040	0.73633	0.77231
0.06	0	0.11110	0.21929	0.41591	0.57039	0.66858	0.70220
0.07	0	0.10020	0.19786	0.37585	0.51647	0.60634	0.63722
0.08	0	0.09059	0.17893	0.34013	0.46780	0.54960	0.57775
0.09	0	0.08200	0.16197	0.30800	0.42378	0.49805	0.52363
0.10	0	0.07426	0.14669	0.27899	0.38393	0.45129	0.47449
0.11	0	0.06727	0.13288	0.25274	0.34784	0.40889	0.42992
0.12	0	0.06094	0.12038	0.22898	0.31515	0.37047	0.38953
0.13	0	0.05521	0.10906	0.20745	0.28553	0.33565	0.35293
0.14	0	0.05002	0.09881	0.18795	0.25869	0.30411	0.31976
0.15	0	0.04532	0.08953	0.17029	0.23438	0.27553	0.28971
0.16	0	0.04106	0.08111	0.15428	0.21235	0.24964	0.26248
0.17	0	0.03720	0.07349	0.13978	0.19240	0.22617	0.23781
0.18	0	0.03371	0.06658	0.12665	0.17431	0.20492	0.21546
0.19	0	0.03054	0.06032	0.11475	0.15793	0.18566	0.19521
0.20	0	0.02767	0.05465	0.10396	0.14309	0.16821	0.17687
0.21	0	0.02507	0.04952	0.09419	0.12964	0.15240	0.16024
0.22	0	0.02271	0.04486	0.08534	0.11746	0.13808	0.14518
0.23	0	0.02058	0.04065	0.07732	0.10642	0.12510	0.13154
0.24	0	0.01864	0.03683	0.07005	0.09642	0.11334	0.11918
0.25	0	0.01689	0.03337	0.06347	0.08736	0.10269	0.10798
0.26	0	0.01530	0.03023	0.05750	0.07915	0.09304	0.09783
0.27	0	0.01386	0.02739	0.05210	0.07171	0.08430	0.08863
0.28	0	0.01256	0.02482	0.04720	0.06497	0.07637	0.08030
0.29	0	0.01138	0.02248	0.04277	0.05886	0.06920	0.07276
0.30	0	0.01031	0.02037	0.03875	0.05333	0.06269	0.06592
∞	0	0	0	0	0	0	0

8.8 Tables of $H_0(x,t)$

t	$H_0(0.0,t)$	$H_0(0.05,t)$	$H_0(0.1,t)$	$H_0(0.2,t)$	$H_0(0.3,t)$	$H_0(0.4,t)$	$H_0(0.5,t)$
0	1	0	0	0	0	0	0
0.0043	1	0.58977	0.28089	0.03103	0.00122	0.00002	0.00000
0.0044	1	0.59403	0.28642	0.03301	0.00138	0.00002	0.00000
0.0045	1	0.59816	0.29184	0.03501	0.00157	0.00002	0.00000
0.0046	1	0.60217	0.29715	0.03706	0.00176	0.00003	0.00000
0.0047	1	0.60606	0.30234	0.03913	0.00197	0.00004	0.00000
0.0048	1	0.60983	0.30743	0.04123	0.00220	0.00004	0.00000
0.0049	1	0.61351	0.31242	0.04335	0.00244	0.00005	0.00000
0.005	1	0.61708	0.31731	0.04550	0.00270	0.00006	0.00000
0.006	1	0.64808	0.36131	0.06789	0.00617	0.00026	0.00000
0.007	1	0.67260	0.39802	0.09097	0.01123	0.00072	0.00002
0.008	1	0.69263	0.42920	0.11385	0.01771	0.00157	0.00008
0.009	1	0.70939	0.45606	0.13604	0.02535	0.00287	0.00019
0.010	1	0.72367	0.47950	0.15730	0.03389	0.00468	0.00041
0.011	1	0.73604	0.50018	0.17753	0.04311	0.00700	0.00075
0.012	1	0.74689	0.51861	0.19671	0.05281	0.00982	0.00125
0.013	1	0.75649	0.53514	0.21485	0.06281	0.01311	0.00193
0.014	1	0.76509	0.55010	0.23200	0.07300	0.01683	0.00281
0.015	1	0.77283	0.56370	0.24821	0.08326	0.02092	0.00389
0.016	1	0.77985	0.57615	0.26355	0.09353	0.02535	0.00519
0.017	1	0.78627	0.58759	0.27808	0.10374	0.03006	0.00670
0.018	1	0.79215	0.59816	0.29184	0.11385	0.03501	0.00841
0.019	1	0.79757	0.60796	0.30490	0.12381	0.04017	0.01032
0.02	1	0.80259	0.61708	0.31731	0.13361	0.04550	0.01242
0.03	1	0.83826	0.68309	0.41422	0.22067	0.10247	0.04123
0.04	1	0.85969	0.72367	0.47950	0.28884	0.15730	0.07710
0.05	1	0.87437	0.75183	0.52709	0.34278	0.20590	0.11384
0.06	1	0.88523	0.77283	0.56370	0.38648	0.24821	0.14890
0.07	1	0.89369	0.78927	0.59198	0.42267	0.28503	0.18139
0.08	1	0.90052	0.80259	0.61707	0.45323	0.31725	0.21112
0.09	1	0.90618	0.81366	0.63733	0.47944	0.34562	0.23819
0.10	1	0.91097	0.82304	0.65466	0.50219	0.37075	0.26276
0.11	1	0.91509	0.83113	0.66969	0.52214	0.39312	0.28504
0.12	1	0.91867	0.83817	0.68286	0.53977	0.41313	0.30523
0.13	1	0.92181	0.84436	0.69449	0.55549	0.43107	0.32354
0.14	1	0.92460	0.84985	0.70482	0.56945	0.44720	0.34012
0.15	1	0.92708	0.85474	0.71404	0.58200	0.46173	0.35515
0.16	1	0.92929	0.85911	0.72231	0.59328	0.47484	0.36876
0.17	1	0.93128	0.86303	0.72974	0.60343	0.48669	0.38109
0.18	1	0.93307	0.86656	0.73643	0.61259	0.49739	0.39227
0.19	1	0.93468	0.86973	0.74240	0.62087	0.50707	0.40239
0.20	1	0.93613	0.87260	0.74791	0.62834	0.51583	0.41157
0.21	1	0.93744	0.87519	0.75283	0.63510	0.52375	0.41988
0.22	1	0.93863	0.87754	0.75728	0.64122	0.53093	0.42741
0.23	1	0.93970	0.87965	0.76131	0.64676	0.53743	0.43423
0.24	1	0.94067	0.88157	0.76495	0.65177	0.54331	0.44041
0.25	1	0.94155	0.88331	0.76825	0.65631	0.54864	0.44601
0.26	1	0.94234	0.88488	0.77124	0.66042	0.55347	0.45109
0.27	1	0.94306	0.88630	0.77394	0.66414	0.55785	0.45568
0.28	1	0.94372	0.88759	0.77639	0.66751	0.56181	0.45985
0.29	1	0.94431	0.88876	0.77861	0.67057	0.56340	0.46362
0.30	1	0.94484	0.88981	0.78062	0.67333	0.56865	0.46704
∞	1	0.95	0.9	0.8	0.7	0.6	0.5

t	$H_0(0.5,t)$	$H_0(0.6,t)$	$H_0(0.7,t)$	$H_0(0.8,t)$	$H_0(0.9,t)$	$H_0(0.95,t)$	$H_0(1,t)$
0	0	0	0	0	0	0	0
0.010	0.00041	0.00002	0.00000	0.00000	0.00000	0.00000	0
0.011	0.00075	0.00005	0.00000	0.00000	0.00000	0.00000	0
0.012	0.00125	0.00011	0.00000	0.00000	0.00000	0.00000	0
0.013	0.00193	0.00020	0.00001	0.00000	0.00000	0.00000	0
0.014	0.00281	0.00034	0.00003	0.00000	0.00000	0.00000	0
0.015	0.00389	0.00053	0.00005	0.00000	0.00000	0.00000	0
0.016	0.00519	0.00080	0.00009	0.00000	0.00000	0.00000	0
0.017	0.00670	0.00114	0.00015	0.00001	0.00000	0.00000	0
0.018	0.00841	0.00157	0.00022	0.00002	0.00000	0.00000	0
0.019	0.01030	0.00208	0.00033	0.00004	0.00000	0.00000	0
0.020	0.01242	0.00270	0.00047	0.00006	0.00000	0.00000	0
0.021	0.01470	0.00341	0.00064	0.00009	0.00001	0.00000	0
0.022	0.01714	0.00423	0.00085	0.00014	0.00002	0.00000	0
0.023	0.01974	0.00515	0.00110	0.00019	0.00003	0.00000	0
0.024	0.02248	0.00617	0.00140	0.00026	0.00004	0.00001	0
0.025	0.02535	0.00729	0.00175	0.00035	0.00006	0.00002	0
0.026	0.02833	0.00851	0.00214	0.00045	0.00008	0.00003	0
0.027	0.03142	0.00982	0.00259	0.00058	0.00011	0.00004	0
0.028	0.03461	0.01123	0.00310	0.00072	0.00014	0.00005	0
0.029	0.03788	0.01273	0.00365	0.00089	0.00018	0.00007	0
0.03	0.04123	0.01431	0.00427	0.00109	0.00023	0.00009	0
0.04	0.07710	0.03389	0.01332	0.00466	0.00136	0.00058	0
0.05	0.11384	0.05777	0.02682	0.01126	0.00392	0.00176	0
0.06	0.14890	0.08321	0.04313	0.02039	0.00788	0.00366	0
0.07	0.18139	0.10863	0.06086	0.03117	0.01287	0.00611	0
0.08	0.21112	0.13315	0.07896	0.04280	0.01849	0.00888	0
0.09	0.23819	0.15633	0.09678	0.05467	0.02437	0.01182	0
0.10	0.26276	0.17797	0.11387	0.06635	0.03027	0.01477	0
0.11	0.28504	0.19799	0.13002	0.07757	0.03599	0.01764	0
0.12	0.30523	0.21640	0.14508	0.08816	0.04145	0.02039	0
0.13	0.32354	0.23328	0.15902	0.09806	0.04657	0.02297	0
0.14	0.34012	0.24869	0.17186	0.10723	0.05134	0.02538	0
0.15	0.35515	0.26274	0.18362	0.11567	0.05574	0.02760	0
0.16	0.36876	0.27552	0.19437	0.12341	0.05978	0.02965	0
0.17	0.38109	0.28714	0.20417	0.13048	0.06348	0.03152	0
0.18	0.39227	0.29769	0.21309	0.13693	0.06686	0.03323	0
0.19	0.40239	0.30727	0.22120	0.14280	0.06994	0.03479	0
0.20	0.41157	0.31596	0.22857	0.14813	0.07274	0.03620	0
0.21	0.41988	0.32385	0.23526	0.15298	0.07529	0.03749	0
0.22	0.42741	0.33099	0.24132	0.15738	0.07760	0.03866	0
0.23	0.43423	0.33747	0.24683	0.16138	0.07970	0.03972	0
0.24	0.44041	0.34334	0.25181	0.16500	0.08160	0.04069	0
0.25	0.44601	0.34866	0.25633	0.16828	0.08333	0.04156	0
0.26	0.45109	0.35349	0.26044	0.17126	0.08489	0.04235	0
0.27	0.45568	0.35786	0.26415	0.17396	0.08631	0.04307	0
0.28	0.45985	0.36182	0.26752	0.17640	0.08760	0.04372	0
0.29	0.46362	0.36540	0.27057	0.17862	0.08876	0.04431	0
0.30	0.46704	0.36865	0.27334	0.18063	0.08982	0.04484	0
∞	0.5	0.4	0.3	0.2	0.1	0.05	0

Bibliography

Abramowitz and Stegun, 1965, *Handbook of Mathematical Functions with Formulas, Graphs, and Mathematical Tables.* National Bureau of Standards, Applied Mathematical Series 55. Fourth Printing. Dover Books of Mathematics. New York USA 1972.

Andrade, 1993, *Calculation of chloride diffusion coefficients in concrete from ionic migration measurements.* Cement and Concrete Research, Vol. 23, 1993.

Ang and Tang, 1984, *Probability Concepts in Engineering, Planning and Design*, Vol. I and II, John Wiley, 1984.

ASTM, 1990, C 856-83: *Standard practice for petrographic examination of hardened concrete.* USA 1990.

ASTM C 114, *Standard test method for chemical analyses for hydraulic cement*, ASTM.

ASTM, 1992, G 109: *Determining the Effects of Chemical Admixtures on the Corrosion of Embedded Steel Reinforcement in Concrete exposed to Chloride Environment, (the Cracked Beam Test).* American Society for Testing and Materials PA, USA 1992.

ASTM, 1994, C 12025: *Standard Test Method for Electrical Indication of Concrete's Ability to resist Chloride Ion Penetration.* American Society for Testing and Materials, Book of ASTM Standards 1994.

Atwood and Johnson, 1924, *The Disintegration of Cement in Sea Water.* Transactions ASCE, Vol. 87, Paper No. 1533, 1924.

Augusti, Baratta and Casiati, 1984, *Probabilistic Methods in Structural Engineering*, Chapman and Hall, 1984.

Benjamin and Cornell, 1970, *Probability, Statistics and Decision for Civil Engineers*, McGraw Hill, 1970.

Bentur, Diamond and Berke, 1997, *Steel Corrosion in Concrete, Fundamental and Civil Engineering Practice.* E & FN SPON, London, England 1997.

Bigas, 1994, *La diffusion des ions chlore dans les morties.* Thèse de Doctorat INSA-Genie Civil, LMDC, INSA de Toulouse, France 1994.

Borges and Castanheta, 1971, *Structural Safety,* Laboratorio Nacional de Engenharia Civil, Lisbon, Portugal 1971.

British Standard Institution, 1993, *Glossary of Building and Civil Engineering Terms.* Blackwell Scientific Publication, London 1993.

Broomfield, 1997, *Corrosion of Steel in Concrete, Understanding, Investigation and Repair,* E & FN SPON, London, England 1997.

Broomfield, Rodriguez, Ortega and Garcia, 1993, Corrosion rate measurement and life prediction for reinforced concrete structures. In *Proceedings of Structural Faults and Repair,* Vol. 2, Engineering Technical Press, University of Edinburgh 1993.

Broomfield, Rodriguez, Ortega and Garcia, 1994, Corrosion rate measurement in reinforced concrete structures by linear polarization device. In *Philip D. Cady Symposium on Corrosion of steel in concrete,* American Concrete Institute SP151, 1994.

Budak, Samarskii and Tikhonov, 1964, *A Collection of Problems in Mathematical Physics,* Pergamon Press, Oxford, London, Edinburgh, New York, Paris, Frankfurt 1964.

Byfors, 1987, *Influence of Silica Fume and Fly Ash on Chloride diffusion and pH-values on Cement Paste.* Cement and Concrete Research, Vol. 17, 1987.

Cannon, 1984, *The One-Dimensional Heat Equation,* Encyclopedia of Mathematics and its Application, Vol. 23, Addision-Wesley Publishing Company, 1984.

Carleson, 1966, *Convergence and growth of partial sums of Fourier series,* Acta Math., 116 (1966), 135–157.

Carslaw and Jaeger, 1947, *Conduction of Heat in Solids,* Oxford at the Clarendon Press, 1947.

Chess *et al.*, 1998, *Cathodic protection of steel in concrete.* E & FN SPON, London, UK 1998.

Clear, 1989, *Measuring the rate of corrosion of steel in field concrete structures.* Transporation Research Board, National Research Counsil, Washington DC 1989.

Collepardi, 1995, *Quick method to determine free and bound chlorides in concrete. Chloride Penetration into Concrete.* Proceedings of the international RILEM Workshop, St-Rémy-les-Chevreuse, France 1995.

Collepardi, Marcialis and Turriziani, 1970, *The Kinetics of Chloride Ions Penetration in Concrete* (in Italian). Il Cemento, Vol. 67, Italy.

Collepardi, Marcialis and Turriziani, 1972, *Penetration of Chloride Ions in Cement Pastes and in Concretes.* Journal American Ceramics Soc. Vol. 55. USA 1972.

Concrete Society, 1989, *Analysis of Hardened Concrete, 9. Microscopical Methods.* Technical Report No. 32, London 1989.

Cornell, 1969, *A Probability-Based Structural Code Format.* ACI Journal Dec. 1969, No. 12, Proceedings Vol. 66, Detroit, USA 1969.

Courant and Hilbert, 1931, *Methoden der mathematischen Physik*, Grundlehren der mathematischen Wissenschaften, Springer-Verlag, Berlin 1931.

Crank, 1956, *The Mathematics of Diffusion*, Second Edition, Clarendon Press, Oxford, UK 1986.

Ditlevsen, 1981, *Uncertainty Modelling with Applications to Multi-dimensional Civil Engineering Systems.* McGraw Hill, 1981.

Ditlevsen and Madsen, 1996, *Structural Reliability Methods.* Danish Institute of Building Research, SBI report 211, Hørsholm, Denmark 1996.

Ditlevsen and Madsen, 1996, *Structural Reliability Methods.* Wiley.

EN 206-1, *Concrete – Execution, production and control.*

EN 1504, 1997, Part 9: *General principles for the use of products and systems. Products and systems for the protection and repair of concrete structures – Definitions, requirements, quality control and evaluation of conformity.* European Prestandard, CEN Brussels, July 1997

EN ISO 8044, 1989, *Corrosion of metals and alloys – Basic terms and definitions.* European committee for standardization, Brussels 1999.

Engell, 1960, *Schnellverfahren zur Auffindung von Kloridgehalten in Beton.* Beton 10, 1960.

EN 206-1, 1990, *Concrete – Performance, Production and Compliance definitions.* European committee for standardization, Brussels 1990.

Eurocode ENV 1992-1-1, *Design of concrete structures. General rules and rules for buildings (Eurocode 2).*

Eydemant, Ostrowski and Demidov, 1993, *Analysis of Diffusion Rate of Migrating Corrosion Inhibitor MCI 2000 in Concrete using Radioactive Isotope tagging Techniques.* Test Report, Institute of Construction Materials and Kurchatov Institute of Nuclear Physics, Moscow, Russia 1993.

Feld, 1968, *Construction Failures.* John Wiley and Sons, Inc., New York, USA 1968.

Fick, 1855, *Poggendorff's Annalen der Physik*, 1855.

Fontana, 1967, *Corrosion Engineering*, Third Edition, McGraw-Hill Book Company, New York, USA 1986.

Frederiksen, Mejlbro and Poulsen, 2000, The HETEK Model of Chloride Ingress into Concrete Made Simpler by Approximations. In *International Rilem Workshop on Testing and Modelling Chloride Ingress into Concrete*, Paris, France 2000.

Frederiksen, Nilsson, Sandberg, Poulsen, Tang and Andersen, 1997, *HETEK, a System of Estimation of Chloride Ingress into Concrete, Theoretical Background.* Report No. 83:1997. The Danish Road Directorate. Copenhagen, Denmark 1997.

Gautefall and Vennesland, 1994, *Methods for Determination of Chloride Diffusivity in Concrete.* REPCON Report No. A16. Statens Vegvesen, Vegdirektoratet, Oslo, Norway 1994.

Germann Petersen, 2003. *Catalogue NTD-2003. NTD-systems, Profile Grinding PF-1100.* Germann Instruments A/S. 102 Emdrupvej, DK-2400 NV Copenhagen. E-mail: germann post6.tele.dk.

Hansen, Seir, 1995, A microstructural study of surface related transformation in concrete pilings, Esbjerg Harbour. *Proceedings of the Nordic Seminar in Lund on Corrosion of Reinforcement, Field and Laboratory Studies for Modelling and Service Life*, Report TVBM–3064. Lund Sweden 1995.

Hoff, 1986, The Service Record of Concrete Off-shore Structures in the North Sea. In *Proceedings, International Conference on Concrete in the Marine Environment*, The Concrete Society, London, UK 1986.

Hope and Thompson, 1995, *Damage to Concrete Induced by Calcium Nitrite.* ACI Materials Journal, September–October 1995, Detroit USA.

John, 1971, *Partial Differential Equations*, Applied Mathematical Sciences, Springer 1071.

Jørsboe and Mejlbro, 1982, *The Carleson-Hunt Theorem on Fourier Series*, Lecture Notes in Mathematics 911, Springer-Verlag, 1982.

Karlsson and Poulsen, 1995, Design of Rebar Concrete Covers in Marine Concrete Structures, Probabilistic Approach. In *Proceedings of the International Rilem Workshop*, St.-Rémy-les-Chevreuse, France 1995.

Kirsh, 1981, *Control of Stray Current on Reinforced Concrete.* Corrosion '81, Paper 51, National Association of Corrosion Engineers, Houston TX, USA 1981.

Klinghoffer and Rislund, 1995, *Determination of chloride diffusivity in matured concrete based on electrical conductivity in the hardening material.* Proceedings of the International Symposium on Non-Destructive testing in Civil Engineering, Berlin, Germany 1995.

Klinghoffer, Mejlbro and Poulsen, 1999, *Prediction of Reinforcement Corrosion by the Macrocell Technique – Evaluation of Observations*, International Conference on Ion and Mass Transport in Cement-Based Materials, University of Toronto, October 4–5, 1999, Canada.

Maage, Helland and Carlsen, 1993, Chloride Penetration in High Performance Concrete Exposed to Marine Environment. In *Proceedings of the Symposium on Utilization of High Strength Concrete*, Lillehammer, Norway 1993.

Maage, Helland and Carlsen, 1995, *Practical non-steady chloride transport as a part of a model for predicting the initiation period*, RILEM Workshop on chloride penetration into concrete, St-Rémy-Les-Chevreuse, France 1995.

Maage, Helland and Carlsen, 1999, *Chloride Penetration into Concrete with Lightweight Aggregates.* Report FoU LIGHTCON, Report No. 3.6, STF22 A98755 SINTEF, Trondheim, Norway 1999.

Maage, Poulsen, Vennesland and Carlsen, 1995, *Service Life Model for Concrete Structures Exposed to Marine Environment – Initiation Period.* LIGHTCON Report No. 2.4, STF70 A94082 SINTEF, Trondheim, Norway 1995.

Madsen, Krenk and Lind, 1986, *Methods of Structural Safety.* Prentice-Hall, 1986.

Mailvaganam, 1992, *Repair and protection of concrete structures.* Chapter 12, Corrosion Protection. CRC Press, Inc. Boca Raton, Florida, USA 1992.

Mangat and Molloy, 1994, *Prediction of Long Term Chloride Concentration in Concrete.* RILEM Materials and Structures, Vol. 27.

Mays *et al.*, 1992, *Durability of concrete structures, investigation, repair, protection.* E & FN SPON, London, UK 1992.

Mejlbro, 1995, *Fick's anden lov for diffusion med stedafhængig diffusionskoefficient*, (Fick's second law with the diffusion coefficient depending on the spatial coordinate; in Danish), MAT-REPORT NR 1995–11, Institute of Mathematics, Technical University of Denmark, 1995.

Mejlbro, 1996, The Complete Solution of Fick's Second Law of Diffusion with Time-dependent Diffusion Coefficient and Surface Concentration. In *Durability of Concrete in Saline Environment.* CEMENTA Special Publication, Danderyd, Sweden 1996.

Mejlbro, 2001, *Initial Study of Fick's Second Law Applied to Concrete Structures with Cracks*, Nordic Mini-Seminar, Göteborg, May 2001 on: Prediction Models for Chloride Ingress and Corrosion Initiation in Concrete Structures.

Mejlbro and Poulsen, 1997, Design of Rebar Covers in Marine Structures by Probabilistic Modelling Applying Cornell's Reliability Index. In *Proceeding of the International Conference on Ion and Mass Transport in Cement-Based Materials*, University of Toronto, Canada – October 1999.

Mejlbro and Poulsen, 2000, *On a model of chloride ingress into concrete exposed to de-icing salt containing chloride.* International RILEM Workshop on testing and modelling chloride ingress into concrete. Paris, France 2000.

Melchers, 1987, *Structural Reliability and Predictions.* Ellis Horwood/J. Wiley, 1987.

Morse and Feshbach, 1953, *Methods of Theoretical Physics*, Vol. I–II, McGraw-Hill, 1953.

Nagayama, 1998, *MCI 2020 Long Term Test, Protection of Rebar in Concrete. Evaluation of 43 Months' Test Results.* Interim Report, General Building Research Corporation of Japan, October 1998. Osaka, Japan 1998.

NCT, 1991, *Specification for Desalination works.* Norwegian Concrete Technologies, Oslo, Norway 1991.

Nielsen, Christoffersen and Frederiksen, 1994, *Danish High Performance Concretes.* ACI SP-159, Proceedings of International Workshop on High Performance Concrete. Bangkok, Thailand 1994.

Nilsson, 2000, *Private communication*, Nødebo 2000.

Nilsson, Massat and Tang, 1994, *The effect of non-linear chloride binding on the prediction of chloride penetration into concrete structures.* ACI SP–145, The 3rd. International Conference on Durability of Concrete. Nice, France 1994.

Nordtest, 1984, NT Build 208: *Concrete hardened: Chloride content*, Edition 2. NORDTEST Register. Espoo, Finland 1992.

Nordtest, 1996a, NT Build 443: *Concrete hardened: Accelerated chloride penetration.* Approved 1995–11. NORDTEST register. Espoo, Finland 1996.

Nordtest, 1996b, NT Build 208: *Concrete hardened: Chloride content by Volhard titration.* Edition 3. NORDTEST Register. Espoo, Finland 1996.

Olver, 1974, *Asymptotics and Special Functions*, Computer Science and Applied Mathematics, Academic Press, New York and London, 1974.

Pedersen and Klinghoffer, 1993, *Factors influencing on the uncertainty in determination of diffusion coefficient by non-linear curve-fitting.* Nordic mini seminar. Chalmers University of Technology, Division of Building Materials, Report P–93:1. Gothenburg, Sweden 1993.

Phanasgaonkar, Cherry and Forsyth, 1997, *Organic Corrosion Inhibitors; How Do They Inhibit and Can They Really Migrate Through Concrete?* Australasian Corrosion Conference Brisbane, AMECR 97/30, Queensland, Australia, December 1997.

Poulsen, 1993, *On a model of chloride ingress into concrete having time-dependent diffusion coefficient*, Nordic Miniseminar on chloride penetration into concrete structures, Göteborg, Sweden 1993.

Poulsen and Frederiksen, 1998, *The method of inverse cores.* NIST/ACI/ASTM Workshop on Models for Predicting the Service life and Life-Cycle Cost of Steel-Reinforced Concrete, Gaithersburg, MD, USA 1998.

prEN 15358, 2002, *Structural timber – Calculation of characteristic 5-percentile value.* European Standard. Management Centre: rue de Stassart 36, B-1050 Brussels, Belgium 2002.

Purvis, Babaei, Clear and Markow, 1994, *Life-cycle Cost Analysis for Protection and Rehabilitation of Concrete Bridges Relative to Reinforcement Corrosion.* Report SHRP–S–377. Appendix B: Time to Deterioration. Strategic Highway Research Program, National Research Council, Washington DC, USA 1994.

Reknes, 1994, *Ringforsøk, felt- og labmetoder for måling av kloridinnhold i betong* (in Norwegian). Project O 6563, Norwegian Institute of Building Research. Oslo, Norway 1994.

Sand, 1993, *The effect of the environmental load on chloride penetration.* Nordic mini seminar. Chalmers University of Technology, Division of Building Materials, Report P–93:1. Gothenburg, Sweden 1993.

Sandberg, 1995, *Critical Evaluation of Factors Affecting Chloride Initiated Reinforcement Corrosion in Concrete.* Report TVBM-7088, LTH-LBM. Lund, Sweden 1995.

Sandberg, 1996, Systematic Collection of Field Data for Service Life Prediction of Concrete Structures. In *Durability of Concrete in Saline Environment*, CEMENTA, Danderyd, Sweden 1996.

Sandberg, 1998, *Chloride initiated reinforcement corrosion in marine concrete.* Report TVBM–1015. Lund University, division of Building Materials. Lund, Sweden 1998.

Schöppel, 1988, *Indication of free chloride ions on concrete surfaces by the IV-test.* Betonwerk+Fertigteil-Technik, 11, 1988.

SHRP–S–658, 1993, *Concrete Bridge Protection and Rehabilitation: Chemical and Physical Techniques – Field Validation.* Strategic Highway Research Council National Research Council, Washington DC, USA 1993.

Stoltzner, 1995, Chloride penetration in Danish bridges. In *Proceedings of the Nordic Seminar in Lund on Corrosion of Reinforcement, Field and Laboratory Studies for Modelling and Service Life*, Report TVBM–3064. Lund, Sweden 1995.

Swamy, Hamada and Laiw, 1995, A Critical Evaluation of Chloride Penetration into Concrete in Marine Environment. In *Proceedings of the Conference on 'Corrosion and Corrosion Protection of Steel in Concrete'*, University of Sheffield, Vol. I, Sheffield, UK 1995.

Sørensen, 1996, *Chloridtransport i hærdnet beton – mekanismer, modeller og målemetoder.* PhD afhandling, AEC, ISBN 87-986767-0-9. (In Danish.)

Takewaka and Mastumoto, 1988, *Quality and Cover Thickness of Concrete Based on the Estimation of Chloride Penetration in Marine Environments.* ACI SP 109–17. American Concrete Institute, Detroit. USA, 1988.

Tang, 1993, *A rapid method for measuring chloride diffusivities by using an electrical field – theory and current experiences.* Nordic mini seminar. Chalmers University of Technology, Division of Building Materials, Report P–93:1, Gothenburg, Sweden 1993.

Tang and Nilsson, 1992, *Rapid determination of chloride diffusivity of concrete by applying an electrical field.* ACI Materials Journal, Vol. 49, No. 1, USA 1992.

Titchmarsh, 1946, *Eigenfunction Expansions Associated with Second-order Differential Equations*, Oxford at the University Press, 1946, also Second Edition, Oxford at the Clarendon Press, 1962.

Toft-Christensen and Baker, 1982, *Structural Reliability Theory and its Applications*. Springer-Verlag, 1982.

Toft-Christensen and Murotsu, 1986, *Application of Structural Systems Reliability Theory*. Springer-Verlag, 1986.

Uji, Matsuoka and Maruya, 1990, Formulation of an Equation for surface Chloride Content due to Permeation of Chloride. In *Proceedings of the Third International Symposium on 'Corrosion of Reinforcement in Concrete Construction'*, Elsevier Applied Science, London, UK 1990.

Ukrainczyk and Bjegovic, 1993, *Diffusion of the MCI 2020 and MCI 2000 corrosion inhibitors into Concrete*. University of Zagreb, Croatia 1993.

Vennesland, 1993, *Chloride penetration into concrete – a description of a test method based on a cyclic chloride environment*. Nordic mini seminar. Chalmers University of Technology, Division of Building Materials, Report P–93:1. Gothenburg, Sweden 1993.

Vincentsen *et al.*, 1999, *Concrete Technology*. The Storebælt Publications. A/S Storebæltsforbindelsen. Denmark 1999.

Vindeløv, 2003, *Kloriddiffusion gennem overfladebehandlet beton, matematisk modellering* (In Danish, Chloride diffusion through surface treated concrete, mathematical modelling). Thesis, BYG-DTU, Technical University of Denmark, 2003.

Watson, 1944, *A Treatise on the Theory of Bessel Functions*, Cambridge at the University Press, Second Edition, 1944.

Wen, 1990, *Structural Load Modelling and Combination for Performance and Safety Evaluation*. Developments in Civil Engineering, Elsevier, Amsterdam 1990.

Østerdal, 1995, *Analyse af chloridprofiler i beton*, (Analysis of chloride profiles in concrete, in Danish), thesis, IMM-EKS-1995-23, Institute of Mathematical Modelling, Technical University of Denmark, 1995.

Index

For Product Safety Concerns and Information please contact our EU
representative GPSR@taylorandfrancis.com
Taylor & Francis Verlag GmbH, Kaufingerstraße 24, 80331 München, Germany

www.ingramcontent.com/pod-product-compliance
Ingram Content Group UK Ltd.
Pitfield, Milton Keynes, MK11 3LW, UK
UKHW020024200926
475702UK00014B/2048

* 9 7 8 0 3 6 7 3 9 1 5 2 2 *